Drug Development Approaches for Poorly Soluble Compounds

Methods and Principles in Medicinal Chemistry

Edited by Helmut Buschmann, Jörg Holenz, Christa Müller

Previous Volumes of the Series

Schneekloth, J., and Pettersson, M. (Eds.)

RNA as a Drug Target: The Next Frontier for Medicinal Chemistry

2024

ISBN: 978-3-527-35100-8
Vol. 84

Daina, A., Przewosny, M., Zoete, V. (Eds.)

Open Access Databases and Datasets for Drug Discovery

2023

ISBN: 978-3-527-34839-8
Vol. 83

Bachhav, Y. (Ed.)

Targeted Drug Delivery

2023

ISBN: 978-3-527-34781-0
Vol. 82

Alza, E. (Ed.)

Flow and Microreactor Technology in Medicinal Chemistry

2022

ISBN: 978-3-527-34689-9
Vol. 81

Rübsamen-Schaeff, H., and Buschmann, H.(Eds.)

New Drug Development for Known and Emerging Viruses

2022

ISBN: 978-3-527-34337-9
Vol. 80

Gruss, M. (Ed.)

Solid State Development and Processing of Pharmaceutical Molecules Salts, Cocrystals, and Polymorphism

2021

ISBN: 978-3-527-34635-6
Vol. 79

Plowright, A.T. (Ed.)

Target Discovery and Validation Methods and Strategies for Drug Discovery

2020

ISBN: 978-3-527-34529-8
Vol. 78

Swinney, D., Pollastri, M. (Eds.)

Neglected Tropical Diseases Drug Discovery and Development

2019

ISBN: 978-3-527-34304-1
Vol. 77

Bachhav, Y. (Ed.)

Innovative Dosage Forms Design and Development at Early Stage

2019

ISBN: 978-3-527-34396-6
Vol. 76

Gervasio, F. L., Spiwok, V. (Eds.)

Biomolecular Simulations in Structure-based Drug Discovery

2018

ISBN: 978-3-527-34265-5
Vol. 75

Sippl, W., Jung, M. (Eds.)

Epigenetic Drug Discovery

2018

ISBN: 978-3-527-34314-0
Vol. 74

Drug Development Approaches for Poorly Soluble Compounds

Edited by
Yogeshwar Bachhav

Series Editors
Helmut Buschmann
Jörg Holenz
Christa Müller

WILEY-VCH

Editor

Yogeshwar Bachhav
Bharatkhand CHS
Tilaknagar
Mumbai

Series Editors

Helmut Buschmann
Sperberweg 15 52076
Aachen
Germany

Jörg Holenz
BIAL – R&D Division
Avenida Siderurgia Nacional
Coronado
Portugal

Christa Müller
Universität Bonn
Pharmazeutisches Institut
An der Immenburg 4
Bonn
Germany

Cover Design: Wiley
Cover Images: © alexkich/Shutterstock, © Photo Oz/Shutterstock

Typesetting Straive, Chennai, India

Library of Congress Card No.: applied for

British Library Cataloguing-in-Publication Data A catalogue record for this book is available from the British Library.

Bibliographic information published by the Deutsche Nationalbibliothek
The Deutsche Nationalbibliothek lists this publication in the Deutsche Nationalbibliografie; detailed bibliographic data are available on the Internet at <http://dnb.d-nb.de>.

Print ISBN: 978-3-527-35478-8
ePDF ISBN: 978-3-527-85075-4
ePub ISBN: 978-3-527-85074-7
oBook ISBN: 978-3-527-85076-1

Printed and bound by
CPI Group (UK) Ltd, Croydon, CR0 4YY

C9783527354788_030726

Contents

Series Editors Preface

Drug discovery and development has become an increasingly costly and inefficient process. The greatest challenge for drug discovery and development is meeting the expectations of the people who will use the drug or the product. The discovery of novel and innovative drugs is a multifaceted process that spans from the initial identification of a biological target to the approval and commercialization of a new medicine. In addition to the intrinsic challenges, drug absorption, sufficient and reproducible bioavailability, and/or pharmacokinetic profile in humans are recognized till today as one of the major challenges in oral delivery of new drug substances. The issue arose especially when drug discovery and medicinal chemistry moved from standard wet chemistry to combinatorial chemistry and high-throughput screening of large compound collections in the mid-1990s. In the past decades, the challenges linked to poor water solubility have been continuously growing without having the warranty that the solutions available today are applicable for a specific lead candidate. It is currently assumed that poorly soluble compounds represent approximately 40% of the top 200 oral drugs marketed in the United States and Europe. In addition, poorly soluble leads represent 90% of new chemical entities, 75% of compounds under development, and around one-third of drugs listed in the United States or European Pharmacopeia.

The primary goal of formulating R&D divisions of a pharmaceutical company is to make the medication accessible at the correct place within the body for its foreseen activity and in the most effective dosage. As a consequence, a variety of pharmaceutical strategies have been designed and developed to address the formulation and delivery challenges presented by poorly soluble drugs in development and medicines already commercialized.

Poor solubility of pharmacologically active compounds remains a major challenge in this field, and volumes addressing novel approaches and solutions will remain an essential part of our book series "Methods and Principles in Medicinal Chemistry."

On behalf of the series editors and the publisher, we would like to express our sincere gratitude to Yogesh Bachhav for compiling valuable contributions and solutions and making these technologies accessible to a broader and interested readership within the medicinal chemistry community.

Yogesh is an enthusiastic scientist who has dedicated his professional life to innovative drug development solutions for almost two decades. His broad and encyclopedic knowledge and his international network have made him an indispensable expert in the field of pharmaceutical formulation of early and late-stage projects. Besides the current volume, he already served as a volume editor for two other volumes of the book series, "Targeted Drug Delivery," a new frontier in drug research that has become a focus for developing novel medications, and "Innovative Dosage Forms: Design and Development at Early Stage," teaching future and current drug developers the latest innovations in drug formulation design and optimization The two volumes have now established themselves as highlights and landmarks of our book series, and the current volume addressing the problem of poor solubility will follow in this sense. The selected chapters of the new book are a valuable addition for pharmaceutical experts and all other disciplines involved in the complex drug discovery and development process.

As with the previous volumes, it was once again very successful to convince leading experts from the pharmaceutical industry and academia authoring various chapters to comprise the basic concepts and practical solutions applying innovative and advanced formulation technologies to address the problem of poor solubility presented in 12 chapters.

The opening chapter, as an introduction to the problem, covers the background, scope, impact, and practical limitations to develop poorly soluble drugs. The subsequent chapters provide a detailed emphasis on the application of the diverse formulation technologies including solid dispersion, nanosizing approaches, cocrystals, self-microemulsifying and self-emulsifying drug delivery systems, liposomes and mixed micelles, the use of mesoporous silica particles, cyclodextrin-based formulations, cosolvent-based formulations, and structural modification options. In addition, two more general chapters address gastrointestinal dissolution and absorption of Biopharmaceutics Classification System (BCS) Class II drugs and formulation considerations of first-in-animal studies. Through the careful selection of topics, this volume is a valuable resource not only for pharmaceutical experts in industry and academia, but also for every scientist involved in the complex development process of new drugs.

The series editors are grateful to Yogesh, who dedicated himself to the problem with great enthusiasm and conviction, to all the contributing authors for their scholarly contributions and commitment, and, last but not least, to Frank Weinreich and Drew Gwilliams from Wiley and all Wiley experts, whose support made this volume possible.

Helmut Buschmann
Jörg Holenz
Christa Müller

Preface

The discovery of new drug molecules is the most challenging and multidisciplinary task in the pharmaceutical industry. The drug discovery process represents several obstacles, and poor solubility of new drugs is one of the major and critical issues to be dealt with, as it can significantly hamper the translation of the safe and efficacious compounds into the commercially available medication.

The literature evidence supports that around 60–70% of the molecules in the modern-day drug discovery pipeline exhibit poor aqueous solubility, which can lead to lower bioavailability and undesired clinical outcomes.

The book, *Drug Development Approaches for Poorly Soluble Compounds*, published by Wiley-VCH GmbH as part of the *Methods and Principles in Medicinal Chemistry Series,* summarizes the exhaustive and comprehensive overview of the existing and model technologies used to tackle the problem of poor solubility and inferior bioavailability in the course of drug development.

The leading experts from the pharmaceutical industry and academia authored various chapters to comprise the basic concepts and practical/operational aspects of advanced formulation technologies to address the problem of poor solubility.

The stage is set by the opening chapter to cover the background, scope, impact, and practical limitations to develop poorly soluble drugs. The subsequent chapters provide a detailed emphasis on the application of the diverse formulation technologies such as solid dispersions, nanosizing, pharmaceutical co-crystals, lipid-based drug delivery systems such as self-microemulsifying drug delivery systems (SMEDDS), liposomes, mesoporous silica particles, and inclusion complexes such as cyclodextrins.

Physiological aspects of drug absorption are also discussed in depth, especially the gastrointestinal dissolution and absorption of BCS Class II drugs.

Furthermore, the volume also comprises the use of the co-solvent-based formulations, structural modifications of the drugs to improve the solubility, and typical considerations to design preclinical formulations of poorly soluble drugs, which are typically barriers to enter first-in-human studies.

This book aims to serve as a powerful tool for researchers, formulators, and students representing pharmaceutical R&D to address the issue of poor solubility of drugs. This book provides a thorough understanding of the formulation technologies along with the practical guide in order to help with the selection of the specific solubility enhancement technology suitable for a specific drug.

The editor is grateful to all the contributing authors for their scholarly contribution and commitment, to the series editors (Dr. Helmut Buschmann, Dr. Jörg Holenz, and Dr. Christa Müller), and to Wiley-VCH for their support in bringing this volume to the fruition. It is expected that this book will promote further innovation in the design and development of advanced drug delivery technologies to formulate poorly soluble compounds and thereby enhance the therapeutic efficacy and patient outcomes.

Yogeshwar Bachhav

1

Impact of Poor Solubility on Drug Development

Preshita Desai

Western University of Health Sciences, College of Pharmacy, Department of Biotechnology and Pharmaceutical Sciences, 309 E. Second Street, Pomona, CA 91766, USA

1.1 Introduction

Poor aqueous solubility remains one of the most significant challenges in modern drug development, impacting nearly every stage of the pharmaceutical product development pipeline. A significantly large proportion of drugs (commonly referred to as active pharmaceutical ingredients [APIs]/new chemical entities [NCEs], lead molecules, etc.), identified through high-throughput screening, exhibit poor solubility, which in turn limits their bioavailability and therapeutic efficacy. This challenge begins as early as the candidate selection phase, where compounds with inadequate solubility may be selected due to their strong and potent pharmacological activity. As development progresses into preclinical and clinical studies, insufficient solubility complicates formulation design, dosing accuracy, and pharmacokinetic profiles, thereby delaying development timelines and increasing costs. Even in post-clinical stages, poor drug solubility can affect long-term storage stability, scalability, etc. (e.g. particle size ripening due to poor drug solubility, loss of product stability due to drug precipitation on storage). This may possibly result in product recalls and, in some cases, complete product withdrawal from the market, hindering the commercial success of the final product.

Over the past decades, numerous strategies have been explored to overcome solubility limitations. These include drug material engineering (salt formation, co-crystallization, prodrugs, etc.), formulation strategies (solubilization techniques, solid dispersions, nanocrystals, lipid-based formulations, cyclodextrin inclusion complexes, etc.), or a combination of them. Selection of the most appropriate solubility enhancement technique, however, is highly dependent on the target product profile, physicochemical properties of the drug, intended route of administration, therapeutic indication, and commercial considerations.

This book gives an in-depth account of such approaches and provides a comprehensive summary of the current strategies available to address poor drug solubility, highlighting their principles, commercial success, advantages, and limitations.

This introductory chapter focuses on understanding the solubility concept and its impact on drug performance. Followed by a brief outline of important and widely reported solubility enhancement strategies. Further, the chapter proposes a possible roadmap to aid researchers and formulators in the systematic selection of solubility enhancement approaches best suited for specific drug candidates. With meticulous integration of these strategies early in development, it is possible to achieve drug bioenhancement, reduce product failures/attrition rates, and ultimately enhance the likelihood of successful translation of the safe, effective, and stable drug product to the market.

1.2 Drug Solubility

Solubility can be defined as the highest amount of solute dissolved in a given solvent to form a homogeneous one-phase saturated solution. Fundamentally, saturation solubility depends innately on the solute (physicochemical properties), solvent (physicochemical properties, composition, volume), and other parameters (temperature, pressure), which are summarized in Figure 1.1 [1, 2].

According to the International Union of Pure and Applied Chemistry (IUPAC), solubility is defined as the analytical composition of a saturated solution expressed as a proportion of a designated solute in a designated solvent. In literature, the solubility for a given solute–solvent system is represented by various units, including

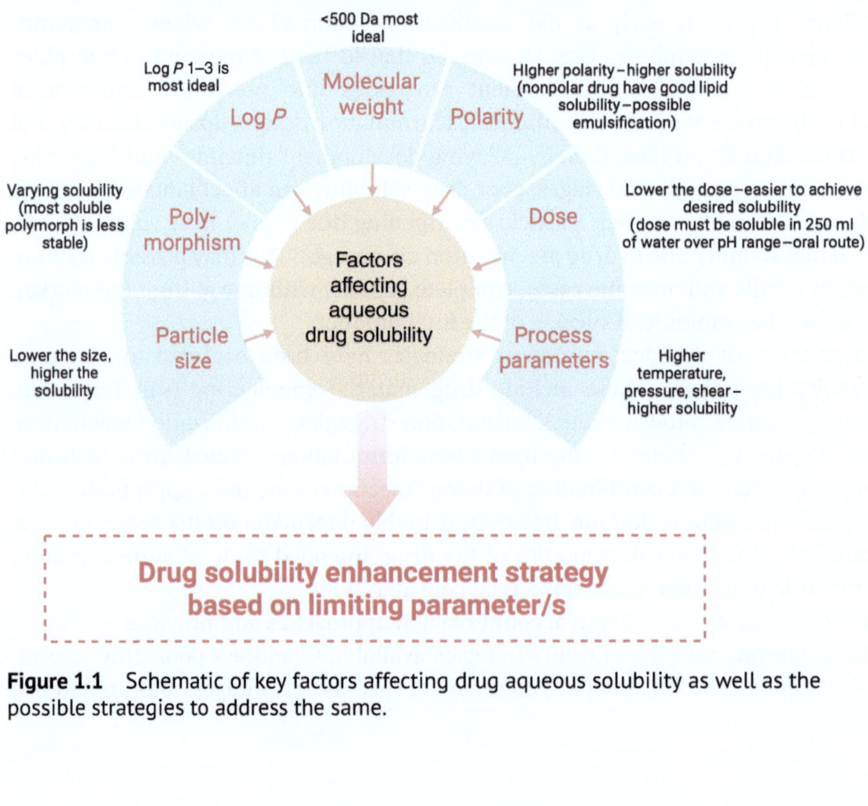

Figure 1.1 Schematic of key factors affecting drug aqueous solubility as well as the possible strategies to address the same.

but not limited to concentration (weight and volume ratio), molality, mole fraction, mole ratio, etc. [1].

Solubility, the commonly described and well-known physical chemistry concept explained above, plays a significant role in the journey of the lead molecule becoming a clinically available drug. Specifically, the overall solubility profile of the drug, more critically, their aqueous solubility exhibits a crucial impact on every stage of drug development, including drug discovery, pharmaceutical product development, and preclinical and clinical success. In simple terms, how a drug moves and/or works in the body depends on how well it dissolves in the body fluids. Good drug solubility in the gastrointestinal (GI) tract enables better absorption, distribution, metabolism, excretion, and toxicity regulation (ADME/Tox). This, in turn, affects the drug safety, bioavailability, effectiveness, and ability to reach the target receptor to produce the desired pharmacological effect.

In this context, knowledge of solubility from a pharmacopoeial perspective and understanding of the Biopharmaceutical Classification System (BCS) are of key importance. The United States Pharmacopoeia (USP) and the British Pharmacopoeia (BP) describe and categorize solubility potential as the saturation solubility of a solute in the given solvent, as described in Table 1.1 [3]. However, for a drug candidate, the solubility concept needs further consideration in relation to its dose. For example, a hypothetical drug candidate "A" that is sparingly soluble but highly potent and administered at a low dose may still dissolve adequately in the GI tract after oral administration. In contrast, a drug candidate "B" that is also sparingly soluble but requires a high dose may not fully dissolve in GI fluids, creating a barrier to absorption.

Furthermore, overall drug absorption is a dynamic process that is not only influenced by drug solubility but also by drug permeability. Drug absorption involves the overall movement of the drug from its site of administration to the systemic circulation. For an orally administered drug, this means that the rate of absorption depends on both the rate and extent of drug dissolution in GI fluids, as well as the ability of the drug to permeate the GI membrane. To amalgamate

Table 1.1 Solubility description per the United States Pharmacopoeia (USP) and the British Pharmacopoeia (BP).

Description forms	Parts of solvent required for one part of solute	Solubility range (mg/ml)
Very soluble (VS)	<1	>1000
Freely soluble (FS)	From 1 to 10	100–1000
Soluble	From 10 to 30	33–100
Sparingly soluble (SPS)	From 30 to 100	10–33
Slightly soluble (SS)	From 100 to 1000	1–10
Very slightly soluble (VSS)	From 1000 to 10,000	0.1–1
Practically insoluble (PI)	>10,000	<0.1

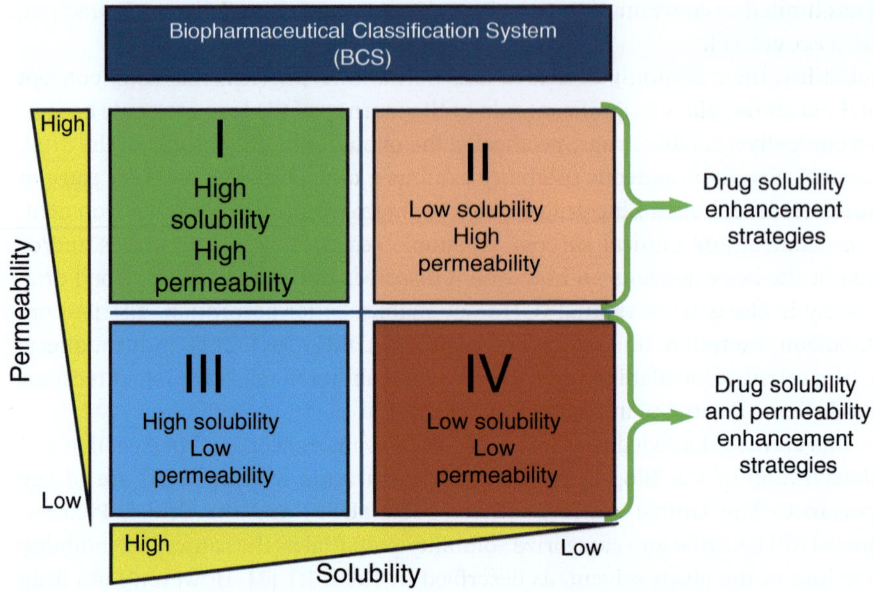

Figure 1.2 Schematic of Biopharmaceutical Classification System (BCS) and the rationale to enhance bioavailability of poorly soluble BCS Class II and IV drugs.

these solubility and permeability parameters, the BCS was established. Per BCS, drug solubility is the ability of a drug to dissolve in an aqueous phase. For oral medications, this is the capacity of the drug to dissolve in GI fluids. As per the classification system, a drug is considered highly soluble if its highest dose dissolves completely in 250 ml of water across the physiological pH range of 1.0–6.8 at $37 \pm 1\,°C$ (encompassing the lowest volume and pH variation across the GI tract). The BCS explains drug permeability as the ability of a drug to pass across the biological membranes (e.g. GI membrane), entering the systemic circulation. As per the classification system, a drug is considered highly permeable if ≥85% of it is absorbed into the systemic circulation. In a nutshell, effective drug absorption requires both high solubility and permeability. Based on variability in solubility and permeability, the drugs are broadly classified into four BCS classes, viz., Class I (high solubility, high permeability), Class II (high permeability, low solubility), Class III (high solubility, low permeability), and Class IV (low solubility, low permeability) (depicted in Figure 1.2) [4].

1.3 Impact of Poor Drug Solubility on Drug-Like Properties

Drug-like properties represent suitable physicochemical and biological characteristics of a lead compound, which make it a potential candidate for consideration as a safe and effective therapeutic drug. For a lead candidate with a potent target effect (pharmacological efficacy), the widely used tool to assess drug-likeness is

Lipinski's rule of 5, which focuses on and describes physicochemical aspects of an orally active molecule to be drug-like based on four parameters, viz., molecular weight less than 500 Da, *n*-octanol/water partition coefficient (Log *P*) not exceeding 5, no more than 5 hydrogen bond donors, and no more than 10 hydrogen bond acceptors [5, 6]. Drug candidates that follow this rule exhibit an acceptable solubility and permeability profile, translating to an advantageous ADME/Tox profile, enhanced drug absorption, and bioavailability.

The challenge of low drug solubility has become increasingly prominent in recent years, with widely preferred screening methods of identifying and shortlisting lead candidates based on high potency at the biological target receptor (drug–ligand interaction). While drug potency will always be a priority, adopting a balanced approach of additionally scrutinizing drug-like properties (ensuring satisfactory ADME/Tox properties) and drugability (potential for commercial translation) will be helpful. Advances in combinatorial chemistry, high-throughput screening, and artificial intelligence-driven drug lead identification have further reinforced this trend, as they often emphasize the discovery of high-potency lead candidates that naturally have higher molecular weight and lipophilicity. This has resulted in the growing prevalence of drug leads with poor aqueous solubility, posing significant challenges to successful drug development and clinical drug product realization. Statistically, this trend is on a significant rise, and as per a report published in 2024, it is estimated that approximately 80–90% of lead drugs under development exhibit poor water solubility (BCS Class II and IV), in contrast to only about 40% of currently marketed oral drugs [7].

The high-risk journey of the drug from discovery to clinic is long and expensive (~10–15 years; ~$1.8–2.2 billion USD) with a high clinical development failure rate of ~90%. Based on analysis of clinical studies conducted between 2010 and 2017, a report published in 2022 states that 40–50% of failures resulted from suboptimal clinical efficacy, 30% from unmanageable toxicity, 10–15% from unacceptable drug-like properties, and 10% due to low commercial need or strategic failure. An increasing number of drugs with poor solubility (BCS Class II and IV) can be a key reason for clinical failure, as poor solubility can negatively impact clinical efficacy, safety (high toxicity), and poor drug-like properties. Hence, it is of key importance to ensure satisfactory drug solubility early on in the drug development process to reduce the rate of attrition and failure during product development and clinical trials. The key considerations to assess the underlying reasons for poor drug aqueous solubility and strategies to address the same are depicted in Figure 1.1.

1.4 Addressing Poor Drug Solubility Challenge

Various approaches have been explored to overcome the challenge of poor aqueous solubility to ensure successful product development. They can be broadly classified into two categories, viz., drug material engineering and formulation approaches, which are described in subsequent sections [4, 8–11].

1.4.1 Drug Material Engineering

The physicochemical properties of the drug itself can be modified to overcome the poor solubility challenge. This approach involves physical and chemical methods as described in the following subsections.

1.4.1.1 Particle Size Engineering

The solubility of any chemical is intrinsically dependent on the particle size. As particle size reduces, surface area and surface area-to-volume ratio increase, allowing better interaction with the solvent and thereby increasing the rate and extent of dissolution. This approach of particle size reduction has been widely explored as a strategy to develop drugs with enhanced dissolution rate, solubility, and bioavailability. To achieve such size reduction, various top-down (comminution, milling, etc.) and bottom-up techniques (spray drying, nano- and microprecipitation, etc.) are reported.

Micronization is one such commonly used method capable of reducing the size of coarse drug particles (preferably less than 5 μm), resulting in a significant enhancement of drug dissolution. However, in some cases, it offers challenges of drug crystal structure interference due to high-energy size reduction processes like pulverization and milling. As a solution to this problem, the use of spray drying and supercritical fluid technology is on the rise [12–14].

In recent years, advancing on this concept, further reduction in drug particle size under 1 μm has gained wide attention, yielding drug nanoparticles. Such extensive particle size reduction can offer the dual benefit of enhancing the drug solubility by extensive reduction in diffusion layer thickness and permeability enhancement due to high proclivity to cross biological membranes (e.g. GI tract lining). Thus, drug nanonization can result in significant bioavailability enhancement compared to drug micronization. For example, danazol, a BCS Class II drug used as a synthetic steroid, exhibited ~13-fold bioavailability improvement using the drug nanoform (size: ~180 nm) compared to the unmilled drug (size: 8 μm) [12–15]. However, particle size reduction to the nanoscale comes with the challenge of stabilizing and maintaining the reduced particle size. Therefore, the drug particle size reduction approach is generally combined with a suitable formulation strategy (nanocrystal, nanosuspensions, solid dispersion, etc.) that is capable of stabilizing the reduced particle size over the product stability period, ensuring retention of low particle size, enhanced solubility, and bioavailability. For a detailed description of this approach, refer to Section 1.4.2.2 and Chapter 3.

1.4.1.2 Modulating Drug Properties: Salts, Solvates, Polymorphs, and Co-crystals

This approach for drug solubility modulation utilized both physical and chemical methods like the development of salts, solvates, specific polymorph selection, and co-crystal technology.

Use of drug salts for poorly soluble drugs is the most widely used and industrially favored approach. This requires an ionizable drug (cationic/anionic/zwitterionic) capable of ionic linking to the counterion. The first salt form of a drug approved by the US Food and Drug Administration (US FDA) back in 1939 was heparin sodium, and since then, over 50% of US FDA-approved drugs are salts. Among these, hydrochloride salts are the most common. This is an interesting approach as the developed salts are ionic in nature, which increases the apparent aqueous solubility but retains the efficacy of the drug without chemical modification [12, 16]. For example, chlorpromazine, a BCS Class IV drug, showed high potency as an antipsychotic, but the original base form was insoluble in water. To overcome the challenge, the drug was converted to chlorpromazine hydrochloride (salt form), which showed enhanced solubility and bioavailability. The drug salt form was successfully translated to a clinically available product (Thorazine®). Further, it was also reported that the crystalline salt form not only increased the solubility and bioavailability but also conferred better stability to the product [17, 18].

Selection of an appropriate polymorph (different crystalline forms), pseudopolymorphs (solvates, hydrates—the US FDA recognized them as polymorphs), or amorphous (non-crystalline) form can also critically modulate the drug solubility. This is one of the key approaches, as 80% of the pharmaceutical compounds are polymorphic in nature [19]. Interestingly, it is reported that the polymorph with low stability (high internal energy) tends to exhibit the highest solubility/dissolution. Therefore, if such a highly soluble polymorph is selected for product development, it comes with a challenge of transformation to a stable polymorph (less soluble) upon storage. Hence, to maintain the highly soluble polymorphic state of a drug (often unstable), a combination with an appropriate drug stabilizing formulation strategy is necessary. The classic example for this is the well-known story of ritonavir (Norvir®), a blockbuster anti-human immunodeficiency virus (anti-HIV) drug that was introduced in the market, wherein the poor drug solubility was enhanced by using a more soluble drug polymorph I. However, upon storage, the drug transformed into stable but less soluble polymorph II, failing to elicit the desired bioavailability and efficacy. Hence, the product was withdrawn from the market, which not only impacted the patients relying on the treatment but also caused severe financial loss to the company. Ultimately, the product was reintroduced in the market by reformulating it with a newer excipient composition that was able to stabilize and retain the soluble polymorph I. This highlights the critical need to approach solubility enhancement strategy through a lens of long-term stability [20].

The new emerging approach in this context is the development of drug co-crystals, which are multimolecular crystalline structures resulting from non-covalent/nonionic interactions. The approach involves the development of a drug–excipient or drug–drug co-crystal for enhancement. Many products (Seglentis®, Depakote®, Entresto®, Lexapro®, etc.) developed with this approach are already available in the market. Seglentis® is a unique drug–drug co-crystal for the co-delivery of two drugs, celecoxib and tramadol, for acute pain in adults. The results indicate enhanced bioperformance, overcoming the individual drug solubility challenge and pharmacokinetic variation [21]. For a detailed description of this approach, refer to Chapter 4.

1.4.1.3 Prodrugs

The concept of prodrugs was first introduced in the late 1950s. A prodrug is a chemically modified derivative of the parent drug. The prodrug is designed with an objective to optimize the physicochemical properties of the parent drug, which, upon administration, can be subsequently converted back to the active parent drug via biotransformation. This unique strategy has been explored for a multitude of drugs and accounts for over 50 US FDA approvals in a 10-year period of 2012–2022 (~13% of US FDA-approved small molecule drugs) [22, 23]. Many such prodrug-based formulations designed for solubility and bioavailability enhancement are available in the market, including Cerebyx®, Solu-Medrol®, Sivextro®, etc. One of the well-known examples of improving drug solubility through a prodrug approach is isavuconazonium sulfate (Cresemba®), a medication prescribed for severe fungal infections (invasive aspergillosis, mucormycosis). Isavuconazole (BCS Class II), the parent drug, exhibited high antifungal potency, but the challenge was to make the drug soluble for intravenous administration. For this, a prodrug, isavuconazonium sulfate, showing significant improvement in solubility, was developed for intravenous use. Interestingly, the solubility enhancement was so significant that ultimately it also enabled the development of an oral version. The prodrug was meticulously designed to be biotransformed in blood (mainly by the enzyme butyrylcholinesterase) upon intravenous administration, while upon oral administration, it is rapidly biotransformed in the gut, liver, and blood. In both cases, the biotransformation to the active drug was highly efficient (about 98% bioavailability) [24]. It is important to note that, for the success of this approach, meticulous design of the prodrug is very important in a way that it should ideally ensure higher solubility and absorption (high bioavailability) followed by prompt biotransformation in systemic circulation to elicit a pharmacological response via the parent drug.

The challenge in this strategy is the possible rapid hydrolysis of such prodrugs prior to absorption. In such a scenario, this transformation would release the water-insoluble parent drug prematurely, resulting in limited absorption and bioavailability. One such example is oral hetacillin (Hetacin®), a prodrug of the widely known antibiotic ampicillin. The prodrug hetacillin showed significant solubility enhancement and was clinically approved for use, but later it was observed that the prodrug was unstable in water, with very quick chemical conversion to ampicillin, and hence did not show significant efficacy enhancement compared to ampicillin itself. Further, during breakdown, it was reported to produce formaldehyde in the GI tract, raising safety concerns. In view of this, the product was withdrawn from the market for human use. However, it is still being used for veterinary purposes via intramammary administration to treat mastitis in cows (PolyMast®). In this context, recently, a new technology, "Sol-Moiety," has been introduced that is capable of delaying and controlling the prodrug hydrolysis, ensuring sufficient solubilization and absorption, leading to enhanced bioavailability [25]. For a detailed description of this approach, refer to Chapter 8.

Overall, poor drug solubility raises issues in product development, but the solubility enhancement strategies based on drug material engineering, chemical form modification, and optimization of crystal form must be explored with caution, ensuring feasibility, efficacy, safety, and stability as pragmatic endpoints.

1.4.2 Drug Formulation Approaches

Drug material engineering offers exceptional approaches to enhancing drug intrinsic solubility. However, stabilization of engineered drugs becomes a key challenge that can be resolved by strategic formulation approaches. Furthermore, unique and novel formulations capable of enhancing drug solubility and permeability are widely reported in the literature with successful clinical translation. These are discussed below.

1.4.2.1 Cosolvency, Hydrotrophy, and Solubilizing Agents

The use of cosolvents is a simple, effective, and widely adopted approach to enhance the solubility of drugs. In simple terms, a cosolvent can be viewed as a bridge between the drug and the aqueous phase. Cosolvents are water-miscible solvents that possess a unique ability to dissolve poorly water-soluble drugs, thereby increasing their solubility in the aqueous phase. The key is to identify a suitable solvent for the given drug, ensuring that it can be used safely at the required dose for the intended route of administration. Further, it is important to critically understand the impact of cosolvent dilution on drug solubility, because upon administration, the drug may precipitate if a poor cosolvent with limited cosolvency and/or miscibility is used.

In this context, several modeling methods such as Hansen solubility parameters, Hildebrand solubility parameters, thermodynamic modeling, and artificial neural networks have been employed as effective tools for identifying suitable cosolvents [26–29].

This strategy is particularly effective with nonionizable drugs having a moderate partition coefficient (log *P*-value: 1–3). Therefore, highly lipophilic drugs may not be suitable candidates for this approach. The use of cosolvency-based formulations of poorly soluble drugs is common in both oral and parenteral products. Well-known examples include Ativan®, Valium®, Nembutal®, Lanoxin®, etc. For instance, Ativan contains lorazepam (a BCS Class II drug, practically insoluble in water), which is dissolved in cosolvents such as propylene glycol, polyethylene glycol, and benzyl alcohol for intravenous and concentrated oral formulations. For a detailed description of this approach, refer to Chapter 10.

Hydrotropes are another category comprising small amphiphilic molecules (nicotinamide, amino acids, sodium benzoate, sodium salicylate, etc.) with small hydrophobic tails (compared to surfactants). Instead of forming micelles like surfactants, hydrotropes form aggregates, reduce water activity, and form complexes with the drug as mechanisms of solubility enhancement [12, 30]. However, the challenge with hydrotropes-based formulations is that they must be used in large amounts, which raises safety concerns. In this context, the new concept of "ionic liquids" is coming into the limelight as powerful, feasible, and safer hydrotropes. In an investigation, ionic liquids of cholinium vanillate, cholinium gallate, and cholinium salicylate were tested to improve the aqueous solubility of two model drugs, ibuprofen and naproxen. The studies revealed that cholinium vanillate- and cholinium gallate-based ionic liquids increased the ibuprofen solubility by ~500-fold and all three ionic liquids increased naproxen solubility by ~600-fold [31, 32].

The next category is surfactant and/or solubilizing agents that are more widely explored. They have the capability of forming micelles at low concentration, thereby reducing surface tension, which is the predominant mechanism for enhanced drug solubilization. This approach is more meaningful for BCS Class II drugs, wherein the drugs exhibit poor solubility but have good permeability. This strategy, if it were to be explored for BCS Class IV drugs, would require further formulation rationale wherein these solubilizing strategies are used in conjunction with permeability enhancement approaches like the development of nanoformulations, discussed in subsequent sections.

1.4.2.2 Cyclodextrin Inclusion Complexes

Inclusion complexes are the supramolecular complexes wherein a solute (drug in this specific scenario) is encapsulated in a host molecule. From a drug solubility enhancement perspective, the host molecule exhibits a hydrophilic outer surface with hydrophobic pockets capable of encapsulating the drug. The macromolecule complex overall allows higher solubilization capacity by virtue of the high solubility of the host molecule. The most common example in the drug delivery spectrum is cyclodextrin-based inclusion complexes [33], with a few already in the market, Sporanox®, Vfend®, Nitropen®, etc. Sporanox is an antifungal oral solution comprising intraconazole, a water-insoluble drug (BCS Class II) that was formulated as an inclusion complex with hydroxypropyl-β-cyclodextrin, showing clinically significant bioenhancement [34]. For a detailed description of this approach, refer to Chapter 9.

1.4.2.3 Solid Dispersions and Solid Solutions

Solid dispersions are unique formulations that overcome poor drug solubility challenges by mixing the drug with hydrophilic excipients at the molecular level using techniques such as melt fusion, hot melt extrusion, and spray drying.

The mechanisms of solubility enhancement with this approach incorporate a multitude of reasons, including particle size reductions, conversion and stabilization of more soluble drug amorphous form, enhanced wettability, reduced drug aggregation, etc. As per the report, between 2012 and 2023, about 48 products based on amorphous solid dispersions have been approved by the US FDA. The most commonly used polymers for such development are copovidone and hypromellose acetate succinate, and the most commonly used methods are spray drying (~50%) and hot melt extrusion (~35%) [35]. However, as discussed in the polymorphism (Section 1.4.1.2), maintaining the molecular form of the drug in solid dispersions is of key importance as phase separation and drug instability are the main challenges in solid dispersion-based dosage forms. Owing to this challenge, a few marketed products like Rezulin®, Isoptin SR®, Incivek®, and Norvir® were withdrawn from the market due to drug instability (physical instability) and drug recrystallization issues impacting drug dissolution rate and bioperformance [36]. Therefore, though solid dispersion is a widely sought approach for solubility enhancement, it must be used with caution for highly lipophilic drugs, ensuring drug stabilization. For a detailed description of this approach, refer to Chapter 2.

1.4.2.4 Nanoformulation Spectrum

Picking up on various concepts of physical and chemical approaches to enhance drug solubility, the current ideal and most explored approach is nanoformulations. To visualize, this approach can be looked upon as the amalgamation of multiple drug solubility enhancement approaches discussed thus far, including but not limited to particle size reduction, solubilization/stabilization with surfactants, cosolvency, encapsulation, etc., ultimately rendering a formulation with particle/globule size in the nanometric range. Nanoformulations increase the solubility of the drug and can also enhance permeability if the drug is encapsulated in a suitable nanocarrier, making them suitable for bioenhancement of BCS Class II and IV drugs.

The simplest of this category is nanosuspensions that aim at creating nanoformulations comprising drug nanocrystals that are stabilized using surfactants and/or steric stabilizers to maintain the nanoparticle size and crystal form. As per a report published in 2024, the first nanocrystal-based tablet product Rapamune® (sirolimus) was approved in 2000, and since then, about 20 products based on nanosuspension/nanocrystal technology have been approved by the US FDA. One of the recent ones in this category is Cabenuva® (cabotegravir/rilpivirine), a combination drug nanosuspension for intramuscular injection for the management of HIV-1 infections [37, 38]. One key technology in this arena is Elan's NanoCrystal® technology, which, in partnership, has entered the market with about nine product approvals. Furthermore, this strategy works best for drugs with high doses. Rather than encapsulating high-dose drugs in a nanoexcipient matrix that increases the bulk of the product, nanosuspensions utilize drug nanocrystals themselves, hence allowing accommodation of higher doses while controlling the product bulk for patient-compliant dosage form.

Lipid-based nanoformulations are another common nanoformulation platform explored for poorly soluble drugs. Many such products are already available in the market that include Sandimmune®, Norvir®, Lamprenes®, Elyxyb®, etc. This strategy harnesses the drug lipophilicity by solubilizing the drug in a lipophilic matrix (lipids/oils), followed by emulsification or vesicle formation that is capable of enhancing drug bioavailability. Self-emulsifying drug delivery systems (micro/nano) are an abundant category explored in this context, which utilizes a unique combination of oil, surfactant, and stabilizers to dissolve the drug in the oily phase with a capability of undergoing self-emulsification upon administration, rendering high drug solubility and permeability via colloidal droplet formation [9, 39]. Importantly, the products are prepared and stored as oily one-phase solutions and are therefore more stable than regular biphasic emulsions. Further, they undergo in situ emulsification in body fluids, allowing the benefits of emulsion-based drug bioenhancement. One such example is the US FDA-approved oral solution of celecoxib (Elyxyb®) for the acute treatment of migraine in adults. Celecoxib is a BCS Class II compound with inherently low water solubility, which was formulated as self-microemulsifying drug delivery system (SMEDDS) to enhance the solubility and bioavailability. This advanced formulation has been clinically proven to enable faster absorption, resulting in a rapid onset of action, which is critically important for patients experiencing sudden and severe migraine attacks. Self-emulsifying systems, though look lucrative and easy to formulate, have the challenge of identifying the uniquely emulsifying oily composition and optimizing drug loading [9, 38, 40]. For a detailed description of this approach, refer to Chapter 5.

Vesicular lipid nanocarriers like liposomes, ethosomes, phytosomes, etc., have also been explored, and their unique feature is the capability of encapsulating both lipophilic and hydrophilic drugs. In the context of enhancing drug solubility, they can be used for both BCS Class II and IV category drugs and can be very advantageous for BCS Class IV drugs. These vesicles have a unique permeability feature owing to their bilayer structure, enabling better bioenhancement for BCS Class IV drugs with both poor solubility and permeability [41, 42]. For a detailed description of this approach, refer to Chapter 6.

Other nanoformulations explored for resolving drug solubility challenges are mesoporous silica-based nanocarriers, dendrimers, drug conjugates, etc., which are used in unique scenarios based on drug physicochemical properties and target product profile. Mesoporous silica has been investigated as an amalgamation of a solid dispersion approach for drug solubility enhancement with nanoformulations, wherein a highly nanoporous excipient like mesoporous silica (mean pore size: 2–50 nm; high surface area: up to 500 m^2/g; high pore volume: >1 cm^3/g) has been used to encapsulate the poorly soluble drug [43]. For a detailed description of this approach, refer to Chapter 7.

A combination of the aforesaid strategies has also been investigated to resolve the challenge of poor drug solubility early on in product development. For example, a study explored a combination of nanocrystal and co-crystal technologies to enhance drug solubility. Co-crystals of itraconazole and indomethacin with various coformers were prepared, and the particle size was reduced to a nanoscale (300–450 nm) using wet milling. Kinetic solubility studies showed that the nano co-crystals significantly increased the solubility compared to raw free drugs and conventional co-crystals. Specifically, the itraconazole–succinic acid nano co-crystal—achieved ~51.5-fold higher solubility compared to the free drug. These results demonstrate that combining multiple formulation strategies can synergistically improve solubility and dissolution, offering a promising approach for enhancing the bioavailability of poorly soluble drugs [44–46]. Further, a choice of drug dissolution enhancement strategy can vary based on the form of the drug, route of administration, etc. For example, Lenacapavir (BCS Class IV) has been approved as Sunlenca® for management of HIV. This unique formulation treatment dose is initiated as an oral therapy comprising a tablet dosage form containing the salt of the drug (lenacapavir sodium) to enhance its oral bioavailability to achieve an initial loading dose. This is followed by a twice-a-year subcutaneous injection given (drug with cosolvent polyethylene glycol depot) to achieve slow and sustained release of the drug over 6 months. Therefore, the dosage form design can vary based on the desired target product profile and therapy regimen [47, 48]. Therefore, the drug solubility

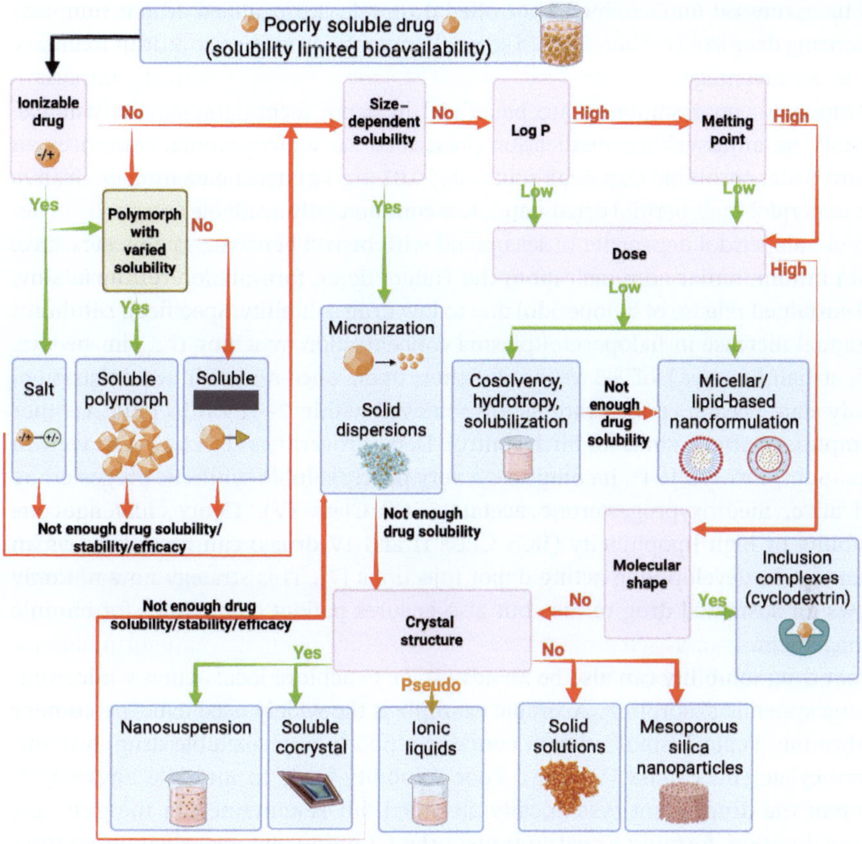

Figure 1.3 Schematic roadmap for selection of appropriate solubility enhancement approaches for successful product development of poorly soluble drugs.

enhancement approach for product development is rather a multipronged approach that requires understanding of the drug profile, dose, dosage form, route of admin istration, efficacy, stability, and commercial viability. Hence, there is no rule of thumb or a gold standard for the use of a specific drug solubility enhancement strategy. Based on extensive literature analysis and our experience in the field, we have developed a possible roadmap to narrow down amicable approaches for product development of poorly soluble drugs. This approach is mainly based on the drug's physicochemical properties and the desired target product profile, which has been depicted in Figure 1.3.

1.5 Poor Drug Solubility–Is It Always a Challenge?

In most cases, poor drug solubility is broadly considered a challenge or a problem, as it limits drug absorption, bioavailability, and overall efficacy. However, in some cases, it can be looked upon as a double-edged sword, wherein the low drug solubility can be harnessed for sustained/controlled drug release, localized drug action, and enhancing drug safety. Therefore, in some therapeutic strategies, low drug solubility can be an advantage.

Drugs with poor solubility are beneficial in depot formulations. For instance, Haldol®, an antipsychotic medication prescribed for schizophrenia, comprises an almost water-insoluble (aqueous solubility: 0.01 mg/kg) decanoate ester derivative of haloperidol (haloperidol decanoate). It is commercially available as a sterile injection of haloperidol decanoate in sesame oil with benzyl benzoate as a preservative. Upon intramuscular administration, the Haldol depot formulation results in slow and sustained release of haloperidol due to low drug solubility. Specifically, it shows a gradual increase in haloperidol plasma concentration, reaching C_{max} in ~6 days, with a half-life ($t_{1/2}$) of ~3 weeks. Further, upon once-a-month administration, steady-state plasma concentrations are achieved within 2–4 months [49]. Another example is quarterly shots for birth control, Depo-Provera®, and its low-dose version Depo-SubQ Provera 104®. It comprises a very poorly soluble synthetic progesterone derivative, medroxyprogesterone acetate (BCS Class IV). Hence, low aqueous solubility or high lipophilicity (BCS Class II and IV drugs) can be utilized as an advantage to develop long-acting depot injections [7]. This strategy now not only allows for sustained drug release but also ensures patient compliance for chronic management.

Poor drug solubility can also be an advantage to achieve local action while minimizing systemic absorption. A simple example is the widely used over-the-counter medication, Pepto Bismol®, which contains a poorly water-soluble drug, bismuth subsalicylate (BCS Class IV drug). Poor solubility leads to multiple advantages, wherein the drug is not systemically absorbed but is contained in the gut for a longer duration, forming a coat to protect the GI lining and also elicit an antibacterial effect. Further, low drug solubility can also be advantageous in a topical product desiring local action without systemic absorption. For example, topical antifungal medication like Monistat® contains a low water-soluble drug, miconazole (BCS Class II), allowing prolonged and effective local and topical antifungal effects. Another interesting example will be hetacillin, which was developed as a water-soluble prodrug of the antibiotic ampicillin for oral use, but was later withdrawn due to rapid transformation to the parent drug with no added benefit. This drug, though withdrawn for human use, was repurposed for veterinary use for mastitis in cows (PolyMast), given via intramammary administration as an oil-based gel. The product so developed allows sustained release of the drug, ensuring efficacy [50, 51]. To summarize, it is advised that the corrective action to enhance drug solubility is not always the preferred approach, but a formulation and product design team must take into consideration the route of administration and the target product profile while designing the product and formulation process.

1.6 Conclusion

To summarize, it is evident that poor aqueous solubility represents a critical barrier in overall pharmaceutical development. Poor drug solubility, especially in aqueous media, plays a vital role in determining ADME/Tox characteristics, thereby influencing therapeutic efficacy and safety and impacting the product stability and commercial success. Multiple strategies have been explored and successfully implemented in developing safe, effective, and stable commercial products of poorly soluble drugs. But there is no single rule of thumb to identify and implement such a strategy. The experts must adopt a multipronged approach of identifying the most suitable drug solubility enhancement approach based on drug physicochemical properties, dose, route of administration, disease state, and commercial feasibility.

References

1 International Union of Pure and Applied Chemistry (IUPAC) (ed.) (2025). Solubility. 5.0.0. https://goldbook.iupac.org/terms/view/S05740 (accessed 24 March 2026).

2 Savjani, K.T., Gajjar, A.K., and Savjani, J.K. (2012). Drug solubility: importance and enhancement techniques. *Int. Sch. Res. Notices* 2012 (1): 195727.

3 SStUN (2012). Description and solubility. http://www.triphasepharmasolutions.com/Private/USP%20DESCRIPTION%20AND%20SOLUBILITY.pdf (accessed 24 March 2026).

4 Desai, P.P., Date, A.A., and Patravale, V.B. (2012). Overcoming poor oral bioavailability using nanoparticle formulations – opportunities and limitations. *Drug Discov. Today Technol.* 9 (2): e87–e95.

5 Walters, W.P. and Murcko, M.A. (2002). Prediction of 'drug-likeness'. *Adv. Drug Deliv. Rev.* 54 (3): 255–271.

6 Lipinski, C.A. (2000). Drug-like properties and the causes of poor solubility and poor permeability. *J. Pharmacol. Toxicol. Methods* 44 (1): 235–249.

7 Alidori, S., Subramanian, R., and Holm, R. (2024). Patient-centric long-acting injectable and implantable platforms — an industrial perspective. *Mol. Pharm.* 21 (9): 4238–4258.

8 Khatri, H., Hussain, M.S., and Tyagi, S. (2022). Solubility enhancement techniques: an overview. *World J. Pharm. Res.* 11 (5): 34–40.

9 Liu, Y., Liang, Y., Yuhong, J. et al. (2024). Advances in nanotechnology for enhancing the solubility and bioavailability of poorly soluble drugs. *Drug Des. Devel. Ther.* 1469–1495.

10 He, Z., Yang, W., Yang, F. et al. (2024). Innovative medicinal chemistry strategies for enhancing drug solubility. *Eur. J. Med. Chem.* 279: 116842.

11 Rocha, B., de Morais, L.A., Viana, M.C., and Carneiro, G. (2023). Promising strategies for improving oral bioavailability of poor water-soluble drugs. *Expert Opin. Drug Discov.* 18 (6): 615–627.

12 Nyamba, I., Sombié, C.B., Yabré, M. et al. (2024). Pharmaceutical approaches for enhancing solubility and oral bioavailability of poorly soluble drugs. *Eur. J. Pharm. Biopharm.* 204: 114513.

13 Kawabata, Y., Wada, K., Nakatani, M. et al. (2011). Formulation design for poorly water-soluble drugs based on biopharmaceutics classification system: basic approaches and practical applications. *Int. J. Pharm.* 420 (1): 1–10.

14 Gulsun, T., Borna, S.E., Vural, I., and Sahin, S. (2018). Preparation and characterization of furosemide nanosuspensions. *J. Drug Delivery Sci. Technol.* 45: 93–100.

15 Kumar, S., Jog, R., Shen, J. et al. (2015). Formulation and performance of danazol nano-crystalline suspensions and spray dried powders. *Pharm. Res.* 32 (5): 1694.

16 Sun, C.C. and Davé, R.N. (2022). Crystal and particle engineering – an indispensable tool for developing and manufacturing quality pharmaceutical products. *Pharm. Res.* 39 (12): 3041–3045.

17 Anagnostis, S., Khehra, N., and Parmar, M.S. (2024). Chlorpromazine. In: *Encyclopedia of Toxicology*, 4e (ed. P. Wexler), 995–1002. Oxford: Academic Press.

18 Gupta, D., Bhatia, D., Dave, V. et al. (2018). Salts of therapeutic agents: chemical, physicochemical, and biological considerations. *Molecules* 23 (7): 1719.

19 Shi, Q., Chen, H., Wang, Y. et al. (2022). Recent advances in drug polymorphs: aspects of pharmaceutical properties and selective crystallization. *Int. J. Pharm.* 611: 121320.

20 Healy, A.M., Worku, Z.A., Kumar, D., and Madi, A.M. (2017). Pharmaceutical solvates, hydrates and amorphous forms: a special emphasis on cocrystals. *Adv. Drug Deliv. Rev.* 117: 25–46.

21 EsteveInc (2021). Esteve announces FDA approval of a novel co-crystal form of celecoxib and tramadol for the management of acute pain. https://www.esteve.com/global/media/news/esteve-announces-fda-approval-of-a-novel-co-crystal-form-of-celecoxib-and-tramadol-for-the-management-of-acute-pain (accessed 24 March 2026).

22 Subbaiah, M.A.M., Rautio, J., and Meanwell, N.A. (2024). Prodrugs as empowering tools in drug discovery and development: recent strategic applications of drug delivery solutions to mitigate challenges associated with lead compounds and drug candidates. *Chem. Soc. Rev.* 53 (4): 2099–2210.

23 Fralish, Z., Chen, A., Khan, S. et al. (2024). The landscape of small-molecule prodrugs. *Nat. Rev. Drug Discov.* 23 (5): 365–380.

24 Markovic, M., Ben-Shabat, S., and Dahan, A. (2020). Prodrugs for improved drug delivery: lessons learned from recently developed and marketed products. *Pharmaceutics* 12 (11): 1031.

25 Karbasi, A.B., Barfuss, J.D., Morgan, T.C. et al. (2024). Sol-moiety: discovery of a water-soluble prodrug technology for enhanced oral bioavailability of insoluble therapeutics. *Nat. Commun.* 15 (1): 8487.

26 Zarei, A., Haghbakhsh, R., and Raeissi, S. (2023). Overview and thermodynamic modelling of deep eutectic solvents as co-solvents to enhance drug solubilities in water. *Eur. J. Pharm. Biopharm.* 193: 1–15.

27 Laksitorini, M.D., Suryani, L.U., Muhammad, F.R., and Purnomo, H. (2023). Application of Hildebrand solubility parameter to identify ethanol-free co-solvent for pediatric formulation. *Indones. J. Pharm.* 34 (2): 6627.

28 Alqarni, M.H., Haq, N., Alam, P. et al. (2021). Solubility data, Hansen solubility parameters and thermodynamic behavior of pterostilbene in some pure solvents and different (PEG-400+ water) cosolvent compositions. *J. Mol. Liq.* 331: 115700.

29 Jouyban, A. (2019). Review of the cosolvency models for predicting drug solubility in solvent mixtures: an update. *J. Pharm. Pharm. Sci.* 22: 466–485.

30 Paul, R., Chattaraj, K.G., and Paul, S. (2021). Role of hydrotropes in sparingly soluble drug solubilization: insight from a molecular dynamics simulation and experimental perspectives. *Langmuir* 37 (16): 4745–4762.

31 Sintra, T.E., Abranches, D.O., Benfica, J. et al. (2021). Cholinium-based ionic liquids as bioinspired hydrotropes to tackle solubility challenges in drug formulation. *Eur. J. Pharm. Biopharm.* 164: 86–92.

32 Zhuo, Y., Cheng, H.-L., Zhao, Y.-G., and Cui, H.-R. (2024). Ionic liquids in pharmaceutical and biomedical applications: a review. *Pharmaceutics* 16 (1): 151.

33 Cid-Samamed, A., Rakmai, J., Mejuto, J.C. et al. (2022). Cyclodextrins inclusion complex: preparation methods, analytical techniques and food industry applications. *Food Chem.* 384: 132467.

34 JanssenPharmaceuticalCompanies (2024). SPORANOX® (itraconazole) oral solution. https://www.accessdata.fda.gov/drugsatfda_docs/label/2024/020657s040lbl.pdf (accessed 24 march 2026).

35 TabletingTechnology (2024). Trends in amorphous solid dispersion drug products approved by the U.S. Food and Drug Administration between 2012 and 2023. https://tabletingtechnology.com/ressources/newsfeeds-page/amorphous-solid-dispersion-drug-products-fda-2012-2023/#:~:text=Between%202012%20and%202023%2C%20the,for%20pediatric%20and%20other%20populations (accessed 24 March 2026).

36 Alshehri, S., Imam, S.S., Hussain, A. et al. (2020). Potential of solid dispersions to enhance solubility, bioavailability, and therapeutic efficacy of poorly water-soluble drugs: newer formulation techniques, current marketed scenario and patents. *Drug Deliv.* 27 (1): 1625–1643.

37 Pardhi, E., Vasave, R., Srivastava, V. et al. (2024). Nanocrystal technologies in biomedical science: from the bench to the clinic. *Drug Discov. Today* 29 (3): 103913.

38 Rossier, B., Jordan, O., Allémann, E., and Rodriguez-Nogales, C. (2024). Nanocrystals and nanosuspensions: an exploration from classic formulations to advanced drug delivery systems. *Drug Deliv. Transl. Res.* 14 (12): 3438–3451.

39 Desai, P., Thakkar, A., Ann, D. et al. (2019). Loratadine self-microemulsifying drug delivery systems (SMEDDS) in combination with sulforaphane for the synergistic chemoprevention of pancreatic cancer. *Drug Deliv. Transl. Res.* 9 (3): 641–651.

40 Gangavarapu, A., Tapia-Lopez, L.V., Sarkar, B. et al. (2024). Lipid nanoparticles for enhancing oral bioavailability. *Nanoscale* 16 (39): 18319–18338.

41 Abouhussein, D.M. (2021). Enhanced transdermal permeation of BCS class IV aprepitant using binary ethosome: optimization, characterization and ex vivo permeation. *J. Drug Delivery Sci. Technol.* 61: 102185.

42 Cheng, X., Han, X., Si, J. et al. (2024). Cationic curcumin nanocrystals liposomes for improved oral bioavailability: formulation development, optimization, in vitro and in vivo evaluation. *Pharmaceutics* 16 (9): 1155.

43 Baumgartner, A. and Planinšek, O. (2021). Application of commercially available mesoporous silica for drug dissolution enhancement in oral drug delivery. *Eur. J. Pharm. Sci.* 167: 106015.

44 Huang, Z., Staufenbiel, S., and Bodmeier, R. (2022). Combination of co-crystal and nanocrystal techniques to improve the solubility and dissolution rate of poorly soluble drugs. *Pharm. Res.* 39 (5): 949.

45 Almansa, C., Frampton, C.S., Vela, J.M. et al. (2019). Co-crystals as a new approach to multimodal analgesia and the treatment of pain. *J. Pain Res.* 12: 2679–2689.

46 Pantwalawalkar, J., Kale, N., Nangare, S. et al. (2025). Pharmaceutical cocrystals: unlocking the potential of challenging drug candidates. *J. Drug Delivery Sci. Technol.* 104: 106572.

47 GileadSciences (2024). Highlights of prescribing information. https://www.gilead.com/-/media/files/pdfs/medicines/hiv/sunlenca/sunlenca_pi.pdf (accessed 24 March 2026).

48 Ma, C., Chang, J., and Yu, B. (2024). Sunlenca® (Lenacapavir): A first-in-class, long-acting HIV-1 capsid inhibitor for treating highly multidrug-resistant HIV-1 infection. *Acta Pharm. Sin. B* 14 (12): 5512–5514.

49 JanssenPharmaceuticals (2025). HALDOL® Decanoate 50 (haloperidol) HALDOL® Decanoate 100 (haloperidol) for IM injection only. https://www.janssenlabels.com/package-insert/product-monograph/prescribing-information/HALDOL+Decanoate-pi.pdf (accessed 24 March 2026).

50 Drugs.com (2025). PolyMast (hetacillin potassium). https://www.drugs.com/vet/polymast-hetacillin-potassium.html#:~:text=Intramammary%20Infusion,any%20alteration%20in%20the%20milk (accessed 24 March 2026).

51 Gingrich, K.F. II, (2023). Boehringer ingelheim udder health 360 coverage advertisement. *Bov. Pract.* 57 (2): 858.

2

Design and Development of Amorphous Solid Dispersion as Enabling Formulation Strategies of Poorly Soluble Drugs

Amrit Paudel[1,2] and Sarah Stryeck[2]

[1]*Graz University of Technology, Institute of Particle and Process Engineering (IPPE), Inffeldgasse 13, Graz 8010, Austria*
[2]*Research Center Pharmaceutical Engineering GmbH, Inffeldgasse 13, Graz 8010, Austria*

2.1 Basics of Amorphous Solid Dispersion (ASD)

2.1.1 Introduction

Over the last decades, significant portions of new chemical entities entering industrial drug development have possessed poor water solubility, leading to enormous absorption challenges when orally administered [1]. Enhancing bioavailability for such drugs is essential, as low solubility limits absorption and, consequently, therapeutic effectiveness. Among the various formulation approaches, amorphous solid dispersions (ASDs) have become a leading solubilization strategy, especially given their ability to improve the thermodynamic solubility and dissolution rate of the drug and maintain supersaturation in vivo [2]. The ASD is a drug–polymer molecular mixture, often referred to as a one-phase glass solution. The basic mechanism of increasing solubility by amorphous state is elimination of crystal lattice contribution to solubilization process, and polymer used as a carrier in ASD stabilizes amorphous state of active pharmaceutical ingredients (APIs) via decreasing molecular mobility and retarding nucleation and crystal growth via stronger drug–polymer intermolecular interaction [3]. Besides solid-state stabilization, the hydrophilic polymers used in ASD also help maintain supersaturated concentration of the solubilized drug during dissolution by acting as the precipitation inhibitor, imparting parachute effect to the supersaturation spring generated by high-energy amorphous state [4]. This growing interest in ASDs reflects their regulatory success and evolving formulation practices, making them integral to modern drug product design and delivery for the poorly soluble drugs requiring higher dosage such as for antiviral, anticancer, and respiratory therapeutics. With the recent US FDA approvals highlighting both product diversity and therapeutic scope, understanding ASD trends is critical for pharmaceutical scientists seeking innovative, patient-centered solutions in drug development.

2.1.2 Basic Design Principles of ASD

The rational ASD design deploys experimental and computational tools that enable the selection of the best polymer carrier and drug loading to balance biopharmaceutics, stability, and processability performances (Figure 2.1). These aspects highly depend upon the intrinsic physicochemical properties of neat drug, for example, intrinsic amorphous stability against crystallization and the ease of amorphization; potential for chemical degradation; and solubility in aqueous, biorelevant, and various organic solvents. The ideal ASD composition provides the extent of drug solubilization during in vitro and in vivo dissolution of the ASD dosage form so that sufficient intestinal absorption is achieved, minimal to no food effect is observed on oral bioavailability, and requires minimal polymer carrier for solubilization and stabilization. From a stability standpoint, the selected polymer and the drug-to-polymer ratio are supposed to prevent phase separation, nucleation, and crystal growth of APIs in ASD during the storage period of designated shelf life under predefined conditions. Since moisture and temperature are two major factors impacting both physical and chemical instability risk, ASD formulations with less sensitivity to these environmental factors are beneficial from a stability standpoint and simple primary packaging. The manufacturing process selection for a given ASD is another important factor and is mainly driven by API and polymer. Common ASD manufacturing processes that are viable and proven at commercial scale are spray drying (SD) and hot melt extrusion (HME). As SD requires single-solution preparation of drug and polymer dissolved in volatile organic solvent(s), sufficient solubility of both components is important. While for the use of HME, desirable thermophysical properties such as lower melting temperatures and melting viscosity and higher thermal stability of API and polymer are the prerequisites. Beyond biopharmaceutics, stability, and manufacturability, the selection of development trajectory of ASD is also heavily influenced by the required dose for the given therapy, downstream processability of ASD intermediates, overall development speed and phase, patient-related factors, as well as sustainability and techno-economical aspects of formulations and processes.

2.1.3 Design Attributes of ASD

For developing ASD, understanding of the basic chemistry and physical structure of ASD and their stabilization mechanisms in solid state, as well as in dissolution environment, is necessary. Amorphous materials at higher temperatures are called supercooled liquid or rubbery state, possessing thermodynamic (free energy, entropy, heat capacity, free volume, etc.) and kinetic (molecular vibration, rotation, diffusion time, viscosity, etc.) properties comparable to molten liquid. The properties of supercool liquids follow temperature dependence. Lowering the temperature of rubbery amorphous state leads, to a certain extent, to breaking the linear relation so that the property versus temperature slope becomes smaller. Crossing this region, called glass transition region, leads to amorphous material to a glassy state, wherein the energy and mobility of molecules are kinetically arrested. The temperature relation to this region is called glass transition temperature (T_g). Glass transition is an important phenomenon of amorphous material, and the T_g is a critical descriptor associated with its structure and dynamics and decisive to several of its thermodynamic and kinetic properties. Due to higher energetic and faster molecular mobility, crystallization timescale in supercooled state is orders of magnitude shorter than in its glassy state. Therefore, the basic mechanism of stabilization of amorphous state of low-molecular-weight API by using a high-molecular-weight polymer carrier is that the lower T_g of the API is significantly elevated by forming a single-phase molecular mixture with a high T_g polymer, the phenomenon called anti-plasticization. As a rule of thumb, the increase in the T_g of ASD by at least 50 °C leads to infinitely slow mobility and diffusivity of molecules, preventing the chance of nucleation and crystal growth. The readers are referred to different books and literature for the details of these aspects [5–8]. Some key attributes governing the ASD design are briefly presented in the following sections (Figure 2.1).

2.1.3.1 Intrinsic Amorphization Potential of API

The possibility of predicting potential success of forming stable ASD of a given API, utilizing the molecular structure and physicochemical properties of API, can save enormous time. To begin with, the inherent amorphization potential of a neat API can already provide some hint in this direction. The glass-forming ability (GFA) of a material is intrinsic propensity to transform into amorphous solid state

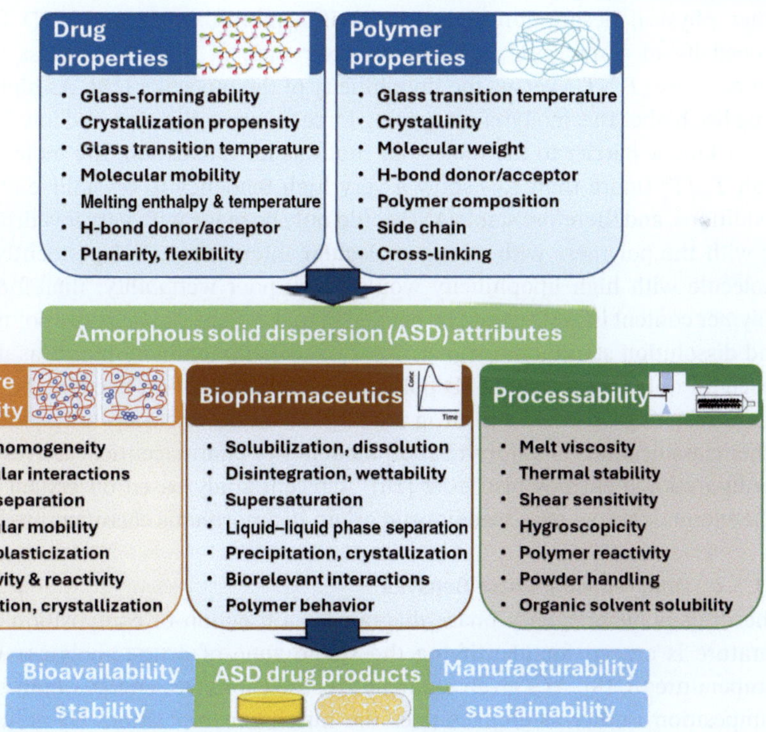

Figure 2.1 Design parameters and design attributes for ASD product development.

when the melt is cooled sufficiently fast. This is determined by index/slope of the temperature-dependent increase in the viscosity of melt, as it approaches the T_g, called fragility. The material with higher fragility values (fragile liquid) tends to crystallize upon cooling from melt, while that with lower fragility (strong liquid) will form amorphous material upon the solidification of the melt. The fragility of organic drug molecules has been structurally related to the number of rotatable bonds they have in their structure [9]. Also, aromaticity of the molecule has been known to increase the molecular rigidity and slow down the local molecular motion [10]. The other features such as molecular conformation-related descriptor such as R-GETAWAY and third-order autocorrelation index weighted by the atomic mass (*R3m*) of an API molecule have recently been proposed as a descriptor to classify whether a molecule can form ASD with a polymer carrier or not [11]. Around 0.6 was found as the threshold value of *R3m* above which drug–polymer melt forms ASD and below which practically none form fully amorphous dispersion. However, this concept has been tested so far only for a single ASD carrier. Widely cited works on GFA of drug molecules classify drugs based on their amorphization characteristics during melt cooling and reheating [12] or variable rate of cooling to find critical cooling rate for amorphization [13]. GFA class I, class II, and class III molecules can be solidified as amorphous at higher, medium, and slow rates of cooling. Evidently, GFA III molecules tend to form stable ASD; however, this GFA class is not always predictive to the behavior for multicomponent ASDs [14]. The other physical description used to classify amorphization and ASD formation propensity of a molecule is based on their ratio of melting to glass transition temperature (T_m/T_g) versus the lipophilicity of the molecule [15]. As high-melting drug has higher thermodynamic driving force for crystallization and low T_g leads to lower kinetic barrier to the molecular motions for nucleation, the molecules with high T_m/T_g (more than 1.4) show a very high tendency to crystallize at ambient conditions, and therefore stable ASD could only be made with very low drug loading or with the polymers with strong molecular interactions. Independently, an API molecule with high lipophilicity would show poor wettability; therefore, higher polymer content in ASD would be needed to make them viable from a solubilization and dissolution standpoint. Again, T_m/T_g versus lipophilicity map was developed based on only one class of ASD polymer; therefore, caution needs to be practiced when selecting polymer and drug loading based on this map only. Besides this, other classifications of amorphization tendency of pharmaceutical actives based on configurational entropy also exist [16]. A recent study based on crystal structure and energetics provides a mechanistic origin to amorphous classification [17].

2.1.3.2 Drug–Polymer Phase Behavior

The drug–polymer binary phase diagram as a function of composition and temperature is crucial for identifying the stable zone of drug-to-polymer ratio and temperature [6, 18]. At a given temperature, the Gibbs free energy of mixing versus composition isotherms characterizes the miscible composition. Namely, exothermic mixing region indicates the stable composition region, while endothermic mixing region is the one with the potential for phase separation and instability.

Therefore, the composition characterizing the minimum of mixing energy versus the composition isotherm represents the most miscible drug-to-polymer ratio. The binary molecular mixture of amorphous drug and polymer can show amorphous–amorphous phase separation with the increase of drug fraction and/or the decrease of temperature, featured by upper critical solution temperature. The phase boundary, splitting stable one-phase amorphous mixture into the zone with the potential of amorphous phase separation, is referred to as binodal line. Increasing the drug fraction further or decreasing the temperature shifts the binary phase to the spinodal region, where a slight change in composition/temperature leads to spontaneous crystallization. The concentration of drug in polymer at which thermodynamically stable drug–polymer complexes exist at the given temperature is defined as the solid solubility of drug in polymer. The ASD with drug content below the solid solubility would show the best stability. However, thermodynamic solid solubility of drug-like molecules in carrier polymer is typically below 5 wt%. Therefore, ASD with such low drug content is practically not beneficial for the drug dose required for most therapies (especially high drug doses). Yet, selecting the polymer with the highest solid solubility of the selected API leads to the lowest thermodynamic driving force for crystallization. Beyond the intrinsic solid solubility limit, the miscible ASD system can still form one-phase amorphous dispersion until certain drug content, called miscibility limit, that is kinetically stable. Such kinetically driven amorphization is triggered by rapid solvent evaporation and droplet-to-particle formation during SD process and rapid solidification of drug–polymer melt mixtures during HME process. The commercially and clinically feasible drug-to-polymer ratio in ASD would be between the solid solubility and miscibility limit within the phase diagram, therefore, are often supersaturated with respect to solid solubility [19, 20]. The extent of solid-state supersaturation is optimized to obtain desired solid-state stability and target pill/capsule size. The better elucidation of ASD in terms of stability and dissolution characteristics would also require understanding the ternary phase diagram, including moisture/humidity or water [21–24]. From a stability standpoint, moisture drastically reduces the drug–polymer solubility and miscibility, thus narrowing down the stable phase zone. This comes from the drastic reduction of T_g due to plasticization by water molecules (each percentage of moisture roughly decreases T_g by 10 °C). This will also have consequences on the behavior of surface dissolution and crystallization interplay, impacting the biopharmaceutics performance. To be further informative to in vitro and in vivo performance, more complex phase diagrams can be generated by including GI fluid components such as bile salts/acids and/or phospholipids and underlying nanoscopic colloidal species that are formed during ASD dissolution.

2.1.3.3 Molecular Mobility in ASD

Molecular mobility in amorphous materials is an order of magnitude faster than that of their crystalline counterparts, and they are often the kinetic drivers for the instability of amorphous state toward crystallization and/or degradation [25, 26]. There are two characteristic types of non-translational molecular motions of amorphous solid-state materials. The timescale taken for a molecule to vibrate entirely around its center of gravity is called primary relaxation time or alpha relaxation time. This is a global molecular motion as entire molecule is involved in the process. This motion process is responsible for most of the reactivity of amorphous states like crystallization or degradation. On the other hand, a segment of molecules (especially in large molecules like polymers), a functional group, or a specific rotational bond exhibits characteristics of relaxation behavior, called secondary relaxation or local molecular motion. At temperatures significantly lower than T_g in glassy state, this secondary motion is still prevalent and responsible for nucleation, crystal growth, or degradation. A characteristic type of secondary motion that is precursor of primary motion is called beta relaxation process or Johari–Goldstein (JG) relaxation process. In a glassy state, the alpha relaxation process is markedly slower, and in a glass transition region, the relaxation time follows the nonlinear temperature dependence. Beyond glass transition region, alpha relaxation time drastically decreases and shows linear temperature dependence. The secondary motions typically show linear temperature dependence with two slopes below and above glass transition. At certain temperature above T_g, called crossover temperature, the overall molecular motions become so fast that all primary and secondary relaxation time converges. Molecules are present as ensembles of short-range order in amorphous domain characterized by cooperatively rearranging region (CRR), wherein molecules exhibit collective motion. Beyond CRR, relaxation time is distributed across amorphous domain, showing nanoscale density heterogeneity and relaxation time distribution called dynamic heterogeneity. In ASD, alpha relaxation time of API is generally (but not necessarily) prolonged by the formation of molecular mixture with high T_g polymer. Such anti-plasticization (or increase in relaxation time) can sometimes be more prominent for secondary molecular motion than alpha relaxation, depending upon the drug and polymer type. Such effects prolong nucleation time beyond shelf life, as relaxation time is related to characteristic nucleation induction time in solid state [27]. An important consideration in interpreting molecular mobility requires the knowledge of the surface-to-volume ratio of the solid, as mobility at the surface can be orders of magnitude faster than in the bulk amorphous material [28]. This can be relevant for ASD of different particle size and thickness generated via SD or different size and shape of ASD particles generated by milling HME strands. The other characteristics related to molecular mobility are viscosity of amorphous state, free volume, and diffusivity, and they show identical temperature dependence. These descriptions are related via various empirical formalisms, such as Vogel–Fulcher–Tammann equation and Adam–Gibbs equation, the details of which are found elsewhere.

2.1.3.4 Drug–Polymer Interactions

Like phase behavior and molecular mobility, drug–polymer intermolecular interactions are crucial molecular mechanisms of ASD stabilization [29]. To form a stable ASD, drug–drug and polymer–polymer homomolecular interactions are to be broken down to form stronger heteromolecular interactions between drug and polymer molecules in solid state [30]. All such interactions are noncovalent in nature; therefore, no chemical changes are supposed to occur due to such interactions. The most prominent and ubiquitous stabilizing interaction is intermolecular hydrogen (H) bonding formed between drug and polymer molecules in ASD [7, 31]. The H-bond is formed by sharing electrons between the donated proton (H) that is covalently bonded to an electron-rich atom (such as oxygen, nitrogen, sulfur) and the proton-accepting electron-rich atom. Generally, drug molecules are rich in H-bond donor group, and most ASD polymers are rich in H-bond acceptor group, while there is often the chance that the molecule contains both H-bond donor and acceptor groups of different strengths. Strong H-bonding between drug and polymer is known to reduce the propensity of amorphous phase separation and retard or prevent drug nucleation from ASD. Such interactions are also known to reduce the overall molecular mobility in ASD. Therefore, T_g can be higher than theoretically predicted using ideal mixing rules for strongly interacting ASD. The stronger drug–polymer interactions in an ASD drastically reduce molecular mobility and potentially delay the crystallization onset [32]. Besides H-bonding, halogen bonding is also emerging and observed as an important stabilizing interaction in ASD, given the fact that most of the drug molecules possess one or more halogen atoms in their structure [33, 34]. The other types of weaker yet notable molecular interactions between drug and polymer molecules contributing to ASD stability can be electrostatic interactions, hydrophobic interactions involving electrons associated with the conjugated/aromatic double bonds, dipole–dipole interactions, etc. While strong molecular interactions such as ionic interactions and H-bonding favor ASD stabilization, they are not the sole prerequisite for stability. Sometimes, anti-plasticization and the reduction in molecular mobility by polymer in ASD can occur without strong molecular interactions, especially for the drug molecules lacking strong H-bond donor/acceptor. Also, as water molecules possess a strong H-bond acceptor and donor, exposure of ASD to moisture can lead to the breaking of drug–molecule H-bonding and the water molecules interacting with drug and polymer molecules. This can be the triggering mechanism underlying the moisture-induced phase separation and crystallization in ASDs [19, 30]. Besides physical stability, the role of drug–polymer interactions on chemical degradation of drug in ASD, as well as on drug release behavior from ASD, is still largely unknown. In recent studies, Que et al. showed that strong drug–polymer H-bonding in ASD can retard drug release [35], while Saraf et al. showed that stronger H-bonding can sometimes be detrimental in terms of chemical stability of drug in ASD [36].

2.1.4 Formulation Basis of ASD Products

The key aspects of ASD formulation development need to encompass the rational selection of polymer stabilizers, balancing the biopharmaceutics, stability, and processability requirements from the product development standpoint. Following the polymer selection, the drug loading or drug-to-polymer ratio is selected also based on the downstream processability, the dose or dose range requirement, the anticipated frequency of administration, the phase of development (early versus late clinical phase), and the dosage form presentation (e.g. tablet or capsules or granules). Also, the selection of non-ASD excipients and their grades required for final dosage form development can influence the performance of ASD in finished dosage forms as compared to ASD intermediates. For example, depending upon biopharmaceutics or processability characteristics, wetting agents, disintegrants, glidants, and flow promoters are to be added in ASD dosage form in amounts sufficient enough that they would not compromise the stability and drug release from ASD.

2.1.4.1 Polymers Used in ASD

The polymers used in ASD are excipients exhibiting the proven track record for safety to be used in pharmaceutical products [37]. While most ASDs are binary drug–polymer systems, sometimes they are also formulated as ternary systems, adding additional component to enable biopharmaceutics, processing, or stability. To name some, such components include surfactants as a wetting agent, plasticizer to decrease viscosity of the drug–polymer melt for HME processing, etc. The most common polymer classes that have been used in commercial and investigational ASDs are vinyl polymers (polyvinyl pyrrolidone [PVP], polyvinyl pyrrolidone vinyl acetate [PVP/VA], polyethylene glycol [PEG], polyvinyl caprolactam–polyvinyl acetate–polyethylene glycol graft copolymer [Soluplus], etc.), cellulose polymers (hydroxypropyl methylcellulose [HPMC], HPMC-acetate/succinate [HPMC-AS], HPMC-phthalate [HPMC-P], hydroxypropyl cellulose [HPC], etc.), and acryl polymers (polyacrylic acid [PAA], polymethacrylate [Eudragit], etc.) [2]. As aforementioned, drug–polymer miscibility, drug solubility in polymer, anti-plasticization, molecular mobility reduction, and strong drug–polymer interactions will be the driver for the selection of the polymer based on the molecular structure of the drug and polymer. These polymers have different solvent solubilities relevant for SD and different thermal stability and melt viscosity relevant for HME that make them unique to be processed by one of the ASD manufacturing technologies, or some are amenable to both processes [38, 39]. For example, HPMC is only soluble in organic solvent mixtures for SD and has very high melt viscosity unfavorable for HME processing. HPMC-AS is soluble in wider types of organic solvents and possesses lower melt viscosity, therefore being suitable for both SD and HME processing. The same goes for PVP/VA; therefore, ASD products based on PVP/VA are found to be prepared via both SD and HME. While PVP is soluble in a wider diversity of solvents and thus suitable for SD, it has limited thermal stability and is therefore less suitable for HME. Polymers like PVP, PEG, and HPMC are available in different molecular weight, providing formulators to select the suitable one in terms of performance. For co-polymers such as PVPVA and substituted polymers like HPMC-AS and HPMC, other subtle structural properties such as degree and sequence of substitutions and polydispersity are equally important for the pharmaceutical performance [40]. The polymers used in ASD formulations are hydrophilic, have linear chain structures, and exhibit different degrees of water uptake and hygroscopicity, which can be decisive for their selection in terms of stability and powder handling/processability standpoint [41]. For example, vinyl polymers like PVP are more hygroscopic than cellulosic polymers like HPMC and HPMC-AS. Also, pH solubility profiles of polymers in aqueous media are an important selection characteristic for ASD polymers.

Polymers that are insoluble in acidic pH in the stomach and soluble in neutral to basic pH in the intestine are called enteric polymers. Enteric polymers like HPMC-AS follow different types of solubilization mechanisms than the polymers, like PVP, possessing pH-independent water solubility.

2.1.4.2 Drug Loading Selection

The selection of an optimal drug-to-polymer ratio is decisive from the performance standpoint of the given ASD; therefore, the right ratio ensures robust balance of biopharmaceutics, stability, and overall processability. From a patient perspective, higher drug-loaded ASD could provide an advantage over the treatment burden. For example, the possibility of producing smaller sizes of finished products can improve swallowability, or high drug load ASD can simply reduce the frequency of pill intake, which overall can reduce the pill burden [42]. Despite these clinical advantages, pushing drug loading excessively high would require a fundamental understanding of associated risks related to stability and biopharmaceutics, and associated rigor and resources for product development. Therefore, it is generally advised to keep drug loading as low as it is to fulfill the required dose. Such a safety margin allows the ASD concept to remain sufficiently far from the edge of performance failure risk, especially to tolerate several unforeseen variations such as API and polymer purity profiles, trace crystals or nuclei, ASD particle properties, or other manufacturing variations. The first driver for the optimal drug loading selection would be the target dissolution performance to achieve the clinical dose required for a given indication of the API. Since the increase in API content in polymer can lead to the lowering of stabilization potential of supersaturated concentration of API generated during dissolution of ASD, the optimal polymer concentration in ASD needs to ensure the prevention of precipitation/crystallization. At a given drug loading, ASD should provide sufficient spring effect of dissolved concentration, both from the initial rate of dissolution and the plateau concentration that is achievable. Also, the concentration achieved of dissolved polymer should be able to act as a strong parachute to maintain the supersaturated drug concentration for a sufficient period relevant to the region and window of GI absorption. Further, the dependence of drug loading on the food effect of drug dissolution and absorption from a given ASD system needs to be duly understood for the payload selection [43]. In the case of treatment needing multiple drugs, there can be biopharmaceutical risks associated with fixed or variable drug combinations on ASDs, as one drug could influence the solubilization and supersaturation of second drug when co-administered [44]. The drug load evidently has a crucial impact on the stability of ASD products. Physical stability failure can be minimal at miscible composition range, which is typical at low drug-loading region for most ASDs. High T_g of the selected polymer and/or strong drug–polymer interaction can reduce crystallization risk even at higher drug loading or prevent crystallinity beyond traces of residual drug crystals. Loading drug content excessively higher than the thermodynamic solubility of the drug in polymer for a given ASD system provides obvious driving force of physical instability, especially coming from various unforeseen variations. Therefore, generally, drug loading beyond one-third fraction in ASD can become a potential physical stability risk. In contrast, due to higher hygroscopicity, lower drug-loaded ASD can be more sensitive to moisture variations. As aforementioned, the inherent GFA class of drug and hygroscopicity are decisive also for achieving certain drug load in ASD. The selected drug load needs to be assessed in terms of risks associated with different manufacturing stages of ASD products. Especially, the estimates of in-process drug nucleation, phase separation, and crystallization of surface or bulk region of ASDs exposed to mechanical forces such as granulation and tableting or water systems such as film coating during the drug product processes are to be accounted for in selecting the robust drug load. While ASD manufacturing processes and involved parameters certainly impact the drug load achievable through kinetic stabilization in ASDs, drug loads need to be considered beyond ASD intermediate stages. Powder and granular properties of ASDs are important for drug product manufacture at a commercial scale and depend on drug load in ASD as well as total ASD content in the powder blend representing final drug product. These properties include powder feedability/flowability, cohesivity, and compactability [45, 46].

2.2 Analytical Characterization of ASDs

The in-depth characterization of physical structures, dynamics, and molecular interactions using a gamut of analytical techniques operating at diverse measurement principles is key to designing successful ASD products. Details about different analytical techniques applicable for the characterization of amorphous API and excipients, ASD intermediates, and ASD products are available elsewhere [47]. Herein, brief accounts and practical aspects of different characterization techniques are described, grouping them into common descriptors, and the features are summarized in Table 2.1.

Table 2.1 Overview of key solid-state characterization techniques for ASDs and descriptors measured.

Techniques	Phase analysis	Molecular interaction	Molecular mobility	Crystallinity
DSC	+++++	+++	+++	++++
DMA	++++	+++	++++	++++
DES	++++	++	+++++	+++
TSC	+++	++	++++	+
XRD	+++	++	—	+++++
IR/Raman	++	++++	+	+++
SS-NMR	++	+++++	+++++	+++++
Microscopy	++	—	—	+++

DES: dielectric spectroscopy, DMA: dynamic mechanical analyzer, DSC: differential scanning calorimetry, IR: infrared, SS-NMR: solid-state nuclear magnetic resonance, TSC: thermally stimulated current, XRD: X-ray diffraction.

+++++: technique of choice, ++++: complementary technique, +++: qualitative analysis, ++: indirectly derived, +: only indicative, and —: no information obtained.

2.2.1 Drug–Polymer Phase Behavior

The phase behavior of drug and polymer is a fundamental characteristic and often is complex. The phase-related descriptors are solid solubility of drug in polymer, glass transition temperature, drug–polymer miscibility, phase homogeneity, phase separation, thermodynamic and energetic properties, etc. Calorimetric techniques such as differential scanning calorimetry (DSC) and isothermal calorimetry (ITC) are cornerstone for the characterization of ASD phase behavior [48]. The DSC measures heat flow or heat capacity changes as a function of heating or cooling at constant rate. The glass transition region is characterized by a step change in heat flow or heat capacity in DSC thermogram. Therefore, T_g can be obtained as a temperature by integrating the transition region. It is important to indicate whether T_g is reported as onset, midpoint, or end of the transition, as the values can vastly differ, especially when transition is wide. Also, it is imperative to report the DSC heating or cooling rate applied during the DSC experiment to obtain the specific T_g, as the values vary linearly with the rate applied. The T_g versus heating rate data can be used to obtain fragility characteristics of amorphous state. An integrated value of the thermal event can also provide heat capacity changes (ΔC_p) related to T_g. The ΔC_p values at T_g of an ASD can be used to derive configuration entropy of mixing [49]. A homogeneous-phase ASD exhibits a single T_g of drug and polymer mixed phase, typically between the T_g of pure drug and pure polymer. A clear phase separation leads to a T_g of drug-rich phase at a lower temperature and that of polymer-rich phase at a higher temperature. The two distinctive glass transitions are visible when the size of the separated phases is larger than 30 nm. When the sizes of the separate phases are smaller, a single glass transition event can be characterized by the increase in the width of the transition zone and the decrease in the heat capacity jump. Therefore, the T_g width also provides an important account of phase homogeneity [30]. For a single amorphous phase, the width of T_g can be related to the variable number of molecular conformations and the dynamic heterogeneity of the system [47]. For ASD samples, the normal DSC thermogram can provide the T_g events overlapping with moisture desolvation endotherm. Thus, to enable better designation, integration, and quantification of thermal events, modulated temperature DSC (mDSC) can be useful. The mDSC decomposes total heat flow signal to resolve thermodynamic events in reversing heat flow signal and kinetic event into non-reversing heat flow signal. This way, T_g event is present in reversing and desolvation endotherm (kinetic event) in non-reversing signal. The thermodynamic solubility of drug in polymer, an important descriptor of ASD design, can also be determined using DSC. The basis of such measurement is analyzing the depression in the melting point of drug with respect to pure drug when polymer is mixed with crystalline drug [19]. This way, the solid–liquid line can be constructed in a plot of melting point versus composition of drug–polymer. On the same plot, the mixed T_g versus composition can be overlaid, representing the liquid–liquid line of the system. The crossing of melting point line and T_g line essentially provides the thermodynamic solution as a drug concentration in polymer at a given temperature. Such experiments need extreme care in sample preparation, and transitions need to be measured at variable heating rates to extrapolate to nearly zero rates (representing equilibrium). The other caveat of this method is that drug crystal can simply dissolve in polymer upon heating when polymer concentration is increased; therefore, melting event can be practically visible only when the drug is present above 60% weight fraction. There are different versions of solid solubility determination using calorimetric methods involving sample preparation via milling, annealing homogeneous ASD to detect phase separation [50]. Apart from scanning versions of calorimetry, ITC presents a unique advantage of studying phase behavior of ASD. Due to high precision and sensitivity of temperature and heat flow change measured by the technique, thermal events like phase separation can be studied in detail and at ambient temperature [47], and enthalpy of mixing of ASD at different compositions can be determined using such experimental data [51]. Different, and less common, thermal methods such as dynamic mechanical thermal analysis (DMTA) and thermally simulated polarizing current (TSPC) analysis can be used to study the phase behavior of ASD. The miscibility limit of drug and polymer in given ASD can also be determined using a viscosity–composition diagram obtained using thermal rheology [52]. While X-ray diffraction (XRD) technique is often reserved for crystalline sample analysis and phase identification, the unique halo pattern obtained in powder X-ray diffraction (pXRD) analysis of ASD samples could be useful for studying the phase homogeneity [47]. By analyzing a high-quality pXRD halo of ASD and comparing it with the individual amorphous component patterns, one can derive the phase homogeneity profile. For that, raw pXRD pattern is often transformed into the derivative plot and analyzed using a pair distribution function (PDF). An ASD with PDF pattern overlaying with pure components and the corresponding physical mixtures is considered phase separated, else is considered as phase homogeneous. Apart from classical pXRD, small-angle X-ray scattering (SAXS) contains a rich information on the nanoscale density heterogeneity that is related to phase homogeneity of ASD [53]. The treatment of SAXS data can also provide information on the domain size of ASD phases and interfaces. A detailed analysis of pXRD pattern and PDF profiles of ASD can yield information on the local molecular environment of the system. Another important analytical tool for studying ASD phase behavior is solid-state nuclear magnetic resonance (ssNMR) relaxometry [54]. The basic principle here is to measure spin–spin and spin–lattice relaxation time in ASD. If a given ASD is characterized by relaxation times comparable to or closer to those of pure drug and polymer, it is referred to as the phase separation system. While the molecularly homogeneous ASD features relaxation times that are uniquely different than the pure components and their physical mixtures [55], the domain sizes of the separated phase could be derived from the NMR relaxation time data. An advantage of NMR method is that phase separation in ASD can be studied at higher resolution than by calorimetric techniques, and the size of the separated domains can be quantitatively derived. Likewise, the phase homogeneity and separation in ASD can be interrogated using vibrational spectroscopic techniques such as infrared (IR) and Raman spectroscopy [21, 56]. Specific to these techniques is that they allow spectroscopic imaging of the sample so that phase distribution in the ASD samples can be spatially mapped. A single T_g ASD has been shown to still show phase separation via

Raman spectroscopic mapping [57, 58]. More advanced techniques such as atomic force microscopy (AFM) and nano-thermal analysis (NTA) can provide a detailed mapping of phase composition in ASD [59, 60]. A detailed review of several analytical methods to study drug–polymer solubility and miscibility can be found in the recent review by Thakore et al. [18].

2.2.2 Molecular Mobility

Molecular motions occurring at various lengths and timescales in ASDs relate closely to their kinetic behavior and stability, and thus are important to be characterized. Some key characteristics associated with molecular motions in ASDs are global relaxation time, local relaxation time, diffusion, diffusivity, and dynamic heterogeneity. Calorimetric techniques such as DSC and ITC are instrumental in characterizing the molecular mobility of ASDs [47, 48]. The common DSC experiment includes annealing the samples below T_g for varying periods to let ASD relax its energy and structure, followed by reheating the sample across T_g. This leads to an endothermic signal about T_g attributable to the recovery of lost enthalpy. In a DSC thermogram obtained via linear heating rate, the enthalpy recovery signal overlaps with the step change in heat capacity due to T_g, therefore difficult to resolve and integrate. Again, here mDSC is useful as enthalpy recovery, a kinetic event, will be clearly separated in non-reversing heat flow, while T_g event remains in reversing heat flow. The plot of enthalpy recovery value as a function of annealing time typically shows an exponential pattern. The average relaxation time and distribution parameters are derived by fitting such an experimental plot using empirical relaxation models such as Kohlrausch–Williams–Watts equation [47]. The relaxation time obtained by DSC is associated with the global or primary molecular mobility and thus is an alpha relaxation time. The dispersity parameter of relaxation time also indicates the dynamic heterogeneity of amorphous systems. Such dynamic heterogeneity is often related to structural heterogeneity and phase composition, and ultimately to both physical and chemical stability behavior. Depending upon the ASD system, a signature of sub-T_g local relaxation can also be observed in some cases. Yet such events show a very weak calorimetric signal and often remain unnoticed in baseline noise. Analogously, ITC is a proven and powerful tool for obtaining the molecular relaxation times of ASD [27]. The advantage of ITC comes from its high sensitivity to heat flow associated with relaxation processes, enabling relaxation time to be obtained at or near room temperature. One of the caveats of ITC is that the total relaxation time is often obtained without being able to recognize the motional processes. Owing to more prominent change in mechanical properties related to the relaxation process, DMTA stands out as a suitable technique to study molecular mobility in ASD [47]. The DMTA acts as a tool for screening and analyzing the thermal transitions associated with molecular relaxation attributable to both sub-T_g local process ($T_g\beta$) and global motional process ($T_g\alpha$) at T_g [61]. The respective relaxation times can be obtained via time–temperature transformation. $T_g\beta$ would have much lower values, and respective relaxation time is much shorter than that of $T_g\alpha$ process.

The beta relaxation process is related to and a precursor of the alpha process and is described via Ngai coupling [62]. The other faster local processes such as gamma and delta processes are secondary molecular motions that are not coupled with the alpha process. The alpha relaxation time is empirically related to the nucleation induction time and used as the descriptor to predict physical stability of ASD [63]. Plasticization of T_g by moisture leads to a marked increase in molecular mobility characterized by the reduction in relaxation time. The JG-type beta relaxation time has been shown to be a stronger predictor of physical stability of amorphous material than alpha relaxation time. Another related technique for characterizing molecular mobility is thermal rheology. The viscosity and shear moduli of dry and wet (humidified) ASD can be determined as a function of temperature using oscillatory rheology to transform into respective molecular mobilities and used to predict crystallization kinetics [64]. Thus, oscillatory rheology is an important technique for molecular mobility characterization of ASD [65]. The core technique for the characterization of different models of molecular mobility of amorphous systems is using dielectric relaxation spectroscopy (DRS). The basic principle of the measurement is measuring the real and imaginary components of dielectric permittivity of the sample sandwiched between two capacitors when exposed to electric fields of varying frequency [47]. This way alpha relaxation distribution will be recorded at the higher-frequency region, and beta and other higher-order relaxation signals will be recorded at the lower-frequency region. The DES spectrum informs about the nature and extent of dynamic heterogeneity in the ASD systems, and the respective relaxation time attributable to distinct modes of molecular mobility can be deduced from the frequency at the peak of DES signal. A common challenge is to resolve the signal of secondary molecular motions of ASDs from the overlapping DC conductivity signal that necessitates special experimental as well as data analysis procedures [66]. The readers are requested to refer to the seminal literature papers regarding the use of DES for the characterization of molecular mobility of ASDs [26, 67]. Another thermal technique, which is relatively less frequently used yet very relevant for measuring molecular mobility, is thermally stimulated current (TSC) spectroscopy [47]. Here, the sample is heated to a certain temperature and cooled at a constant rate under the application of current, leading to the polarization of the dipoles in the sample. Subsequently, the sample is heated using constant rate without electric field, which generates the relaxation/depolarization of frozen dipoles. This mode of analysis is known as thermally stimulated depolarization current (TSDC) spectroscopy [47]. The TSDC spectroscopic signal constitutes global relaxation profiles of both sub-T_g and T_g-related processes when sample is first heated beyond T_g before polarization during cooling. In addition, thermal windowing experiments are performed with a procedure of repeatedly collecting TSDC signals by first heating to several different temperatures before cooling polarization. This way, individual relaxation signals are deconvoluted from global relaxation signals, and this provides information on relaxation time distribution as well as dynamic heterogeneity in ASD. Another variant of TSC is the measurement of polarization current during the heating of sample under electric field, known as TSPC spectroscopy [68].

NMR relaxometry is an important analytical technique for profiling relaxation behavior in ASDs. Based on solid-state NMR spectroscopic profiles, the magnetization inversion or recovery profile of a particular peak attributable to special functional group(s) of drug and/or polymer of ASD samples is recorded using special pulse sequences and analyzed using exponential analytical expressions to obtain molecular site-specific relaxation time. The spin–lattice relaxation time measured under static condition (T_1) and constant rotation (T_{1p}) is related to local (such as β) and global (α) molecular motions, respectively. The T_1 relaxation times are distinctly different between the drug and polymer, such that proton (H1) T_1/T_{1p} relaxation time profiles of ASDs can provide information on the overall miscibility within the domain size below 30 nm. The general interpretation is that if the relaxation times of ASDs are identical to those of individual components, then there is phase separation, while the averaging of relaxation times to a new value means molecular mixing within the given spatial dimension. The spin–spin relaxation time (T_2) can be obtained from NMR line broadening and provides information on the miscibility in ASD within smaller domain size (<10 nm) as well as on the diffusion coefficient. Overall, NMR relaxometry is a prominent tool for comprehensive speciation of relaxation dynamics and phase homogeneity in ASD. Variable temperature–NMR relaxometry can relate the change in relaxation times to various phase transitions such as glass transitions, crystallization, and melting of API in the sample. For details, the readers are referred to the review article by Paudel et al. [54]. Like NMR relaxometry, terahertz (THz) time domain spectroscopy is also used to profile local and global molecular mobility in ASD [69]. An interesting technique for studying global molecular mobility and diffusion rate in ASD is by using AFM, wherein the rate of the change in nanoscopic surface roughness created by nano-grating is measured at the relevant temperature and analyzed [70]. An indirect account of molecular motion in ASD can be probed by measuring the overall molar free volume in the sample using positron annihilation lifetime spectroscopy (PALS) [71]. The technique can provide a wealth of information; however, the sample preparation can be complex, and PALS source is not widely accessible.

2.2.3 Characterization of Molecular Interactions in ASD

The inter- and intramolecular interactions between drug–drug, polymer–polymer, and drug–polymer interactions are best characterized using spectroscopic techniques. Mid-IR spectroscopy is most common for characterizing molecular interactions in ASD owing to its ease of interpretation and sample preparation. The IR spectrum of an ASD constitutes the characteristic vibrational bands of API and polymer functional group as well as peak shifts attributable to API–polymer interactions [33]. The fingerprint region represents vibration bands' relation to double bonds (C=X, where X can be O, N, S) of API and/or polymer (mostly C=O), and vibration bands attributable to stretching H—X bonds (where X can be O, N, S) lie in the region beyond 3000 cm^{-1}. The H-bonding between the donor group of drugs (OH, NH, SH) and the acceptor group of polymers (mostly C=O, for example, in vinyl polymers like PVP, PVPVA) leads to the shift of polymer band to the lower wavenumber, called red shift [30]. The OH/NH stretching bands of H-bonded fraction show the red shift as well. For the latter vibrations, the change is distinctive using near-IR (NIR) spectroscopy and thus has been used to analyze ASD in hyphenation with water sorption [72]. In case of ASDs, wherein there could be salt formation between the functional group of drug and polymer, a clear appearance of bands related to the ionic interaction can be noticed. The IR spectra of such ASDs of the drug with carboxylic acid function and polymer with basic amine function group like methacrylate derivatives or acid polymer like PAA and amine drug [73] exhibit a peculiar band of the ionic carboxylate species significantly red-shifted with respect to the C=O band of the neutral counterpart. In phase-separated ASDs, the IR bands of strongly and weakly H-bonded species can be observed that are attributable to drug-rich and polymer-rich fractions [21, 30]. The IR spectroscopic method is best suited for qualitative analysis of molecular interactions, although the peak deconvolution or relative intensities of the bands of interest can provide semi-quantitative information. The exact quantitative insight is difficult to extract due to the difficulty in generating reference to the interacting motifs. A caveat of IR method is that it is highly sensitive to moisture in the sample, as water can act as both H-bond donor and acceptor. To this end, Raman spectroscopy is advantageous for studying molecular interactions in ASD, since water is a weak Raman scatterer [74]. Also, Raman spectroscopy is highly sensitive toward detecting weaker nonpolar interactions such as aromatic interactions, dipolar bond formation [47] as well as halogen bonding between halogen of API and acceptor group of polymers [34]. Though less often, UV and fluorescence spectroscopy [75] are also utilized to analyze molecular interactions in ASD. The spectroscopic techniques are also available as imaging setup to use as a hyperspectral imaging tool for the characterization of ASDs. Extensive reports on the use of Fourier transform infrared (FTIR) microscopy [57], confocal Raman imaging [58], and UV/Vis imaging of ASD are published, deciphering the spatial and temporal mapping of interactions, phase separations, and domain/cluster sizes in ASDs, even at elevated temperatures and humidity conditions, as well as under dissolution environment. Furthermore, these spectral techniques are used as process analytical technologies (PATs) for monitoring ASD manufacturing and production [76, 77]. For in-depth study of molecular interactions in ASDs, solid-state NMR is considered a gold standard [54]. As in vibrational spectroscopy, SS-NMR spectra of an ASD will be characterized by peaks attributable to drug–polymer interactions such as H-bonding, ionic interactions, and dipolar interactions. The advantage of SS-NMR is that it can directly speciate on exact functional group involved in the formation of such interactions and provide quantitative insight into the strength of such interaction (H-bond length, bond angle, etc.) [78]. Such information is achieved via the change in SS-NMR shift of proton (H1) or carbon (C13) or even using abundant heteroatoms in APIs (N15, F19) in ASD in comparison to individual components. In addition, correlation spectroscopy obtained from 2D SS-NMR data can provide precise information on the structural motifs of drug and polymer involved in such stabilizing molecular interactions in ASDs and detailed regioselectivity of substituent groups in complex polymer structure like HPMC-AS [79].

2.2.4 Crystallinity Analysis in ASD

The crystallinity analysis of ASD is crucial to support different stages of ASD product development [80]. During the early development and screening phases, qualitative methods are employed to detect residual drug crystals, whereas the late-phase development necessitates methods to quantify drug crystals in ASD intermediates as well as drug products to support drug product process development and stability studies [47]. Polarized optical microscopy is a ubiquitous tool for facile analysis of any residual crystals remaining in ASD samples, as the presence of crystals is detected by the birefringent pattern, and the lack of birefringence confirms fully amorphous nature of an ASD. The analysis can also be performed using a heating or cooling program, as well as isothermal condition, to study crystal nucleation and growth in ASD. DSC analysis leads to the identification of melting event of any residual API crystal in an ASD sample generated during accelerated stability studies, dissolution tests, or suboptimal manufacturing process. The melting point of the drug crystal present in ASD is often lower than in pure form, and the ratio of enthalpy of melting endotherm of drug in ASD to that in pure form can be used to quantify the relative crystallinity in an ASD. The drug load in ASD is used to weigh the crystallinity quantitation. While recording a DSC thermogram, ASD sample can show a glass transition event, followed by a cold crystallization exotherm and a melting event of the drug fraction crystallized in situ during DSC run. In such case, the enthalpy of crystallization would be closer or equal to the melting enthalpy (after correcting the heat capacity difference) [47]. If the melting enthalpy is excess to the crystallization enthalpy, the difference will provide an estimate of residual crystallinity in ASD. As melting process in ASD is heterogeneous than in pure state, the melting endotherm is broader for drug crystal in ASD and thus difficult to accurately obtain the melting enthalpy for the trace crystallinity. Fast heating rates can increase sensitivity to detect small fractions of crystallinity. Crystallinity quantitation becomes complex when analyzing ASD drug product with several additional excipients potentially showing thermal events overlapping with drug crystal melting. For suitable ASD systems, DSC analysis enables modeling of sub-T_g crystal growth [81]. The technique of choice for measuring crystallinity is pXRD. The full amorphous nature of ASD is characterized by the diffuse halo pattern without any distinct Bragg peak in the diffractogram [47]. The trace crystal in ASD can be detected by the presence of discrete Bragg peaks in pXRD data. Moreover, the peak position enables the polymorphic identity of the trace crystal present in ASD, whereas the width of peak can be used to derive the size of the crystallites in the sample. The simple method to obtain relative crystallinity based on pXRD data is from the ratio of Bragg peaks of interest in ASD to those in pure crystal. However, due to the nonlinear baseline due to amorphous halo in ASD, the deconvolution process is often applied to resolve the accurate Bragg peak intensity and area under the peak. The area ratio method represents more information than the intensity ratio alone for crystallinity quantification. In the case of multiple Bragg peaks observed in ASD, the multivariate method can be developed for the precise quantification of crystallinity. The use of strong X-ray source in synchrotron XRD can drastically enhance the limit of detection and quantification of crystallinity in ASD [47]. Besides DSC and pXRD, spectroscopic techniques such as IR, NIR, THz, and Raman spectroscopy are also extensively used for crystallinity analysis, as amorphous and crystalline states possess distinct spectroscopic signatures. The NIR method in combination with the chemometric model is shown to provide comparable quantitative power to pXRD for the analysis of crystallinity in ASD [82], and due to specificity and speed of quantitation, NIR spectroscopy is used for the quantification of crystallinity in drug product processes as PAT tool or offline analysis of ASD drug products. Likewise, Raman spectroscopy is widely used for crystallinity quantitation, including monitoring of crystallinity during ASD dissolution [83]. Especially, low-frequency (LF) Raman spectroscopy is highly sensitive to the presence of phonon peak of crystals, therefore amenable for the detection and quantification of low amount of crystal in ASD [84]. LF-Raman is proven to be applicable for crystallinity detection in ASD screening phase for nanogram sample [85]. Also, detection of trace crystallinity using transmission electron microscopy has been shown for ASD [86]. For trace crystallinity detection in amorphous samples like ASD, second harmonic generation (SHG) spectroscopy, a nonlinear imaging technique, is extremely sensitive to crystal (parts per million) [87]. The quantitative power of such method is yet to be fully established, especially for the ASD drug product with several excipients. Several method development aspects such as sample preparation, calibration procedures, calibration mix preparation, and crystal spiking are to be duly considered for robust crystallinity quantification, especially in view of regulatory-compliant and validated method development and deployment in GMP setting and transfer to commercial production environments.

2.2.5 Surface Analysis of ASD

Surface characteristics of ASD impact biopharmaceutics, stability, and manufacturability performance. Surface chemical composition of ASD can be analyzed using surface-sensitive techniques such as X-ray photoelectron spectroscopy (XPS), which provides quantitative insight and resolution into the elemental composition of a few nanometers' surface of the solid [47]. The information on the phase separation process at ASD surface can be obtained using X-ray fluorescence (XRF) spectroscopy [88]. Likewise, time-of-flight secondary ion mass spectrometry (ToF-SIMS) enables the analysis of the molecular composition of ASD particle surface based on molecular mass of the species and has been applied to analyze ASD surface [89] as well as ASD tablets [47]. Besides spectroscopy, high-resolution microscopies such as scanning and transmission electron microscopy (SEM and TEM) can provide detailed information of ASD particle surfaces as well as nanoscopic surface topology [47]. In addition, AFM can provide information about the surface phase composition and topology, as well as the molecular mobility of the surface of ASD [60]. Nano-thermal analysis (Nano-TA), a combination of AFM and thermal analysis, is a powerful technique to map thermal transition at high spatial resolution, yielding information on phase homogeneity in ASDs [90]. Moreover, annealing-induced structural relaxation at the surface of ASD particles has been studied using inverse gas chromatography (IGC) [91]. Also, IGC could provide information on surface energy, surface crystallization [92], and surface miscibility in ASD [93]. The surface energy of an ASD relates to the powder properties, flowability, and cohesivity, as well as wettability, an important metric to drug dissolution performance. The wettability of ASD can also be estimated via sessile drop contact angle method [94]. An important metric to analyze for ASD is its interaction with moisture, and it is studied using sorption method, specifically using dynamic vapor sorption (DVS) [47]. As most moisture interaction at lower water activity occurs through physisorption, DVS reveals the surface phenomena in ASDs like plasticization, drug–polymer–water interactions, crystallization, and surface coalescence [95] that are crucial for rational selection of manufacturing, packaging, and storage conditions. For example, the change in T_g as a function of the equilibration of ASD at certain water activity provides information on drug–polymer–water interactions and miscibility [96]. Beyond certain water activity, moisture sorption, however, becomes a bulk phenomenon.

2.2.6 Bulk and Powder Characterization of ASD

The appropriate characterization of particle and powder properties of ASD intermediates is necessary for selecting downstream manufacturability of ASDs. Foremost, the particle size and shape distribution of ASD powder can be measured using laser diffraction, microscopy, and their packing arrangement in granules or tablets or surface topography and roughness of ASDs can be measured via advanced SEM or TEM. The microstructure of spray-dried ASD was investigated in-depth using focused ion beam SEM (FIB-SEM) and X-ray micro-computed tomography (XRCT) combined with an AI-based image analysis algorithm [97]. The use of synchrotron X-ray source for such workflow is shown to yield detailed information on the particle surface area, wall thickness of hollow particles, and overall spray-dried particle microstructure generated using varying SD process parameters [98]. ASD particles manufactured via SD and HME are typically fine and thus possess poor powder flow and high cohesivity. The particles generated via SD are often hollow and wrinkled. The shell thickness of these particles, important in terms of their compressibility, has been characterized in depth using X-ray computed microtomography and advanced image analysis. The surface area of ASD powders is analyzed using nitrogen physisorption techniques, providing insight into their dissolution kinetics. The surface area of ASD powders manufactured via HME is often too low to be quantified using nitrogen physisorption, while krypton physisorption can be used to analyze such low surface area. Also, porosity of particles can be characterized using several techniques, the most common being mercury porosimetry and gas physisorption. Besides microscopic surface and porosity, nanoscopic surface and interface of ASD samples can be analyzed in detail using SAXS, providing unambiguous connection of surface and interface heterogeneity and dispersity of ASD [53]. Connected to porosity, the true density of ASD particles and powders is analyzed using helium pycnometry. The characteristics related to bulk handling and processability of ASDs include tapped and bulk density, which can be characterized using the method specified in pharmacopeia. The tapped and bulk density values are used to derive power compressibility and flowability. The flowability of powder is also characterized by using angle of repose. The other powder flow-related parameters, such as flow function coefficient, cohesivity, and powder bed permeability, can be obtained using powder rheometry and ring shear tester. Related to powder flow and cohesion, electrostatic properties of ASD powders can be characterized using tools like Faraday cup, as powder tribo-charging can lead to process variations and safety concerns [99]. Besides micromeritics and flowability, the characterization of mechanical properties of ASD is crucial for selecting their processability. The data on attributes such as hardness and mechanical moduli (Young's modulus, elastic constant, etc.) of HME strand can guide the optimal milling process selection to obtain powder. Furthermore, mechanical properties of ASD intermediates, granules, and blends are important to be characterized for assessing tabletability of ASDs as a function of polymer types and drug loading. At bulk level, the change in powder volume and the tensile strength of resulting compact as a function of compaction pressure are characterized by using single punch tablet press or compaction simulator, providing insight into elastoplastic behavior, Young's modulus, compressibility, compactability, and tabletability [100]. While such characterization is routine for oral drug product development, mechanical characteristics and deformation behavior of ASDs are not widely characterized.

2.2.7 Miscellaneous Analysis of ASD

The chemical characterization of ASD in terms of drug assay, appearance, moisture content, purity, and batch homogeneity of the manufacturing batches is essential to ensure the loading of target drug dose. Thus, quantitative chromatography (such as liquid chromatography) or spectroscopy (UV/Vis or IR) method is developed for the drug assay in ASD including that also for the dissolution testing samples. Also, any formulation and process-related impurities can not only be detrimental for patient safety, but also can alter physical stability and biopharmaceutics performance, thus accounting for the alteration in the product quality [36]. For this purpose, impurity profiling and stability-indicating analytical methods are developed and employed for ASD product development using liquid chromatography hyphenated to mass spectrometry. Moreover, polymers being a prominent part of ASDs, analytical methods are developed to ensure the chemical identity and molecular and structural integrity of polymers during formulation and processing of ASDs. However, chemical analysis of polymers in ASD is not routinely performed during development. Analytical methods like size exclusion chromatography and quantitative NMR spectroscopy are utilized for this purpose. To investigate the formation of reactive impurities such as acids, oxidizing peroxide, and free radicals formed of polymers during processing, quantitative NMR spectroscopy as well as electron spin resonance spectroscopy are employed [36, 101].

2.3 Preformulation and Screening of ASD

Preformulation and screening investigations are inevitable prerequisites for obtaining the crucial data and knowledge into design and development attributes of

ASDs [102]. These studies are conducted at an earlier stage prior to formulation and process development of ASDs. The objectives are to obtain key insight into biopharmaceutical performance, stability, and processability of ASD formulations that enable the selection of the best-performing ASD polymers for given API and the drug loadings with optimal performance using small amount of API. Therefore, experiments at this stage are performed using miniaturized setup and often in high-throughput fashion in well plates/vials, applying automation, robotics, and in situ analysis wherever relevant, with an aspiration to obtain data on the maximum number of conditions and compositions.

2.3.1 Biopharmaceutics Screening

The screening is conducted using several ASD polymers to investigate their solubilization potential in biorelevant conditions and temperature, using a few milligrams of API. By starting with excess undissolved API in pre-dissolved polymer solution using simulated physiological medium such as fasted and fed-state gastric and intestinal fluids, the increase in solubilization of API can be assessed in comparison to API alone. As the next step, similar experiments can be performed by introducing supersaturated drug solutions (with starting drug concentration higher than equilibrium drug solubility) into the pre-dissolved polymer solution and observing the extent and the rate of loss of supersaturation via precipitation. The polymers enabling the highest supersaturation and for the longest duration before drug precipitation are considered to be the best ones. Such supersaturation screenings are conducted using a few different polymer concentrations. In recent years, the emphasis has also been given to the studies for obtaining amorphous drug solubility, identifying the tendency of forming liquid–liquid phase-separated (LLPS) systems in biorelevant media [103]. The concentration and size of drug-rich and polymer-rich droplets formed in physiological media as a function of drug-to-polymer ratios are obtained by using UV/Visible spectroscopy, dynamic light scattering, NMR spectroscopy, etc., enabling the construction of a complete phase diagram [104].

2.3.2 Composition Screening

This is often the succeeding step to biopharmaceutics ranking and utilizes greater API amounts. Here, the selected polymers are used to prepare miniaturized ASD specimen with several drug loadings each and characterize their physical state, namely, T_g, miscibility, crystallinity, etc. Manual screening can be performed using mini powder blends, heating and cooling in DSC to obtain solid-state information, or drug–polymer solutions can be directly dried inside DSC pans to form film samples [30]. Alternatively, robotics is used for high-throughput ASD preparations using liquid dispensing unit that dispenses several ratios of drug and polymer solutions from the stock solutions to well plates in industrial setting. The drug–polymer solutions in organic solvents are dried to obtain films [105]. The well plates containing ASD films are amenable to high-throughput analysis via microscopy for crystallinity detection or via IR, Raman, or THz spectroscopy for crystallinity and molecular interaction analysis. Some well plates are X-ray transparent, enabling XRD analysis of screening samples. Some studies have demonstrated the use of inkjet printing [106] and microarray systems [107] for ASD screening. The miniaturized setup using picoliter-volume inkjet printing coupled to polarized microscopy is shown to enable high-throughput screening of ASD at ng scale [108]. Using the same sample plate or by generating extra plates only with amorphous and homogeneous-phase composition, the biorelevant media can be introduced directly onto the samples to conduct miniaturized dissolution studies. Though such studies are not fully representative of traditional in vitro dissolution, the intention here is to obtain relative ranking of dissolution, wetting, and supersaturation behavior. Also, the residual sample remained in such study can be subjected to high-throughput solid-state analysis. The screening samples are exposed to accelerated stress conditions including elevated temperature and/or humidity to interrogate their relative stability behavior.

2.3.3 Process-Relevant Screening

For the narrowed-down ASD candidates based on biopharmaceutics and stability screening, the process-relevant screening is the next logical step to conduct when several milligrams of API are available, yet prior to small-scale processing using SD or HME. For ASDs' processability via HME, melt viscosity and melt flow indices of the selected compositions are obtained via such screens, and they can be conducted using melt rheology [109]. The recovered samples exposed to variable shear, torque, and temperature history can be subjected to solid-state analysis and drug assay. An intuitive ASD screening using the heating of sample filled in the syringe plunged through a bent needle has been reported [39]. In recent years, vacuum compression molding has been used as the screening tool for ASDs, owing to its ability to generate specimens with well-defined geometry and surface using different melting temperatures and heating rates [110]. SD itself can be conducted on a small scale using laboratory-scale equipment, requiring small API amount to run screening experiments compared to HME. At screening phase, actual process performance such as yields and efficiency is not strictly in the scope rather to generate samples under varying solvent and drying conditions. For such purposes, an automated robotic platform has been integrated into SD for ASD screening, enabling high-throughput experiments [111]. An experiment reports an interesting custom setup including atomization nozzle and a drying assembly utilized for ASD screening relevant to SD [112]. Also, there have been reports on SD-on-chip and ultrasonic levitated single droplet drying having potential to be developed and implemented as a screening tool [38].

2.4 Manufacturing Processes of ASD

The manufacturing processes for ASD are employed to convert the molecularly mixed-state drug–polymer solution or melt into solid by applying rapid evaporation in the case of solution or fast cooling of the melt. This way, the mixed drug–polymer

Table 2.2 A comparative overview of different aspects of ASD manufacturing processes.

ASD process	Biopharmaceutics	Stability	Processability	Downstream
Spray drying	Particles with high surface and dissolution rates	High surface area impacts moisture sorption and surface crystallization	APIs and polymers need to be soluble in common volatile organic solvents	The particles often need flow aid; tableting can be favorable
Hot melt extrusion	The extrudates are compact, thus dissolution and wetting can be affected	Low surface and high density lead to superior stability and lower sorption	APIs and polymers need to be thermostable	Particles often need granulation, as compressibility is lower
Co-precipitation	High surface can increase wetting and dissolution	Stability might be impacted by residual solvent	The pH solubility in solution stability of drug and polymer needs to be optimal	Larger and porous particles are produced with improved tabletability
Fluid bed layering	Favorable surface of dissolving film	Surface crystallization can be issue	Only possible for low drug loading	Direct filling into hard capsules is possible
Ball milling	High surface for dissolution	Residual nuclei or crystals can be deleterious	Scale-up is an issue	Particles might need flow aid and/or granulation

solid state is generated, often forming the supersaturated single-phase ASD that is kinetically stabilized. The process selection is made based on the properties of API and polymer, the development stage, intended dosage form, and sustainability aspects [38, 39]. Table 2.2 provides an overview of different ASD manufacturing processes, and schematics of the processes are depicted in Figure 2.2.

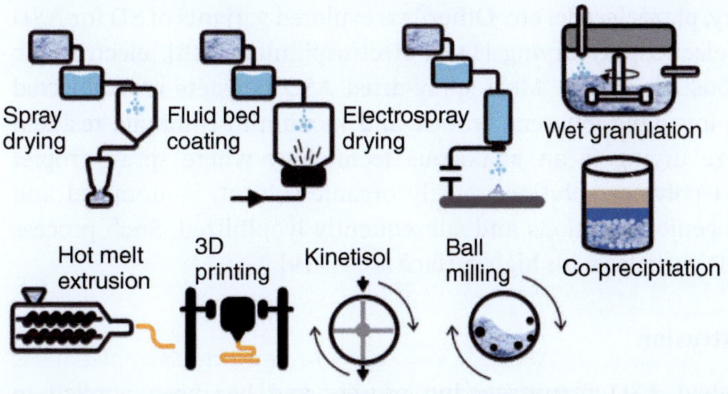

Figure 2.2 Schematic overview of industrially relevant manufacturing processes of ASD.

Different ASD manufacturing processes can be categorized broadly into solvent- and melt-based technologies. Below, a brief account of the process and formulation considerations for different manufacturing technologies is presented.

2.4.1 Spray Drying

SD is a widely used process for manufacturing ASDs with maximum number of approved ASDs [2]. It offers a continuous and commercially scalable drying process which enables particle engineering opportunities [38]. The process consists of several steps. First, the feed solution containing the drug and the polymer (and possibly other additives) is prepared in volatile organic solvent or solvent blend. The solvents used for SD must be of International Council for Harmonisation (ICH) of Technical Requirements for Pharmaceuticals for Human Use class III (acetone, ethanol, butyl acetate) or class II (dichloromethane, methanol) in order to comply with the residual solvent limit in the drug product as per regulatory guidance [38]. The feed solution is then pumped at a controlled rate to the atomization nozzle that generates spray of feed solution droplets inside a drying chamber, wherein a stream of hot gas aids an instantaneous evaporation of the solvent to convert droplets into particles. The gas flow also carries the generated ASD particles to collection chamber that separates particles from gas stream and collects particles via cyclone mechanism. Finding a common solvent system to completely dissolve selected drug and polymer is necessary. Also, care should be taken to ensure that no crystal solvate with low solubility is formed in feed solvent. The solvent selection for SD is critical as it directly impacts atomization, evaporation, and thereby particle properties as well as phase homogeneity of the generated spray-dried dispersion (SDD) [113]. The inlet temperature of drying gas acts as an important parameter controlling the drying kinetics of the droplets and the outlet temperature at which the exiting particles are exposed to. Thus, it determines both solid-state [114] and particle properties of ASDs [98]. A very useful descriptor in SD is Peclet number, the ratio of drying kinetics to diffusion kinetics [115]. The other driving process parameter in SD is drying gas flow rate, as it determines the heat and mass transfer to and from the droplets. The balance of drying gas rate to spray (feed) rate ensures the right product quality [116]. Different atomization nozzles are available including twin fluid, pressure, rotary, piezoelectric, etc. Other less explored variants of SD for ASD manufacturing are electrospray drying [117], electrospinning [118], electrostatic SD, and pulse combustion drying. Most spray-dried ASD powders are subjected to secondary drying involving low temperature and vacuum to eliminate residual solvents. Spray-freeze drying is an analogous technology where spray droplet containing drug and polymer solution, mostly organic solvent, is atomized and collected under cryogenic conditions and subsequently lyophilized. Such process produces porous ASD particles with high surface area [119].

2.4.2 Hot Melt Extrusion

HME is an established ASD manufacturing process and has been applied to numerous commercial ASD products. HME is a continuous and solvent-free manufacturing process with several downstream possibilities [39]. In HME process, the powder blend of drug and polymer is fed into the extruder, consisting of a heated barrel enclosing the co-rotating twin screw. At manufacturing scale, the screw elements are designed to perform different functions across the processing zones [120]. The conveying element of screw in powder intake zone possesses higher pitch to convey the powder mass further, so higher volume is available. As powder moves to the kneading zone, the screw pitch is reduced to enable powder densification and air elimination. In melting zone of extruder, the angular pitches of screw with different inclinations ensure stagnant flow and sufficient residence time of the feed to ensure thorough melting of drug and polymer. This zone is followed by a melting mixing zone, wherein more aggressive pitch design of screw enables melt-filled barrels and efficient mixing. The mixed melt is further conveyed to die exit through which extrudate is discharged in the form of strand or fiber. The set barrel temperatures are higher as powder moves from conveying zone to melting and mixing zones of the screw, and then again lower toward the exit die. The temperature at melting zone is kept sufficiently higher than melting temperature of drug crystal and glass transition temperature of amorphous polymer to obtain optimal melt viscosity for mixing. The prerequisite for HME processing of ASD is thermal/melt stability of API and polymer, as well as optimal viscosity of the melt (often driven by polymer). The viscosity below 10^3 Pa·S is often required for robust processing in HME. ASD polymers like PVP/VA, HPMC-AS, and several methacrylate polymers possess favorable melt viscosity and melt flow, thus suitable for extrusion. The melt viscosity of polymers such as HPMC is higher than necessary for HME. Often, molten drug plasticizes the polymer to make it less viscous and extrudable. Alternatively, liquid (e.g. triethyl citrate, glycerol), gas (e.g. carbon dioxide), or solid (surfactants like vitamin E derivatives, polysorbates, polyoxy polymers) are added as plasticizers to aid the processing and production of ASD via HME. Care should be taken here, as plasticizers can reduce T_g and increase molecular mobility to trigger phase separation and crystallization. The important process parameters in HME are powder feed rate, screw rotation rate, and temperature of different processing zones. Compared to SDD, the dense ASD produced by HME is often physically more stable and comparatively slower [121]. These parameters determine the torque and back pressure experienced by the system as well as the residence time of powder and melting in different zones, melting, mixing, and homogeneity of resulting ASDs. The existing HME strands are often milled to transform into ASD powder. The milled ASD powder is sieved to obtain homogeneous particle size range to further process into tablet. There are several other downstream possibilities that can be hyphenated to HME such as pelletization, calendaring, strand cutting, or even use in additive manufacturing such as injection molding [122] and 3D printing for personalized drug delivery [123]. To use the extrudate obtained via HME as input material for 3D printing via fused deposition modeling (FDM), the strands need to have optimal mechanical properties, spoolability, and flexibility [124]. Alternatively, direct powder extrusion to convert powder to printed ASD tablets was explored in the context of rapid formulation prototyping to expedite preclinical to early clinical supply [125].

2.4.3 Other Technologies

While SD and HME are the major process technologies of most commercial ASDs, there are several other interesting processes that are used for producing few commercial ASD products or emerging at research stages with future potential. Fluid bed coating/layering has been used to prepare a commercial ASD of itraconazole [126], which involves spraying drug–polymer solution in organic solvent onto fluidized cores like inert cellulose beads and simultaneously. Since ASD is loaded as film, the technology is suitable only for lower drug dose. The advantage of the process is that the ASD-coated pellets can directly act as the finished dosage form or, after filling into the hard capsule, thus requiring non-to-minimal downstream operation. The drying can operate at lower temperatures, making it suitable for thermolabile actives. An interesting solvent-based ASD processing is co-precipitation technology, and commercial HPMC-AS-based ASD produced by this technology is vemurafenib [127]. The process is rather old, simple, and, in recent years, applied to ASD with ionic polymers. A drug–polymer solution in organic solvent is mixed with aqueous solvent to trigger rapid precipitation of ASD particles. The lower pH of aqueous anti-solvent drastically reduces the solubility of enteric polymer like HPMC-AS. This is often accompanied by agitation and high-shear processing to aid intimate mixing and solvent exchange of ASD components and prevent drug crystallization in the solvent system. The organic drug–polymer solvent was mixed with aqueous surfactant anti-solvent under rotor--stator mixing to form instant emulsion and subsequently isolated and dried to obtain ASD particles, a process named controlled API–polymer solidification [128]. The resulting suspension is filtered to recover ASD particles; the wet solid mass is washed to remove residual organic solvent and dried to obtain final dry ASD. The cold and pH-controlled anti-solvent of drug and polymer is used to induce controlled precipitation. The temperature is set below the expected T_g of ASD to ensure particle hardening. Other variants like flash precipitation and evaporative precipitation are also used to prepare ASDs. The produced particles are often much larger than spray-dried ASD particles, easing the drug product process such as direct compression. On the other hand, a modified version of co-precipitation has been used to prepare high drug-loaded ASD nanoparticles [129]. Moreover, hierarchical particles with controlled properties and surfaces of ASD produced by co-precipitation result in ASDs with improved wettability [130]. The group of Merck scientists has published several papers in recent years on ASD manufacturing via co-precipitation [131]. The hot anti-solvent wash of the wet mass leads to an increase in bulk density from the annealing of ASD powder, thereby improving the downstream manufacturability. The thin film drying of co-precipitate also densifies the resulting ASD particulates [132]. Also, microfluidic technique has been applied to conduct co-precipitation under flow to create ASD. Supercritical carbon dioxide ($scCO_2$) technology has also been reported as a process for ASD manufacturing. Elevated pressure and $scCO_2$ lead to drug–polymer melting and mixing, and when the system is decompressed, $scCO_2$ is released, yielding ASD. Owing to complexity and scalability challenges,

$scCO_2$ technology has not reached commercial product manufacturing. An innovative process reported over the last year is atomic layering technology to coat ASD particles via aerosol technology for reducing moisture sensitivity [133]. An emerging melt-based ASD process is KinetiSol® that amorphizes via high-shear melting and mixing of drug–polymer upon subsequent quenching [134]. The melting and mixing process is extremely rapid. Thus, the process can be suitable for thermolabile API and polymer with very short exposure to thermal stress [135]. Several ASDs' manufacturing processes are also reported via mechanochemical milling of the powder blend of crystalline drug and polymer [136]. The mechanical and impact force breaks API crystal lattice to form amorphous state that concomitantly mixes with the polymer. The process is not yet available for industrial ASD manufacturing scale. Another emerging ASD manufacturing technology is via microwave heating, including in situ amorphization within tablet [137]. Some specialized granulation processes are also used to manufacture ASD directly [138, 139], for example, commercial tacrolimus amorphous formulation was manufactured using melt granulation technology and PEG as a semi-crystalline stabilizer, called MeltDose® technology [140]. Besides process efficiency, current strides are also moving toward environmental friendliness and sustainability for ASD manufacturing.

2.4.4 Downstream Processing of ASD

As much as primary ASD manufacturing process, ASD drug product manufacturing process is equally important for the critical quality attributes of the final product and to meet the target product profiles [141]. Most ASDs are intended to be developed as oral solid dosage, with tablet being the major dosage form, followed by hard-filled capsules. As milled HME extrudate or SDD powder possesses fine particle size range and is rather cohesive [45], ASD powder feeding and direct tableting can be challenging, often necessitating flow aids to be added. Likewise, ASD powder is hygroscopic, and therefore, tight control of environmental humidity around the production line needs to be ensured. Also, often wet granulation process is avoided to prevent solvent- and drying-induced phase separation and crystallization of ASD [142]. This leads to dry granulation via roller compaction as a preferred drug product process for ASD. From both roller compaction and tableting, the mechanical properties of ASD play a pivotal role in producing tablets with optimal tensile strength and disintegration properties. The deformation properties of ASD are mostly dominated by the polymer property. The mechanical properties of polymer such as Young's modulus, hardness, elasticity, and plasticity are dictated by the molecular structures such as T_g, molecular weight, molar volume, and intra- and intermolecular H-bonding [143]. The hardness of ASD was found not related with their T_g, as T_g was decreasing with increased drug loading in contrast to the increase in the hardness, which impacted their tableting behavior. Compaction behavior was also found to improve with high drug load [144]. A recent work by Marcozzi et al. investigated the effects of ASD polymer content, filler, disintegrant type and quantity, glidant, and tablet manufacturing on tablet properties [145]. Mannitol-based tablets were found to possess the best balance of tensile strength and disintegration time. The deformation behavior is also influenced by particle properties such as wall thickness of spray-dried hollow particles. The mechanical stress imparted during compaction of ASD is shown to induce amorphous–amorphous phase separation in some cases [146, 147], while in some other cases, compaction induced better miscibility of ASD [148]. The roller-compacted ASD is reported to show lower physical stability than direct compression counterpart, as roller stress might lead to phase separation or nuclei generation [149]. The understanding of compaction impact and tableting excipient on ASD attributes is still limited and deserves further proportionate research as much as ASD intermediates. Another important unit operation that can have deleterious impact on physical state and stability is film coating process of ASD tablets, as solvent involved in coating can lead to API migration to coating surface and crystallization [149, 150]. The moisture or solvent, as well as plasticizer, used in film composition can lead to such instability and needs attention during product development [151]. Similar investigations need to be performed for other types of dosage forms such as filling into hard capsules or preparation of ASD as dermal films [152]. The transfer of a non-GMP ASD formulation and process to a GMP facility requires establishing clear relationships among critical process parameters (CPP), critical material attributes (CMA), and critical quality attributes (CQA), while the manufacturing location must be selected early based on capabilities. Controlling the process based on a rational design model facilitates scale-up and technology transfer, and accommodates potential variability of API and polymer.

2.5 Biopharmaceutics of ASD

Biopharmaceutics attributes of a drug substance and drug product relate their physical and chemical interactions within the physiological environment to the absorption, pharmacokinetics, and bioavailability of drug. ASD is a solubilizing formulation technology for drug with poor aqueous solubility and with or without poor gastrointestinal membrane permeability (BCS class II or BCS class IV drugs). In this context, thorough biopharmaceutics evaluation at different stages of development is pivotal to the selection of ASD formulation performing the best in vivo [153]. Various in vitro and in silico tools implemented to investigate the in vivo fate of drug delivered as ASD are summarized below, and schematic overview of key biopharmaceutics aspects of ASD is given in Figure 2.3.

2.5.1 Solubility, Supersaturation, and Nano-colloids

ASD formulation is a supersaturating drug delivery system [154]. This means that the dissolved drug concentration in aqueous solution can significantly increase with respect to the equilibrium solubility of crystalline drug. This is in part that the polymer available in ASD solid can prevent the crystallization of drug at the surface in contact with solvent medium and prevent the dissolved drug from precipitating/crystallizing from the supersaturation solution or at least prolong the

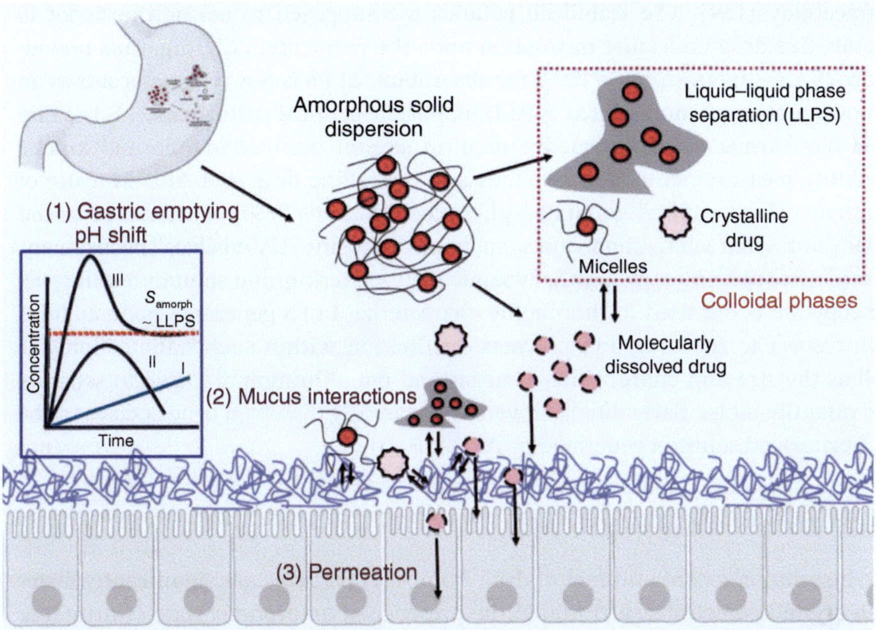

Figure 2.3 Schematics of typical biopharmaceutical behavior of ASD. The dissolution curves in the inset represent (I) slow dissolution without crystallization, (II) fast dissolution followed by drop in concentration due to surface crystallization, and (III) fast supersaturation followed by drop in concentration to form LLPS (dotted line is LLPS or amorphous solubility of drug).

supersaturation. On the other hand, the component present in gastrointestinal fluid or simulated gastric or intestinal fluid can significantly impact the equilibrium as well as kinetic solubility of drug delivered as ASD. This is factored by the interaction among ASD components (drug, polymer, surfactant) and biosurfactants (bile salts and phospholipids) in solutions. Bile salt and phospholipid form colloidal self-assembly, mixed micelles of different sizes. The dissolved API can interact with free bile salt or mixed micelles depending upon their affinity, and this can significantly increase drug solubilization [155]. In the context of ASD, drug solubilization by physiological components is also influenced by the affinity of polymers to these micelles [156]. ASD polymers that bind to micelles impact flux through membrane for the drug that interacts with micelles, but not for the drug having no interaction with micelles, and the polymers that do not interact with micelles have no impact on drug flux [157]. In recent years, the phase behavior of drugs, polymers, and biorelevant components has been extensively investigated in lieu of understanding the in vivo behavior of ASDs [158]. Thermodynamically, the supersaturated solution of drug and polymer in aqueous solution undergoes LLPS beyond certain drug concentration, often found to be closer to amorphous solubility of drug in the given medium. The LLPS results in drug-rich and polymer-rich nano-colloids, and these nano-species are found to be stabilized by bile salt–phospholipid mixed micelles in biorelevant medium, often responsible for drug supersaturation and permeability [159]. The stabilized colloids are supposed to act as the depot to liberate free drug molecules in solution upon the permeation of drug, thus providing constant supersaturation drive for absorption. LLPS is now characterized as an important phenomenon related to ASD biopharmaceutical performance. Enriching ASD biopharmaceutics knowledge requires several thermodynamic and kinetic solubility measurements of amorphous and crystalline drug and ASD in water of different pH, as well as simulated physiological fluids [159]. Further, advanced biophysical characterization tools such as viscosity, UV/visible spectroscopy, fluorescence spectroscopy, SAXS, dynamic light scattering, and solution NMR spectroscopy are being used to thoroughly characterize LLPS-generated nano-colloids with respect to the drug and polymer distribution within such nano-colloids as well as the size and charge. Microdialysis and nanofiltration are used to separate and quantify molecularly dissolved versus apparently dissolved drug species in the supersaturated solution generated by ASD [160, 161].

2.5.2 Drug Dissolution from ASD

The dissolution extent and rate of drug from ASD particles are significantly faster than crystalline counterparts due to the presence of amorphous state of drug. For ASDs with ionizable drug and polymer, drug release and supersaturation profiles become pH-dependent [162]. The in vitro dissolution testing of ASD can be performed using sink method, wherein total dissolved drug concentration achievable is lower than the solubility value of drug in the given medium. The sink method is often used for quality control purposes to identify the variations in drug release kinetics of ASD in terms of manufacturing or storage condition changes [163]. The method can detect the change due to initial composition change or due to the presence of residual crystallinity in ASD. However, to have a discriminatory method at the screening and development phases of ASD, a non-sink method is preferred, wherein the total dissolved drug is expected to be higher than the solubility of drug in the given medium [164]. Often, non-sink conditions are employed as biorelevant dissolution testing of ASD. A wide diversity of non-sink methods can be created by changing dissolution conditions such as medium volume and type (pH and other components), ASD dose used, and dissolution setup. The intention of non-sink method is to obtain entire dissolution, supersaturation, pH ionization status of API and polymer, and precipitation profile of given ASD (spring-parachute effect) [165]. Such non-sink method can be accompanied by a monitoring tool for surface/matrix crystallization of drug at ASD surface interfacing with the dissolution medium. More elaborate and in vivo relevant form of dissolution testing starts with dissolution testing in gastric medium and dynamic transfer of the dissolution system to intestinal medium, either by pH shift or transfer of first medium to second. This is also referred to as physiologically based dissolution testing (PBDT) [166]. PBDT can provide rich data on the mechanistic insight into pH solubility, supersaturation, LLPS, precipitation, and surface crystallization interplay for ASD. Further to this, experimental setups integrating dissolution and membrane flux are used to evaluate in vitro performance of ASD in a bio-predictive manner so that continuous de-supersaturation induced by drug permeation can be accounted [167, 168]. The dissolution in donor compartment is coupled with permeating artificial lipid membrane or cellulose membrane or their combination [169]. More complex in vitro settings also simulate gastric motility, emptying, and transfer rate in the experiments in addition to dissolution and permeation. The simpler one-vessel biphasic dissolution setup includes aqueous medium as a donor dissolution compartment layered with immiscible organic layer (e.g. Decanol) as the acceptor compartment for elucidating dissolution and partitioning process of ASD [170]. An important aspect to consider the dissolution kinetics of ASD is the relationship between the relative release rate of API and polymer and drug loading in ASD. At lower drug loading, drug and polymer release rate is congruent; therefore, supersaturation can be maintained duly in the solution. At a certain drug loading and beyond, drug release rate drastically drops with respect to polymer release [104]. This is defined as the limit of congruency (LoC) for an ASD system, an important consideration with respect to pharmaceutical performance. The wet T_g of the system with respect to dissolution temperature, as well as polymer molecular weight and drug–polymer interactions in ASD, is reported to affect the LoC [171]. At and beyond LoC, drug release is postulated to be hindered by the formation of drug-rich layer at ASD and dissolution medium interface. Furthermore, stronger drug–polymer interaction in solid state is shown to reduce dissolution kinetics of ASD, pointing to the need to balance dissolution and stabilization interplay [35]. Also, the microstructure and the domain size of different amorphous phases formed as a function of drug loading impact their drug release kinetics [90]. Relative to drug dissolution, there is a limited characterization done so far on the disintegration properties of ASD products [172]. At high polymer content, ASD polymers often lead to poor disintegration, swelling, and gel formation in ASD. Apart from physical states, particle properties and wettability can impact the dissolution behavior of ASD. To disregard these effects, intrinsic dissolution testing is performed, preparing the ASD with defined surface exposed to the dissolution medium.

2.5.3 In Vivo and Ex Vivo Aspects of ASD

Several effects drive the dynamic of physicochemical, physiological, and hydrodynamic effects on the ASD products administered in vivo, of which there is still partial understanding at the mechanistic level. This still makes the in vitro–in vivo correlation (IVIVC) exercise quite challenging for strategic decision-making on ASD development. There are cases wherein ASD shows very fast dissolution in vitro and markedly slow dissolution and absorption in vivo [173]. However, there have been some exciting findings over the past few years on our understanding of the mechanism behind the in vivo outcome of ASDs. One of them is evolution toward the impact of LLPS occurring in the formulation in vivo, and how it contributes to drug absorption. The carefully designed quantitative in vivo studies in rodents have shown that drug-rich nanodroplets formed upon LLPS are directly taken by intestinal enterocytes [174]. Even though the flux of nanodroplets is slower than that of the molecularly dispersed drug, the overall higher concentration of them contributes to the higher exposure for ASDs forming nano-colloids prior to absorption in vivo. When the dissolved concentration from ASD exceeds several magnitudes of amorphous solubility, the rich nano-colloids formed in vivo are shown to significantly enhance the absorption to even compensate efflux of the P-glycoprotein (P-gp) substrate drug [175]. The nanoscale drug-rich droplets effectively diffuse into the unstirred water layer [176]. For some ASD, the region with higher concentration generated in vivo leads to a higher flux in the case of HPLC-AS/HF polymer than HPMC-AS/LF. Thus, nano-colloids formed upon LLPS tend to increase both solubility and permeability in vivo. Various in vitro membrane and ex vivo cell models were implemented to study transport of the amorphous nanoparticle formed upon ASD and compared with intestinal tissue results [177]. The other emerging concept in this direction is particle drift phenomenon of nano-colloids generated in vivo [178]. This describes the mechanism of how nano-colloids overcome the resistance to drug diffusion across the unstirred water layer to intestinal wall and release free drug near mucosal surface. The variation in amorphous nanoparticle drifting effect has been observed between the in vivo experiments in mice and rats, pointing toward the possible difference coming from the different bile salt composition in two animal species. The difference in GI luminal composition among different animal models, as well as humans, can pose distinct challenges in translating the learning of nano-colloidal behavior of ASDs in vivo. ASD formulation performance in vivo can also be substantially affected by co-administration of feed together with drug products. Food components can either increase (positive) or decrease (negative) oral absorption and therefore need to be studied thoroughly [43]. The food effect of ASD is still less predictable using the in vitro tests, matching standard US FDA-recommended diet and calories. Further to food, the digestion process taking place in vivo can significantly affect the solubilization behavior, resulting in nano-colloids, and overall oral absorption behavior of drug delivered as ASD. in vitro lipolysis, a common digestion test, shows variable outcome in terms of vivo predictability for ASDs, and the effect has been found to be slightly drug dependent [179]. Likewise, with the effect of other physiological and pharmacological conditions such as achlorhydria, GI phospholipid disorder, or co-administration of proton-pump inhibitors, the in vivo fate of ASD can become even more unpredictable [180]. Also, gut microbiota-mediated bile acid metabolism can alter the in vivo solubility of drug and solubilization capacity of ASD. The other crucial factor often overlooked is the presence and impact of mucus layer on the performance of ASD, especially how a certain polymer- or drug-rich nano-colloids formed in vivo interact with mucus components such as glycoproteins is currently unknown [181]. These aspects deserve further research.

2.6 Stability Considerations of ASD

The stability of ASD formulation is one of the critical quality attributes ensuring the safety and efficacy of the drug product. As ASD is physically metastable, it tends to phase-separate or crystallize over time or upon exposure to various stresses introduced during drug product development. Also, amorphous state being more energetic and mobile makes it more reactive toward degradation conditions in drug formulation or process environment. Therefore, ensuring both physical and chemical stability of ASD requires thorough experimental investigations under a wide range of conditions representing manufacturing and storage conditions of the products. Often, these testing conditions expose ASD to elevated temperatures, humidity, mechanical stress, oxygen, or other reactive environments so that instability kinetics are accelerated. These stress tests provide understanding of the potential mode of instability and help mitigate stability-related risk during manufacturing, formulation, packaging, storage, and usage. These accelerated stability data, together with formal ambient conditions stability, form the data for shelf-life determination and setting expiry date of the product. Below are brief accounts of different stability considerations.

2.6.1 Physical Stability Assessment of ASD

The physical stability assessment of ASD includes the investigation of phase separation and crystallization under accelerated storage conditions. The conditions used often are ICH-recommended temperature and humidity combinations, e.g. 25 °C/60%RH and 40 °C/75%RH. Further, the dry-state elevated temperature data can be used as a reference to understand the effect of temperature alone. The samples are pulled at different conditions to conduct solid-state analysis such as amorphous phase heterogeneity, crystallinity, moisture content, and molecular interactions using a gamut of techniques described above [182]. The analytical methods need to be quantitative and robust to monitor changes. The accelerated stability data, on the one hand, elucidate the instability mechanism and, on the other hand, with further kinetic modeling and statistical analysis, enable the long-term stability prediction to the best or at least the relative ranking of stability during the screening and development stage. Further, more extreme and elaborate combinations of storage conditions are employed for the solid-state stability study (called accelerated predictive stability study) of ASD and further kinetic modeling for stability prediction [182]. Importantly, kinetics of solid-state transition like phase separation, nucleation, and crystallization often exhibit non-Arrhenius behavior with respect to temperature and humidity. The kinetic analysis of stability mostly uses T_g plasticization and corresponding molecular mobility to connect to long-term stability [63]. The data generated at highly accelerating stress storage conditions below T_g of ASD and subsequent analysis using kinetic model and modified Arrhenius equation with both temperature and humidity factors are shown to predict long-term crystallinity levels accurately for ASDs [183]. Most stability investigations use crystallinity as the primary response for kinetic modeling and stability extrapolation [184], while sometimes crystal size/volume is also used for the stability data analysis to model crystal growth [185]. More elaborate thermodynamic and kinetic modeling approach to experimental data on moisture sorption and viscosity has been used to construct an entire phase diagram for ASD in view of shelf-life prediction [186]. On the practical side, solid-state transitions are kinetically controlled and can show quite some variability. Stability samples for

ASD can often be moist and pose problems in reproducible sample preparation and transfer. Precise control of sample preparation, transfer time, and storage temperature after sample is pulled from the stability chamber until analysis is crucial to obtain a reproducible result [187]. Another type of physical stability study of ASD is performed in suspension in the scope of preclinical enabling formulation to be tested in rodents for acute toxicology studies. For this purpose, only limited polymer type is used, e.g. HPMC or HPMC-AS being the most common ones, and the vehicle is typically either aqueous buffer medium, surfactant, or cyclodextrin-based system [188]. The sensitive analytical method for crystallinity or other instability detection on such samples is challenging due to the heterogeneity of the suspension sampling and limited study period. In the context of oral solid dosage form of ASD, the additional excipients present in tablet, such as filler, disintegrant, and binder, can have a dramatic effect on physical stability. Moreover, excipient compatibility study for the ASD intermediate is suitable to perform using the compact of the blend instead of loose powder blend, as the latter lacks the intimate contact surface between the ASD particles and excipient particles [189]. This also works, inferring that hygroscopic excipients are not necessarily deleterious to physical stability, but rather less hygroscopic crystalline fillers like lactose and mannitol leave the sorbed water in free state to induce surface crystallization of API from the ASD in the compact upon exposure to accelerated stability.

2.6.2 Chemical Stability Assessment of ASD

While physical stability is duly assessed using solid-state analytical techniques during accelerated and long-term stability studies, the evaluation of chemical stability of ASD is equally crucial. Owing to the reactive amorphous state and enhanced molecular mobility, drug in ASD has inherently higher tendency to degrade compared to the crystalline drug. Also, ASD formulations are intrinsically more hygroscopic, leading to higher accessibility of water for degradation. So, the stability samples need to be analyzed using HPLC for drug assay as well as full impurity profiling and structural elucidation of degradation products are obtained using liquid chromatography-mass spectrometry (LC-MS) analysis. The ASDs containing enteric and acidic polymers, when exposed to accelerated conditions, lead to the degradation of drug due to the generation of acidic microenvironmental pH within the solid interface [190]. The organic acid byproducts generated upon hydrolysis of cellulose ester-based ASD carriers at elevated temperature and humidity are shown to induce chemical degradation of acid-labile drug in physically stable ASDs [191, 192]. Such residual acids can also form upon ASD processing such as HME of HPMC-AS-based ASD. Also, it has been observed that drug dehydration impurity formed in ASD upon isothermal heating or during HME process catalyzes the hydrolysis of protic polymer like HPMC-AS [193]. There is also a report of direct drug–polymer chemical interaction occurring during HME, leading to the formation of drug degradation products in HPMC-based ASD [194]. The degradation data obtained in DSC measurement, combined with kinetic analysis, is shown to help anticipate drug degradation during HME processing of ASD [195]. Also, many ASD polymers contain synthetic impurities that are reactive to drugs such as peroxides, free radicals, and formaldehydes [101]. Formaldehyde present in polymer can cause drug degradation at higher temperatures; therefore, excipient compatibility for ASD needs to take these into account accordingly. A challenging situation has been reported in ASDs where physically stabilizing intermolecular H-bonding interactions between drug and polymer increased the autoxidation susceptibility of drug, leading to increased fraction of impurity formed via free radicals present in PVP [36]. For an HPMC-AS-based ASD, the intimate noncovalent molecular interactions led to an increase in physical stability, and at the same time, such proximity led to drug esterification by the succinyl group [196]. These observations suggest that polymer, formulation, and process selection for ASD necessitates the balance between physical and chemical stability. For oxidative susceptibility test of amorphous drug with free radicals present in ASD carrier like PVP, mechano-activation tool is found to be fast and suitable [197]. For solid-state prescreening of oxidative sensitivity of API in ASD, a special version of excipient enriched in peroxide can be used, or experimental setup with elevated oxygen environment can be useful [36]. Besides ASD, additional tablet excipients can also be deleterious to the chemical stability of drug in ASD [198]. An important yet relatively less considered avenue is about the interplay between impurity and physical stability in ASD. The presence of certain impurities, even at trace level, can alter molecular interactions and/or mobility in the formulation to the extent that it can notably impact the kinetics and extent of phase separation and crystallization or even induce variation in biopharmaceutics performance [199]. This concerted effect deserves deconvolution through future research in ASD stability science.

2.6.3 Product Performance Stability Assessment of ASD

It is important to evaluate the physical and chemical changes introduced by exposure of ASD products to high temperature and humidity in relation to possible changes in their performance such as dissolution, disintegration, and mechanical properties. An obvious change in dissolution can be inferred based on the physical state evolution during storage such as increase in crystallinity can reduce dissolution and, likewise, oral absorption [200]. However, loss of solubility advantage and increase in crystallinity in stability samples in ASD are not necessarily proportional, and there is limited in vitro and even rarer in vivo data. A recent study on human volunteers reports that drug crystallinity generated in commercial tacrolimus ASD products exposed to accelerated storage condition makes them non-bioequivalent, with prominent impact on maximum blood concentration achieved [201]. Non-sink dissolution test is found to be able to discriminate the alteration in performance due to the storage-induced crystallinity [202]. A less obvious situation can be the dissolution behavior of the stability sample of ASDs with amorphous–amorphous phase separation. In some cases, phase-separated ASDs are found to perform superiorly during non-sink dissolution than a single T_g ASD of the same composition [114]. The moisture sorption by ASD products can significantly alter their mechanical properties owing to the moisture-induced plasticization [203]. The storage at humidity can drastically reduce the tensile strength of ASD tablets triggered by plasticization as well as relaxation of the system [204]. Furthermore, disintegration time of ASD tablets is found to increase upon storage at elevated temperatures and humidity. The effect can be stemming from premature surface dissolution and swelling of disintegrants induced by moisture and reduction in porosity of the tablet microstructure due to particle coalescence and sintering [205]. While these changes in drug products can originate from increased molecular mobility of ASD and excipient surfaces, they are not strictly solid-state transitions and are much less studied in the context of ASD product stability. Future research direction in ASD stability is expected to pave the way for further molecular-level understanding of these stability failure modes.

2.7 Computational Tools for ASD Development

The use of computational tools is prevalent for screening, development, processing, stability, and biopharmaceutics performance of ASDs [206]. With the increased understanding of scientific domain knowledge as well as computational power, ASD-related modeling tools have extensive applications in recent years [206, 207]. In combination with experimental data, various in silico investigations of ASD attributes can reduce overall experimental efforts and streamline the development timelines. While the predictions made of computational tools are approximate, they can provide insight into the mechanisms underlying the observed stabilization of ASD by certain polymers, stability trends, behavior in biorelevant media, process-related changes, and biopharmaceutics. As shown in Figure 2.4, modeling and simulation covering a wide range of length, time, and energy scales are applicable for various aspects of ASD [208]. A summary of different categories of computation tools applied to ASDs and cases is provided below.

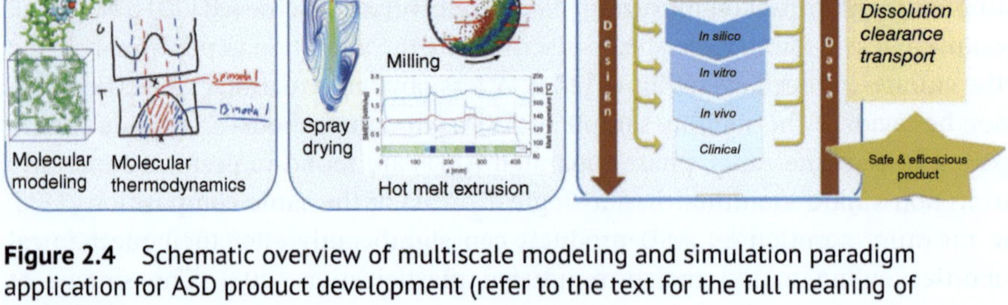

Figure 2.4 Schematic overview of multiscale modeling and simulation paradigm application for ASD product development (refer to the text for the full meaning of acronyms).

2.7.1 Thermodynamic and Phenomenological Models

Thermodynamic assessment and modeling play foundational role in designing ASD. These semi-empirical models provide qualitative guidance to polymer selection for miscible ASD [18, 208]. A simpler way of learning about the contribution of enthalpy versus entropy on drug–polymer mixing is by comparing experimental T_g of an ASD to theoretical T_g predicted by different ideal mixing expressions mathematically derived from the glass transition temperature of corresponding drug and polymer, their relative molar fraction, and physical constant (Gordon–Taylor equation) or heat capacity change in T_g (Couchman and Karasz equation) [49]. The higher value of the predicted T_g than experimental T_g infers the prevalence of enthalpy contribution to the mixing, and the lower value indicates entropy contribution. Several other complex expressions have also been used for predicting T_g of ASD [209]. A ubiquitously used evaluation is based on the Hildebrand or Hansen solubility parameter calculation [19]. Solubility parameter is a molecular descriptor, a square root of cohesion energy density, the latter being the ratio of enthalpy of evaporation to molar volume. Total solubility parameter or fractional solubility parameter including dispersive, polar, and H-bonding can be calculated using the molecular structure and group contribution method. The difference in the solubility parameter of drug and polymer provides insight into relative miscibility of drug and polymer; the lower the difference, the better the miscibility, a typical "like dissolves like principle" [18]. The solubility parameter difference is also used to calculate interaction parameters for mixing. This method is suitable for initial virtual screening, but would not provide all types of mixing possibilities. Next, the popular model for constructing a theoretical temperature–composition phase diagram is using Flory–Huggins's (FH) theory [19]. Basically, Gibbs free energy difference of drug–polymer mixing is estimated as a function of drug-to-polymer ratio using FH theory according to the combinatorial entropy and enthalpy contributions, the latter also involving FH drug–polymer interaction parameter (χ). The negative value of χ means stronger mixing (heteromolecular) interaction in ASD, and the positive value indicates stronger homomolecular interaction (drug–drug, polymer–polymer) and the potential of phase separation. The negative χ results in negative free energy of mixing, pointing to higher miscibility. The χ value as a function of composition and temperature can be obtained via thermal analysis of drug–polymer mixture. The first and second derivatives of FH expression provide binodal and spinodal points, and having these data as a function of temperature will provide the entire phase diagram [210]. A ternary phase diagram including drug, polymer, and water can also be constructed to account for the impact of moisture on the phase miscibility [19]. Modified expressions of classical FH theory, including mechanical, mixing, and melting or including H-bond contribution via Hansen H-bonding solubility parameters, have been applied to ASD phase modeling [211, 212]. Despite that, the main caveat of FH theory for ASD phase modeling is that it is unable to account for directionality of interaction such as H-bonding and the impact on conformational entropy and enthalpy [213]. A more elaborate thermodynamic equation-of-state model accounting for molecular association is based on perturbed-chain statistical associating fluid theory (PC-SAFT), pioneered by Sadowski and her colleagues in ASDs and relevant molecular systems [214]. Here, drug and polymer molecules are represented as spherical segments in chains of certain size, with their segments interacting through dispersive as well as H-bonding, forming self- or inter-association of molecules at a definite size. With respect to ASD, PC-SAFT modeling has been extensively utilized to estimate thermodynamic activity, miscibility gap, phase behavior, and solid-state solubility that leads to the prediction of the physical stability of ASD of certain composition range in response to a certain combination of temperature and humidity [23, 24, 64, 65, 104, 186]. The PC-SAFT phase modeling has also been used as a thermodynamic predictive tool and combined with heat and energy balance model to support the ternary phase modeling in context of SD process design of ASD [215]. Furthermore, the drug/polymer/aqueous medium phase diagrams constructed via PC-SAFT help understanding the thermodynamic mechanisms underlying the drug dissolution from ASD as a function of drug loading [104]. The growing PC-SAFT modeling is yet to evolve further for ASD systems with complex molecular interactions, also with polymers with heterogeneous microstructure of semi-crystalline polymers like PVA and PEG. The model refinement to account for long-range polar interactions, selection of accurate number of association sites, ionization (pH effect), and H-bond cooperativity in PC-SAFT can further improve predictivity. Broadly, advanced thermodynamic estimation of solubility and phase behavior alone cannot suffice to predict stability of ASD. In addition to thermodynamic descriptors, models accounting for kinetic factors like molecular mobility, viscosity, and diffusivity decoupling, as well as crystal growth rate expression, have been employed for ASD [185]. Such models need wider adoption and testing for diverse ASD cases. Also, a model relates the polymer overlap concentration (c^*) in ASD, where polymer chains overlap via coiling, to the nucleation induction time [216].

2.7.2 Multiscale Molecular Simulation and Modeling

Several first-principle molecular modeling methods were shown to mechanistically understand the design attributes, predict and optimize the structures, interactions, and dynamics in relation to the properties of ASD [217]. The most elaborate models applied are quantum chemical and mechanical (QC/QM) calculations such as density functional theory (DFT). It involves the study of the change in chemical bonds, conformations, the electron configurations (band gaps, orbitals), total energy and thermodynamics, frequency and spectra calculations, noncovalent interactions, molecular binding, etc. [208]. QM modeling has been applied to compute the strength, site, and type of intermolecular interactions between the drug and dimer version of polymer in ASD [218]. The QM/DFT calculation is also shown to support the identification of optical hydrocarbon chain length of ASD polymer for favorable solvation-free energy [219]. In an ASD with HPMC-P, formation of ion-paired drug–polymer complex between the protonated drug bound to the carboxylate group of the polymer was elucidated using DFT calculation and correlated with experimental results [220]. Further, DFT results deciphered in a study of ASD that stronger interaction between the drug and PVP was formed than between drug molecules themselves [221]. The QM/QC method for structurally realistic ASD systems is still computationally intensive, and model accuracy depends upon the theoretical basis set selected for calculations. Faster QM methods combined with statistical thermodynamics, such as the conductor-like screening model (COSMO) for real solvents (COSMO-RS) and COSMO segment activity coefficient (COSMO-SAC), are being used as efficient in silico tools for the computation of activity coefficient of drug in polymer, thereby drug solubility in polymer, solid–liquid equilibrium, and phase diagram to assist ASD polymer screening with respect to physical stability and supersaturation [222–224]. A widely used molecular mechanics (MM) modeling for ASD is via molecular dynamics (MD) simulation, wherein molecular simulation is executed on an energy-minimized and equilibrated molecular system applying a force field (FF). The FF defines all bonded (bond length and angle, torsion) and non-bonded (Coulombic, van der Waals, H-bonding, etc.) interactions of molecular species. MD simulation provides the trajectory of atomic position, velocity, interactions, energies, etc. at every simulation step executed based on Newtonian mechanics. The results obtained can be analyzed to extract information on structural stability, solvation, and spatial atomic distribution. MD simulations of ASDs lead to the molecular-level understanding of molecular interactions, thermodynamics, kinetics, molecular mobility, diffusivity, etc. in relation to their performance [225]. MD simulations enable solubility parameter calculation and eventually the miscibility of ASD [226, 227] and drug–polymer interaction strengths [228, 229] and the T_g of ASD including plasticizer [230]. Moreover, water clustering and diffusion, along with interactions with drug and polymer in ASD, have also been studied using MD simulations in relevance to its physical stability and dissolution performance [230, 231]. The radial distribution function profile obtained from MD elaborates on molecular proximity and neighboring environments in ASD [232]. A quantitative MD simulation of ASDs reports intuitive method of searching for pre-crystallite drug assemblies as a precursor of instability [232]. Due to simulation box size and time limitation, MD simulations often represent ASD polymers in the form of few covalent monomers; however, a recent study has simulated entire polymer length [233]. Despite advantages, MD simulations are computationally resource-intensive and challenging to accurately account for interactions and structures in complex environments such as the presence of water and charged structures. A coarse-grained MD simulation built on the full atomistic MD simulation can be explored to increase the speed of computation [234, 235]. The other MM approaches such as molecular docking are also reported for ASD systems [230, 236]. Future development in MD simulation is expected to make it more powerful for ASD modeling, with methods such as molecular glass activation and relaxation, machine-learned FF, and interatomic potentials [237].

2.7.3 Macroscale Modeling

The macroscale modeling methods applied to ASD are mostly relevant to the various phenomena related to manufacturing process and drug dissolution. The models at this scale address heat, mass, and energy balances and treat a system as continuous medium within distinction of atoms and molecules. An important model is computational fluid dynamics (CFD) modeling applied to SD process. CFD models can be used to understand the interaction, velocity, and hydrodynamics of atomized droplets with drying gas at a set of temperatures and flow rates and optimize the drying process to obtain ASD powder with desired solid-state and micromeritic properties [238, 239]. CFD modeling with experimental data is used as a rational approach for predicting solute precipitation and motion of droplet ejecting from atomizer nozzle during SD of ASD [115]. The CFD and flowsheet models enable SD process scale-up [116]. CFD has also been applied for modeling co-rotating HME process of ASD for profiling transport capacity of screws and the pressure buildup [240]. The caveats of mesh-based CFD modeling of HME are in modeling fluids with complex viscosity like drug–polymer melts with spatial heterogeneity of filling levels in HME regions. These limitations are overcome by a mesh-free variant of CFD, known as smoothed particle hydrodynamics (SPH). SPH modeling has been extensively and successfully applied in understanding, optimizing, and scaling up HME process [39, 120]. Besides process modeling, CFD models are also useful for drug dissolution modeling [241]. The other relevant process modeling at macroscale is performed using a contact mechanics model called discrete element modeling (DEM). DEM addresses granular flow, and particle interactions are therefore useful for modeling ASD manufacturing via mechanical milling [242]. CFD, DEM, and hybrid CFD–DEM models as well as other macroscopic models such as finite element method (FEM) and finite volume method (FVM) are extensively used as high-fidelity digital tools for pharmaceutical process simulation, including granulation, coating, capsule filling, tableting, etc., therefore, equally applicable for ASD drug product process understanding [243, 244].

2.7.4 Biopharmaceutics Modeling

The applications of models that enable the prediction of the in vivo behavior and overall biopharmaceutics performance of ASD are rapidly evolving as a valuable tool [245]. They are referred to as physiologically based pharmacokinetic modeling (PBPK) or physiologically based biopharmaceutics modeling (PBBM). Based on the input from the detailed in vitro biorelevant solubility, dissolution, and permeability characterization and basic in vivo PK data (clearance, volume of distribution, etc.) from intravenous administration, the mechanistic PBBM models predict the in vivo drug concentration–time profile, providing insight into the fraction of dose absorbed, maximum plasma concentration, time to maximum concentration, etc. The PBBM model essentially accounts for absorption, clearance, and transit by constitutive equations. The model can utilize the entire in vitro dissolution profile of ASD as an input, and amorphous solubility of API in the presence of polymer can be accounted [246]. Another element such as the effect of drug–polymer nano-colloid formed during ASD dissolution on the permeability can be accounted in PBBM model [247]. A simulation covering the physiology of wider population can be useful for ASD, especially including the in vivo precipitation risk into the model [248]. The full speciation of solubilized, micellized drugs, LLPS, etc. related to ASD dissolution provides confidence to the PBBM model prediction. For example, a good IVIVC for ASD through PBBM is reported to be obtained using only the molecularly dissolved drug in the model, while including entire dissolved species leads to overprediction [249]. The case of PBBM enabled prediction of the complex interplay of food effect, acid-reducing agent, and intersubject variability of clinical performance of various commercial ASDs of itraconazole ASD, which was recently published [43]. The best quantitative models are the validated version using clinical data. Overall, PBBM is currently an important industrial practice in ASD development, aiding speedup of formulation design, reducing preclinical testing in animals, and predicting the clinical bridging, virtual bioequivalence assessment, and generic product development [250].

2.7.5 Data-Driven and Hybrid Modeling

Statistical methods like design of experiments are widely used in ASD development for exploring the design space and identifying optimal parameters to yield the desired ASD characteristics. The recent adoption of data-driven models such as machine learning (ML) or AI in ASD research has shown their application in formulation design [251–253]. With the ability of AI/ML algorithm to identify patterns in complex datasets and extract valuable information from experimental data, such tools can rapidly uncover relationships between formulation parameters like excipient type, processing conditions, and final product properties like dissolution rate and stability. A type of ML model, random forest (RF) model trained with small-scale experimental high-throughput screening data, is shown to predict the large-scale ASD formulations for both binary and ternary systems [251]. The RF regressor is used to capture structure–property relationship of novel polymers amenable to stable ASD design [254]. Artificial neural network (ANN), a deep learning (DL) model, is being extensively used to predict physical and chemical stability, dissolution performance, and composition of ASD [255]. Besides using a sole data-driven model, hybrid models combine AI/ML/DL with mechanistic or physics-driven models to predict ASD performances. For example, the use of MD simulation-enriched ML models is shown to improve the accuracy of stability predictions of amorphous pharmaceuticals [256]. To empower mechanistic models for predicting ASD performances, data-driven models can generate virtual datasets to augment or parameterize models like PC-SAFT or FF used for MD simulations [257]. Furthermore, AI models integrated with PBPK modeling have been used to model in vitro and in vivo performance of ASDs [258].

2.8 Emerging Innovations in ASD

ASD has grown into an established drug formulation technology that is extensively used, starting from preclinical stage up to commercial production of enabled formulations of poorly soluble drugs in pharmaceutical industry. While being a mature and flexible formulation platform, there are still some critical hurdles for the seamless adoption of ASD as a universal solubilizing formulation. An important avenue to explore is ASD formulations for high-dose therapies to circumvent the pill burden [259]. In this direction, a renewed interest in ASD-containing drug and polymer-forming salt is emerging as stable ASD with high drug-loading potential [260]. Using enteric polymer like HPMC-P in the form of salt in ASDs is proposed as a promising way of improving drug solubility [261], while in situ salt formation of weakly basic drug in ASD matrix using acid additive led to the formation of high T_g but less dissolving form [262]. Other promising approaches can be combining ASD with second solubilization mechanisms such as in combination with the self-emulsifying lipid-based system [263, 264], porous inorganic carriers [265], or phospholipids [266]. Using small carriers such as amino acids or sugar derivatives to prepare so-called co-amorphous systems of poorly soluble drugs is increasingly reported in the literature [57]. While these approaches can offer significant drug load, the overall lower T_g as compared to traditional polymeric ASD could pose challenges in their stability. The use of polymer blends or adding surfactants as ASD carrier can provide advantage of different mechanisms of solubilization and stabilization; however, it overall increases the development complexity [267, 268]. The molecular mass and flexibility of modern synthetic molecules with poor solubility such as protein degraders are extremely large, thereby creating hurdles in obtaining high-dose ASD [269]. These drug molecules will be intrinsically stable in amorphous state as such, yet ASD is preferred for improved wettability. In contrast, their chemical stability or even intestinal permeability can be low. There is also interest toward developing ASDs including absorption enhancers for increasing both solubility and permeability of BCS class IV drugs [270]. Leveraging the learning from synthetic drug molecules, there is an enormous opportunity to explore solubilization potential of ASD to increase oral absorption of phytopharmaceuticals and natural product compounds. While synthetic and semi-synthetic polymers are the most common ASD carriers, there is also future for hydrophilic natural polymers like shellac and natural proteins/peptides as ASD carriers from a sustainability viewpoint [108, 271]. From the process innovation perspective, HME processability is limited for emerging high-melting drug candidates, and SD process uses excessive organic solvent. The acid–base complexation without salt formation of acid drugs and base additives combined with ASD is shown to improve solubility as well as drastically reduce the HME processing temperature [272, 273]. For environmentally green SD process, the increase in feed solution temperature in combination with the addition of volatile additive in aqueous feed is proposed for ASD manufacturing [274]. Overall, delivery of existing poorly soluble crystalline drugs as ASD formulation is a great way to reduce the dose to the patient and burden of environmental exposure [275].

References

1 Stegemann, S., Moreton, C., Svanbäck, S. et al. (2023). Trends in oral small-molecule drug discovery and product development based on product launches before and after the rule of five. *Drug Discov. Today* 28: 103344.

2 Moseson, D.E., Tran, T.B., Karunakaran, B. et al. (2024). Trends in amorphous solid dispersion drug products approved by the U.S. Food and Drug Administration between 2012 and 2023. *Int. J. Pharm. X* 7: 100259.

3 Van Den Mooter, G. (2012). The use of amorphous solid dispersions: a formulation strategy to overcome poor solubility and dissolution rate. *Drug Discov. Today Technol.* 9: e79–e85.

4 Brouwers, J., Brewster, M.E., and Augustijns, P. (2009). Supersaturating drug delivery systems: the answer to solubility-limited oral bioavailability? *J. Pharm. Sci.* 98: 2549–2572.

5 Laitinen, R., Priemel, P.A., Surwase, S. et al. (2014). Theoretical considerations in developing amorphous solid dispersions. In: *Amorphous Solid Dispersions: Theory and Practice* (ed. S. Navnit, H. Sandhu, D.S. Choi, et al.), 35–90. New York: Springer.

6 Janssens, S. and Van den Mooter, G. (2009). Review: physical chemistry of solid dispersions. *J. Pharm. Pharmacol.* 61: 1571–1586.

7 Taylor, L.S. and Zografi, G. (2025). From unwanted annoyances to oral delivery saviors: the rollercoaster journey of amorphous drugs. *Mol. Pharm.* 22: 5125–5138.

8 Zografi, G., Newman, A., and Shalaev, E. (2025). Structural features of the glassy state and their impact on the solid-state properties of organic molecules in pharmaceutical systems. *J. Pharm. Sci.* 114: 40–69.

9 Kawakami, K. (2019). Ultraslow cooling for the stabilization of pharmaceutical glasses. *J. Phys. Chem. B* 123: 4996–5003.

10 Yao, B., Wojnarowska, Z., and Paluch, M. (2021). Effect of structure on molecular dynamics in glass-forming liquids. The case of aromaticity. *J. Mol. Liq.* 344: 117757.

11 DeBoyace, K., Bookwala, M., Zhou, D. et al. (2024). Understanding the influence of API conformations on amorphous dispersion formation potential predictions using the R3m molecular descriptor. *Mol. Pharm.* 21: 770–780.

12 Baird, J.A., Van Eerdenbrugh, B., and Taylor, L.S. (2010). A classification system to assess the crystallization tendency of organic molecules from undercooled melts. *J. Pharm. Sci.* 99: 3787–3806.

13 Blaabjerg, L.I., Lindenberg, E., Löbmann, K. et al. (2016). Glass forming ability of amorphous drugs investigated by continuous cooling and isothermal transformation. *Mol. Pharm.* 13: 3318–3325.

14 Panini, P., Rampazzo, M., Singh, A. et al. (2019). Myth or truth: the glass forming ability class III drugs will always form single-phase homogenous amorphous solid dispersion formulations. *Pharmaceutics* 11: 529.

15 Vodak, D.T. and Morgen, M. (2014). Design and development of HPMCAS-based spray-dried dispersions. In: *Amorphous Solid Dispersions: Theory and Practice* (ed. N. Shah, H. Sandhu, D.S. Choi, et al.), 303–322. New York: Springer.

16 Zhou, D., Schmitt, E.A., Law, D. et al. (2019). Assessing physical stability risk using the amorphous classification system (ACS) based on simple thermal analysis. *Mol. Pharm.* 16: 2742–2754.

17 Fenwick, N.W., Telford, R., Bowen, R.D., and Seaton, C.C. (2024). Crystal or glass? Chemical and crystallographic factors in the amorphization of molecular materials. *Cryst. Growth Des.* 24: 1124–1135.

18 Thakore, S.D., Akhtar, J., Jain, R. et al. (2021). Analytical and computational methods for the determination of drug-polymer solubility and miscibility. *Mol. Pharm.* 18: 2835–2866.

19 Paudel, A., Van Humbeeck, J., and Van den Mooter, G. (2010). Theoretical and experimental investigation on the solid solubility and miscibility of naproxen in poly (vinylpyrrolidone). *Mol. Pharm.* 7: 1133–1148.

20 Janssens, S., De Zeure, A., Paudel, A. et al. (2010). Influence of preparation methods on solid state supersaturation of amorphous solid dispersions: a case study with Itraconazole and Eudragit E100. *Pharm. Res.* 27: 775–785.

21 Rumondor, A.C.F., Wikström, H., Van Eerdenbrugh, B., and Taylor, L.S. (2011). Understanding the tendency of amorphous solid dispersions to undergo amorphous–amorphous phase separation in the presence of absorbed moisture. *AAPS PharmSciTech* 12: 1209–1219.

22 Rumondor, A.C.F. and Taylor, L.S. (2010). Effect of polymer hygroscopicity on the phase behavior of amorphous solid dispersions in the presence of moisture. *Mol. Pharm.* 7: 477–490.

23 Grönniger, B., Kimpe, K., Singh, A., and Sadowski, G. (2024). Simultaneous Water sorption and crystallization in ASDs 1: stability studies lasting for two years. *Mol. Pharm.* 21: 957–969.

24 Grönniger, B., Fritschka, E., Kimpe, K. et al. (2024). Simultaneous water sorption and crystallization in ASDs 2: modeling long-term stabilities. *Mol. Pharm.* 21: 2908–2921.

25 Bhattacharya, S. and Suryanarayanan, R. (2009). Local mobility in amorphous pharmaceuticals—characterization and implications on stability. *J. Pharm. Sci.* 98: 2935–2953.

26 Mehta, M., McKenna, G.B., and Suryanarayanan, R. (2016). Molecular mobility in glassy dispersions. *J. Chem. Phys.* 144: 204506.

27 Bhugra, C., Shmeis, R., Krill, S.L., and Pikal, M.J. (2008). Prediction of onset of crystallization from experimental relaxation times. II. Comparison between predicted and experimental onset times. *J. Pharm. Sci.* 97: 455–472.

28 Yu, L. (2016). Surface mobility of molecular glasses and its importance in physical stability. *Adv. Drug Deliv. Rev.* 100: 3–9.

29 Cools, L. and Van den Mooter, G. (2025). A comprehensive overview of the role of intermolecular interactions in amorphous solid dispersions. *Int. J. Pharm.* 674: 125441.

30 Paudel, A., Nies, E., and Van den Mooter, G. (2012). Relating hydrogen-bonding interactions with the phase behavior of naproxen/PVP K 25 solid dispersions: evaluation of solution-cast and quench-cooled films. *Mol. Pharm.* 9: 3301–3317.

31 Singh, D. (2025). A pioneer review on intermolecular hydrogen bonding in drug-polymer mixtures: implication of a stable delivery system. *J. Macromol. Sci. Part B* 64: 1510–1526.

32 Mistry, P., Mohapatra, S., Gopinath, T. et al. (2015). Role of the strength of drug–polymer interactions on the molecular mobility and crystallization inhibition in ketoconazole solid dispersions. *Mol. Pharm.* 12: 3339–3350.

33 Bookwala, M. and Wildfong, P.L.D. (2023). The implications of drug-polymer interactions on the physical stability of amorphous solid dispersions. *Pharm. Res.* 40: 2963–2981.

34 Bookwala, M. (2023). The implications of the halogen atom on the physical stability of amorphous solid dispersions. Doctoral dissertation. Duquesne University.

35 Que, C., Deac, A., Zemlyanov, D.Y. et al. (2021). Impact of drug–polymer intermolecular interactions on dissolution performance of copovidone-based amorphous solid dispersions. *Mol. Pharm.* 18: 3496–3508.

36 Saraf, I., Roskar, R., Modhave, D. et al. (2022). Forced solid-state oxidation studies of nifedipine-PVP amorphous solid dispersion. *Mol. Pharm.* 19: 568–583.

37 Zhang, J., Guo, M., Luo, M., and Cai, T. (2023). Advances in the development of amorphous solid dispersions: the role of polymeric carriers. *Asian J. Pharm. Sci.* 18: 100834.

38 Paudel, A., Worku, Z.A., Meeus, J. et al. (2013). Manufacturing of solid dispersions of poorly water soluble drugs by spray drying: formulation and process considerations. *Int. J. Pharm.* 453: 253–284.

39 Matić, J., Paudel, A., Bauer, H. et al. (2020). Developing HME-based drug products using emerging science: a fast-track roadmap from concept to clinical batch. *AAPS PharmSciTech* 21: 176.

40 Bhadoriya, T.S., Chakraborty, S., Pariskar, A., and Bansal, A.K. (2026). Excipient variability critically impacts crystallization propensity of spray-dried amorphous solid dispersions. *Eur. J. Pharm. Sci.* 216: 107294.

41 Patel, N.G. and Serajuddin, A.T.M. (2022). Moisture sorption by polymeric excipients commonly used in amorphous solid dispersion and its effect on glass transition temperature: I. Polyvinylpyrrolidone and related copolymers. *Int. J. Pharm.* 616: 121532.

42 Stegemann, S., Sheehan, L., Rossi, A. et al. (2022). Rational and practical considerations to guide a target product profile for patient-centric drug product development with measurable patient outcomes – a proposed roadmap. *Eur. J. Pharm. Biopharm.* 177: 81–88.

43 Rudolph, N., Charbe, N., Plano, D. et al. (2025). A physiologically based biopharmaceutics modeling (PBBM) framework for characterizing formulation-dependent food effects: paving the road towards fed state virtual BE studies for itraconazole amorphous solid dispersions. *Eur. J. Pharm. Sci.* 209: 107047.

44 Arca, H.Ç., Mosquera-Giraldo, L.I., Dahal, D. et al. (2017). Multidrug, anti-HIV amorphous solid dispersions: nature and mechanisms of impacts of drugs on each other's solution concentrations. *Mol. Pharm.* 14: 3617–3627.

45 Leane, M., Pitt, K., and Reynolds, G. (2015). A proposal for a drug product Manufacturing Classification System (MCS) for oral solid dosage forms. *Pharm. Dev. Technol.* 20: 12–21.

46 Hsiao, W.K., Hörmann, T.R., Toson, P. et al. (2020). Feeding of particle-based materials in continuous solid dosage manufacturing: a material science perspective. *Drug Discov. Today* 25: 800–806.

47 Paudel, A., Meeus, J., and Van den, M.G. (2014). Structural characterization of amorphous solid dispersions. In: *Amorphous Solid Dispersions: Theory and Practice* (ed. N. Shah, H. Sandhu, D.S. Choi, et al.), 421–485. New York: Springer.

48 Baird, J.A. and Taylor, L.S. (2012). Evaluation of amorphous solid dispersion properties using thermal analysis techniques. *Adv. Drug Deliv. Rev.* 64: 396–421.

49 Pinal, R. (2008). Entropy of mixing and the glass transition of amorphous mixtures. *Entropy* 10: 207–223.

50 Larsen, B.S., Meiland, P., Tzdaka, E. et al. (2024). A unifying approach to drug-in-polymer solubility prediction: streamlining experimental workflow and analysis. *Eur. J. Pharm. Biopharm.* 203: 114478.

51 Alin, J., Setiawan, N., Defrese, M. et al. (2018). A novel approach for measuring room temperature enthalpy of mixing and associated solubility estimation of a drug in a polymer matrix. *Polymer (Guildf)* 135: 50–60.

52 Song, S., Wang, C., Zhang, B. et al. (2023). A rheological approach for predicting physical stability of amorphous solid dispersions. *J. Pharm. Sci.* 112: 204–212.

53 Laggner, P. and Paudel, A. (2018). Density fluctuations in amorphous pharmaceutical solids. Can SAXS help to predict stability? *Colloids Surf. B Biointerfaces* 168: 76–82.

54 Paudel, A., Geppi, M., and Van Den Mooter, G. (2014). Structural and dynamic properties of amorphous solid dispersions: the role of solid-state nuclear magnetic resonance spectroscopy and relaxometry. *J. Pharm. Sci.* 103: 2635–2662.

55 Sarpal, K. and Munson, E.J. (2021). Amorphous solid dispersions of felodipine and nifedipine with Soluplus®: drug-polymer miscibility and intermolecular interactions. *J. Pharm. Sci.* 110: 1457–1469.

56 Kushwah, V., Succhielli, C., Saraf, I., and Paudel, A. (2024). Amorphous solid dispersions: implication of method of preparation and physicochemical properties of API and excipients. *Pharmaceutics* 16: 1035.

57 Kilpeläinen, T., Ervasti, T., Uurasjärvi, E. et al. (2022). Detecting different amorphous – amorphous phase separation patterns in co-amorphous mixtures with high resolution imaging FTIR spectroscopy. *Eur. J. Pharm. Biopharm.* 180: 161–169.

58 Qian, F., Huang, J., Zhu, Q. et al. (2010). Is a distinctive single T_g a reliable indicator for the homogeneity of amorphous solid dispersion? *Int. J. Pharm.* 395: 232–235.

59 Lauer, M.E., Siam, M., Tardio, J. et al. (2013). Rapid assessment of homogeneity and stability of amorphous solid dispersions by atomic force microscopy—from bench to batch. *Pharm. Res.* 30: 2010–2022.

60 Chakraborty, S. and Bansal, A.K. (2025). Application of atomic force microscopy in the development of amorphous solid dispersion. *J. Pharm. Sci.* 114: 70–81.

61 Kissi, E.O., Grohganz, H., Löbmann, K. et al. (2018). Glass-transition temperature of the β-relaxation as the major predictive parameter for recrystallization of neat amorphous drugs. *J. Phys. Chem. B* 122: 2803–2808.

62 Ngai, K.L. (2017). Universal properties of relaxation and diffusion in condensed matter. *Chin. Phys. B* 26: 18105.

63 Greco, S., Authelin, J.-R., Leveder, C., and Segalini, A. (2012). A practical method to predict physical stability of amorphous solid dispersions. *Pharm. Res.* 29: 2792–2805.

64 Grönniger, B., Danzer, A., Kimpe, K. et al. (2023). Viscoelastic behavior of supercooled and glassy ASDs at humid conditions can be predicted. *Mol. Pharm.* 20: 2568–2578.

65 Wolbert, F., Stecker, J., Luebbert, C., and Sadowski, G. (2020). Viscosity of ASDs at humid conditions. *Eur. J. Pharm. Biopharm.* 154: 387–396.

66 Bhardwaj, S.P. and Suryanarayanan, R. (2011). Subtraction of DC conductivity and annealing: approaches to identify Johari–Goldstein relaxation in amorphous trehalose. *Mol. Pharm.* 8: 1416–1422.

67 Grzybowska, K., Adrjanowicz, K., and Paluch, M. (2016). Application of broadband dielectric spectroscopy to study molecular mobility in pharmaceutical systems. In: *Disordered Pharmaceutical Materials* (ed. M. Descamps), 301–360. Weinheim: Wiley.

68 Péter-Haraszti, A., Záhonyi, P., Farkas, A. et al. (2024). Thermal investigation of relaxations of interacting and non-interacting amorphous solid dispersions. *J. Therm. Anal. Calorim.* 149: 8067–8083.

69 Santitewagun, S. and Zeitler, J.A. (2025). Probing the influence of water on the molecular mobility of PVP/VA using terahertz spectroscopy. *Mol. Pharm.* 22: 5928–5938.

70 Bannow, J., Karl, M., Larsen, P.E. et al. (2020). Direct measurement of lateral molecular diffusivity on the surface of supersaturated amorphous solid dispersions by atomic force microscopy. *Mol. Pharm.* 17: 1715–1722.

71 Shokoya, M.M., Benkő, B.-M., Süvegh, K. et al. (2023). Positron annihilation lifetime spectroscopy as a special technique for the solid-state characterization of pharmaceutical excipients, drug delivery systems, and medical devices—a systematic review. *Pharmaceuticals* 16: 252.

72 Moran, A. and Buckton, G. (2009). Studies of the crystallization of amorphous trehalose using simultaneous gravimetric vapor sorption/near IR (GVS/NIR) and "Modulated" GVS/NIR. *AAPS PharmSciTech* 10: 297–302.

73 Gui, Y., McCann, E.C., Yao, X. et al. (2021). Amorphous drug–polymer salt with high stability under tropical conditions and fast dissolution: the case of clofazimine and poly (acrylic acid). *Mol. Pharm.* 18: 1364–1372.

74 Paudel, A., Raijada, D., and Rantanen, J. (2015). Raman spectroscopy in pharmaceutical product design. *Adv. Drug Deliv. Rev.* 89: 3–20.

75 van Drooge, D.J., Braeckmans, K., Hinrichs, W.L.J. et al. (2006). Characterization of the mode of incorporation of lipophilic compounds in solid dispersions at the nanoscale using fluorescence resonance energy transfer (FRET). *Macromol. Rapid Commun.* 27: 1149–1155.

76 Hitzer, P., Bäuerle, T., Drieschner, T. et al. (2017). Process analytical techniques for hot-melt extrusion and their application to amorphous solid dispersions. *Anal. Bioanal. Chem.* 409: 4321–4333.

77 Winck, J., Daalmann, M., Berghaus, A., and Thommes, M. (2022). In-line monitoring of solid dispersion preparation in small scale extrusion based on UV–vis spectroscopy. *Pharm. Dev. Technol.* 27: 1009–1015.

78 Pugliese, A., Tobyn, M., Hawarden, L.E. et al. (2022). New development in understanding drug–polymer interactions in pharmaceutical amorphous solid dispersions from solid-state nuclear magnetic resonance. *Mol. Pharm.* 19: 3685–3699.

79 Cosquer, R.P., Pinon, A.C., Šoltésová, M. et al. (2025). Characterizing 13C spectral assignments and substituent distributions of hydroxypropylmethylcellulose acetyl succinate using dynamic nuclear polarization nuclear magnetic resonance spectroscopy. *Mol. Pharm.* 22: 5870–5878.

80 Moseson, D.E. and Taylor, L.S. (2023). Cystallinity: a complex critical quality attribute of amorphous solid dispersions. *Mol. Pharm.* 20: 4802–4825.

81 Svoboda, R. (2025). Calorimetric monitoring of the sub-T_g crystal growth in molecular glasses: the case of amorphous nifedipine. *Molecules* 30: 1679.

82 Rumondor, A.C.F. and Taylor, L.S. (2010). Application of Partial Least-Squares (PLS) modeling in quantifying drug crystallinity in amorphous solid dispersions. *Int. J. Pharm.* 398: 155–160.

83 Xie, T. and Taylor, L.S. (2016). Dissolution performance of high drug loading celecoxib amorphous solid dispersions formulated with polymer combinations. *Pharm. Res.* 33: 739–750.

84 Bordos, E., Islam, M.T., Florence, A.J. et al. (2019). Use of terahertz-Raman spectroscopy to determine solubility of the crystalline active pharmaceutical ingredient in polymeric matrices during hot melt extrusion. *Mol. Pharm.* 16: 4361–4371.

85 Haku, R., Suzuki, N., Ohashi, K. et al. (2025). Nanogram-scale screening of amorphous solid dispersion and its stability with monitoring phase transition using low-frequency Raman spectroscopy. *J. Drug Deliv. Sci. Technol.* 112: 107237.

86 S'ari, M., Blade, H., Cosgrove, S. et al. (2021). Characterization of amorphous solid dispersions and identification of low levels of crystallinity by transmission electron microscopy. *Mol. Pharm.* 18: 1905–1919.

87 Smith Casey, J., Janny, D., Schmitt Paul, D. et al. (2018). Calibration-free second harmonic generation (SHG) image analysis for quantification of trace crystallinity within final dosage forms of amorphous solid dispersions. *Appl. Spectrosc.* 72: 1594–1605.

88 Shi, C., Li, L., Zhang, G.G.Z., and Borchardt, T.B. (2019). Direct visualization of drug–polymer phase separation in ritonavir–copovidone amorphous solid dispersions using in situ synchrotron X-ray fluorescence imaging of thin films. *Mol. Pharm.* 16: 4751–4754.

89 Meeus, J., Scurr, D.J., Chen, X. et al. (2015). Combination of (M)DSC and surface analysis to study the phase behaviour and drug distribution of ternary solid dispersions. *Pharm. Res.* 32: 1407–1416.

90 Li, N. and Taylor, L.S. (2019). Microstructure formation for improved dissolution performance of lopinavir amorphous solid dispersions. *Mol. Pharm.* 16: 1751–1765.

91 Hasegawa, S., Ke, P., and Buckton, G. (2009). Determination of the structural relaxation at the surface of amorphous solid dispersion using inverse gas chromatography. *J. Pharm. Sci.* 98: 2133–2139.

92 Miyanishi, H., Nemoto, T., Mizuno, M. et al. (2013). Evaluation of crystallization behavior on the surface of nifedipine solid dispersion powder using inverse gas chromatography. *Pharm. Res.* 30: 502–511.

93 Školáková, T., Smržová, D., Pekárek, T. et al. (2022). Investigation of tadalafil molecular arrangement in solid dispersions using inverse gas chromatography and Raman mapping. *Int. J. Pharm.* 623: 121955.

94 Yu, D., Fiddler, F., Ibrahim, A. et al. (2022). Surface characterization as a tool for identifying the factors affecting the dissolution rate of amorphous solid dispersion tablets. *AAPS PharmSciTech* 23: 282.

95 Kubat, K., Strojewski, D., Majda, D. et al. (2023). Molecular phenomena related to moisture uptake in amorphous solid dispersion loaded with bosentan monohydrate and copovidone. *Powder Technol.* 430: 119020.

96 Hancock, B.C. and Zografi, G. (1994). The relationship between the glass transition temperature and the water content of amorphous pharmaceutical solids. *Pharm. Res.* 11: 471–477.

97 Xi, H., Zhu, A., Klinzing, G.R. et al. (2020). Characterization of spray dried particles through microstructural imaging. *J. Pharm. Sci.* 109: 3404–3412.

98 Marcozzi, T., Baviriseaty, S., Yawman, P. et al. (2025). Synchrotron computed tomography combined with AI-based image analysis for the advanced characterization of spray dried amorphous solid dispersion particles. *J. Pharm. Sci.* 114: 530–543.

99 Asare-Addo, K., Šupuk, E., Al-Hamidi, H. et al. (2015). Triboelectrification and dissolution property enhancements of solid dispersions. *Int. J. Pharm.* 485: 306–316.

100 Sun, C.C. (2009). Materials science tetrahedron—a useful tool for pharmaceutical research and development. *J. Pharm. Sci.* 98: 1671–1687.

101 Saraf, I., Kushwah, V., Werner, B. et al. (2025). Quantification of hydrogen peroxide in PVP and PVPVA using 1H qNMR spectroscopy. *Polymers (Basel)* 17: 739.

102 Hu, Q., Wyttenbach, N., Shiraki, K., and Choi, D.S. (2014). Miniaturized screening tools for polymer and process evaluation. In: *Amorphous Solid Dispersions: Theory and Practice* (ed. N. Shah, H. Sandhu, D.S. Choi, et al.), 165–195. New York: Springer.

103 Ilevbare, G.A. and Taylor, L.S. (2013). Liquid–liquid phase separation in highly supersaturated aqueous solutions of poorly water-soluble drugs: implications for solubility enhancing formulations. *Cryst. Growth Des.* 13: 1497–1509.

104 Deac, A., Luebbert, C., Qi, Q. et al. (2024). Dissolution mechanisms of amorphous solid dispersions: application of ternary phase diagrams to explain release behavior. *Mol. Pharm.* 21: 1900–1918.

105 Senniksen, M.B., Wyttenbach, N., Page, S., and Dressman, J. (2025). Combining high-throughput ASD screening with the rDCS to streamline development of poorly soluble drugs. *Eur. J. Pharm. Sci.* 211: 107130.

106 Papakostas, G., Corner, P.A., Hook, A.L. et al. (2025). Miniaturized high-throughput amorphous solid dispersion screening via picoliter volume 2D-inkjet printing of formulation microarrays. *Mol. Pharm.* 22: 2040–2052.

107 Sodhi, I. and Sangamwar, A.T. (2018). Microarray plate method for estimation of precipitation kinetics of celecoxib under biorelevant conditions and precipitate characterization. *Mol. Pharm.* 15: 2423–2436.

108 Clark, E.F., Howard, A., Feliu, S.D.M. et al. (2024). D-Xylose oxetane copolymers as bioderived and tuneable polyesters for amorphous solid dispersions. *RSC Appl. Polym.* 2: 1104–1112.

109 Bochmann, E.S., Üstüner, E.E., Gryczke, A., and Wagner, K.G. (2017). Predicting melt rheology for hot-melt extrusion by means of a simple T_g-measurement. *Eur. J. Pharm. Biopharm.* 119: 47–55.

110 Shadambikar, G., Kipping, T., Di-Gallo, N. et al. (2020). Vacuum compression molding as a screening tool to investigate carrier suitability for hot-melt extrusion formulations. *Pharmaceutics* 12: 1019.

111 Klimša, V., Ruphuy, G., Jonáš, J. et al. (2023). Spray drying robot for high-throughput combinatorial fabrication of multicomponent solid dispersions. *Powder Technol.* 428: 118872.

112 Ousset, A., Bassand, C., Chavez, P.-F. et al. (2019). Development of a small-scale spray-drying approach for amorphous solid dispersions (ASDs) screening in early drug development. *Pharm. Dev. Technol.* 24: 560–574.

113 Paudel, A. and Van den Mooter, G. (2012). Influence of solvent composition on the miscibility and physical stability of naproxen/PVP K 25 solid dispersions prepared by cosolvent spray-drying. *Pharm. Res.* 29: 251–270.

114 Paudel, A., Loyson, Y., and Van den Mooter, G. (2013). An investigation into the effect of spray drying temperature and atomizing conditions on miscibility, physical stability, and performance of naproxen–PVP K 25 solid dispersions. *J. Pharm. Sci.* 102: 1249–1267.

115 Boel, E., Koekoekx, R., Dedroog, S. et al. (2020). Unraveling particle formation: from single droplet drying to spray drying and electrospraying. *Pharmaceutics* 12: 625.

116 Dobry, D.E., Settell, D.M., Baumann, J.M. et al. (2009). A model-based methodology for spray-drying process development. *J. Pharm. Innov.* 4: 133–142.

117 Smeets, A., Clasen, C., and Van den Mooter, G. (2017). Electrospraying of polymer solutions: study of formulation and process parameters. *Eur. J. Pharm. Biopharm.* 119: 114–124.

118 Verreck, G., Chun, I., Peeters, J. et al. (2003). Preparation and characterization of nanofibers containing amorphous drug dispersions generated by electrostatic spinning. *Pharm. Res.* 20: 810–817.

119 Kožák, J., Bühlbecker, P., Rautenberg, A. et al. (2025). Hydroxypropyl cellulose-based lyospheres accelerate dissolution of aprepitant combined with its supersaturation for oral delivery. *Eur. J. Pharm. Sci.* 213: 107220.

120 Matić, J., Alva, C., Eder, S. et al. (2021). Towards predicting the product quality in hot-melt extrusion: pilot plant scale extrusion. *Int. J. Pharm. X* 3: 100084.

121 Martynek, D., Ridvan, L., Sivén, M., and Šoóš, M. (2025). Stability and recrystallization of amorphous solid dispersions prepared by hot-melt extrusion and spray drying. *Int. J. Pharm.* 672: 125331.

122 Wood, C.C., Patel, K.G., Weber, V.L. et al. (2025). Development of impact resistant immediate release amorphous solid dispersion via hot-melt extrusion and injection molding. *Int. J. Pharm.* 680: 125746.

123 Roche, A., Sanchez-Ballester, N.M., Bataille, B. et al. (2024). Fused deposition modelling 3D printing and solubility improvement of BCS II and IV active ingredients – a narrative review. *J. Control. Release* 365: 507–520.

124 Alva, C., Goetzinger, E., Matić, J. et al. (2025). Tailoring the release of highly loaded amorphous solid dispersions via additive manufacturing. *J. Control. Release* 382: 113723.

125 Fratini, C., Moroni, S., De, A.D. et al. (2025). Direct powder extrusion (DPE) 3D-printing of mini-tablets for preclinical studies in rodents. *Int. J. Pharm.* 675: 125542.

126 Gilis, P.M.V., De Condé, V.F.V., and Vandecruys, R.P.G. (1997). Beads having a core coated with an antifungal and a polymer. US Patent 5633015.

127 Shah, N., Iyer, R.M., Mair, H.-J. et al. (2013). Improved human bioavailability of vemurafenib, a practically insoluble drug, using an amorphous polymer-stabilized solid dispersion prepared by a solvent-controlled coprecipitation process. *J. Pharm. Sci.* 102: 967–981.

128 Zhang, T., Jankovsky, C., Blattner, S.M. et al. (2026). Controlled API-polymer solidification (CAPS): a novel manufacturing process approach to produce amorphous solid dispersion with enhanced physical properties. *J. Pharm. Sci.* 115: 104028.

129 Yu, M., Zhou, D., Oberoi, H.S. et al. (2025). Scale-up and clinical bioavailability assessment of a 45% drug loaded amorphous nanoparticle formulation of a BCS IV compound for oral delivery. *J. Pharm. Sci.* 114: 383–393.

130 Hiew, T.N., Solomos, M.A., Kafle, P. et al. (2025). The importance of surface composition and wettability on the dissolution performance of high drug loading amorphous dispersion formulations. *J. Pharm. Sci.* 114: 289–303.

131 Solomos, M.A., Punia, A., Saboo, S. et al. (2023). Evaluating spray drying and co-precipitation as manufacturing processes for amorphous solid dispersions of a low T_g API. *J. Pharm. Sci.* 112: 2087–2096.

132 Potter, C., Tian, Y., Walker, G. et al. (2015). Novel supercritical carbon dioxide impregnation technique for the production of amorphous solid drug dispersions: a comparison to hot melt extrusion. *Mol. Pharm.* 12: 1377–1390.

133 Benson, E.G., Moseson, D.E., Bhalla, S. et al. (2025). Dissolution of copovidone-based amorphous solid dispersions: influence of atomic layer coating, hydration kinetics, and formulation. *J. Pharm. Sci.* 114: 323–335.

134 Ellenberger, D.J., Miller, D.A., and Williams, R.O. (2018). Expanding the application and formulation space of amorphous solid dispersions with KinetiSol®: a review. *AAPS PharmSciTech* 19: 1933–1956.

135 DiNunzio, J.C., Brough, C., Hughey, J.R. et al. (2010). Fusion production of solid dispersions containing a heat-sensitive active ingredient by hot melt extrusion and Kinetisol® dispersing. *Eur. J. Pharm. Biopharm.* 74: 340–351.

136 Dedroog, S., Huygens, C., and Van den Mooter, G. (2019). Chemically identical but physically different: a comparison of spray drying, hot melt extrusion and cryo-milling for the formulation of high drug loaded amorphous solid dispersions of naproxen. *Eur. J. Pharm. Biopharm.* 135: 1–12.

137 Doreth, M., Löbmann, K., Priemel, P. et al. (2018). Influence of PVP molecular weight on the microwave assisted in situ amorphization of indomethacin. *Eur. J. Pharm. Biopharm.* 122: 62–69.

138 Tanaka, R., Hattori, Y., Horie, Y. et al. (2019). Characterization of amorphous solid dispersion of pharmaceutical compound with pH-dependent solubility prepared by continuous-spray granulator. *Pharmaceutics* 11: 159.

139 Chen, Y.C., Ho, H.O., Da, C.J., and Sheu, M.T. (2014). Physical and dissolution characterization of cilostazol solid dispersions prepared by hot melt granulation (HMG) and thermal adhesion granulation (TAG) methods. *Int. J. Pharm.* 473: 458–468.

140 Baraldo, M. (2016). Meltdose tacrolimus pharmacokinetics. *Transplant. Proc.* 48: 420–423.

141 Démuth, B., Nagy, Z.K., Balogh, A. et al. (2015). Downstream processing of polymer-based amorphous solid dispersions to generate tablet formulations. *Int. J. Pharm.* 486: 268–286.

142 Jijun, F., Lishuang, X., Xiaoli, W. et al. (2011). Nimodipine (NM) tablets with high dissolution containing NM solid dispersions prepared by hot-melt extrusion. *Drug Dev. Ind. Pharm.* 37: 934–944.

143 Patel, S., Kou, X., Hou, H.H. et al. (2017). Mechanical properties and tableting behavior of amorphous solid dispersions. *J. Pharm. Sci.* 106: 217–223.

144 Yu, D., Nie, H., and Hoag, S.W. (2024). Comprehensive evaluation of polymer types and ratios in spray-dried dispersions: compaction, dissolution, and physical stability. *Int. J. Pharm.* 650: 123674.

145 Marcozzi, T., Otava, M., Andersen, S.K. et al. (2025). Investigating the effects of formulation and manufacturing routes on the tensile strength and disintegration time of tablets containing an HPMC AS-based spray dried powder. *Int. J. Pharm.* 686: 126314.

146 Ayenew, Z., Paudel, A., and Van Den Mooter, G. (2012). Can compression induce demixing in amorphous solid dispersions? A case study of naproxen–PVP K25. *Eur. J. Pharm. Biopharm.* 81: 207–213.

147 Singh, A., Bharati, A., Frederiks, P. et al. (2016). Effect of compression on the molecular arrangement of itraconazole–soluplus solid dispersions: induction of liquid crystals or exacerbation of phase separation? *Mol. Pharm.* 13: 1879–1893.

148 Singh, A., Van Humbeeck, J., and Van den Mooter, G. (2014). A new twist in the old story-can compression induce mixing of phase separated solid dispersions? a case study of spray-dried miconazole-PVP VA64 solid dispersions. *Pharm. Res.* 31: 3191–3200.

149 Leane, M.M., Sinclair, W., Qian, F. et al. (2013). Formulation and process design for a solid dosage form containing a spray-dried amorphous dispersion of ibipinabant. *Pharm. Dev. Technol.* 18: 359–366.

150 Raciborska, L., Buczkowska, E.M., Kukuls, K. et al. (2025). Active tablet coating with amorphous solid dispersion of ibuprofen–HPMCAS from organic solution. *Pharmaceutics* 17: 1514.

151 Chatterjee, K., Punia, A., Confer, A.M., and Lamm, M.S. (2025). Understanding the effect of plasticizers in film coat materials on the physical stability of amorphous solid dispersions. *J. Pharm. Sci.* 114: 468–476.

152 Wolff, H.-M., Arth, C., Quere, L., and Müller, W. (2018). Polyvinylpyrrolidone for the stabilization of a solid dispersion of the non-crystalline form of rotigotine. Google Patents.

153 Boyd, B.J., Bergström, C.A.S., Vinarov, Z. et al. (2019). Successful oral delivery of poorly water-soluble drugs both depends on the intraluminal behavior of drugs and of appropriate advanced drug delivery systems. *Eur. J. Pharm. Sci.* 137: 104967.

154 Kawakami, K. (2017). Supersaturation and crystallization: non-equilibrium dynamics of amorphous solid dispersions for oral drug delivery. *Expert Opin. Drug Deliv.* 14: 735–743.

155 Hammad, M.A. and Müller, B.W. (1998). Increasing drug solubility by means of bile salt–phosphatidylcholine-based mixed micelles. *Eur. J. Pharm. Biopharm.* 46: 361–367.

156 Pigliacelli, C., Belton, P., Wilde, P., and Qi, S. (2019). Probing the molecular interactions between pharmaceutical polymeric carriers and bile salts in simulated gastrointestinal fluids using NMR spectroscopy. *J. Colloid Interface Sci.* 551: 147–154.

157 Schlauersbach, J., Hanio, S., Lenz, B. et al. (2021). Leveraging bile solubilization of poorly water-soluble drugs by rational polymer selection. *J. Control. Release* 330: 36–48.

158 Kawakami, K. (2025). Roles of supersaturation and liquid–liquid phase separation for enhanced oral absorption of poorly soluble drugs from amorphous solid dispersions. *Pharmaceutics* 17: 262.

159 Zhao, P., Han, W., Shu, Y. et al. (2023). Liquid–liquid phase separation drug aggregate: merit for oral delivery of amorphous solid dispersions. *J. Control. Release* 353: 42–50.

160 Ricarte, R.G., Van Zee, N.J., Li, Z. et al. (2019). Recent advances in understanding the micro- and nanoscale phenomena of amorphous solid dispersions. *Mol. Pharm.* 16: 4089–4103.

161 Tres, F., Posada, M.M., Hall, S.D. et al. (2018). Mechanistic understanding of the phase behavior of supersaturated solutions of poorly water-soluble drugs. *Int. J. Pharm.* 543: 29–37.

162 Wang, B., Nethercott, M.J., Narula, A. et al. (2022). pH-Dependent supersaturation from amorphous solid dispersions of weakly basic drugs. *Pharm. Res.* 39: 2919–2936.

163 Hermans, A., Milsmann, J., Li, H. et al. (2022). Challenges and strategies for solubility measurements and dissolution method development for amorphous solid dispersion formulations. *AAPS J.* 25: 11.

164 Sun, D.D., Wen, H., and Taylor, L.S. (2016). Non-sink dissolution conditions for predicting product quality and in vivo performance of supersaturating drug delivery systems. *J. Pharm. Sci.* 105: 2477–2488.

165 de Alencar Danda, L.J., de Carvalho Amaral, A.R., Soares-Sobrinho, J.L., and de La Roca Soares, M.F. (2024). Optimizing nonsink dissolution testing for amorphous solid dispersions: exploring sample handling variables. *Mol. Pharm.* 21: 1861–1871.

166 Jankovic, S., O'Dwyer, P.J., Box, K.J. et al. (2019). Biphasic drug release testing coupled with diffusing wave spectroscopy for mechanistic understanding of solid dispersion performance. *Eur. J. Pharm. Sci.* 137: 105001.

167 Holzem, F.L., Jensen, I.H., Petrig Schaffland, J. et al. (2023). Combining in vitro dissolution/permeation with microdialysis sampling: capabilities and limitations for biopharmaceutical assessments of supersaturating drug formulations. *Eur. J. Pharm. Sci.* 188: 106533.

168 Oktay, A.N. and Polli, J.E. (2025). Formulation optimization of ritonavir tablets containing spray dried dispersion from HPMCAS-L or PVP-VA and comparative dissolution/permeation performance. *Int. J. Pharm.* 685: 126211.

169 Sitovs, A., Cernisova, V., and Mohylyuk, V. (2025). Dissolution-permeation systems in biopharmaceutical research: a graphical review on the current state-of-the-art. *J. Drug Deliv. Sci. Technol.* 107: 106814.

170 Naing, M.D. and Tsume, Y. (2025). Incorporating the biphasic system to GIS-α Improves in vivo prediction for low solubility drugs. *Eur. J. Pharm. Biopharm.* 211: 114724.

171 Deac, A., Que, C., Cousineau, M.L. et al. (2025). Dissolution mechanisms of amorphous solid dispersions: role of polymer molecular weight and identification of a new failure mode. *J. Pharm. Sci.* 114: 486–496.

172 Zhang, W., Thool, P., Weitz, B.W., and Hou, H.H. (2025). Investigating the effects of formulation variables on the disintegration of spray dried amorphous solid dispersion tablets. *J. Pharm. Sci.* 114: 304–312.

173 Oktay, A.N. and Polli, J.E. (2025). Comparison of in vitro and in vivo dissolution of Norvir® oral powder: in vivo relevance of a too rapid in vitro dissolution test. *Turk. J. Pharm. Sci.* 22: 270.

174 Yoshikawa, E., Ueda, K., Hakata, R. et al. (2024). Quantitative investigation of intestinal drug absorption enhancement by drug-rich nanodroplets generated via liquid–liquid phase separation. *Mol. Pharm.* 21: 1745–1755.

175 Beig, A., Fine-Shamir, N., Lindley, D. et al. (2017). Advantageous solubility-permeability interplay when using amorphous solid dispersion (ASD) formulation for the BCS class IV P-gp substrate rifaximin: simultaneous increase of both the solubility and the permeability. *AAPS J.* 19: 806–813.

176 Ueda, K., Takemoto, S., Higashi, K., and Moribe, K. (2025). Impact of colloidal drug-rich droplet size and amorphous solubility on drug membrane permeability: a comprehensive analysis. *J. Pharm. Sci.* 114: 136–144.

177 Sabra, R., Narula, A., Taylor, L.S., and Li, N. (2022). Comparisons of in vitro models to evaluate the membrane permeability of amorphous drug nanoparticles. *Mol. Pharm.* 19: 3412–3428.

178 Sharma, M., Yang, D.H., Adhikari, S. et al. (2025). The particle drifting effect in vivo: impact of dose and animal species. *Int. J. Pharm.* 683: 126060.

179 Patel, R.P., Cristofoletti, R., Wu, F. et al. (2024). in vitro lipolysis model to predict food effect of poorly water-soluble drugs itraconazole, rivaroxaban, and ritonavir. *J. Pharm. Sci.* 113: 2361–2373.

180 Wu, D., Liu, J., Paragas, E.M. et al. (2025). Assessing and mitigating pH-mediated DDI risks in drug development – formulation approaches and clinical considerations. *Drug Metab. Rev.* 57: 291–310.

181 Kordorwu, V., Castleberry, S., Lustig, S., and Carrier, R.L. (2025). Mucin mimics and impacts the function of polymeric inhibitors in stabilizing drug supersaturation. *Mol. Pharm.* 22: 1396–1407.

182 O'Connell, P., Yoon, J.H., Geever, L.M. et al. (2025). Accelerated predictive stability (APS) strategies applied to screening pharmaceutical formulations: A comparison of spray dried and hot melt extruded nifedipine amorphous solid dispersions. *Int. J. Pharm.* 683: 126012.

183 Leon, A.S.C., Waterman, K.C., Wang, G. et al. (2024). Accelerated stability modeling of recrystallization from amorphous solid dispersions: a griseofulvin/HPMC-AS case study. *Int. J. Pharm.* 657: 124189.

184 Liu, B., Theil, F., Lehmkemper, K. et al. (2021). Crystallization risk assessment of amorphous solid dispersions by physical shelf-life modeling: a practical approach. *Mol. Pharm.* 18: 2428–2437.

185 Ojo, A.T. and Lee, P.I. (2021). A mechanistic model for predicting the physical stability of amorphous solid dispersions. *J. Pharm. Sci.* 110: 1495–1512.

186 Wolbert, F., Luebbert, C., and Sadowski, G. (2023). The shelf life of ASDs: 2. Predicting the shelf life at storage conditions. *Int. J. Pharm.* X 6: 100207.

187 Miyazaki, T., Mizoguchi, R., Ueda, K. et al. (2023). Crystallization of amorphous nifedipine under isothermal conditions: inter-laboratory reproducibility and investigation of the factors affecting reproducibility. *J. Pharm. Sci.* 112: 2703–2716.

188 Nespi, M., Ly, J., Fan, Y. et al. (2025). Vehicle effect on in-vitro and in-vivo performance of spray-dried dispersions. *J. Pharm. Sci.* 114: 566–576.

189 Yu, D., Li, M., Hoag, S.W., and Nie, H. (2025). Understanding excipient-induced crystallization of spray-dried amorphous solid dispersion. *J. Pharm. Sci.* 114: 234–244.

190 Yu, D., Li, J., Wang, H. et al. (2022). Role of polymers in the physical and chemical stability of amorphous solid dispersion: a case study of carbamazepine. *Eur. J. Pharm. Sci.* 169: 106086.

191 Li, J., Yu, D., Zeng, C. et al. (2025). Hydrolysis of cellulose acetate phthalate and hydroxypropyl methylcellulose phthalate in amorphous solid dispersions. *J. Pharm. Sci.* 114: 89–97.

192 Peng, T., She, Y., Zhu, C. et al. (2018). Influence of polymers on the physical and chemical stability of spray-dried amorphous solid dispersion: dipyridamole degradation induced by enteric polymers. *AAPS PharmSciTech* 19: 2620–2628.

193 de Alvarenga, B.R. Jr., Moseson, D.E., Carneiro, R.L., and Taylor, L.S. (2022). Impact of polymer type on thermal degradation of amorphous solid dispersions containing ritonavir. *Mol. Pharm.* 19: 332–344.

194 Mittal, A., Aras, G., Yuvaraaj, V.K. et al. (2025). Identification and characterization of interaction product impurities and degradation products of atorvastatin hot melt extrusion formulation using LC-HRMS/MS and ATR-IR. *Arch. Pharm. (Weinheim)* 358: e3129.

195 Liu, T. and Zhang, F. (2023). Modelling drug degradation of amorphous solid dispersion during twin-screw extrusion. *Eur. J. Pharm. Biopharm.* 190: 197–205.

196 Kojima, T., Takeda, J., Song, Y. et al. (2023). Polymer-inducing chemical degradation of amorphous solid dispersions driven by drug-polymer interactions for physical stabilization. *Int. J. Pharm.* 647: 123504.

197 Iyer, J., Barbosa, M., Saraf, I. et al. (2023). Mechanoactivation as a tool to assess the autoxidation propensity of amorphous drugs. *Mol. Pharm.* 20: 1112–1128.

198 Saraf, I., Jakasanovski, O., Stanić, T. et al. (2024). Investigation of the influence of copovidone properties and hot-melt extrusion process on level of impurities, in vitro release, and stability of an amorphous solid dispersion product. *Mol. Pharm.* 21: 5703–5715.

199 Huzjak, T., Jakasanovski, O., Berginc, K. et al. (2024). Overcoming drug impurity challenges in amorphous solid dispersion with rational development of biorelevant dissolution-permeation method. *Eur. J. Pharm. Sci.* 192: 106655.

200 Trasi, N.S., Purohit, H.S., and Taylor, L.S. (2017). Evaluation of the crystallization tendency of commercially available amorphous tacrolimus formulations exposed to different stress conditions. *Pharm. Res.* 34: 2142–2155.

201 Taylor, L.S., Trasi, N.S., Purohit, H.S. et al. (2025). Changes in drug crystallinity in a commercial tacrolimus amorphous formulation result in variable pharmacokinetics. *J. Pharm. Sci.* 114: 313–322.

202 Purohit, H.S., Trasi, N.S., Sun, D.D. et al. (2018). Investigating the impact of drug crystallinity in amorphous tacrolimus capsules on pharmacokinetics and bioequivalence using discriminatory in vitro dissolution testing and physiologically based pharmacokinetic modeling and simulation. *J. Pharm. Sci.* 107: 1330–1341.

203 Okwuosa, T.C., Sadia, M., Isreb, A. et al. (2021). Can filaments be stored as a shelf-item for on-demand manufacturing of oral 3D printed tablets? An initial stability assessment. *Int. J. Pharm.* 600: 120442.

204 Pantazos, I., Poimenidou, M., Kouskouridas, D. et al. (2025). Influence of drug properties, formulation composition, and processing parameters on the stability and dissolution performance of amorphous solid dispersions-based tablets. *Polymers (Basel)* 17: 2484.

205 Ibrahim, I., Mann, J., Abbott, A. et al. (2025). Linking powder to tablet stability: length- and time-scale prediction of moisture sorption. *Int. J. Pharm.* 684: 126154.

206 Bhatta, H.P., Han, H.-K., Maharjan, R., and Jeong, S.H. (2025). Recent techniques to improve amorphous dispersion performance with quality design, physicochemical monitoring, molecular simulation, and machine learning. *Pharmaceutics* 17: 1249.

207 Sadowski, G., Kontogeorgis, G.M., Artusio, F. et al. (2025). The thermodynamic pharma challenge: a cross-cutting perspective. *Eur. J. Pharm. Sci.* 212: 107178.

208 Chougule, S., Manchanda, K., Sood, M. et al. (2025). Pharmacoinformatics in drug–excipient interactions. *Mol. Pharm.* 22: 7210–7238. doi: 10.1021/acs.molpharmaceut.5c00801.

209 Kalogeras, I.M. (2011). A novel approach for analyzing glass-transition temperature vs. composition patterns: application to pharmaceutical compound + polymer systems. *Eur. J. Pharm. Sci.* 42: 470–483.

210 Liu, H., Zhang, X., Suwardie, H. et al. (2012). Miscibility studies of indomethacin and Eudragit®; E PO by thermal, rheological, and spectroscopic analysis. *J. Pharm. Sci.* 101: 2204–2212.

211 Huang, J., Li, Y., Wigent, R.J. et al. (2011). Interplay of formulation and process methodology on the extent of nifedipine molecular dispersion in polymers. *Int. J. Pharm.* 420: 59–67.

212 Bellantone, R.A., Patel, P., Sandhu, H. et al. (2012). A method to predict the equilibrium solubility of drugs in solid polymers near room temperature using thermal analysis. *J. Pharm. Sci.* 101: 4549–4558.

213 Anderson, B.D. (2018). Predicting solubility/miscibility in amorphous dispersions: it is time to move beyond regular solution theories. *J. Pharm. Sci.* 107: 24–33.

214 Wu, H., Wang, L., Zhang, H. et al. (2025). Application of the perturbed-chain statistical associating fluid theory to predicting drug-polymer miscibility and stability in amorphous solid dispersions: a comprehensive overview. *Front. Chem. Sci. Eng.* 19: 117.

215 Dohrn, S., Rawal, P., Luebbert, C. et al. (2021). Predicting process design spaces for spray drying amorphous solid dispersions. *Int. J. Pharm. X* 3: 100072.

216 Song, S., Cui, S., Sun, C.C. et al. (2025). Crystallization inhibition in molecular liquids by polymers above the overlap concentration (c*): delay of the first nucleation event. *J. Pharm. Sci.* 114: 82–88.

217 Walden, D.M., Bundey, Y., Jagarapu, A. et al. (2021). Molecular simulation and statistical learning methods toward predicting drug–polymer amorphous solid dispersion miscibility, stability, and formulation design. *Molecules* 26: 182.

218 Maniruzzaman, M., Pang, J., Morgan, D.J., and Douroumis, D. (2015). Molecular modeling as a predictive tool for the development of solid dispersions. *Mol. Pharm.* 12: 1040–1049.

219 Mosquera-Giraldo, L.I., Borca, C.H., Meng, X. et al. (2016). Mechanistic design of chemically diverse polymers with applications in oral drug delivery. *Biomacromolecules* 17: 3659–3671.

220 Nie, H., Mo, H., Zhang, M. et al. (2015). Investigating the interaction pattern and structural elements of a drug–polymer complex at the molecular level. *Mol. Pharm.* 12: 2459–2468.

221 Wang, B., Wang, D., Zhao, S. et al. (2017). Evaluate the ability of PVP to inhibit crystallization of amorphous solid dispersions by density functional theory and experimental verify. *Eur. J. Pharm. Sci.* 96: 45–52.

222 Zeneli, E., Bohets, H., Mebenga, F.N. et al. (2025). Guiding excipient selection for physically stable amorphous solid dispersions: a combined in-vitro in-silico approach. *Eur. J. Pharm. Sci.* 212: 107152.

223 Zeneli, E., Bohets, H., Mebenga, F.N. et al. (2025). Guiding excipient selection for amorphous solid dispersions by combining an in vitro-in-silico approach – II: supersaturation and drug release. *Int. J. Pharm.* 682: 125976.

224 Lin, X.C., Huang, H.C., and Hsieh, C.M. (2025). Prediction of drug solubility in polymer with COSMO-SAC. *J. Taiwan Inst. Chem. Eng.* 173: 106177.

225 Brunsteiner, M., Khinast, J., and Paudel, A. (2018). Relative contributions of solubility and mobility to the stability of amorphous solid dispersions of poorly soluble drugs: a molecular dynamics simulation study. *Pharmaceutics* 10: 101.

226 Gupta, J., Nunes, C., Vyas, S., and Jonnalagadda, S. (2011). Prediction of solubility parameters and miscibility of pharmaceutical compounds by molecular dynamics simulations. *J. Phys. Chem. B* 115: 2014–2023.

227 Xiang, T.-X. and Anderson, B.D. (2013). Molecular dynamics simulation of amorphous indomethacin-poly (vinylpyrrolidone) glasses: solubility and hydrogen bonding interactions. *J. Pharm. Sci.* 102: 876–891.

228 Gupta, K.M., Chin, X., and Kanaujia, P. (2023). Molecular interactions between APIs and enteric polymeric excipients in solid dispersion: insights from molecular simulations and experiments. *Pharmaceutics* 15: 1164.

229 Aulifa, D.L., Al Shofwan, A.A., Megantara, S. et al. (2024). Elucidation of molecular interactions between drug–polymer in amorphous solid dispersion by a computational approach using molecular dynamics simulations. *Adv. Appl. Bioinforma. Chem.* 17: 1–19.

230 Barmpalexis, P., Karagianni, A., Katopodis, K. et al. (2019). Molecular modelling and simulation of fusion-based amorphous drug dispersions in polymer/plasticizer blends. *Eur. J. Pharm. Sci.* 130: 260–268.

231 Walter, S., Mileo, P.G.M., Afzal, M.A.F. et al. (2024). Predicting the release mechanism of amorphous solid dispersions: a combination of thermodynamic modeling and in silico molecular simulation. *Pharmaceutics* 16: 1292.

232 Zhong, H., Lu, T., Wang, R., and Ouyang, D. (2024). Quantitative analysis of physical stability mechanisms of amorphous solid dispersions by molecular dynamic simulation. *AAPS J.* 27: 9.

233 Budiman, A., Fakih, T.M., Megantara, S. et al. (2025). Leveraging non-covalent adaptable networks to stabilize drug-polymer systems in supersaturated solutions: a computational and experimental approach. *Appl. Sci.* 15: 307.

234 Sun, M., Li, B., Li, Y. et al. (2017). Experimental observations and dissipative particle dynamic simulations on microstructures of pH-sensitive polymer containing amorphous solid dispersions. *Int. J. Pharm.* 517: 185–195.

235 Otto, D.P., Otto, A., and De Villiers, M.M. (2013). Experimental and mesoscale computational dynamics studies of the relationship between solubility and release of quercetin from PEG solid dispersions. *Int. J. Pharm.* 456: 282–292.

236 Gangurde, A.B., Kundaikar, H.S., Javeer, S.D. et al. (2015). Enhanced solubility and dissolution of curcumin by a hydrophilic polymer solid dispersion and its insilico molecular modeling studies. *J. Drug Deliv. Sci. Technol.* 29: 226–237.

237 Madanchi, A., Azek, E., Zongo, K. et al. (2025). Is the future of materials amorphous? challenges and opportunities in simulations of amorphous materials. *ACS Phys. Chem. Au* 5: 3–16.

238 Milanesi, A., Rizzuto, F., Rinaldi, M. et al. (2022). Thermodynamic balance vs. computational fluid dynamics approach for the outlet temperature estimation of a benchtop spray dryer. *Pharmaceutics* 14: 296.

239 Lebedev, E.A. and Menshutina, N.V. (2024). Experimental study and modeling of dispersion in spray drying for producing powdered pharmaceutical compositions. *Theor. Found. Chem. Eng.* 58: 1980–1990.

240 Düphans, V., Kimmel, V., Messing, L. et al. (2024). Experimental and numerical characterization of screw elements used in twin-screw extrusion. *Pharm. Dev. Technol.* 29: 675–683.

241 Mattusch, A.M., Schaldach, G., Bartsch, J., and Thommes, M. (2025). Intrinsic dissolution modeling: interdependence between dissolution rate, solubility, and boundary layer thickness. *Pharmaceutics* 17: 570.

242 Pas, T., Smeets, B., Ramon, H. et al. (2020). Mechanodegradation of polymers: a limiting factor of mechanochemical activation in the production of amorphous solid dispersions by cryomilling. *Mol. Pharm.* 17: 2987–2999.

243 Boehling, P., Jacevic, D., Detobel, F. et al. (2020). Validating a numerical simulation of the ConsiGma(R) coater. *AAPS PharmSciTech* 22: 10.

244 Siegmann, E., Forgber, T., Toson, P. et al. (2020). Powder flow and mixing in different tablet press feed frames. *Adv. Powder Technol.* 31: 770–781.

245 Kambayashi, A., Kiyota, T., Fujiwara, M., and Dressman, J.B. (2019). PBPK modeling coupled with biorelevant dissolution to forecast the oral performance of amorphous solid dispersion formulations. *Eur. J. Pharm. Sci.* 135: 83–90.

246 Pepin, X., Arora, S., Borges, L. et al. (2024). Parameterization of physiologically based biopharmaceutics models: workshop summary report. *Mol. Pharm.* 21: 3697–3731.

247 Stewart, A.M. and Grass, M.E. (2020). Practical approach to modeling the impact of amorphous drug nanoparticles on the oral absorption of poorly soluble drugs. *Mol. Pharm.* 17: 180–189.

248 Arora, S., Pansari, A., Kilford, P.J. et al. (2022). A mechanistic absorption and disposition model of ritonavir to predict exposure and drug–drug interaction potential of CYP3A4/5 and CYP2D6 substrates. *Eur. J. Drug Metab. Pharmacokinet.* 47: 483–495.

249 Lynnerup, J.T., Lillotte, T., Feldmueller, M. et al. (2025). Untangling "dissolved" drug species from various formulations of a poorly soluble drug: sampling methods, mechanistic insights, and IVIVC. *J. Pharm. Sci.* 114: 103884.

250 Karakitsios, E., Angelerou, M.-F.-G., Kapralos, I. et al. (2025). Integrating in vitro dissolution and physiologically based pharmacokinetic modeling for generic drug development: evaluation of amorphous solid dispersion formulations for tacrolimus. *Pharmaceutics* 17: 227.

251 Lu, T., Wu, Y., Xiong, P. et al. (2025). Combining high-throughput screening and machine learning to predict the formation of both binary and ternary amorphous solid dispersion formulations for early drug discovery and development. *Pharm. Res.* 42: 697–709.

252 Jiang, J., Lu, A., Ma, X. et al. (2023). The applications of machine learning to predict the forming of chemically stable amorphous solid dispersions prepared by hot-melt extrusion. *Int. J. Pharm. X* 5: 100164.

253 Ge, K., Huang, Y., and Ji, Y. (2024). Machine learning with active pharmaceutical ingredient/polymer interaction mechanism: prediction for complex phase behaviors of pharmaceuticals and formulations. *Chin. J. Chem. Eng.* 66: 263–272.

254 Di Mare, E.J., Punia, A., Lamm, M.S. et al. (2024). Data-driven design of novel polymer excipients for pharmaceutical amorphous solid dispersions. *Bioconjug. Chem.* 35: 1363–1372.

255 Barmpalexis, P., Koutsidis, I., Karavas, E. et al. (2013). Development of PVP/PEG mixtures as appropriate carriers for the preparation of drug solid dispersions by melt mixing technique and optimization of dissolution using artificial neural networks. *Eur. J. Pharm. Biopharm.* 85: 1219–1231.

256 Barnard, T. and Sosso, G.C. (2023). Combining machine learning and molecular simulations to predict the stability of amorphous drugs. *J. Chem. Phys.* 159: 014503.

257 Habicht, J., Sadowski, G., and Brandenbusch, C. (2025). An adaptive ML framework to predict PC-SAFT parameters for mixtures. *Ind. Eng. Chem. Res.* 64: 22413–22427.

258 Zhu, J., Xiong, P., Wang, W. et al. (2025). Integrating artificial intelligence and physiologically based pharmacokinetic modeling to predict in vitro and in vivo fate of amorphous solid dispersions. *J. Control. Release* 386: 114123.

259 Newman, A. and Zografi, G. (2023). Considerations in the development of physically stable high drug load API- polymer amorphous solid dispersions in the glassy state. *J. Pharm. Sci.* 112: 8–18.

260 Yao, X., Neusaenger, A.L., and Yu, L. (2021). Amorphous drug-polymer salts. *Pharmaceutics* 13: 1271.

261 Qi, Q. and Taylor, L.S. (2022). Improved dissolution of an enteric polymer and its amorphous solid dispersions by polymer salt formation. *Int. J. Pharm.* 622: 121886.

262 Hiew, T.N. and Taylor, L.S. (2022). Combining drug salt formation with amorphous solid dispersions – a double edged sword. *J. Control. Release* 352: 47–60.

263 Zupančič, O., Matić, J., Doğan, A. et al. (2024). Comparing low-dose carvedilol continuous manufacturing by solid and liquid feeding in self-emulsifying delivery Systems via Hot Melt EXtrusion (SEDEX). *Pharmaceuticals* 17: 1290.

264 Nora, G.I., Venkatasubramanian, R., Strindberg, S. et al. (2022). Combining lipid based drug delivery and amorphous solid dispersions for improved oral drug absorption of a poorly water-soluble drug. *J. Control. Release* 349: 206–212.

265 Vidiyala, N., Sunkishala, P., Parupathi, P. et al. (2025). High drug loading of amorphous solid dispersion by hot melt extrusion: the role of magnesium aluminometasilicate (Neusilin® US2). *Sci. Pharm.* 93: 30.

266 de Carvalho, D.M., Lourenço, A.C.M., Moreira, G.G. et al. (2025). Role of phospholipids on drug dissolution in polymer solid dispersions prepared by hot-melt extrusion. *ACS Omega* 10: 31501–31508.

267 Li, J., Wang, Y., and Yu, D. (2023). Effects of additives on the physical stability and dissolution of polymeric amorphous solid dispersions: a review. *AAPS PharmSciTech* 24: 175.

268 Hörmann, T.R., Jäger, N., Funke, A. et al. (2018). Formulation performance and processability window for manufacturing a dual-polymer amorphous solid dispersion via hot-melt extrusion and strand pelletization. *Int. J. Pharm.* 553: 408–421.

269 Screen, M.A., Askin, S., McCabe, J.F. et al. (2025). Developing multi-component solid formulation strategies for PROTAC dissolution enhancement. *Mol. Pharm.* 22: 7052–7067.

270 Indulkar, A.S., Hanouch, L., Gignac, N. et al. (2025). Investigating the effect of permeation enhancers on oral absorption of a BCS IV compound, ombitasvir, utilizing animal models. *J. Pharm. Sci.* 114.

271 Kapote, D.N. and Wagner, K.G. (2021). Influence of shellac on the improvement of solubility and supersaturation of loratadine amorphous solid dispersion using a new grade of HPMC. *J. Drug Deliv. Sci. Technol.* 61: 102116.

272 Raje, V., Lu, I., Keir, R.S., and Serajuddin, A.T.M. (2025). Development of indomethacin amorphous solid dispersion by applying acid-base supersolubilization (ABS) principle to enhance solubility and enable low-temperature hot melt extrusion. *Int. J. Pharm.* 126405.

273 Syed, M.I., Kandagatla, H.P., and Serajuddin, A.T.M. (2026). Lowering melt extrusion temperature and increasing solubility and dissolution rate of telmisartan amorphous solid dispersion by applying acid-base supersolubilization principle. *Int. J. Pharm.* 687: 126404.

274 Zheng, W., Ke, J., Liu, K. et al. (2025). Aqueous solution spray drying preparations of binary amorphous solid dispersions. *Crystals (Basel)* 15: 323.

275 Panda, R. and Lankalapalli, S. (2025). Bioavailability enhancement and polymorphic stabilization of one BCS class IV metastable drug through novel formulation approach. *Turk. J. Pharm. Sci.* 22: 333.

3

Nanosizing of Drugs to Improve the Apparent Dissolution and Bioavailability

Yalamanda Varla[1], Vishvesh Joshi[2], Rikin Patel[3], and Derajram Benival[1]

[1]*National Institute of Pharmaceutical Education and Research–Ahmedabad (NIPER–A), Department of Pharmaceutics, Opposite the Air Force Station in Palaj, Gandhinagar, Gujarat 382355, India*
[2]*Chartwell Pharmaceuticals LLC, 77 Brenner Dr, Congers, NY 10920, USA*
[3]*Intas Pharmaceuticals Ltd., Near Sola Bridge, S. G. Highway, Thaltej, Ahmedabad, Gujarat 380054, India*

3.1 Introduction

The effectiveness of an oral drug depends on its ability to dissolve in biological fluids and reach systemic circulation in therapeutic concentrations [1–3]. Solubility and dissolution rate are crucial parameters determining how efficiently the body absorbs and utilizes a drug [4, 5]. Poorly water-soluble drugs often suffer from solubility and dissolution-limited bioavailability, leading to inconsistent therapeutic outcomes and inter-patient variability [6, 7].

3.1.1 Importance of Solubility and Dissolution in Drug Bioavailability

Solubility refers to the maximum concentration of a solute that can be dissolved in a solvent at a specific temperature, while dissolution is the process by which a substance forms a solution to become available for absorption [4, 8, 9]. Both solubility and dissolution are key factors that influence a drug's bioavailability, which refers to the fraction of an administered dose that successfully enters the systemic circulation in its active form [10]. The extent to which a drug dissolves in the biological environment directly influences its absorption and therapeutic effectiveness. Poor solubility can lead to inadequate drug absorption, resulting in variability in drug response and suboptimal treatment outcomes [6, 7, 11]. Furthermore, insufficient dissolution may fail to reach the desired plasma concentration, potentially requiring dose modifications. This can, in turn, increase the risk of side effects and reduce patient compliance, further impacting therapeutic effectiveness [12]. Since dissolution is the rate-limiting step in the bioavailability of poorly water-soluble drugs, regulating the dissolution process is essential to ensure appropriate drug absorption and pharmacological activity [10, 13]. Therefore, maintaining a balance between solubility and dissolution is fundamental to achieving the desired drug concentration in systemic circulation and ensuring consistent therapeutic performance.

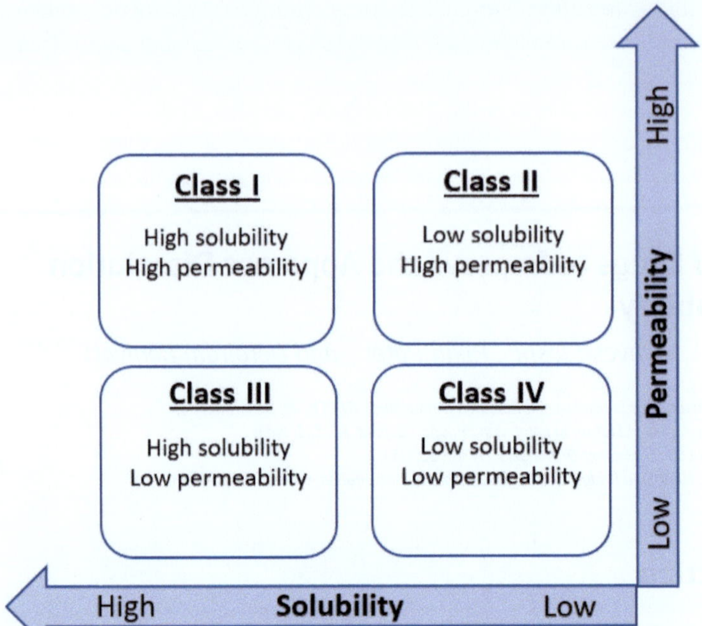

Figure 3.1 Schematic representation of the Biopharmaceutics Classification System (BCS).

The Biopharmaceutics Classification System (BCS) classifies drugs based on their solubility and permeability, as shown in Figure 3.1. From the four classes, BCS Class II drugs (low solubility, high permeability) and Class IV drugs (low solubility, low permeability) face significant bioavailability challenges due to their poor dissolution rates [14]. Approximately 40% of marketed drugs and 90% of new drug candidates fall under these categories, making solubility enhancement a critical aspect of pharmaceutical development [12, 15, 16]. Various formulation strategies have been explored to address this, with nanosizing emerging as one of the most effective approaches for improving drug dissolution and bioavailability [17–19].

3.1.2 Challenges in Formulating Poorly Water-Soluble Drugs

Developing formulations for poorly water-soluble drugs presents multiple challenges, including slow drug absorption, dose escalation requirements, food-dependent absorption, formulation instability, and poor patient compliance [20, 21]. These drugs dissolve at a much slower rate, leading to delayed absorption and reduced bioavailability, which, in turn, can compromise their therapeutic efficacy. Higher doses are often required to achieve the desired plasma concentration to compensate for this limitation, potentially increasing the risk of adverse events and toxicity [7, 22–25]. Additionally, many BCS Class II drugs exhibit significant food-dependent bioavailability, either positive or negative, which results in unpredictable plasma drug concentrations when taken with or without meals. A high-fat diet may increase drug absorption compared to the fasted state, potentially enhancing bioavailability and altering pharmacokinetic profiles. This inconsistency makes it difficult to establish optimal dosing regimens [26–28]. Gu et al. conducted a comprehensive analysis of the food effect on drug absorption by evaluating clinical data from 90 marketed drugs (92 total entries). Their study categorized food effects into three groups based on statistical significance: positive food effect (31 drugs), where food significantly increased drug absorption (area under the curve [AUC]); negative food effect (24 drugs), where food significantly decreased absorption; and no food effect (37 drugs), where no significant change in AUC was observed. The study also examined the relationship between clinical dose and maximum absorbable dose (MAD) for many drugs from BCS Class I and IV. Among them, hydrochlorothiazide (BCS Class III) was assessed at two different doses, demonstrating dose-dependent food effects. At a lower dose (MAD > clinical dose), absorption was not limited, ensuring consistent bioavailability. In contrast, at a higher dose (MAD < clinical dose), absorption was incomplete, highlighting food-dependent variability. By analyzing physicochemical properties, Gu et al. developed a logistic regression model to predict food effects based on solubility, permeability, and dose number. Their findings provide valuable insights into how food alters drug absorption, emphasizing the importance of dose optimization and formulation strategies for drugs with food-dependent bioavailability [29, 30]. Traditional solubility enhancement strategies, such as salt formation and co-solvency, can potentially improve solubility to some extent; however, they also have limitations. Stability concerns primarily hinder salt formation, while co-solvency is often a slow and costly process, further constrained by toxic effects, especially at high concentrations [21, 31–33]. Furthermore, frequent dosing regimens required for poorly soluble drugs contribute to patient non-adherence, reducing treatment effectiveness [7, 12]. Addressing these challenges requires advanced formulation approaches that enhance solubility and dissolution while ensuring stability, scalability, and patient compliance. Various formulation approaches have been employed to overcome these challenges, including salt formation, solid dispersions, lipid-based formulations, and complexation techniques. However, particle size reduction is a widely used and effective strategy to enhance drug performance. This reduction significantly increases the surface area available for dissolution, thereby enhancing the dissolution rate, promoting better drug absorption, and ultimately improving bioavailability [18, 34, 35].

3.1.3 Noyes–Whitney Equation and Surface Area Relationship

The dissolution rate of a solid drug particle into a finite volume is described by the Nernst–Brunner form of the Noyes–Whitney equation:

$$\frac{dC}{dt} = \frac{DA}{h}(C_s - C)$$

where D is the diffusion coefficient of drug, A the solid's surface area, h the stagnant boundary-layer thickness, C_s the saturation solubility, and C the bulk concentration. Nanosizing increases A dramatically and can reduce the effective h via improved wetting and dispersion, thereby increasing dC/dt. When nanosizing also elevates

C_s (e.g. via surface energy effects or amorphous content), dissolution is further accelerated. These mechanistic levers—$\uparrow A$, $\downarrow h$, and sometimes $\uparrow C_s$—explain why nanocrystals and nanosuspensions consistently enhance dissolution-limited bioavailability [29, 35–39].

3.1.4 Overview of Particle Size Reduction Technologies

Reducing the particle size of a drug enhances its surface area, thereby improving its dissolution rate, as explained by the Noyes–Whitney equation [40, 41]. Traditional micronization techniques, such as jet milling and ball milling, reduce particle size to the micrometer range. However, micronized drugs often fail to improve the saturation solubility, dissolution, and bioavailability for very low aqueous-soluble drugs [42–45]. This limitation has driven interest in nanosizing, which further reduces particle size to the nanometer range (typically <1000 nm) thereby increasing surface area, dissolution rate, and in some cases, enhancing apparent saturation solubility or supersaturation, leading to improved dissolution kinetics and bioavailability [46, 47].

3.1.5 Role of Nanosizing in Enhancing Dissolution and Bioavailability

Nanosizing has emerged as a highly effective strategy that involves reducing drug particles to nanometer dimensions to improve their solubility, dissolution rate, and bioavailability [48]. As mentioned earlier, by leveraging the principles of the Noyes–Whitney equation, nanosized drug formulations can achieve significantly faster dissolution rates, leading to better absorption, reduced dose variability, and improved therapeutic outcomes. This is particularly beneficial for BCS Class II and IV drugs, where dissolution serves as the primary barrier to effective absorption [49–51]. Additionally, nanosized particles improve supersaturation solubility due to increased surface free energy, allowing for better gastrointestinal absorption [52, 53]. Faster dissolution ensures that a greater proportion of the drug is available for absorption, leading to higher and more predictable plasma concentrations [13, 54, 55]. However, not all the dissolved drugs can get absorbed, this also depends on the permeability of the drug, especially when it comes to BCS Class IV. It may be problematic for the basic drugs, which tend to precipitate when they transit through the small intestine, especially if the medication is administered orally. Along with low permeability, if a medication undergoes P-gp efflux, its absorption remains low despite its higher solubility [7, 56–59]. Furthermore, it helps in minimizing the food-related variations in drug absorption, as nanoparticles dissolve more uniformly in gastric fluids, ensuring consistent bioavailability regardless of fed or fasted states [60, 61]. The homogeneous particle size distribution enhances formulation stability and dosage uniformity, reducing the variability in therapeutic outcomes [62, 63].

3.1.6 Traditional Versus Nanosizing Techniques in Solubility Enhancement

One of the biggest challenges in drug formulation is improving the solubility of poorly water-soluble drugs, as their low dissolution rate directly impacts bioavailability [64]. Several traditional techniques have been employed to enhance solubility, but each comes with significant limitations, especially regarding long-term stability [7]. Nanosizing, which involves the reduction of particle size to the nanometer range, has emerged as one of the alternatives, addressing many of these issues [18, 41].

Salt formation is the most straightforward, economical, and widely used strategy to enhance the aqueous solubility of poorly soluble drugs. This process involves the conversion of the drug into its more soluble ionic form by reacting it with an appropriate acid or base. Salts are formed when a compound that is ionized in solution forms a strong ionic interaction with an oppositely charged counterion [65–68]. While this method can significantly improve solubility, it is majorly applicable to ionizable drugs, meaning that non-ionizable compounds cannot benefit from this approach. Salts often suffer from stability issues, such as hygroscopicity, which can lead to premature degradation or moisture-induced instability [69–71]. Moreover, active pharmaceutical ingredients (APIs) with pH-dependent solubility can rapidly precipitate when transitioning from the acidic environment of the stomach to the more alkaline conditions of the intestine [6, 72]. Besides this, replacing a suboptimal salt form with a more suitable one during drug development can be time-consuming and resource-intensive, as it often requires re-evaluation of various biological, toxicological, formulation, and stability assessments [66]. In contrast, nanosizing improves dissolution by increasing the surface area available for drug-medium interaction, irrespective of the drug's ionizability. Unlike salt formation, nanosizing does not require conversion of the drug into an ionized salt form, makes nanosizing a more broadly applicable approach than salt formation. However, the drug's pH-dependent solubility may still influence its dissolution behaviour [35, 73].

Another common approach is co-solvency, where water-miscible organic solvents like ethanol, propylene glycol, or polyethylene glycol are used to improve the solubility of poorly dissolved drugs [74, 75]. Co-solvents reduce the interfacial tension between the drug and water, allowing greater solubilization [76]. However, a major drawback of this technique is solubility dependency on solvent concentration, meaning that upon administration, the solvent may get diluted and the drug may precipitate out, leading to diminished absorption. Additionally, co-solvents pose toxicity and vascular occlusion concerns, especially when used in high concentrations [12, 77–79]. Nanosizing, on the other hand, eliminates the risk of precipitation upon dilution, as most nanosizing techniques do not require the use of organic solvents [80].

Micellar solubilization, using surfactants and amphiphilic polymers, is another technique where the drug is entrapped within the hydrophobic core of micelles, allowing it to dissolve in an aqueous environment [79, 81, 82]. At and above the CMC, surfactants organize into micelles, consisting of a hydrophobic core and hydrophilic layers, which can effectively incorporate drugs. While this significantly improves solubility, these systems suffer from instability, as micelles tend to dissociate upon dilution, leading to premature release, and may precipitate the drug [79, 83–86]. Additionally, many surfactants are associated with toxicity, making them less desirable for pharmaceutical applications [79, 87–89]. Nanosizing will increase the surface area of nanoparticles and facilitate direct dissolution in water, bypassing the requirement for carrier molecules.

Cyclodextrins (CDs) are cyclic oligosaccharides widely used in pharmaceutical formulations to improve the aqueous solubility, stability, and bioavailability of poorly soluble drugs, basically through complexation. They consist of six, seven, or eight dextrose units, forming the so-called parent molecules: α-CD, β-CD, and γ-CD, respectively. These molecules have a hydrophobic internal cavity and a hydrophilic outer surface, which allows them to encapsulate poorly aqueous-soluble drugs via inclusion complexation [90, 91]. This mechanism enhances drug solubility in aqueous environments and eases formulation development for poorly aqueous-soluble drugs. The choice of cyclodextrin majorly depends on the drug's properties, the route of administration, and safety considerations. α-CD is relatively safe for parenteral use; however, its small cavity size and potential to cause renal toxicity when administered parenterally limit its applicability to only small molecules. For the same reason, it is not suitable for bulky or aromatic compounds. Conversely, γ-CD offers the largest cavity, capable of accommodating bulky drug molecules with high water solubility. However, concerns about systemic build-up and adverse events, particularly at high doses, restrict its use in parenteral applications [92, 93]. β-CD, although effective for some compounds, has lower water solubility compared to α- and γ-CDs, and is associated with nephrotoxicity, which limits its suitability for injectable formulations. However, modified β-CDs such as sulfobutylether-β-CD (Captisol®) and hydroxypropyl-β-CD are highly soluble and exhibit improved safety profiles, making them more appropriate for parenteral administration. Despite their enhanced solubility, these derivatives can still cause adverse effects, particularly in patients with renal impairment who are more vulnerable due to reduced baseline glomerular filtration rates [91, 94–96]. Randomly methylated β-CD, while enhancing drug permeability in nasal or rectal formulations, has been shown to cause tissue irritation and potential toxicity at high concentrations [91]. Therefore, selecting the appropriate CD requires careful consideration of the drug's properties, the administration route, the goals of solubility enhancement, and established safety thresholds. In contrast, nanosizing increases the drug's dissolution rate without requiring additional excipients, making it a more scalable and cost-effective solution.

Solid dispersions, where the drug is dispersed within a hydrophilic polymer matrix, can also improve solubility by preventing drug crystallization and enhancing wettability [97–99]. However, these formulations often suffer from physical instability, as amorphous drug forms tend to recrystallize over time, leading to reduced solubility and dissolution rates [100]. Additionally, key challenges in formulating solid dispersions include determining the miscibility between the selected drug(s) and carrier polymers, preserving the structural integrity of the matrix immediately after processing, and the absence of reliable predictive models for miscibility and long-term stability [69]. Nanosizing circumvents this issue by breaking down drug particles into their stable crystalline or amorphous form while maintaining high dissolution rates, thereby providing long-term solubility enhancement without the risk of phase transformation.

Amorphization is another strategy used to enhance the solubility of poorly water-soluble drugs by converting them into high-energy amorphous forms. These forms offer improved solubility and dissolution rates, ultimately leading to better bioavailability. However, their inherent thermodynamic instability poses a significant limitation, as amorphous drugs tend to revert to their more stable, low-energy crystalline forms over time, so-called recrystallization [69, 101].

In recent decades, pharmaceutical science has actively explored versatile nanosystems [102], leading to the successful development and commercialization of several United States Food and Drug Administration (US FDA)-approved nanosized drug formulations, like fenofibrate (TriCor®) [103], sirolimus (Rapamune®) [104, 105], and megestrol acetate (Megace ES®) [106], highlighting the potential of nanotechnology in transforming drug delivery. These market successes promote more advanced nanotechnology-based formulations with improved properties in the future. The next sections of this chapter will explore the various nanosizing techniques, their applications, and the challenges associated with implementation, providing a comprehensive overview of this transformative drug delivery approach.

3.2 Techniques for Nanosizing

As comprehended earlier, nanosizing is a crucial strategy in pharmaceutical formulation to improve the solubility and bioavailability of poorly aqueous-soluble drugs. Nanosizing can be done using top-down and bottom-up technologies. Top-down approaches, such as milling, involve breaking down larger drug particles into the nanometer range. In contrast, bottom-up methods create nanoparticles by assembling them from individual drug molecules in solution, typically through processes like precipitation. These are well-established and commonly used approaches to reduce particle size to the desired range. The method of preparation significantly impacts the development and stability of the final formulation, influencing its overall performance and effectiveness [107–109]. Hybrid technologies that combine both top-down and bottom-up approaches have attracted significant attention for their potential to optimize nanosizing processes, as they may overcome the limitations of each method by integrating their complementary strengths [110]. The choice of an appropriate technique depends on the drug's properties, stability concerns, scalability, and the intended route of administration. The method of manufacture can significantly impact the formation and stability of nanoformulations, thereby affecting their overall performance.

As illustrated in Figure 3.2, nanosizing approaches are broadly grouped into **top-down** (mechanical comminution) and **bottom-up** (controlled precipitation/crystallization) methods, with emerging **hybrid** processes combining both.

3.2.1 Top-Down Approaches

Top-down approaches involve the mechanical breakdown of larger drug particles into nanosized dimensions using high-energy forces such as shear, impact, or cavitation. These methods typically begin with bulk drug powders and reduce their size through techniques like milling, homogenization, or ultrasonication. The primary advantage of top-down techniques is their ability to produce crystalline nanoparticles for drugs without altering their chemical composition. However, these methods

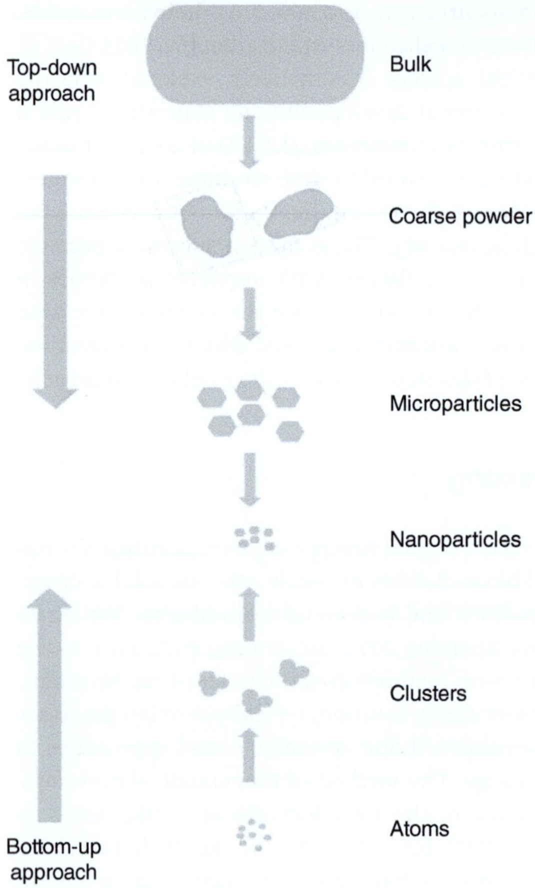

Figure 3.2 Schematic representation of top-down and bottom-up approaches for nanosizing. *Source:* Abid et al. [111] / with permission of Elsevier.

may result in particle aggregation, high energy consumption, equipment wear, and potential changes in the drug's polymorphic form. Additionally, heat generated during processing may degrade the drug [107, 112–114].

3.2.1.1 Wet Media Milling

Milling techniques in the pharmaceutical industry are broadly classified into two categories, wet and dry milling. Among them, wet media milling is the most widely used technique in the pharmaceutical industry for reducing particle size, primarily through abrasion, cleavage, and fracturing. The media milling process involves dispersing the API in a liquid medium, such as water or a suitable buffer, along with stabilizers, and milling is performed using high-shear rotating beads made up of zirconium silicate, magnesium oxide–stabilized zirconia, glass, or highly cross-linked polystyrene resin, which help break down particles and maintain nano-sized dispersions, typically 400 nm [115–117]. In contrast, dry milling techniques like jet milling can produce micron-sized particles without the use of solvents, but they have notable limitations. Dry milling is generally not preferred for very potent compounds because it generates fine airborne particles that pose inhalation and handling risks. Additionally, the high-energy impact and friction involved in dry milling can introduce amorphous regions into the drug, potentially affecting its stability. Moreover, electrostatic charges generated during dry milling can cause particles to adhere to equipment surfaces, leading to material loss and inconsistent dosing, critical concerns when dealing with highly potent drugs that require precise microgram or nanogram dosing. Wet media milling overcomes these limitations by maintaining the drug in a dispersed liquid phase throughout the process, minimizing airborne exposure, reducing electrostatic issues, and improving drug recovery. This method allows for safer handling, better uniformity, and enhanced control over particle size distribution, making it the preferred approach for producing nanosized pharmaceutical formulations with improved dissolution and bioavailability [45, 118]. The process can be carried out in either batch or recirculation mode, depending on the desired particle size and distribution. In batch mode, achieving dispersions with unimodal distribution profiles and mean particle diameters below 200 nm typically requires 30–60 minutes. In contrast, the recirculatory mode allows for continuous processing, optimizing milling time to less than 5 hours to obtain the desired particle size, with $d90 \leq 400$ nm. Throughout the process, the movement of milling media, driven by mechanical agitation, generates shear and impact forces leading to particle size reduction [119]. The resulting particle size is determined by the stress intensity and the number of contact points. In particular, stress intensity depends on the kinetic energy of the grinding beads, while using smaller beads can increase the number of contact points, enhancing the efficiency of particle size reduction [120]. Wet media milling allows the processing of heat-sensitive drugs by carefully selecting parameters such as bead loading, bead size, and stirrer speed. This technique is highly preferred due to its ability to process hydrophobic drugs, ease of scalability from laboratory to commercial production, and relatively simple operation. Additionally, it ensures a narrow particle size distribution and provides flexibility in handling drug concentrations ranging from 1 to 400 mg/ml, making it suitable for both dilute and highly concentrated formulations. Despite its advantages, wet media milling presents certain challenges, including particle aggregation, media contamination, and temperature sensitivity. Although wet media milling can process heat-sensitive drugs with careful parameter selection, extremely heat-sensitive drugs may still require alternative approaches. A key concern is the potential introduction of milling media residues into the final product, which is particularly critical for long-term therapies. However, advancements in polystyrene resin-based beads have significantly reduced this risk. These beads release minimal contaminants, with monomer residues as low as 50 ppb and milling-induced residues kept below 0.005% w/w of the final product [121, 122].

3.2.1.2 High-Pressure Homogenization

High-pressure homogenization (HPH) is widely used as an effective technique in achieving fine particle size, particularly in pharmaceutical and biotechnological formulations. It works on the principle by forcing the sample through a narrow gap or valve under high pressure, typically ranging from 50 to 500 MPa, depending on the type of equipment available. As a result of high pressure, the piston pump creates a pulsating flow that forces the premix through the narrow valve or aperture, which produces high local stress, responsible for a drastic reduction in particle size. A complete homogenization cycle occurs when the processed material is recirculated through the narrow gap under the same pressure conditions. With each cycle, droplet size continues to decrease, especially as the fluid re-experiences the same mechanical stresses. The number of cycles directly correlates with the mean particle size, assuming constant pressure. Notably, the time required to complete a cycle depends on the liquid's viscosity. As droplet size reduces over successive cycles, the viscosity of the emulsion also decreases accordingly, which decreases the time required to complete the cycle. An HPH system typically comprises key components such as pumps, pistons, a homogenizing valve, and extremely narrow capillaries comparable to hair strands that are crucial for achieving size reduction. Lab-scale systems usually contain a single piston, whereas pilot and production-scale homogenizers often include three to five pistons for enhanced throughput. Systems capable of operating above 200 MPa are referred to as ultra-high-pressure homogenizers (UHPHs). Sometimes the term "microfluidizer" is used in literature instead of high-pressure homogenizer with counter-jet dispergator. However, the microfluidizer is a proprietary technology developed by Microfluidics that incorporates a specialized counter-jet design. While its working principle shares similarities with that of HPH, the design and mechanism, particularly the counter-jet dispergator, offer distinct performance characteristics and operational advantages. Therefore, the two technologies should not be confused with each other, as the microfluidizer represents a unique system with specific design features and processing behavior [123–125]. HPH has been successfully employed in commercial nanosized formulations such as Rapamune® (sirolimus) and Triglide® (fenofibrate) [126, 127]. A key advantage of HPH is its ability to produce stable nanoparticles with a narrow size distribution. However, multiple cycles of homogenization are often required to reach the desired particle size, and temperature-sensitive drugs may degrade due to the heat generated during the process [125, 128].

3.2.1.3 Ultrasonication

Ultrasonication alone or combined with other methods has become a well-recognized and established approach for enhancing the solubility of poorly water-soluble drugs by reducing particle size to the nanometer range, either for the pure drug or for drug-loaded formulations. Ultrasonication is a technique that utilizes high-frequency sound waves (typically above 20 kHz) to agitate particles in a liquid medium, facilitating processes such as particle size reduction, emulsification, and dispersion. The core mechanism behind ultrasonication is acoustic cavitation, wherein ultrasonic waves create microscopic bubbles that rapidly collapse, releasing significant energy. This phenomenon exerts strong shear forces capable of breaking apart particles and droplets, leading to nanosizing and homogenization. There are two primary methods of ultrasonication, namely direct and indirect. Direct ultrasonication involves the immersion of an ultrasonic probe directly into the sample, transmitting energy without any barriers, making it highly effective for strong cavitation, which leads to particle breakdown. However, it carries the risk of sample contamination due to potential probe erosion. On the other hand, indirect ultrasonication uses a water bath to transmit ultrasonic waves through the container holding the sample. This method provides more gentle and uniform energy distribution, although it is less intense due to energy loss through the container walls. The outcome of these methods and the application mode depend on the volume, the characteristics of the sonicated dispersion, and the desired level of size reduction [129–134].

3.2.2 Bottom-Up Approaches

Bottom-up approaches involve the controlled assembly of nanoparticles from molecular or atomic levels, utilizing precipitation or crystallization techniques. Unlike top-down methods, bottom-up techniques build nanoparticles rather than breaking down larger particles, leading to more uniform size distribution and better control over drug properties. These are low-energy methods, require simple instruments, are cost-effective, and can be operated at a low temperature, making them particularly suitable for thermolabile drugs. However, these methods often require precise solvent selection and stabilizers to prevent particle aggregation and stability issues [107, 135].

3.2.2.1 Precipitation Techniques

Precipitation is a widely adopted bottom-up technique, owing to its simplicity, low energy requirements, ease of implementation, tunability, reproducibility, and versatility [136]. The precipitation involves the drug being dissolved in an organic solvent and subsequently mixed with an antisolvent, leading to rapid nucleation and formation of nanoparticles. Surfactants or stabilizers are added to prevent aggregation [137, 138]. However, challenges such as the difficulty in precisely controlling particle size and achieving uniform size distribution arise, as the kinetics of nucleation and particle growth are highly sensitive to process parameters such as mixing speed, solvent-to-antisolvent ratio, and supersaturation level. Inadequate control can lead to broad particle size distributions, which compromise formulation stability. Additionally, complete removal of the organic solvent remains a significant challenge, as the regulatory agencies impose strict limits on residual solvents in the final drug product [139].

3.2.2.2 Supercritical Fluid Technology

Supercritical fluid (SCF) methods utilize a SCF, which is a highly compressed gas that combines the properties of gases and liquids. The most commonly used SCF is carbon dioxide (SCO_2), as it is inert, non-toxic, easily available, cheap, and a highly solvating fluid for nonpolar compounds. CO_2 exhibits unique solvent properties above its critical temperature (the temperature beyond which a substance cannot be liquefied regardless of the applied pressure) and critical pressure (the minimum pressure required to liquefy a substance at its critical temperature) to dissolve and precipitate nanosized drug particles [140–143]. SCO_2, due to its low polarity, can dissolve hydrophobic drugs to form a solution. When this solution is rapidly expanded through a capillary into the ambient environment, it results in the formation of fine particles. This process is known as rapid expansion of supercritical solution (RESS). For compounds with limited solubility in SCO_2, the solvent can instead act as an antisolvent to induce precipitation from a polar drug solution. This technique is referred to as the supercritical antisolvent (SAS) process. It involves either injecting the drug solution into the SCF or introducing the SCF into the drug solution, which leads to solvent expansion and subsequent drug precipitation. The main challenges associated with the SCF method include expensive instruments, complex operations, and time-consuming [144, 145].

3.2.2.3 Solvent Evaporation

The solvent evaporation technique is one of the most widely adopted methods to prepare polymeric nanoparticles, especially drug-loaded polymer-based systems used in pharmaceutical formulations. Typically, the polymer is dissolved in a volatile organic solvent, along with the drug. This organic phase is then emulsified into an aqueous phase containing a surfactant using high-speed homogenization, leading to the formation of an emulsion. Once a stable emulsion is achieved, the organic solvent is removed either by applying heat under reduced pressure or by continuous stirring, resulting in a nanoscale dispersion of drug-loaded polymeric particles. Despite its simplicity, completely removing the organic solvent is often difficult. As mentioned earlier, the US FDA imposes strict limits on residual solvents in the final drug product [146].

3.2.3 Hybrid and Advanced Approaches

Hybrid techniques integrate both top-down and bottom-up approaches for particle size reduction. They leverage the benefits of each method to improve efficiency and control. This combination approach can overcome the limitations of either approach when used alone.

3.2.3.1 Combination Techniques (Microprecipitation and Homogenization)

A combination of nanosizing techniques that integrate bottom-up and top-down approaches has emerged as an effective strategy to overcome the limitations of individual methods. Microprecipitation followed by HPH is a widely studied method. This combinative technique involves initially dissolving the drug in a suitable organic solvent and then rapidly precipitating it by adding it to an aqueous antisolvent containing stabilizers. This process results in the formation of nanocrystals, which are subsequently subjected to mechanical size reduction via HPH. The synergy between controlled precipitation and mechanical processing allows for more efficient particle size reduction compared to either method alone, while also enhancing control over crystallinity and particle size distribution. Industrial implementations of this method, such as the NANOEDGE™ and H69™ (cavi-precipitation) technologies, have been successfully applied to drugs like itraconazole [41, 147, 148]. Similarly, Shelar et al. demonstrated the application of this combinative method for the poorly soluble antihypertensive drug isradipine, reporting a significant reduction in particle size and a marked improvement in solubility, dissolution rate, and oral bioavailability [149].

3.3 Formulation Strategies for Nanosized Drugs

Nanosizing is a widely explored technique to improve the aqueous solubility of poorly water-soluble drugs. By reducing drug particles to the nanometer scale, typically less than 1 μm [41]. However, for practical and effective administration, these nanosized drug particles must be converted into suitable formulations. Among the various formulation strategies available, the most widely preferred and extensively researched include nanocrystals, nanosuspensions, and nanoemulsions [150, 151]. These formulations not only stabilize the nanosized particles but also enable their administration through different routes such as oral, injectable, and topical delivery. Each of these systems brings unique advantages in terms of particle size, stability, and scalability, making them prominent choices in the field of nanomedicine and drug delivery [113, 152, 153]. Currently, there is a controversial discussion regarding the size of nanoparticles. Most commonly, nanoparticles are defined as particles smaller than 100 nm. However, some researchers in colloid and polymer sciences suggest that only particles smaller than 20 nm truly represent the nanoscale. Regulatory agencies such as the US FDA, the British Standards Institution (BSI), and European cosmetic authorities use the 100 nm size limit in their guidelines. In Europe, cosmetic products must be labeled as "nano" if they contain particles below this threshold. This definition is closely tied to safety concerns, as particles smaller than 100 nm can enter most types of cells through endocytosis, increasing the potential for systemic exposure and toxicity. In contrast, larger particles are usually taken up only by macrophages, which limits their distribution in the body. To address these differences, a Nanotoxicological Classification System (NCS) has been proposed. Similar to the BCS, it classifies nanoparticles based on size (above or below 100 nm) and whether they are biodegradable or non-biodegradable, as depicted in Figure 3.3, helping in the evaluation of potential risks and guiding formulation development [156]. The NCS categorizes nanoparticles into four primary

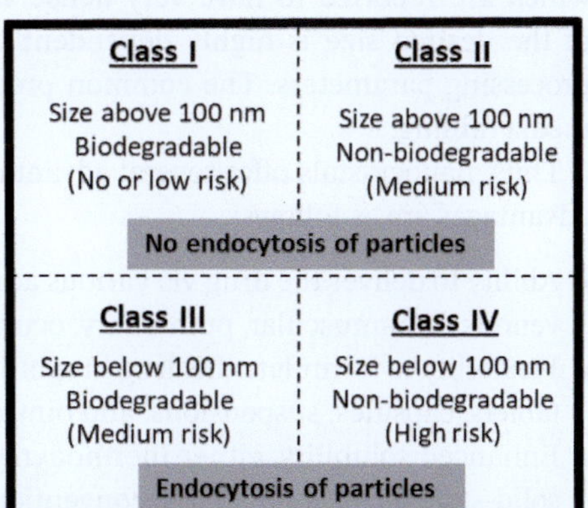

Figure 3.3 Nanotoxicological Classifications System. *Source:* Keck and Müller [154] and Müller et al. [155].

classes (I–IV) based on size and biodegradability, which are critical determinants of their interaction with biological systems and potential toxicity. Class I includes particles larger than $100 \leq 1000$ nm that are biodegradable, typically posing the lowest risk as they are less likely to enter cells and can be broken down by biological systems. Class II represents large (>100 nm) but non-biodegradable particles, which are more persistent and less compatible with biological clearance mechanisms, making them associated with moderate toxicity. Class III consists of smaller (<100 nm) biodegradable particles, which are small enough to be internalized by most cells via endocytosis but are more easily broken down, hence presenting a moderate to high toxicity risk due to their increased cellular access. Class IV, the highest risk category, includes small (<100 nm) and non-biodegradable particles that can both be internalized by most cells and persist in the body, leading to higher toxicity potential. Each class is further subdivided based on surface compatibility into biodegradable (B) or non-biodegradable (NB), resulting in a total of eight subclasses (e.g. I-B and I-NB to IV-B and IV-NB), which help predict the risk level of nanoparticles in pharmaceutical and biomedical applications [156].

3.3.1 Nanocrystals and Nanosuspension

Nanocrystals are pure drug solid particles with a regular, repeating lattice structure, typically having an average diameter of <1 μm and a crystalline character. The nanocrystals solely contain drug particles without any matrix material and a minimum of excipients, and the majority of these nanocrystals are formulated as tablets [126, 151, 157]. Where nanocrystals are dispersed in a liquid medium and stabilized by a minimum of surfactant or polymer are known as a nanosuspension [158]. Nanocrystals ideally lie in the range of 200–800 nm, which may be optimal to improve the dissolution and prevent stability issues like aggregation due to increased surface energy [159]. But in some cases, the desired size range will be preferred based on several factors, for instance, improved wound healing properties have been observed in the range of 10–400 nm [160], while the size range between 100 and 300 nm is ideal to avoid capillary blockage and for the enhanced permeability and retention effect (EPR effect) to achieve drug concentration in solid tumors, which are reported to have very dense vasculature [161, 162]. The achievement of the desired size is highly dependent on the type of preparation method and processing parameters. The common preparation methods include HPH and wet media milling.

Thus, nanocrystals offer several advantages and disadvantages [158]. Among the advantages are as follows:

- Ability to deliver the drug via various administration routes, including oral, intravenous, intramuscular, pulmonary, ocular, and dermal.
- Flexibility to formulate the drug in multiple pharmaceutical dosage forms such as tablets, capsules, suspensions, and ointments.
- Enhanced solubility, either thermodynamic or kinetic, depending on the drug's solid-state form, compared to conventional particles.
- Faster dissolution rate relative to conventional drug particles.

- Nanocrystals with prolonged circulation time in the bloodstream.
- Potential for both passive and active drug targeting mechanisms.
- Lower risk of tissue irritation following subcutaneous or intramuscular injection.
- Feasibility of developing hybrid nanocrystals for combined diagnostic and therapeutic use.

Fewer are the disadvantages, such as:

- Physicochemical-related stability problems, and particularly nanosuspension agglomeration, sedimentation, or Ostwald ripening, are the major concern.
- High-energy input required for preparation and a need for stabilizers.
- Risk of crystal growth during storage, leading to reduced efficacy.

3.3.2 Nanoemulsions

Nanoemulsions are colloidal particulate systems that typically lie in the submicron size, less than 1 μm. These are kinetically stable, homogeneous systems in which two immiscible liquids, typically oil and water, are mixed to form a fine colloidal dispersion with the aid of surfactants and co-surfactants. Nanoemulsions generally have droplet sizes in the range of 20–600 nm. This size range is ideal because it ensures optical transparency, stability against sedimentation or creaming, and enhanced bioavailability. Droplets larger than this may lead to instability and reduced bioavailability [163, 164]. Nanoemulsions can be prepared using high-energy methods like ultrasonication and HPH or low-energy methods such as phase inversion composition and phase inversion temperature [165]. Nanoemulsions offer several advantages and disadvantages [163].

Among the advantages are:

- Excellent physical stability over time (resistant to sedimentation, flocculation, and creaming).
- Enhanced solubilization of lipophilic drugs.
- Increased absorption and bioavailability due to nanoscale droplet size.
- Potential for controlled or sustained drug release.
- Useful for both hydrophilic and hydrophobic drugs depending on the phase composition.
- Easily manufactured by scalable methods (e.g. HPH, ultrasonication).
- Applicable for multiple administration routes (oral, dermal, ocular, parenteral).
- Potential for targeting through surface modification.
- Less irritation potential in topical formulations due to mild composition and small droplet size.

Disadvantages include:

- Thermodynamic instability may undergo phase separation over time.
- Limited solubilization capacity for high-dose hydrophobic drugs.
- Use of high surfactant concentrations may lead to toxicity or irritation, especially in parenteral and ocular formulations.

3.3.3 Marketed Nanosized Drug Formulations

The commercial success of nanosizing technologies clearly demonstrates their effectiveness in enhancing the dissolution rate and bioavailability of poorly water-soluble drugs. This is reflected in the number of nanosized drug formulations that have reached the market, as summarized in Table 3.1. These marketed products highlight the practical applicability and therapeutic advantages of nanosizing in overcoming solubility-related challenges in drug delivery.

3.4 Challenges and Limitations in Nanosizing

3.4.1 Stability Issues

One of the primary concerns in nanosizing is the stability of nanoparticles. Aggregation occurs due to high surface energy, leading to particle growth, which can significantly affect drug dissolution and bioavailability. To mitigate aggregation, stabilizers such as surfactants and polymers are commonly employed. The mechanism of stability is explained by the Derjaguin, Landau, Verwey, and Overbeek (DLVO) theory, which accounts for the balance of attractive and repulsive forces between particles [48, 179]. Ostwald ripening is another challenge, where larger particles grow at the expense of smaller ones, resulting in an increase in particle size over time. This phenomenon is driven by the solubility differences between small and large particles, as described by the Ostwald–Freundlich equation [180–183]. Ostwald ripening can be partially controlled by reducing the particle size distribution, which helps to minimize differences in saturation solubility and drug concentration gradients within the medium. Additionally, stabilizers play a crucial role by lowering the interfacial tension between the solid particles and the surrounding liquid, thereby preventing the Ostwald ripening process [184].

3.4.2 Scale-Up and Manufacturing Constraints

All the methods described above can be broadly classified into bottom-up and top-down processes. The latter one has been widely adopted by the pharmaceutical industry. The transition from laboratory-scale to commercial production presents significant challenges [185]. However, the bottom-up technique is less common at the industrial scale, as this technique requires the removal of the traces of the organic solvent, which is a very complex process. Reasons may include the need for solvent removal, the difficulty in controlling the process, and the fact that many poorly soluble drugs are poorly soluble not only in aqueous media but also in organic solvents [186, 187]. Numerous products developed using top-down technologies have been successfully commercialized, demonstrating the scalability and robustness of these methods for industrial production while complying with stringent pharmaceutical quality standards [188]. Scaling up nanomedicine from laboratory to market involves several critical considerations, including the careful selection of materials, solvents, and formulation procedures and ensuring that the

Table 3.1 Comprehensive table of marketed nanosized drug formulations.

Product name	API	Nanocarrier type	Formulation type	Preparation method	Indication	Company	References
Emend®	Aprepitant	Nanocrystal	Oral capsule and suspension	Elan's NanoCrystal® technology	Chemotherapy-induced nausea/vomiting	Merck	[166, 167]
Rapamune®	Sirolimus	Nanocrystal	Oral suspension and tablet	Elan's NanoCrystal® technology	Immunosuppressant (transplants)	Wyeth (Pfizer)	[166, 168]
TriCor®	Fenofibrate	Nanocrystals	Oral tablet	Elan's NanoCrystal® technology	Dyslipidemia	Abbott Laboratories	[168]
Megace ES®	Megestrol acetate	Nanocrystals	Oral suspension	Elan's NanoCrystal® technology	Appetite stimulant, anorexia, cachexia, and weight loss in patients with HIV	Par Pharma (Endo)	[169]
Invega Sustenna®	Paliperidone palmitate	Nanosuspension	Extended release injectable (IM)	Elan's NanoCrystal® technology	Schizophrenia, schizoaffective disorder	Johnson & Johnson	[168, 170]
Estrasorb™	Estradiol	Nanoemulsion	Topical lotion	Phase inversion temperature method	Moderate to severe vasomotor symptoms due to menopause	Novavax	[171, 172]
Restasis®	Cyclosporine A	Nanoemulsion	Ophthalmic emulsion	Emulsification (oil-in-water)	Chronic dry eye/inflammation	Allergan	[173, 174]
Cleviprex®	Clevidipine butyrate	Nanoemulsion	Injectable emulsion (IV)	Emulsification (oil-in-water)	Acute hypertension in perioperative settings	The Medicines Company (Novartis)	[175]
Diprivan®	Propofol	Nanoemulsion	Injectable emulsion (IV)	High-shear mixing and emulsification	Induction and maintenance of general anesthesia	AstraZeneca/Zeneca Pharmaceuticals	[176–178]

final product is acceptable to both clinicians and patients [186]. However, during the scale-up process, the desirable characteristics of nanoparticles may be compromised sometimes. For example, in a study of the scale-up of nanoparticles prepared using the emulsion method, it was observed that an increase in impeller speed and agitation time decreased particle size, although entrapment efficiency was not altered [189]. Furthermore, large-scale production demands the development of cost-effective and reproducible manufacturing processes. High-pressure melt homogenization and milling have shown potential for commercial viability, as they utilize well-established instrumentation. However, precise control over process parameters, such as the homogenization pressure, temperature, number of cycles, volume of milling media, solid loading, hardness of the particle, and processing heat-sensitive drugs, remains a limitation due to the temperature increase during homogenization and milling [119, 186, 190–192].

3.4.3 Regulatory Considerations and Safety Concerns

Regulatory agencies such as the United States Food and Drug Administration (US FDA), European Medicines Agency (EMA), and Central Drugs Standard Control Organization (CDSCO) require rigorous evaluation of nanosized drug formulations to ensure safety and efficacy [193–197]. Key concerns include potential toxicity arising from increased surface area and altered biodistribution. Nanosized particles often accumulate in tissues and may exhibit inefficient clearance, raising the risk of long-term retention and associated toxicity. Therefore, the behavior of nanoparticles in biological systems, including their absorption, metabolism, and clearance, requires thorough investigation to establish robust safety profiles [198, 199]. Furthermore, variability in particle size distribution, surface charge, and crystallinity can significantly influence drug performance, necessitating stringent quality control measures during production [200, 201]. Additionally, ensuring batch-to-batch consistency is crucial for regulatory approval [202].

3.5 Future Perspectives

Nanosizing technologies are set to revolutionize drug delivery systems, especially for BCS Class II and IV drugs. Looking ahead, one of the most promising directions is the integration of nanosizing with targeted delivery systems and personalized medicine, where nanoparticles can be designed for site-specific drug release using ligands or stimuli-responsive coatings. Nanosuspensions offer expanded potential for site-specific delivery and reduced systemic side effects due to their ability to modify surface characteristics and enhance permeability [48, 203]. Emerging hybrid approaches, such as combining precipitation and homogenization, are gaining traction for their scalability and efficiency. Technologies like NANOEDGE™ and H69™ showcase the ability to produce highly stable nanocrystals with tailored release profiles [147, 204]. Furthermore, regulatory agencies are increasingly emphasizing the toxicity of nanoparticles and biocompatibility evaluations, pushing the pharmaceutical industry to refine formulation strategies to align with evolving safety standards [205]. Advancements may also focus on solid-state nanoparticle engineering, enabling long-acting formulations for chronic diseases [113]. With the rise of AI and machine learning in pharmaceutical development, predictive modeling could optimize nanosizing parameters, speeding up drug development timelines and reducing the attrition rate of new molecules [206].

3.6 Conclusion

Nanosizing has emerged as a highly promising strategy for enhancing the solubility, dissolution, and bioavailability of poorly water-soluble drugs, particularly for BCS Class II and IV. It addresses key limitations of traditional solubility enhancement methods like salt formation, co-solvency, and solid dispersion by increasing the surface area and improving dissolution kinetics. Products like TriCor® and Rapamune® demonstrate that technologies including wet media milling, HPH, and hybrid approaches are effective in both laboratory and commercial settings. However, challenges remain, particularly with scale-up, regulatory review, long-term stability, and safety evaluation. Among these, toxicity concerns are increasingly important, as nanosizing can lead to altered biodistribution, tissue accumulation, and potentially prolonged retention in the body. These issues are expected to be addressed through further advancements in material science and nanotechnology. As pharmaceutical research advances toward personalized medicine and targeted therapies, nanosizing is likely to serve as a core platform technology, driving next-generation drug development and improving therapeutic outcomes for diverse patient populations.

References

1 Dasgupta, A. (2019). Issues of interferences in therapeutic drug monitoring. *Biot. Other Interf. Immunoassays* 1: 99–112. doi: 10.1016/B978-0-12-816429-7.00006-X.

2 Stielow, M., Witczyńska, A., Kubryń, N. et al. (2023). The bioavailability of drugs—the current state of knowledge. *Molecules* 28 (24): 8038. doi: 10.3390/MOLECULES28248038.

3 Mantri, R.V., Sanghvi, R., and Zhu, H.J. (2017). Solubility of pharmaceutical solids. In: *Developing Solid Oral Dosage Forms*, 2e (ed. Y. Qiu, Y. Chen, G.G.Z. Zhang, et al.), 3–22. Elsevier. doi: 10.1016/B978-0-12-802447-8.00001-7.

4 Lu, J.X., Tupper, C., Gutierrez, A.V, and Murray, J. (September 12, 2022). Biochemistry, dissolution and solubility. *StatPearls*. https://www.ncbi.nlm.nih.gov/books/NBK431100/. (accessed 12 March 2025).

5 Savjani, K.T., Gajjar, A.K., and Savjani, J.K. (2012). Drug solubility: importance and enhancement techniques. *ISRN Pharm.* 2012: 195727. doi: 10.5402/2012/195727.

6 Zhuo, Y., Zhao, Y.G., and Zhang, Y. (2024). Enhancing drug solubility, bioavailability, and targeted therapeutic applications through magnetic nanoparticles. *Molecules* 29 (20): 4854. doi: 10.3390/MOLECULES29204854.

7 Bhalani, D.V., Nutan, B., Kumar, A., and Chandel, A.K.S. (2022). Bioavailability enhancement techniques for poorly aqueous soluble drugs and therapeutics. *Biomedicines* 10 (9): 2055. doi: 10.3390/BIOMEDICINES10092055.

8 USP. (2025). Dissolution testing and drug release tests | USP. https://www.usp.org/small-molecules/dissolution. (accessed 12 March 2025).

9 Surasarang, S.H. and Williams, R.O. (2016). Pharmaceutical cryogenic technologies. In: *Formulating Poorly Water Soluble Drugs* (ed. R.O. Williams III, A.B. Watts, and D.A. Miller). Springer. https://link.springer.com/chapter/10.1007/978-3-319-42609-9_11. (accessed 15 March 2025).

10 Jambhekar, S.S. and Breen, P.J. (2013). Drug dissolution: significance of physicochemical properties and physiological conditions. *Drug Discov. Today* 18 (23–24): 1173–1184. doi: 10.1016/J.DRUDIS.2013.08.013.

11 Wang, Z., Xu, W., Liu, D. et al. (2025). Impact of food physical properties on oral drug absorption: a comprehensive review. *Drug Des. Devel. Ther.* 19: 267–280. doi: 10.2147/DDDT.S497515.

12 Xie, B., Liu, Y., Li, X. et al. (2024). Solubilization techniques used for poorly water-soluble drugs. *Acta Pharm. Sin. B* 14 (11): 4683–4716. doi: 10.1016/J.APSB.2024.08.027.

13 FDA. (1997). Office of training and communications division of communications management. The Drug Information Branch. 20857:301–827. https://www.fda.gov/media/70936/download.

14 FDA. (2021). M9 biopharmaceutics classification system-based biowaivers guidance for industry FDA-2018-D-3614. https://www.fda.gov/vaccines-blood-biologics/guidance-compliance-regulatory-information-biologics/biologics-guidances (accessed 12 March 2025).

15 Price, D.J. (2022). The developability classification system (DCS): enabling an optimized approach for formulation of poorly soluble molecules. https://www.pharmaexcipients.com/news/developability-classification-system-dcs/ 7.

16 Pantwalawalkar, J., Kale, N., Nangare, S. et al. (2025). Pharmaceutical cocrystals: unlocking the potential of challenging drug candidates. *J. Drug Delivery Sci. Technol.* 104: 106572. doi: 10.1016/J.JDDST.2024.106572.

17 Rodriguez-Aller, M., Guillarme, D., Veuthey, J.L., and Gurny, R. (2015). Strategies for formulating and delivering poorly water-soluble drugs. *J. Drug Delivery Sci. Technol.* 30: 342–351. doi: 10.1016/J.JDDST.2015.05.009.

18 Lai, F., Schlich, M., Pireddu, R. et al. (2015). Production of nanosuspensions as a tool to improve drug bioavailability: focus on topical delivery. *Curr. Pharm. Des.* 21 (42): 6089–6103. doi: 10.2174/1381612821666151027152350.

19 Jain, S., Patel, N., and Lin, S. (2015). Solubility and dissolution enhancement strategies: current understanding and recent trends. *Drug Dev. Ind. Pharm.* 41 (6): 875–887. doi: 10.3109/03639045.2014.971027.

20 Meola, T.R., Bremmell, K.E., Williams, D.B. et al. (2022). Bio-enabling strategies to mitigate the pharmaceutical food effect: a mini review. *Int. J. Pharm.* 619: 121695. doi: 10.1016/J.IJPHARM.2022.121695.

21 Kumari, L., Choudhari, Y., Patel, P. et al. (2023). Advancement in solubilization approaches: a step towards bioavailability enhancement of poorly soluble drugs. *Life (Basel)* 13 (5): 1099. doi: 10.3390/LIFE13051099.

22 Kalepu, S. and Nekkanti, V. (2015). Insoluble drug delivery strategies: review of recent advances and business prospects. *Acta Pharm. Sin. B* 5 (5): 442–453. doi: 10.1016/j.apsb.2015.07.003.

23 Price, G. and Patel, D.A. (2023). *Drug Bioavailability*, 1–2. StatPearls Publishing. doi: 10.1016/B978-008055232-3.60035-2.

24 May, S. and Fahr, A. (2013). Self-assembled delivery vehicles for poorly water-soluble drugs: basic theoretical considerations and modeling concepts. In: *Drug Delivery Strategies for Poorly Water-Soluble Drugs* (ed. D. Douroumis and A. Fahr), 1–35. Wiley. doi: 10.1002/9781118444726.CH1.

25 Sun, D., Gao, W., Hu, H., and Zhou, S. (2022). Why 90% of clinical drug development fails and how to improve it? *Acta Pharm. Sin. B* 12 (7): 3049–3062. doi: 10.1016/J.APSB.2022.02.002.

26 Raman, S. and Polli, J.E. (2016). Prediction of positive food effect: bioavailability enhancement of BCS class II drugs. *Int. J. Pharm.* 506 (1–2): 110–115. doi: 10.1016/J.IJPHARM.2016.04.013.

27 Haser, A., DiNunzio, J.C., Martin, C. et al. (2016). Melt extrusion. *AAPS Adv. Pharm. Sci. Ser.* 22: 383–435. doi: 10.1007/978-3-319-42609-9_9.

28 Shargel, L., Wu-Pong, S., and Yu, A.B.C. (2012). *Applied Biopharmaceutics and Pharmacokinetics*, 811. McGraw Hill Professional.

29 Warnken, Z., Smyth, H.D.C., and Williams, R.O. (2022). Route-specific challenges in the delivery of poorly water-soluble drugs. *AAPS Adv. Pharm. Sci. Ser.* 50: 1–31. doi: 10.1007/978-3-030-88719-3_1.

30 Gu, C.H., Li, H., Levons, J. et al. (2007). Predicting effect of food on extent of drug absorption based on physicochemical properties. *Pharm. Res.* 24 (6): 1118–1130. doi: 10.1007/s11095-007-9236-1.

31 Millard, J.W., Alvarez-Núñez, F.A., and Yalkowsky, S.H. (2002). Solubilization by cosolvents: establishing useful constants for the log–linear model. *Int. J. Pharm.* 245 (1–2): 153–166. doi: 10.1016/S0378-5173(02)00334-4.

32 He, M., Zheng, W., Wang, N. et al. (2022). Molecular dynamics simulation of drug solubilization behavior in surfactant and cosolvent injections. *Pharmaceutics* 14 (11): 2366. doi: 10.3390/PHARMACEUTICS14112366/S1.

33 Nie, H., Byrn, S.R., Zhou, Q., and (Tony). (2017). Stability of pharmaceutical salts in solid oral dosage forms. *Drug Dev. Ind. Pharm.* 43 (8): 1215–1228. doi: 10.1080/03639045.2017.1304960.

34 Kumar, R., Thakur, A.K., Chaudhari, P., and Banerjee, N. (2022). Particle size reduction techniques of pharmaceutical compounds for the enhancement of their dissolution rate and bioavailability. *J. Pharm. Innov.* 17 (2): 333–352. doi: 10.1007/S12247-020-09530-5.

35 Dizaj, S.M., Vazifehasl, Z., Salatin, S. et al. (2015). Nanosizing of drugs: effect on dissolution rate. *Res. Pharm. Sci.* 10 (2): 95. https://www.researchgate.net/publication/281873785_Nanosizing_of_drugs_Effect_on_dissolution_rate (accessed 23 March 2025).

36 Nernst, W. (1904). Theorie der reaktionsgeschwindigkeit in heterogenen systemen. *Z. Phys. Chem.* 47U (1): 52–55. doi: 10.1515/ZPCH-1904-4704/PDF.

37 Brunner, E. (1904). Reaktionsgeschwindigkeit in heterogenen systemen. *Z. Phys. Chem.* 47U (1): 56–102. doi: 10.1515/ZPCH-1904-4705/PDF.

38 Dokoumetzidis, A. (2006). A century of dissolution research: from Noyes and Whitney to the biopharmaceutics classification system. *Int. J. Pharm.* 321 (1–2): 1–11. https://www.sciencedirect.com/science/article/pii/S0378517306005813?casa_token=_P1PAQQvjPYAAAAA:fukDwf9VYFS7qdEB6SJofvFPNZ09Engm_k9_Q707TyvsG8I_skUSwcOIL2M6jtK4J2HSuHw1Og (accessed 25 March 2025).

39 Noyes, A.A. and Whitney, W.R. (1897). The rate of solution of solid substances in their own solutions. *J. Am. Chem. Soc.* 19 (12): 930–934. doi: 10.1021/JA02086A003/ASSET/JA02086A003.FP.PNG_V03.

40 Gao, Y., Glennon, B., He, Y., and Donnellan, P. (2021). Dissolution kinetics of a BCS class II active pharmaceutical ingredient: diffusion-based model validation and prediction. *ACS Omega* 6 (12): 8056–8067. https://pmc.ncbi.nlm.nih.gov/articles/PMC8014923/ (accessed 16 March 2025).

41 Al-Kassas, R., Bansal, M., and Shaw, J. (2017). Nanosizing techniques for improving bioavailability of drugs. *J. Control. Release* 260: 202–212. doi: 10.1016/J.JCONREL.2017.06.003.

42 Junyaprasert, V.B. and Morakul, B. (2015). Nanocrystals for enhancement of oral bioavailability of poorly water-soluble drugs. *Asian J. Pharm. Sci.* 10 (1): 13–23. doi: 10.1016/J.AJPS.2014.08.005.

43 Saleem, I.Y. and Smyth, H.D.C. (2010). Micronization of a soft material: air-jet and micro-ball milling. *AAPS PharmSciTech* 11 (4): 1642. doi: 10.1208/S12249-010-9542-5.

44 Dhiman, A. and Prabhakar, P.K. (2021). Micronization in food processing: a comprehensive review of mechanistic approach, physicochemical, functional properties and self-stability of micronized food materials. *J. Food Eng.* 292: 110248. doi: 10.1016/J.JFOODENG.2020.110248.

45 Loh, Z.H., Samanta, A.K., and Sia Heng, P.W. (2015). Overview of milling techniques for improving the solubility of poorly water-soluble drugs. *Asian J. Pharm. Sci.* 10 (4): 255–274. doi: 10.1016/J.AJPS.2014.12.006.

46 Scheler, S. (2013). Micro- and nanosizing of poorly soluble drugs by grinding techniques. In: *Drug Delivery Strategies for Poorly Water-Soluble Drugs* (ed. D. Douroumis and A. Fahr), 509–550. Wiley. doi: 10.1002/9781118444726.CH16.

47 Sutradhar, K.B., Khatun, S., and Luna, I.P. (2013). Increasing possibilities of nanosuspension. *J. Nanotechnol.* 2013 (1): 346581. doi: 10.1155/2013/346581.

48 Jacob S, Nair AB, Shah J 2020. Emerging role of nanosuspensions in drug delivery systems. *Biomater. Res.* 24(1):3. https://pmc.ncbi.nlm.nih.gov/articles/PMC6964012/. (accessed 17 March 2025).

49 Liu, Y., Liang, Y., Yuhong, J. et al. (2024). Advances in nanotechnology for enhancing the solubility and bioavailability of poorly soluble drugs. *Drug Des. Devel. Ther.* 18: 1469. doi: 10.2147/DDDT.S447496.

50 Dahlgren, D., Sjögren, E., and Lennernäs, H. (2020). Intestinal absorption of BCS class II drugs administered as nanoparticles: a review based on in vivo data from intestinal perfusion models. *ADMET DMPK* 8 (4): 375. doi: 10.5599/ADMET.881.

51 Jacob, S., Kather, F.S., Boddu, S.H.S. et al. (2025). Nanosuspension innovations: expanding horizons in drug delivery techniques. *Pharmaceutics* 17 (1): 136. doi: 10.3390/PHARMACEUTICS17010136.

52 Gan, Y., Baak, J.P.A., Chen, T. et al. (2023). Supersaturation and precipitation applicated in drug delivery systems: development strategies and evaluation approaches. *Molecules* 28 (5): 2212. doi: 10.3390/MOLECULES28052212.

53 Kellogg, R.M. and Leeman, M. (2012). 9.16 crystallization as a tool in industrial applications of asymmetric synthesis. *Compr. Chirality* 9: 367–399. doi: 10.1016/B978-0-08-095167-6.00918-6.

54 Zhang, X., Xing, H., Zhao, Y., and Ma, Z. (2018). Pharmaceutical dispersion techniques for dissolution and bioavailability enhancement of poorly water-soluble drugs. *Pharmaceutics* 10 (3): 74. doi: 10.3390/PHARMACEUTICS10030074.

55 Talevi, A. and Bellera, C.L. (2024). Drug absorption. In: *ADME Processes in Pharmaceutical Sciences Dosage, Design, and Pharmacotherapy Success* (ed. A. Talevi and P.A.M. Quiroga), 11–96. Springer. doi: 10.1007/978-3-319-99593-9_2.

56 Raines, K., Agarwal, P., Augustijns, P. et al. (2023). Drug dissolution in oral drug absorption: workshop report. *AAPS J.* 25 (6): 1–20. doi: 10.1208/S12248-023-00865-8/TABLES/1.

57 Nguyen, T.T.L., Duong, V.A., and Maeng, H.J. (2021). Pharmaceutical formulations with p-glycoprotein inhibitory effect as promising approaches for enhancing oral drug absorption and bioavailability. *Pharmaceutics* 13 (7): 1103. doi: 10.3390/PHARMACEUTICS13071103.

58 Wagner, C., Jantratid, E., Kesisoglou, F. et al. (2012). Predicting the oral absorption of a poorly soluble, poorly permeable weak base using biorelevant dissolution and transfer model tests coupled with a physiologically based pharmacokinetic model. *Eur. J. Pharm. Biopharm.* 82 (1): 127–138. doi: 10.1016/J.EJPB.2012.05.008.

59 Shekhawat, B., P. and Pokharkar, B., V. (2017). Understanding peroral absorption: regulatory aspects and contemporary approaches to tackling solubility and permeability hurdles. *Acta Pharm. Sin. B* 7 (3): 260–280. doi: 10.1016/J.APSB.2016.09.005.

60 Rangaraj, N., Sampathi, S., Junnuthula, V. et al. (2022). Fast-fed variability: insights into drug delivery, molecular manifestations, and regulatory aspects. *Pharmaceutics* 14 (9): 1807. doi: 10.3390/PHARMACEUTICS14091807.

61 Miao, Y., Zhao, S., Zuo, J. et al. (2022). Reduced the food effect and enhanced the oral bioavailability of ivacaftor by self-nanoemulsifying drug delivery system (SNEDDS) using a new oil phase. *Drug Des. Devel. Ther.* 16: 1531–1546. doi: 10.2147/DDDT.S356967.

62 Danaei, M., Dehghankhold, M., Ataei, S. et al. (2018). Impact of particle size and polydispersity index on the clinical applications of lipidic nanocarrier systems. *Pharmaceutics* 10 (2): 57. doi: 10.3390/PHARMACEUTICS10020057.

63 Huang, C.Y. and Sherry, K.M. (2010). Prediction of drug particle size and content uniformity in low-dose solid dosage forms. *Int. J. Pharm.* 383 (1–2): 70–80. doi: 10.1016/J.IJPHARM.2009.09.009.

64 Rocha, B., de Morais, L.A., Viana, M.C., and Carneiro, G. (2023). Promising strategies for improving oral bioavailability of poor water-soluble drugs. *Expert Opin. Drug Discov.* 18 (6): 615–627. doi: 10.1080/17460441.2023.2211801.

65 Pindelska, E., Sokal, A., and Kolodziejski, W. (2017). Pharmaceutical cocrystals, salts and polymorphs: advanced characterization techniques. *Adv. Drug Deliv. Rev.* 117: 111–146. doi: 10.1016/J.ADDR.2017.09.014.

66 Hughey, J.R., Huang, S., and Williams, R.O. (2016). Solid-state techniques for improving solubility. *AAPS Adv. Pharm. Sci. Ser.* 22: 121–163. doi: 10.1007/978-3-319-42609-9_3.

67 Gupta, D., Bhatia, D., Dave, V. et al. (2018). Salts of therapeutic agents: chemical, physicochemical, and biological considerations. *Mol. A J. Synth. Chem. Nat. Prod. Chem.* 23 (7): 1719. doi: 10.3390/MOLECULES23071719.

68 Kumar, L., Amin, A., and Bansal, A. (2008). Salt selection in drug development. *Pharm. Technol.* 32. https://www.pharmtech.com/view/salt-selection-drug-development (accessed 23 March 2025).

69 Li, S., Jones, D.S., and Andrews, G.P. (2013). Hot melt extrusion: a process overview and use in manufacturing solid dispersions of poorly water-soluble drugs. In: *Drug Delivery Strategies for Poorly Water-Soluble Drugs* (ed. D. Douroumis and A. Fahr), 325–358. Wiley. doi: 10.1002/9781118444726.CH11.

70 Lüdeker, D. and Briel, T. (2023). Improving API solubility by salt and cocrystal formation.

71 Serajuddin, A.T.M. (2007). Salt formation to improve drug solubility. *Adv. Drug Deliv. Rev.* 59 (7): 603–616. doi: 10.1016/J.ADDR.2007.05.010.

72 Ellenberger, D., O'Donnell, K.P., and Williams, R.O. (2016). Optimizing the formulation of poorly water-soluble drugs. *AAPS Adv. Pharm. Sci. Ser.* 22: 41–120. doi: 10.1007/978-3-319-42609-9_2.

73 Mou, D., Chen, H., Wan, J. et al. (2011). Potent dried drug nanosuspensions for oral bioavailability enhancement of poorly soluble drugs with pH-dependent solubility. *Int. J. Pharm.* 413 (1–2): 237–244. doi: 10.1016/J.IJPHARM.2011.04.034.

74 Shah, S.M., Jain, A.S., Kaushik, R. et al. (2014). Preclinical formulations: insight, strategies, and practical considerations. *AAPS PharmSciTech* 15 (5): 1307. doi: 10.1208/S12249-014-0156-1.

75 Surasarang, S.H. and Williams, R.O. (2016). Co-solvent and complexation systems. *AAPS Adv. Pharm. Sci. Ser.* 22: 215–256. doi: 10.1007/978-3-319-42609-9_5.

76 Nayak, A.K. and Panigrahi, P.P. (2012). Solubility enhancement of etoricoxib by cosolvency approach. *Int. Sch. Res. Not.* 2012 (1): 820653. doi: 10.5402/2012/820653.

77 Dick, F.D. (2006). Solvent neurotoxicity. *Occup. Environ. Med.* 63 (3): 221. doi: 10.1136/OEM.2005.022400.

78 Jörgensen, A.M., Friedl, J.D., Wibel, R. et al. (2020). Cosolvents in self-emulsifying drug delivery systems (SEDDS): do they really solve our solubility problems? *Mol. Pharm.* 17 (9): 3236–3245. doi: 10.1021/ACS.MOLPHARMACEUT.0C00343/ASSET/IMAGES/LARGE/MP0C00343_0008.JPEG.

79 Biswas, S., Vaze, O.S., Movassaghian, S., and Torchilin, V.P. (2013). Polymeric micelles for the delivery of poorly soluble drugs. In: *Drug Delivery Strategies for Poorly Water-Soluble Drugs* (ed. D. Douroumis and A. Fahr), 411–476. Wiley. doi: 10.1002/9781118444726.CH14.

80 Ahmadi Tehrani, A., Omranpoor, M.M., Vatanara, A. et al. (2019). Formation of nanosuspensions in bottom-up approach: theories and optimization. *DARU J. Pharm. Sci.* 27 (1): 451. doi: 10.1007/S40199-018-00235-2.

81 Perumal, S., Atchudan, R., and Lee, W. (2022). A review of polymeric micelles and their applications. *Polymers (Basel)* 14 (12): 2510. doi: 10.3390/POLYM14122510.

82 Khan, K.U., Minhas, M.U., Badshah, S.F. et al. (2022). Overview of nanoparticulate strategies for solubility enhancement of poorly soluble drugs. *Life Sci.* 291. doi: 10.1016/j.lfs.2022.120301.

83 Naguib, Y.W., O'Mary, H.L., Cui, Z., and Watts, A.B. (2016). Injectable formulations of poorly water-soluble drugs. *AAPS Adv. Pharm. Sci. Ser.* 22: 257–293. doi: 10.1007/978-3-319-42609-9_6.

84 Polat, H., Kutluay, G., and Polat, M. (2020). Analysis of dilution induced disintegration of micellar drug carriers in the presence of inter and intra micellar species. *Colloid Surf. A Physicochem. Eng. Asp.* 601: 124989. doi: 10.1016/J.COLSURFA.2020.124989.

85 Trivedi, R. and Kompella, U.B. (2010). Nanomicellar formulations for sustained drug delivery: strategies and underlying principles. *Nanomedicine (Lond.)* 5 (3): 485. doi: 10.2217/NNM.10.10.

86 Lu, Y., Zhang, E., Yang, J., and Cao, Z. (2018). Strategies to improve micelle stability for drug delivery. *Nano Res.* 11 (10): 4985. doi: 10.1007/S12274-018-2152-3.

87 Inácio, Â.S., Mesquita, K.A., Baptista, M. et al. (2011). In vitro surfactant structure-toxicity relationships: implications for surfactant use in sexually transmitted infection prophylaxis and contraception. *PLoS One* 6 (5): doi: 10.1371/JOURNAL.PONE.0019850.

88 Perinelli, D.R., Cespi, M., Lorusso, N. et al. (2020). Surfactant self-assembling and critical micelle concentration: one approach fits all? *Langmuir* 36 (21): 5745–5753. doi: 10.1021/ACS.LANGMUIR.0C00420/ASSET/IMAGES/LARGE/LA0C00420_0006.JPEG.

89 Ríos, F., Lechuga, M., Lobato-Guarnido, I., and Fernández-Serrano, M. (2023). Antagonistic toxic effects of surfactants mixtures to bacteria pseudomonas putida and marine microalgae *Phaeodactylum tricornutum*. *Toxics* 11 (4): 344. doi: 10.3390/TOXICS11040344.

90 Nicolaescu, O.E., Belu, I., Mocanu, A.G. et al. (2025). Cyclodextrins: enhancing drug delivery, solubility and bioavailability for modern therapeutics. *Pharmaceutics* 17 (3): 288. doi: 10.3390/PHARMACEUTICS17030288.

91 European Medicines Agency. (2017). Committee for human medicinal products (CHMP). www.ema.europa.eu/contact (accessed 7 July 2025).

92 Ferreira, L., Campos, J., Veiga, F. et al. (2022). Cyclodextrin-based delivery systems in parenteral formulations: a critical update review. *Eur. J. Pharm. Biopharm.* 178: 35–52. doi: 10.1016/J.EJPB.2022.07.007.

93 Donaubauer, H.H., Fuchs, H., Langer, K.H., and Bär, A. (1998). Subchronic intravenous toxicity studies with γ-cyclodextrin in rats. *Regul. Toxicol. Pharmacol.* 27 (2): 189–198. doi: 10.1006/RTPH.1998.1224.

94 Musuc, A.M. (2024). Cyclodextrins: advances in chemistry, toxicology, and multifaceted applications. *Molecules* 29 (22): doi: 10.3390/MOLECULES29225319.

95 European Medicines Agency. (2017). Committee for Human Medicinal Products (CHMP) Questions and answers on cyclodextrins used as excipients in medicinal products for human use End of consultation (deadline for comments). www.ema.europa.eu/contact (accessed 7 July 2025).

96 Mentzafos, D. and Bethanis, K. (2015). β-Cyclodextrin dimeric inclusion complexes: a review of some crystal structure aspects. https://www.researchgate.net/publication/297767688_b-Cyclodextrin_dimeric_inclusion_complexes_A_review_of_some_crystal_structure_aspects (accessed 7 July 2025).

97 Sharma, K.S., Sahoo, J., Agrawal, S., and Kumari, A. (2019). Solid dispersions: a technology for improving bioavailability. *J. Anal. Pharm. Res.* 8 (4): 127–133. doi: 10.15406/JAPLR.2019.08.00326.

98 Choi, M.J., Woo, M.R., Choi, H.G., and Jin, S.G. (2022). Effects of polymers on the drug solubility and dissolution enhancement of poorly water-soluble rivaroxaban. *Int. J. Mol. Sci.* 23 (16): 9491. doi: 10.3390/IJMS23169491.

99 Zhang, J., Guo, M., Luo, M., and Cai, T. (2023). Advances in the development of amorphous solid dispersions: The role of polymeric carriers. *Asian J. Pharm. Sci.* 18 (4): 100834. doi: 10.1016/J.AJPS.2023.100834.

100 Moroni, A.B., Mayoral, E.P., Lionello, D.F. et al. (2023). Solid-state properties of nifurtimox. Preparation, analytical characterization, and stability of an amorphous phase. *Eur. J. Pharm. Biopharm.* 184: 25–35. doi: 10.1016/J.EJPB.2023.01.008.

101 Kim, D.H., Kim, Y.W., Tin, Y.Y. et al. (2021). Recent technologies for amorphization of poorly water-soluble drugs. *Pharmaceutics* 13 (8): 1318. doi: 10.3390/PHARMACEUTICS13081318.

102 Mazayen, Z.M., Ghoneim, A.M., Elbatanony, R.S. et al. (2022). Pharmaceutical nanotechnology: from the bench to the market. *Futur. J. Pharm. Sci.* 8 (1): 1–11. doi: 10.1186/S43094-022-00400-0.

103 Jain, D., Thakur, P.S., Thakore, S.D. et al. (2020). Impact of differential particle size of fenofibrate nanosuspensions on biopharmaceutical performance using physiologically based absorption modeling in rats. *J. Drug Delivery Sci. Technol.* 60: 102040. doi: 10.1016/J.JDDST.2020.102040.

104 Shen, L.J. and Wu, F.L.L. (2007). Nanomedicines in renal transplant rejection – focus on sirolimus. *Int. J. Nanomedicine* 2 (1): 25. doi: 10.2147/NANO.2007.2.1.25.

105 Haeri, A., Osouli, M., Bayat, F. et al. (2018). Nanomedicine approaches for sirolimus delivery: a review of pharmaceutical properties and preclinical studies. *Artif. Cells Nanomed. Biotechnol.* 46 (sup1): 1–14. doi: 10.1080/21691401.2017.1408123.

106 Jang, K., Yoon, S., Kim, S.E. et al. (2014). Novel nanocrystal formulation of megestrol acetate has improved bioavailability compared with the conventional micronized formulation in the fasting state. *Drug Des. Devel. Ther.* 8: 851. doi: 10.2147/DDDT.S62176.

107 Mahesh, K.V., Singh, S.K., and Gulati, M. (2014). A comparative study of top-down and bottom-up approaches for the preparation of nanosuspensions of glipizide. *Powder Technol.* 256: 436–449. doi: 10.1016/J.POWTEC.2014.02.011.

108 Khan, I., Saeed, K., and Khan, I. (2019). Nanoparticles: properties, applications and toxicities. *Arab. J. Chem.* 12 (7): 908–931. doi: 10.1016/J.ARABJC.2017.05.011.

109 Paramasivam, G., Palem, V.V., Sundaram, T. et al. (2021). Nanomaterials: synthesis and applications in theranostics. *Nanomaterials (Basel)* 11 (12): 3228. doi: 10.3390/NANO11123228.

110 Böhringer, C. and Rutherford, T.F. (2008). Combining bottom-up and top-down. *Energy Econ.* 30 (2): 574–596. doi: 10.1016/J.ENECO.2007.03.004.

111 Abid, N., Khan, A.M., Shujait, S. et al. (2022). Synthesis of nanomaterials using various top-down and bottom-up approaches, influencing factors, advantages, and disadvantages: a review. *Adv. Colloid Interf. Sci.* 300: 102597. doi: 10.1016/J.CIS.2021.102597.

112 Narayan, R., Pednekar, A., Bhuyan, D. et al. (2017). A top-down technique to improve the solubility and bioavailability of aceclofenac: in vitro and in vivo studies. *Int. J. Nanomedicine* 12: 4921–4935. doi: 10.2147/IJN.S141504.

113 Lhaglham, P., Jiramonai, L., Jia, Y. et al. (2024). Drug nanocrystals: surface engineering and its applications in targeted delivery. *iScience* 27 (11): 111185. doi: 10.1016/J.ISCI.2024.111185.

114 Verma, S., Gokhale, R., and Burgess, D.J. (2009). A comparative study of top-down and bottom-up approaches for the preparation of micro/nanosuspensions. *Int. J. Pharm.* 380 (1–2): 216–222. doi: 10.1016/J.IJPHARM.2009.07.005.

115 Date, A.A. and Patravale, V.B. (2004). Current strategies for engineering drug nanoparticles. *Curr. Opin. Colloid Interface Sci.* 9 (3–4): 222–235. doi: 10.1016/J.COCIS.2004.06.009.

116 De Castro, L. (1995). Method of preparing stable drug nanoparticles. https://patents.google.com/patent/US5534270A/en (U.S. Patent No. US5534270A).

117 Shegokar, R. (2016). Wet media milling: an effective way to solve drug solubility issue. In: *Handbook of Nanoparticles* (ed. M. Aliofkhazraei), 385–406. Springer. doi: 10.1007/978-3-319-15338-4_20.

118 Grigorov, P., Long, C.P., and Lekhal, A. (2011). High pressure homogenization as particle size reduction technology in the pharmaceutical industry: optimization, modeling and isolation techniques. *2011 AIChE Annual Meeting*, Minnepolis, MN (16–21 October 2011). https://www.researchgate.net/publication/267312700_High_Pressure_Homogenization_As_Particle_Size_Reduction_Technology_In_the_Pharmaceutical_Industry_Optimization_Modeling_and_Isolation_Techniques. (accessed 14 April 2025).

119 Catlin, E.J., Fandiño, O.E., Lopez-Vidal, L. et al. (2024). A novel temperature-controlled media milling device to produce drug nanocrystals at the laboratory scale. *Int. J. Pharm.* 666: 124780. doi: 10.1016/J.IJPHARM.2024.124780.

120 Rabinow, B.E. (2004). Nanosuspensions in drug delivery. *Nat. Rev. Drug Discov.* 3 (9): 785–796. doi: 10.1038/NRD1494.

121 Guner, G., Seetharaman, N., Elashri, S. et al. (2022). Analysis of heat generation during the production of drug nanosuspensions in a wet stirred media mill. *Int. J. Pharm.* 624: 122020. doi: 10.1016/J.IJPHARM.2022.122020.

122 Patravale, V.B., Date, A.A., and Kulkarni, R.M. (2004). Nanosuspensions: a promising drug delivery strategy. *J. Pharm. Pharmacol.* 56 (7): 827–840. doi: 10.1211/0022357023691.

123 Malik, T., Sharma, R., Ameer, K. et al. (2023). Potential of high-pressure homogenization (HPH) in the development of functional foods. *Int. J. Food Prop.* 26 (1): 2509–2531. doi: 10.1080/10942912.2023.2249262.

124 Håkansson, A. (2025). A low-cost method for characterizing the inception and extent of cavitation in high-pressure homogenizers. *Ind. Eng. Chem. Res.* doi: 10.1021/ACS.IECR.5C00512/ASSET/IMAGES/LARGE/IE5C00512_0013.JPEG.

125 Yadav, K.S. and Kale, K. (2020). High pressure homogenizer in pharmaceuticals: understanding its critical processing parameters and applications. *J. Pharm. Innov.* 15 (4): 690–701. doi: 10.1007/S12247-019-09413-4/TABLES/4.

126 Jarvis, M., Krishnan, V., Mitragotri, S. et al. (2018). Nanocrystals: a perspective on translational research and clinical studies. *Bioeng. Transl. Med.* 4 (1): 5. doi: 10.1002/BTM2.10122.

127 Chen, Y., Liu, H., and Pan, M. (2016). Sirolimus external preparation, preparation method and the usage. https://patents.google.com/patent/CN105663027B/en (China, Patent No, CN105663027B).

128 Pandiselvam, R., Özaslan, Z.T., Sahni, P. et al. (2024). High pressure homogenization for preservation of liquid foods – mechanisms, molecular modifications and recent developments. *Futur. Foods* 10: 100488. doi: 10.1016/J.FUFO.2024.100488.

129 Sandhya, M., Ramasamy, D., Sudhakar, K. et al. (2021). Ultrasonication an intensifying tool for preparation of stable nanofluids and study the time influence on distinct properties of graphene nanofluids – a systematic overview. *Ultrason. Sonochem.* 73: 105479. doi: 10.1016/J.ULTSONCH.2021.105479.

130 Sharifalhoseini, Z., Entezari, M.H., and Jalal, R. (2015). Direct and indirect sonication affect differently the microstructure and the morphology of ZnO nanoparticles: optical behavior and its antibacterial activity. *Ultrason. Sonochem.* 27: 466–473. doi: 10.1016/J.ULTSONCH.2015.06.016.

131 Yun, J.S., Hwangbo, S.A., and Jeong, Y.G. (2023). Preparation of uniform nano liposomes using focused ultrasonic technology. *Nanomaterials* 13 (19): 2618. doi: 10.3390/NANO13192618/S1.

132 Betts, J.N., Johnson, M.G., Rygiewicz, P.T. et al. (2013). Potential for metal contamination by direct sonication of nanoparticle suspensions. *Environ. Toxicol. Chem.* 32 (4): 889–893. doi: 10.1002/ETC.2123.

133 Sesis, A., Hodnett, M., Memoli, G. et al. (2013). Influence of acoustic cavitation on the controlled ultrasonic dispersion of carbon nanotubes. *J. Phys. Chem. B* 117 (48): 15141–15150. doi: 10.1021/JP410041Y.

134 Partheniadis, I., Shah, R.R., and Nikolakakis, I. (2022). Application of ultrasonics for nanosizing drugs and drug formulations. *J. Dispers. Sci. Technol.* 43 (11): 1587–1602. doi: 10.1080/01932691.2021.1878035.

135 Kumari, S., Raturi, S., Kulshrestha, S. et al. (2023). A comprehensive review on various techniques used for synthesizing nanoparticles. *J. Mater. Res. Technol.* 27: 1739–1763. doi: 10.1016/J.JMRT.2023.09.291.

136 Kuddushi, M., Kanike, C., Bin, X.B., and Zhang, X. (2025). Recent advances in nanoprecipitation: from mechanistic insights to applications in nanomaterial synthesis. *Soft Matter* 21 (15): 2759–2781. doi: 10.1039/D5SM00006H.

137 Matteucci, M., Hotze, M., Johnston, K., and Langmuir, R.W. (2006). Drug nanoparticles by antisolvent precipitation: mixing energy versus surfactant stabilization. *Langmuir* 22 (21): 8951–8959. doi: 10.1021/la061122t.

138 Verma, V., Ryan, K.M., and Padrela, L. (2021). Production and isolation of pharmaceutical drug nanoparticles. *Int. J. Pharm.* 603: 120708. doi: 10.1016/J.IJPHARM.2021.120708.

139 Sinha, B., Müller, R.H., and Möschwitzer, J.P. (2013). Bottom-up approaches for preparing drug nanocrystals: formulations and factors affecting particle size. *Int. J. Pharm.* 453 (1): 126–141. doi: 10.1016/J.IJPHARM.2013.01.019.

140 Girotra, P., Singh, S.K., and Nagpal, K. (2013). Supercritical fluid technology: a promising approach in pharmaceutical research. *Pharm. Dev. Technol.* 18 (1): 22–38. doi: 10.3109/10837450.2012.726998.

141 Smith, R.M. (1993). Nomenclature for supercritical fluid chromatography and extraction. *Pure Appl. Chem.* 65 (11): 2397–2403. doi: 10.1351/PAC199365112397/MACHINEREADABLECITATION/RIS.

142 Clifford, A.A. and Williams, J.R. (2000). Introduction to supercritical fluids and their applications. In: *Supercritical Fluid Methods and Protocols* (ed. J.R. Williams and A.A. Clifford), 1–16. Springer. doi: 10.1385/1-59259-030-6:1.

143 Yildiz-Ozturk, E. and Yesil-Celiktas, O. (2019). Supercritical fluid technology in bioseperation. In: *Comprehensive Biotechnology* (ed. M. Moo-Young), 713–724. Elsevier. doi: 10.1016/B978-0-444-64046-8.00122-1.

144 Chan, H.K. and Kwok, P.C.L. (2011). Production methods for nanodrug particles using the bottom-up approach. *Adv. Drug Deliv. Rev.* 63 (6): 406–416. doi: 10.1016/J.ADDR.2011.03.011.

145 Li, W., Yuan, H., Liu, Y. et al. (2024). Current analytical strategies for the determination of resveratrol in foods. *Food Chem.* 431. doi: 10.1016/j.foodchem.2023.137182.

146 Cheaburu-Yilmaz, C.N., Karasulu, H.Y., and Yilmaz, O. (2019). Nanoscaled dispersed systems used in drug-delivery applications. In: *Polymeric Nanomaterials in Nanotherapeutics* (ed. C. Vasile), 437–468. Elsevier. doi: 10.1016/B978-0-12-813932-5.00013-3.

147 Salazar, J., Müller, R.H., and Möschwitzer, J.P. (2014). Combinative particle size reduction technologies for the production of drug nanocrystals. *J. Pharm.* doi: 10.1155/2014/265754.

148 Rabinow, B., Kipp, J., Papadopoulos, P. et al. (2007). Itraconazole IV nanosuspension enhances efficacy through altered pharmacokinetics in the rat. *Int. J. Pharm.* 339 (1–2): 251–260. doi: 10.1016/J.IJPHARM.2007.02.030.

149 Shelar, D.B., Pawar, S.K., and Vavia, P.R. (2013). Fabrication of isradipine nanosuspension by anti-solvent microprecipitation-high-pressure homogenization method for enhancing dissolution rate and oral bioavailability. *Drug Deliv. Transl. Res.* 3 (5): 384–391. doi: 10.1007/S13346-012-0081-3/TABLES/2.

150 Tanuku, S., Velisila, D., Thatraju, D. et al. (2024). Nanoemulsion formulation strategies for enhanced drug delivery. *J. Pharma Insights Res.* 2 (4): 125–138. doi: 10.69613/3F8M9151.

151 Rossier, B., Jordan, O., Allémann, E., and Rodríguez-Nogales, C. (2024). Nanocrystals and nanosuspensions: an exploration from classic formulations to advanced drug delivery systems. *Drug Deliv. Transl. Res.* 14 (12): 3438. doi: 10.1007/S13346-024-01559-0.

152 Pınar, S.G., Oktay, A.N., Karaküçük, A.E., and Çelebi, N. (2023). Formulation strategies of nanosuspensions for various administration routes. *Pharmaceutics* 15 (5): 1520. doi: 10.3390/PHARMACEUTICS15051520.

153 Helmy, S.A., El-Morsi, R.M., Helmy, S.A.M., and El-Masry, S.M. (2022). Towards novel nano-based vaccine platforms for SARS-CoV-2 and its variants of concern: advances, challenges and limitations. *J. Drug Deliv. Sci. Technol.* 76: 103762. doi: 10.1016/J.JDDST.2022.103762.

154 Keck, C.M. and Müller, R.H. (2013). Nanotoxicological classification system (NCS) – a guide for the risk-benefit assessment of nanoparticulate drug delivery systems. *Eur. J. Pharm. Biopharm.* 84 (3): 445–448.

155 Müller, R.H., Gohla, S., and Keck, C.M. (2011). State of the art of nanocrystals – special features, production, nanotoxicology aspects and intracellular delivery. *Eur. J. Pharm. Biopharm.* 78 (1): 1–9. doi: 10.1016/J.EJPB.2011.01.007.

156 Keck, C.M., Kobierski, S., Mauludin, R., and Müller, R.H. (2012). Novel top-down technologies: effective production of ultra-fine. In: *Drug Delivery Strategies for Poorly Water-Soluble Drugs* (ed. D. Douroumis and A. Fahr), 247–263. Wiley. doi: 10.1002/9781118444726.CH8;CTYPE:STRING:BOOK.

157 Chary, P.S., Shaikh, S., Bhavana, V. et al. (2024). Emerging role of nanocrystals in pharmaceutical applications: a review of regulatory aspects and drug development process. *Appl. Mater. Today* 40: 102334. doi: 10.1016/J.APMT.2024.102334.

158 Gigliobianco, M.R., Casadidio, C., Censi, R., and Di Martino, P. (2018). Nanocrystals of poorly soluble drugs: drug bioavailability and physicochemical stability. *Pharmaceutics* 10 (3): 134. doi: 10.3390/PHARMACEUTICS10030134.

159 Peltonen, L., Strachan, C., Rades, T. et al. (2015). Understanding critical quality attributes for nanocrystals from preparation to delivery. *Molecules* 20 (12): 22286–22300. doi: 10.3390/MOLECULES201219851.

160 Jari Litany, R.I. and Praseetha, P.K. (2022). Tiny tots for a big-league in wound repair: tools for tissue regeneration by nanotechniques of today. *J. Control. Release* 349: 443–459. doi: 10.1016/J.JCONREL.2022.07.005.

161 Ahire, E., Thakkar, S., Darshanwad, M., and Misra, M. (2018). Parenteral nanosuspensions: a brief review from solubility enhancement to more novel and specific applications. *Acta Pharm. Sin. B* 8 (5): 733–755. doi: 10.1016/J.APSB.2018.07.011.

162 Wu, J. (2021). The enhanced permeability and retention (EPR) effect: the significance of the concept and methods to enhance its application. *J. Pers. Med.* 11 (8): 771. doi: 10.3390/JPM11080771.

163 Jaiswal, M., Dudhe, R., and Sharma, P.K. (2014). Nanoemulsion: an advanced mode of drug delivery system. *3 Biotech* 5 (2): 123. doi: 10.1007/S13205-014-0214-0.

164 Choudhury, H., Pandey, M., Gorain, B. et al. (2019). Nanoemulsions as effective carriers for the treatment of lung cancer. In: *Nanotechnology-Based Targeted Drug Delivery Systems for Lung Cancer* (ed. P. Kesharwani), 217–247. Elsevier. doi: 10.1016/B978-0-12-815720-6.00009-5.

165 Kumar, M., Bishnoi, R.S., Shukla, A.K., and Jain, C.P. (2019). Techniques for formulation of nanoemulsion drug delivery system: a review. *Prev. Nutr. Food Sci.* 24 (3): 225. doi: 10.3746/PNF.2019.24.3.225.

166 Goel, S., Sachdeva, M., and Agarwal, V. (2019). Nanosuspension technology: recent patents on drug delivery and their characterizations. *Recent Pat. Drug Deliv. Formul.* 13 (2): 91. doi: 10.2174/1872211313666190614151615.

167 Mayo Clinic. (2025). Aprepitant (oral route). https://www.mayoclinic.org/drugs-supplements/aprepitant-oral-route/description/drg-20066897. 1 April 2026. (accessed 16 April 2025).

168 Malamataris, S. (2018). Pharmaceutical nanocrystals: production by wet milling and applications. *Drug Discov. Today*. https://www.academia.edu/56324213/Pharmaceutical_nanocrystals_production_by_wet_milling_and_applications (accessed 16 April 2025).

169 NA (2007). Megestrol acetate NCD oral suspension–Par Pharmaceutical: megestrol acetate nanocrystal dispersion oral suspension, PAR 100.2, PAR-100.2. *Drugs R D* 8 (6): 403–406. doi: 10.2165/00126839-200708060-00009.

170 INVEGA SUSTENNA®. (2006). (paliperidone palmitate) extended-release injectable suspension, for intramuscular use highlights of prescribing information, U.S. Food and Drug Administration. Available from: https://www.accessdata.fda.gov/drugsatfda_docs/label/2024/022264s037lbledt.pdf.

171 Tadros, T., Izquierdo, P., Esquena, J., and Solans, C. (2016). Cosmetic nanoemulsion. *Adv. Colloid Interf. Sci.* 108–109: 303–318. doi: 10.1016/J.CIS.2003.10.023.

172 ESTRASORB® (estradiol topical emulsion). (2025). Highlights of prescribing information, U.S. Food and Drug Administration; 1975. Available from: https://www.accessdata.fda.gov/drugsatfda_docs/label/2024/021371s009lbl.pdf (accessed 16 April 2025).

173 Lallemand, F., Schmitt, M., Bourges, J.L. et al. (2017). Cyclosporine A delivery to the eye: a comprehensive review of academic and industrial efforts. *Eur. J. Pharm. Biopharm.* 117: 14–28. doi: 10.1016/J.EJPB.2017.03.006.

174 RESTASIS MULTIDOSE®. (cyclosporine ophthalmic emulsion) 0.05%, for topical ophthalmic use, highlights of prescribing information full prescribing information, U.S. Food and Drug Administration; 1983. Available from: https://www.accessdata.fda.gov/drugsatfda_docs/label/2025/050790s031lbl.pdf pdf.

175 Motheram, R. and Williams, G.C. (2013). Clevidipine emulsion formulations containing antimicrobial agents. https://patents.google.com/patent/US11103490B2/en.

176 FDA. (2001). Drug approval package: Diprivan (Propofol) NDA #019627s035. Available from: https://www.accessdata.fda.gov/drugsatfda_docs/nda/96/019627_S027_Diprivan_APPROVAL_PACKAGE.pdf (accessed 16 April 2025).

177 Cèbe, A., Dessane, B., Gohier, P. et al. (2023). Hospital production of sterile 2% propofol nanoemulsion: proof of concept. *Pharmaceutics* 15 (3): 905. doi: 10.3390/PHARMACEUTICS15030905.

178 Pramanick S., Gurjar M.K., and Mehta S.S. (2014). Pharmaceutical composition of propofol. WO patent WO2014033751A2. 2014 Mar 6. Available from: https://patentimages.storage.googleapis.com/2c/71/25/52ec98f1cec1cb/WO2014033751A2.pdf.

179 Abdelrasoul, A., Doan, H., Lohi, A., and Cheng, C.H. (2018). The effect of contaminated particle sphericity and size on membrane fouling in cross flow ultrafiltration. *Environ. Technol.* 39 (2): 203–220. doi: 10.1080/09593330.2017.1297490.

180 Eslami, F. and Elliott, J.A.W. (2014). Role of precipitating solute curvature on microdrops and nanodrops during concentrating processes: the Nonideal Ostwald-Freundlich equation. *J. Phys. Chem. B* 118 (50): 14675–14686.

181 Yamashita, Y., Miyahara, R., and Sakamoto, K. (2017). Emulsion and emulsification technology. In: *Cosmetic Science and Technology Theoretical Principles and Applications* (ed. K. Sakamoto, R.Y. Lochhead, H.I. Maibach, and Y. Yamashita), 489–506. Elsevier. doi: 10.1016/B978-0-12-802005-0.00028-8.

182 Sun, S., Zhang, X., Yang, Q. et al. (2018). Cuprous oxide (Cu_2O) crystals with tailored architectures: a comprehensive review on synthesis, fundamental properties, functional modifications and applications. *Prog. Mater. Sci.* 96: 111–173. doi: 10.1016/J.PMATSCI.2018.03.006.

183 Zhang, Z., Wang, Z., He, S. et al. (2015). Redox reaction induced Ostwald ripening for size- and shape-focusing of palladium nanocrystals. *Chem. Sci.* 6 (9): 5197. doi: 10.1039/C5SC01787D.

184 Deshpande, A., Mohamed, M., Daftardar, S.B. et al. (2017). Solid lipid nanoparticles in drug delivery: opportunities and challenges. In: *Emerging Nanotechnologies for Diagnostics, Drug Delivery and Medical Devices* (ed. A.K. Mitra, K. Cholkar, and A. Mandal), 291–330. Elsevier. doi: 10.1016/B978-0-323-42978-8.00012-7.

185 Fernandes, C., Jathar, M., Sawant, B.K.S., and Warde, T. (2023). Scale-up of nanoparticle manufacturing process. In: *Pharmaceutical Process Engineering and Scale-up Principles* (ed. A.B. Jindal), 173–203. Springer. doi: 10.1007/978-3-031-31380-6_12.

186 Paliwal, R., Babu, R.J., and Palakurthi, S. (2014). Nanomedicine scale-up technologies: feasibilities and challenges. *AAPS PharmSciTech* 15 (6): 1527. doi: 10.1208/S12249-014-0177-9.

187 Junghanns, J.U.A.H. and Müller, R.H. (2008). Nanocrystal technology, drug delivery and clinical applications. *Int. J. Nanomedicine* 3 (3): 295. doi: 10.2147/IJN.S595.

188 Zhang, X., Li, L.C., and Mao, S. (2014). Nanosuspensions of poorly water soluble drugs prepared by top-down technologies. *Curr. Pharm. Des.* 20 (3): 388–407. doi: 10.2174/13816128113199990401.

189 Colombo, A.P., Briançon, S., Lieto, J., and Fessi, H. (2001). Project, design, and use of a pilot plant for nanocapsule production. *Drug Dev. Ind. Pharm.* 27 (10): 1063–1072. doi: 10.1081/DDC-100108369.

190 Negi, A., Nimbkar, S., and Moses, J.A. (2023). Engineering inhalable therapeutic particles: conventional and emerging approaches. *Pharmaceutics* 15 (12): 2706. doi: 10.3390/PHARMACEUTICS15122706.

191 Wang, M., Chen, S., Zhou, W. et al. (2020). Algal cell lysis by bacteria: a review and comparison to conventional methods. *Algal Res.* 46. doi: 10.1016/J.ALGAL.2020.101794.

192 Desai, N. (2012). Challenges in development of nanoparticle-based therapeutics. *AAPS J.* 14 (2): 282. doi: 10.1208/S12248-012-9339-4.

193 Gupta, Y.K., Singht, G.N., Dinda, A.K., and Pradhan, A.K. (2019). Guidelines for evaluation of nanopharmaceuticals in India. *Guidel. Nanopharm.* 12: 1–49.

194 Sainz, V., Conniot, J., Matos, A.I. et al. (2015). Regulatory aspects on nanomedicines. *Biochem. Biophys. Res. Commun.* 468 (3): 504–510. doi: 10.1016/J.BBRC.2015.08.023.

195 Foulkes, R., Man, E., Thind, J. et al. (2020). The regulation of nanomaterials and nanomedicines for clinical application: current and future perspectives. *Biomater. Sci.* 8 (17): 4653–4664. doi: 10.1039/D0BM00558D.

196 Gaspar, R. (2007). Regulatory issues surrounding nanomedicines: setting the scene for the next generation of nanopharmaceuticals. *Nanomedicine* 2 (2): 143–147. doi: 10.2217/17435889.2.2.143.

197 Soares, S., Sousa, J., Pais, A., and Vitorino, C. (2018). Nanomedicine: principles, properties, and regulatory issues. *Front. Chem.* 6 (AUG): 360. doi: 10.3389/FCHEM.2018.00360.

198 Khlebtsov, N. and Dykmana, L. (2011). Biodistribution and toxicity of engineered gold nanoparticles: a review of in vitro and in vivo studies. *Chem. Soc. Rev.* 40 (3): 1647–1671. doi: 10.1039/C0CS00018C.

199 Manke, A., Wang, L., and Rojanasakul, Y. (2013). Mechanisms of nanoparticle-induced oxidative stress and toxicity. *Biomed. Res. Int.* 2013: doi: 10.1155/2013/942916.

200 Peltonen, L. (2018). Practical guidelines for the characterization and quality control of pure drug nanoparticles and nano-cocrystals in the pharmaceutical industry. *Adv. Drug Deliv. Rev.* 131: 101–115. doi: 10.1016/J.ADDR.2018.06.009.

201 Haripriyaa, M. and Suthindhiran, K. (2023). Pharmacokinetics of nanoparticles: current knowledge, future directions and its implications in drug delivery. *Futur. J. Pharm. Sci.* 9 (1): 1–26. doi: 10.1186/S43094-023-00569-Y.

202 El-Tanani, M., Satyam, S.M., Rabbani, S.A. et al. (2025). Revolutionizing drug delivery: the impact of advanced materials science and technology on precision medicine. *Pharmaceutics* 17 (3): 375. doi: 10.3390/PHARMACEUTICS17030375.

203 Hajebi, S., Chamanara, M., Nasiri, S.S. et al. (2024). Advances in stimuli-responsive gold nanorods for drug-delivery and targeted therapy systems. *Biomed. Pharmacother.* 180: 117493. doi: 10.1016/J.BIOPHA.2024.117493.

204 Srivalli, K.M.R. and Mishra, B. (2014). Drug nanocrystals: a way toward scale-up. *Saudi Pharm. J.* 24 (4): 86–404. doi: 10.1016/j.jsps.2014.04.007.

205 De Jong, W.H., Geertsma, R.E., and Borchard, G. (2022). Regulatory safety evaluation of nanomedical products: key issues to refine. *Drug Deliv. Transl. Res.* 12 (9): 2042–2047. doi: 10.1007/s13346-022-01208-4.

206 SEHS. (2024). View of AI-driven design and optimization of nanoparticle-based drug delivery systems. https://li01.tci-thaijo.org/index.php/sehs/article/view/264345/179293 (accessed 16 April 2025).

4

Pharmaceutical Cocrystals

Ghaida Mustafa[1,2], *Shahzeb Khan*[1], *and Anant Paradkar*[1]

[1] *University of Bradford, Centre for Pharmaceutical Engineering and Skin Sciences, School of Pharmacy, Richmond Road, West Yorkshire, Bradford, BD7 1DP England, UK*
[2] *University of Khartoum, Department of Pharmaceutics, Faculty of Pharmacy, Al-Qasr Avenue, Khartoum State, Khartoum, Sudan*

4.1 Introduction

4.1.1 Pharmaceutical Cocrystals and Importance in the Pharmaceutical Industry

Pharmaceutical cocrystals are molecular complexes of drug/s and suitable generally recognized as safe (GRAS) coformer/s that are solids at ambient temperatures, in stochiometric ratio which are joined by noncovalent, nonionic intermolecular interactions especially hydrogen bonding [1].

Pharmaceutical cocrystals are innovative crystalline structures engineered to tailor physicochemical and biopharmaceutical attributes of drugs to develop stable, efficacious, and robust pharmaceutical products. Cocrystallization approach can significantly enhance the aqueous solubility and ultimately bioavailability of poorly water-soluble drugs, which represents approximately half of the orally administered drugs [1]. In addition, cocrystals have been reported to enhance drugs' stability and to tailor drug release profiles [1]. Moreover, formation of drug–coformer cocrystal results in modification of physical properties of the parent drug such as melting point; therefore, it can ease drug processing and manufacturing [2]. Furthermore, these innovative crystalline structures can allow generation of cost-effective products while maintaining regulatory requirements. Regulatory bodies such as US Food and Drug Administration (USFDA) consider them intermediate pharmaceutical products rather than new drugs [3, 4].

4.1.2 Fundamentals of Pharmaceutical Cocrystals

4.1.2.1 Historical Background

The origin of cocrystals dates to 1844, when Wöhler, during the study in quinones, synthesized quinhydrone complex. Upon the introduction of X-ray analysis,

Quinone (**1**)
(mp 113–115 °C)

Hydroquinone (**2**)
(mp 172–175 °C)

Quinhydrone (**3**)
(mp 170–172 °C)

Figure 4.1 Chemical structures of quinone, hydroquinone, and quinhydrone. *Source:* Stahly [5] / with permission of American Chemical Society.

the complex was found to be quinone and hydroquinone in a stochiometric ratio of 1 (Figure 4.1) [6].

Initially, different designations are given to these complexes such as adducts, molecular complexes, and solid-state complexes [1]. Even though the first cocrystal patent was in 1937 [7], the term cocrystal was first used in 1967 as a suggestion for description of 9-methyl adenine and 1-methyl thymine complex [1]. It's worth noting that, at the beginning of twenty-first century, the debate on cocrystal definition was initiated, and many definitions were provided. Karagianni et al., in their review, addressed these arguments [1].

In 2013, USFDA defined cocrystals as "solids which are crystalline materials composed of two or more molecules in the same crystal lattice" [8]. Based on this definition, cocrystals are considered as "drug product intermediates" and not new active pharmaceutical ingredients (APIs), limiting the scope of coformer to the pharmacologically inert compounds [2]. Recently, the European Medicines Agency (EMA) has provided a reflection paper to consider cocrystals as "new active substance" under the condition that they are safe and efficacious. This EMA consideration opens new doors for the development of such class of compounds [9]. They have become a delicious dish on the table of the pharmaceutical industry; extensive research has been conducted over the last 25 years, and many patents have been granted [10].

4.1.2.2 Basic Concept and Definitions

During the pre-formulation studies, physicochemical properties of APIs such as physical form, crystallinity, solubility, dissolution rate, and stability are characterized to select the dosage form in which API can be administered. Most drugs are crystalline solids that are more stable and easily obtained in high purity with reproducible properties compared to amorphous drugs. A major challenge in the development of crystalline pharmaceuticals is the limited aqueous solubility observed in a significant proportion of APIs. Notably, 40% of the marketed drugs and those under development, and 90% of APIs that are in the production line, suffer from limited aqueous solubility [11]. These poorly soluble drug molecules captured the attention of researchers to overcome challenges associated with their poor bioavailability, particularly when administered orally. Therefore, several strategies have been established to address the low aqueous solubility issues including micronization, solubilization in cosolvents (cosolvents–surfactant mixtures), micellar solutions, amorphous solid dispersions, mesoporous silica particles, complexation with cyclodextrins, salt formation, and the application of lipidic delivery systems for lipophilic drugs [12]. However, the successful implementation of such solubility-enhancing approaches relies on specific physicochemical nature of the drug molecules [13]. Although micronization of APIs improves their dissolution rate via increased surface area, it has nothing to do with equilibrium solubility, and for very slightly soluble drugs, this approach is often efficient to significantly enhance the bioavailability. Furthermore, the resultant ultrafine particles are more prone to van der Waals and electrostatic interactions, leading to a reduction of the effective surface area for dissolution and limitation of improvement in drug's bioavailability. Cosolvency is a suitable solubilization approach for nonionizable drugs via reduction of system polarity. However, this method has got limitations related to the added solvent/s such as cellular toxicity. Amorphous solid dispersions are dispersed systems in which the amorphous drug is distributed in a polymeric matrix to enhance its aqueous solubility. These dispersions have the drawback of susceptibility to recrystallization during storage, leading to short-term storage stability [14]. Solubilization methods such as micellar systems require reasonable miscibility and compatibility of APIs in a restricted scope of pharmaceutically permissible excipients, whereas many APIs entering development have weak ionization or neutral nature that constrains salt formation capability to improve their dissolution rate [13]. Other solubilization methods have their limitations that have been addressed nicely in the cited review [12].

Crystal engineering strategies emerge as a surrogate and potentially exuberant method for enhancing solubility, dissolution rate, and consequent bioavailability of poorly soluble drugs, while retaining suitable physicochemical stability of them. The technique is based on the tailoring of crystalline solids in terms of physicochemical characteristics to develop efficacious, stable, and robust pharmaceutical products. The recent crystal engineering science originated as a technique for investigation of solid materials' reactions at the molecular level. This field has progressed rapidly, especially with the introduction of the modern crystallographic techniques that enable understanding, monitoring, and controlling of the crystal design. Crystal engineering encloses two approaches (Figure 4.2). Single-component methods manipulate the crystalline material attributes via converting crystal shape (habit) or crystal structure (polymorph) [13]. The second crystal engineering approach is a multicomponent strategy through formation of salts, solvates, hydrates, and cocrystals [13].

Pharmaceutical cocrystallization is a feasible surrogate crystal engineering approach to achieve the desired physicochemical properties of APIs when other crystal manipulation strategies are not applicable. Pharmaceutical cocrystals are molecular complexes of API/s and suitable GRAS coformer in stochiometric ratio which are connected mainly by hydrogen bonding (86% of cocrystals). At ambient temperatures, cocrystals exist as solids [15]. USFDA defines coformer as "a component that interacts non-ionically with the API in the crystal lattice, that is not a solvent (including water) and is typically non-volatile" [16].

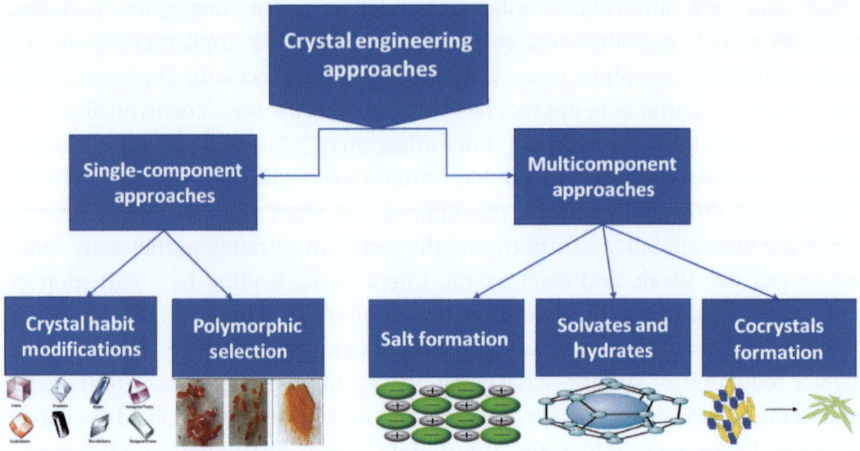

Figure 4.2 Crystal engineering approaches.

4.1.2.3 Differences Between Cocrystals, Salts, and Polymorphs

Cocrystals are unique crystalline solids; the neutral nature of their constituents distinguishes them from salts whose crystal lattices are composed of ionized molecules. They are also distant from polymorphs, as their crystal lattices are made up of API and suitable coformers instead of API alone. Figure 4.3 shows the distinction between cocrystals and other material engineering approaches.

4.1.2.4 Selection of Coformers

The critical parameter for cocrystal formation is the presence of functional groups in API and coformer with capability of hydrogen bond donation and acceptance. As per USFDA guidance, the pK_a difference of 1 or >1 with the proton transfer between API and excipient is associated with salt formation, and pK_a difference of <1 without

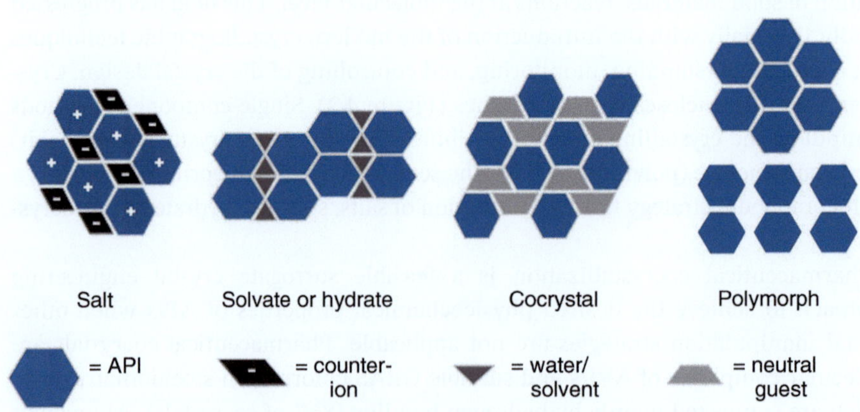

Figure 4.3 Pictures showing difference between cocrystals and other common solid-state pharmaceutical strategies. *Source:* Schultheiss and Newman [10] / with permission of American Chemical Society.

proton transfer is related to cocrystal formation [17]. However, the process of proton transfer is not only pK_a dependent, but also affected by temperature and crystal environment; thus, no clear boundary exists between salt and cocrystals, i.e. they form a continuum as reported in the case of escitalopram oxalate that has ability to cocrystallize with oxalic acid and can behave as a salt or as a cocrystal [18].

As a rule of thumb, compounds that exhibit polymorphism have a great tendency to form cocrystals due to their structural flexibility. Structural flexibility of API alone is not a guarantee for cocrystal formation, as other parameters are needed, and these include number and assortment of hydrogen bonds, solubility parameters, small molecular weight, lattice energy, and pK_a. The coformer is selected based on capability of hydrogen bond formation, potential to elucidate it at the molecular level, and its safety profile [19]. Additional information can be found in the cited review [20].

4.2 Synthon Theory and Prediction Tools in Cocrystal Design

4.2.1 Definition, Importance, and Types of Synthons in Pharmaceutical Cocrystals

A synthon is an essential unit in crystal engineering, demonstrating a repeated structural pattern that assists in predicting and constructing the overall crystal structure. The synthon occurs in an enormous number of cases in which a particular set of molecular functions is a good or robust synthon [21]. When special regular repetition of crystal configuration occurs, the intermolecular interactions can be stated as supramolecular synthons. They are classified into two types (Figure 4.4) [23]:

1) **Supramolecular homosynthons:** They entail arrangements of similar functional groups (for instance, acid–acid).
2) **Supramolecular heterosynthons:** They are united by distinct but complementary functionalities (for instance, acid–amide).

In pharmaceutical cocrystal structures, the principal supramolecular synthons imply interactions such as C=O· · ·H—O, C=O· · ·H, C=O· · ·H—N hydrogen bonds. These intermolecular interactions are observed with functional groups such as carboxylic acids, alcohols, carbohydrates, and amides [22, 24].

4.2.2 Computational Approaches and Software Tools for Cocrystal Screening

For successful design of pharmaceutical cocrystals, it is crucial to select appropriate coformers and understand the interactions between cocrystal components. Some cocrystal preparation methods are used as screening tools to predict cocrystal formations such as solid- and liquid-assisted grinding [25, 26]. These techniques have led to development of cocrystals with enhanced formulation properties;

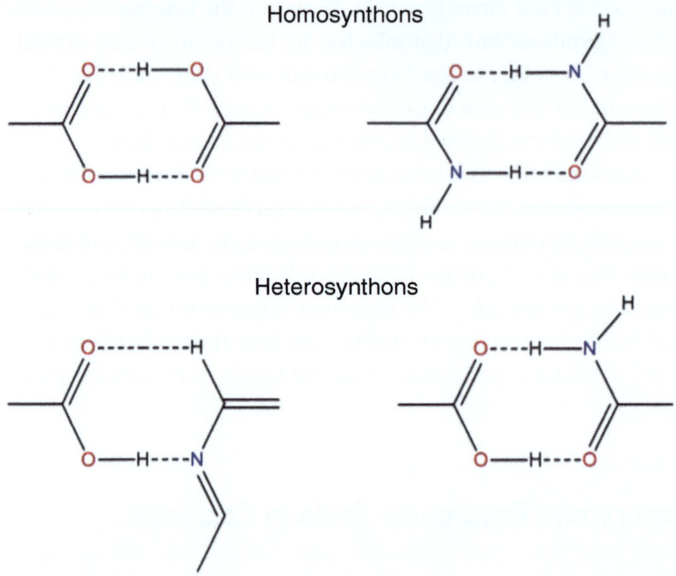

Figure 4.4 Common homosynthons and heterosynthons utilized in cocrystal formation. *Source:* Banerjee et al. [22] / with permission of Elsevier.

however, they have drawbacks of requiring more practical work, relying on trial-and-error approach, and often failure. One of the suggested strategies to boost the area of cocrystals prediction is the combination of conventional prediction methods with high-throughput screening tools and software [27–29]. Adoption of such approach will assist researchers to focus on the most promising candidates for synthesis and improve drug development process. Figure 4.5 shows a suggested roadmap for rational cocrystal design.

A range of in silico tools is presently available for predicting cocrystal formation. Some of these methods not only assess the probability of cocrystal formation but also offer insights into the potential crystal structures of the resulting cocrystals, as demonstrated in Figure 4.6. This subsection provides a summary of the most applied computational methods along with their prediction approach and software packages and algorithms that support them. It is worth noting that there is an overlap between these prediction strategies, and multiple approaches are often adopted for more reliable results.

4.2.2.1 Quantum-Mechanical Methods

Quantum-mechanical (QM) methods are vital tools for predicting cocrystal formation that rely on the assessment of intermolecular interactions such as hydrogen bonding and electrostatic forces [31, 32]. QM methods calculations are based on techniques such as density functional theory (DFT) and molecular electrostatic potential. These calculations are used to estimate stability and crystal structures via evaluation of the interaction energies and prediction of molecular complementarity [33, 34]. Prediction capabilities of QM methods are further enhanced by periodic

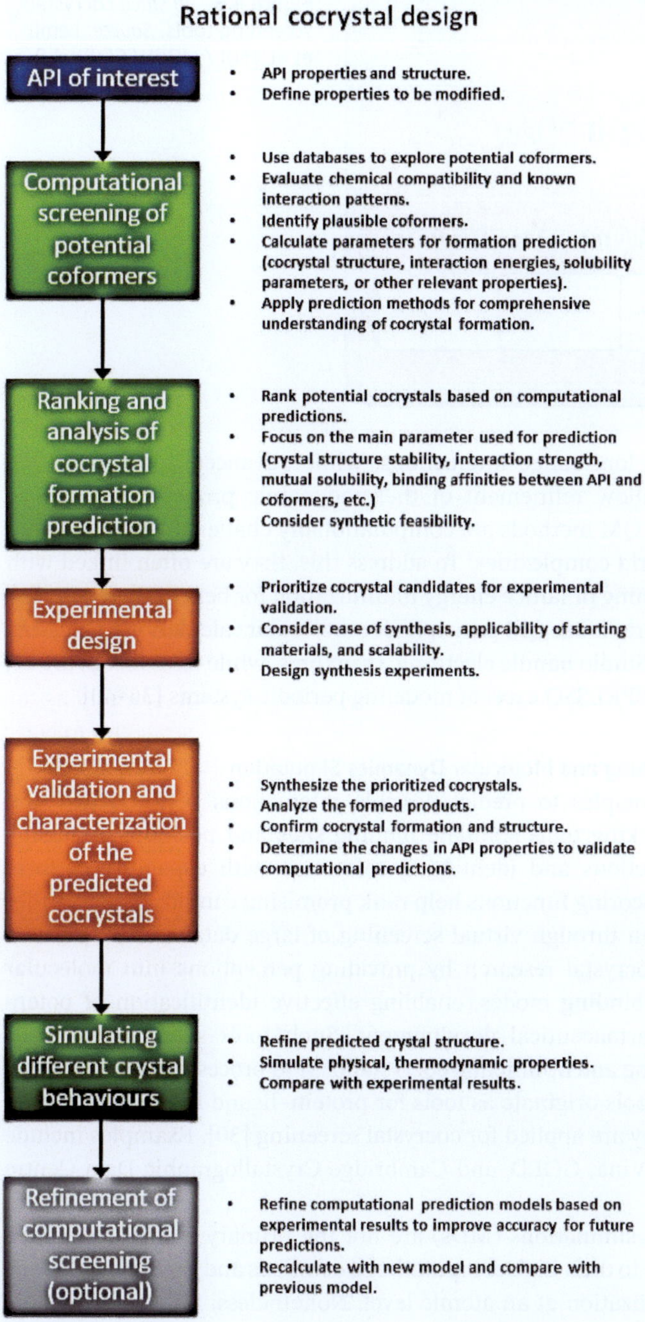

Figure 4.5 A roadmap for rational cocrystal design. *Source:* Lemli et al. [30] / MDPI / CC BY 4.0.

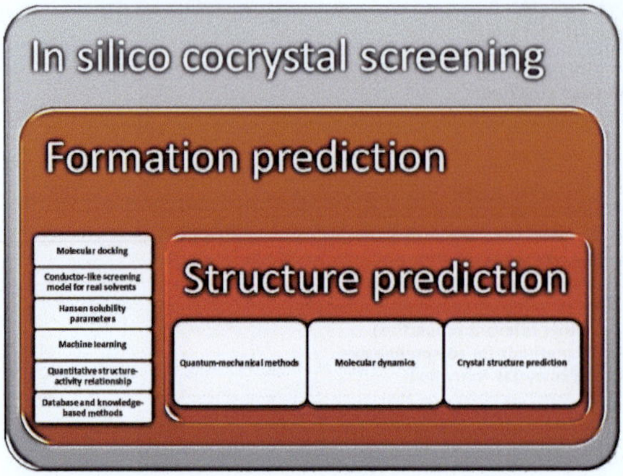

Figure 4.6 *In silico* cocrystals screening tools. *Source:* Lemli et al. [30] / MDPI / CC BY 4.0.

DFT, which considers long-range interactions, while advanced approaches like phonon calculations allow refinement of thermodynamic property predictions. Despite their accuracy, QM methods are computationally challenging and may not account for all real-world complexities. To address this, they are often linked with tools like machine learning or lattice energy minimization for better efficiency [35].

Several software programs use QM principles for molecular calculations. Gaussian and BIOVIA Materials Studio handle electronic structures, while tools like CASTEP, VASP, and Quantum ESPRESSO excel at modeling periodic systems [36–40].

4.2.2.2 Molecular Docking and Molecular Dynamics Simulation

They rely on QM principles to predict possible interactions between API and coformer [41, 42]. Docking tools simulate binding sites and positions, assessing the strength of interactions and identifying coformers with capability to form stable cocrystals [43]. Scoring functions help rank promising candidates for further experimental evaluation through virtual screening of large databases [44]. These methods modernize cocrystal research by providing perceptions into molecular complementarity and binding modes, enabling effective identification of potential coformers for pharmaceutical development. Such tools significantly assist researchers in evaluating and optimizing cocrystallization processes [45].

Molecular docking tools originate as tools for protein–ligand interaction predictions, but currently they are applied for cocrystal screening [30]. Examples include AutoDock, AutoDock Vina, GOLD, and Cambridge Crystallographic Data Centre (CCDC) [46–49].

Molecular dynamics simulations (MDs) are not the primary method for direct cocrystal screening due to their high computational demands and time requirements for modeling cocrystallization at an atomic level. Nonetheless, they become basic for recognizing potential cocrystal structures by applying methods such as lattice energy minimization or docking simulations. Throughout analysis of parameters such as docking poses, conformational flexibility, and intermolecular interactions, MDs refine these crystalline structures. Effective MD simulations can be achieved by tools such as Desmond and LAMMPS, which particularly offer understandings into the dynamic behavior and stability of cocrystal structures [50, 51]. These tools supplement other prediction methods by assisting researchers in picking the most promising candidates for further study and experimental validation.

4.2.2.3 Crystal Structure Prediction and Lattice Energy Minimization

Crystal structure prediction (CSP) uses chemical formula of API and coformer to predict possible crystal structures [52]. They are then ranked according to their lattice energies to identify the structure with the minimum lattice energy as the most stable form. An example of CSP software is DMACRYS [53, 54].

It is worth noting that, Cambridge Crystallographic Data Centre (CCDC) organizes a blind test, as a standard initiative aimed at evaluating and comparing the performance of CSP methods. The primary objective of this investigation is to judge the accuracy and effectiveness of various computational approaches in predicting the crystal structures of previously uncharacterized molecules [55].

4.2.2.4 Semi-empirical Screening Methods

COSMO-RS (conductor-like screening model for real solvents) is one of the semi-empirical approaches, which integrates quantum chemical calculations with empirically derived parameters to estimate molecular interactions [56]. The COSMO-RS prediction relies on evaluation of critical parameters such as solubility and miscibility using statistical thermodynamics. This allows screening of coformer/s without demanding extensive molecular simulations.

Remarkably, COSMO-RS is suitable for large-scale predictive assessments of cocrystal formation through computing the excess enthalpy of mixture between API and coformer [57]. This enthalpy acts as a key sign of the probability of successful cocrystallization, therefore rationalizing the process of identifying potential candidates.

For prediction of intermolecular compatibility, a semi-empirical principle known as Hansen solubility parameters (HSPs) is used [58]. The method evaluates the key intermolecular interactions affecting solubility. These critical parameters are dispersion, polarity, and hydrogen bonding. By integrating experimental data with theoretical concepts, such as the geometric mean derived from corresponding states theory, HSPs provide insights into the miscibility of API and potential coformers. Components with similar HSP values are more liable to demonstrate compatibility and form stable cocrystals.

Identification of suitable cocrystal components is facilitated by HSPiP software that supports the computation and visualization of HSPs [59]. The reliability of HSPs predictions is affected by the quality and scope of data in available databases, which may vary in content and regularity of updates. Despite these restrictions, HSPs remain a valuable tool for reviewing the potential for successful cocrystallization.

4.2.2.5 Machine Learning and Data-Driven Techniques

Machine learning (ML) performs cocrystal prediction by utilization of existing database for cocrystal structures and properties, as well as failed experiments. Algorithms such as Random Forests extract patterns and relationships between molecular properties and cocrystal structures, allowing researchers to predict the crystal structures of new materials based only on their molecular composition [60]. Scikit-learn (Machine Learning in Python) is an example of open-access database that allows implementations of commonly used algorithms for developing and applying ML models [61].

QSAR (quantitative structure–activity relationship) models are conventionally used in drug activity prediction. They can be applied to estimate cocrystal formation potential based on relationships between cocrystal components' molecular structures [62]. KNIME Analytics Platform is an example of free database that allows QSAR modeling [63].

An effective cocrystal prediction strategy can be achieved by integrating ML and QSAR methods [64]. QSAR offers the initial understanding of molecular structure and the factors influencing cocrystal formation, while ML utilizes the generated data to develop advanced predictive models, enhancing the accuracy and efficiency of the cocrystal prediction process.

4.2.2.6 Database and Knowledge-Based Approaches

Virtual screening is a computational approach in cocrystal discovery that identifies coformers capable of forming stable cocrystals with a specific API. This process assesses molecular attributes of cocrystal components such as shape and size that are necessary to ensure efficient packing in cocrystal lattice. Chemical functional groups on the API and coformer are also analyzed for their ability to enable intermolecular interactions such as hydrogen bonding and π–π stacking, which are essential for cocrystal stability. Other properties, such as solubility and hygroscopicity, may be considered to refine the screening results further. These techniques provide a rational method for ranking coformers with favorable characteristics and rationalize the identification process in cocrystal design studies.

CCDC software package, containing Mercury, ConQuest, and CSD Python API, is a vital tool for identifying cocrystal candidates by virtual screening [49, 65–67]. There are valuable databases to assist in ranking potential coformers for cocrystal discovery. For instance, DrugBank allows identification of biologically active coformers, while Crystallography Open Database (COD) contains crystal structures of organic and inorganic substances obtained by several elucidation techniques [68, 69].

4.3 Synthesis of Pharmaceutical Cocrystals

Several innovative methods have been reported in the literature for cocrystal preparations. Generally, these methods can be divided into solid-based (e.g. grinding and sonication), solvent-based (e.g. slurry, spray drying, and solution crystallization), and supercritical fluids-based approaches, as shown in Figure 4.7. Other methods include thermal, mechanical, and precipitation-based approaches. Each of these approaches has its advantages, limitations, and challenges. For instance, solvent evaporation method is a simple method, but the selection of the

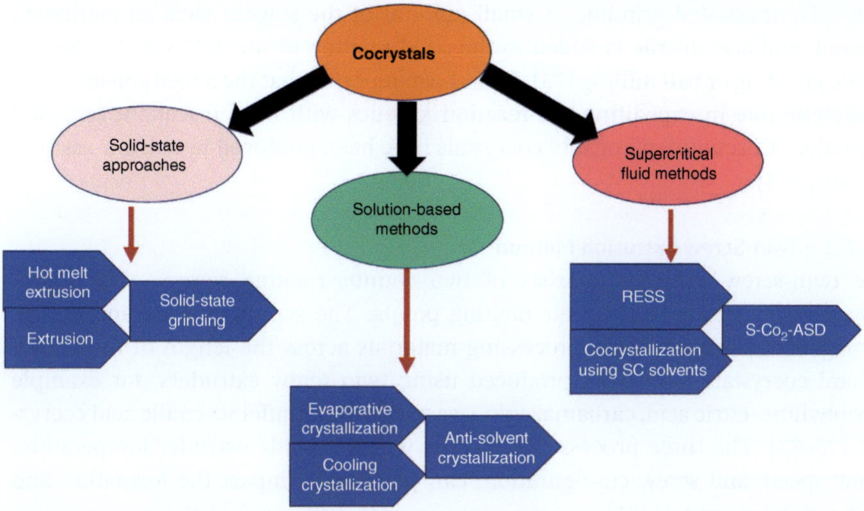

Figure 4.7 Approaches for production of cocrystals. RESS: rapid expansion of supercritical solutions. ASD: assisted spray drying.

solvent is crucial. Another example is applying mechanical cocrystal preparation approaches can result in an amorphous formation. This section provides details of such cocrystallization methods.

4.3.1 Solid-State Approaches

These methods are based on the principle of mixing of two APIs or API with the coformer under controlled experimental and process conditions. The cocrystals can spontaneously be produced through mixing. However, in some cases, pre-grinding of the individual components to achieve the adequate particle size, followed by mixing, can result in stable cocrystals [70]. The process parameters that impact the production of cocrystals depend on the physicochemical attributes of the processing materials. In contrast to spontaneous formation of the cocrystals such as isoniazid and benzoic acid, some of the crystallization reactions have been impacted by humidity and a range of temperatures [71, 72]. The following methods are based on the principle of solid-state processing of the API and coformers.

4.3.1.1 Dry and Solvent-Assisted Grinding

In dry grinding methods, the API and the coformer are properly ground by mechanical pressure being applied using pestle and mortar or through ball mill. Based on the physicochemical attributes of the API and coformer, the cocrystal formation depends on the grinding time. Several cocrystals have been produced through solid-state grinding or neat grinding. For instance, solid-state grinding was used for preparation of pyrazinamide–diflunisal cocrystals [58]. This is a simple and quick method for production of cocrystals. However, some of the issues associated with this approach include incomplete formation of cocrystals, crystal defects, and purification.

In solvent-assisted grinding, a small amount of the solvent such as methanol, ethanol, and acetonitrile is added to the solid mixture of the API/s and coformer before grinding or ball milling [73]. It has been reported that the added solvent plays a catalytic role in expediting the reaction kinetics with the subsequent cocrystal formation. Curcumin–pyrogallol cocrystals have been produced by solvent-assisted grinding [74].

4.3.1.2 Twin-Screw Extrusion Method

The twin-screw extruder consists of two counter-rotating screws, which processes the materials below their melting points. The screws provide and ensure homogeneous mixing of the processing materials across the length of the barrel. Several cocrystals have been produced using twin-screw extruders, for example theophylline–citric acid, carbamazepine–saccharin, and caffeine–oxalic acid cocrystals [75–77]. The three process parameters, which include extruder temperature, screw speed, and screw configuration, can potentially impact the formation and purity of the cocrystals [77].

4.3.1.3 Hot Melt Extrusion Method

Hot melt extrusion (HME) has a noticeable attraction to produce cocrystals. This technique facilitates simultaneous melting and mixing of the added raw materials [78]. The coformer and API are mixed in appropriate stoichiometric ratio and fed into the extruder (Figure 4.8). The high temperature of the extruder melts the materials, followed by mixing and nucleation that leads to the formation of cocrystals. The main advantages of this approach are continuous engineering of the cocrystals, fast formation of cocrystals, and no use of organic solvent. The key process parameters for producing stable cocrystals using HME include temperature profile of the processing materials, feeding rate and speed of the screw. There are numerous reports of production of cocrystals using HME such as ibuprofen: nicotinamide, carbamazepine: nicotinamide, and hydrochlorothiazide: nicotinamide cocrystals [78, 80, 81].

4.3.2 Cocrystals from Solutions

Solution crystallization has been a well-established approach for controlling polymorphs and crystal habits of many drugs. Solubility profile of the compounds plays an important role in establishing supersaturation and nucleation of the specific compound in a solvent. Owing to the different solubility attributes of the coformers and API, the underlying mechanism of cocrystal formation from solution is different compared to formation of the single component's crystals from solution. Crystallization is a solubility- and supersaturation-driven process, which could be impacted by various factors including concentration of the solutes in solution and temperature. The solubility of a cocrystal system is most accurately represented in a ternary phase diagram. Ternary phase diagram represents the relationship between solubility of

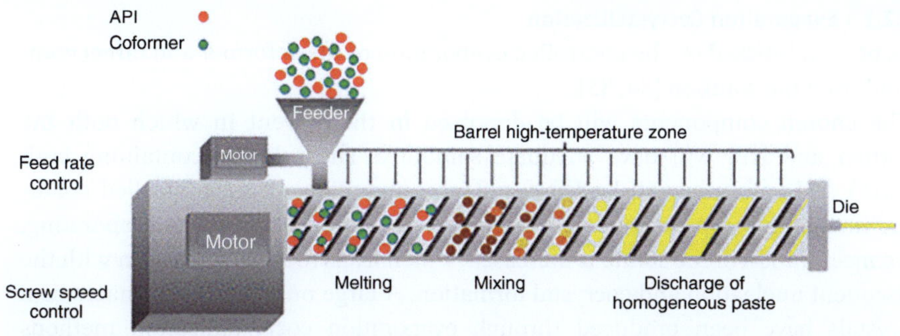

Figure 4.8 Schematic representation of a typical HME instrument that involves feeding drug and coformer physical mixture, melting, mixing, and molding. *Source:* Karimi-Jafari et al. [79] / with permission of American Chemical Society.

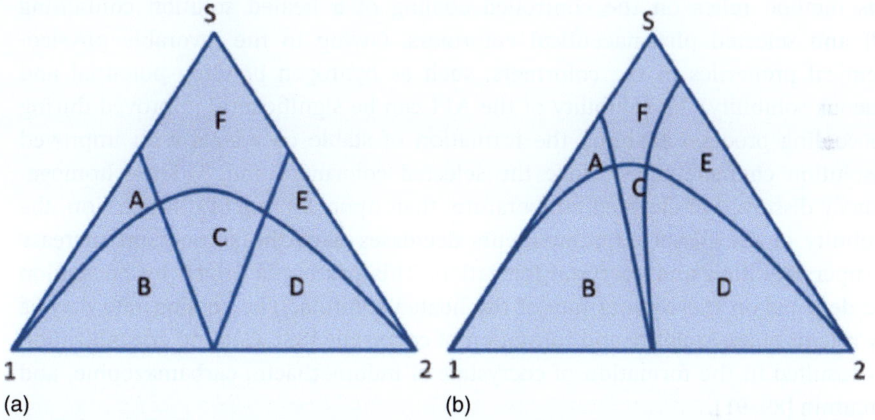

Figure 4.9 Schematic representation of isothermal ternary phase diagram. (a) Similar solubilities between API and coformer (1 and 2) in solvent S and (b) different solubilities of 1 and 2 in S. Region A, component 1 and solvent; B, component 1 + cocrystal; C, cocrystal; D, component 2 + cocrystal; E, component 2 and solvent; F, solution. *Source:* Karimi-Jafari et al. [79] / with permission of American Chemical Society.

the chosen solute in the specific solvent as a function of temperature and pressure (Figure 4.9). As depicted in Figure 4.9, a multicomponent ternary phase diagram, which represents the three-phase behavior of the API, coformer, and solvent, tracking the conditions that lead to cocrystals formation.

Solution crystallization is one of the established operation processes to produce one-component stable crystals that can potentially be scaled for cocrystal formation as well. There are numerous reports summarizing the production of different cocrystals, which include carbamazepine–nicotinamide and caffeine–dapsone using solution crystallization approaches [82, 83]. The following are the main solutions-based cocrystallization methods.

4.3.2.1 Evaporation Cocrystallization

This process is based on the controlled evaporation of the coformer and target compound from the solution [84, 85].

The chosen components will be dissolved in the solvent in which both the coformer and API will have adequate solubility. The solution containing both the API and coformer will be left for slow evaporation under controlled conditions including temperature and humidity. Once the solvent starts evaporating, the concentration of the solute is increased, which leads to supersaturation with the subsequent nucleation and cocrystal formation. A large number of pharmaceutical cocrystals have been produced through evaporation cocrystallization methods such as isoniazid–curcumin, ibuprofen–nicotinamide, and aspirin–benzoic acid cocrystals [79, 86, 87].

4.3.2.2 Cooling Crystallization

This method relies on the controlled cooling of a heated solution containing API and selected pharmaceutical coformers. Owing to the favorable physicochemical properties of the coformers, such as hydrogen bonding potential and aqueous solubility, the solubility of the API can be significantly improved during the cooling process, enabling the formation of stable cocrystals with improved dissolution characteristics. Once the selected coformers and API are homogeneously dissolved at elevated temperature, then upon cooling of this solution, the solubility of the dissolved components decreases with the subsequent increase in supersaturation and cocrystal formation. This process is driven by nucleation and depends on the cooling rate of the heated solution. The cooling rate plays a key role in phase stability and formation of cocrystals [88]. Cooling crystallization has resulted in the formation of cocrystals of indomethacin, carbamazepine, and curcumin [89–91].

4.3.2.3 Slurry Method

This method involves the mixing of the coformer and API in a fixed molar ratio in a suitable solvent. This could also be achieved by addition of the API or any target molecule to the coformer solution. Zhang et al. proposed this method as an efficient screening method for cocrystal formation [92].

The slurry formation depends on the concentration of the target molecule and coformer. Additionally, the formation of cocrystals from the slurry depends on the nucleation and kinetics of the chosen systems and nature of the chosen solvent as well. Despite the high amount of the initial materials required and loss of the materials during this process, this method has a noticeable attraction. The common steps involved in cocrystal formation using this method include slurry preparation, agitation of the suspension, interaction of the components within slurry, and drying. The formation of the cocrystals using this method could be controlled by the process parameters such as temperature, agitation rate, solvent composition, and processing time [93]. Some of the cocrystals prepared by this approach include sulfamerazine–salicylamide, sulfamethazine–nicotinamide, and sulfamerazine–anthranilic acid [94].

4.3.2.4 Anti-solvent Precipitation Method

In contrast to the cooling and evaporative crystallization methods, anti-solvent precipitation method is a low-energy bottom-up approach can be used for a wide range of poorly water-soluble drugs. The method is based on the solubility attributes of the coformers and target molecules in the chosen solvent system [95]. The purpose of the addition of anti-solvent is to reduce solubility of coformers and target API that will potentially lead to nucleation and cocrystallization. This is pertinent to mention that both solvent and anti-solvent should be miscible with each other. The key factors that can impact cocrystallization of the target molecules using the anti-solvent cocrystallization approach include selection of the solvent system, solvent–anti-solvent ratio, solubility of the coformers and target API in the mixture of the solvents, and agitation rate [96]. This method could be exploited for the formation of nano-cocrystals (NCCs). Carbamazepine–nicotinamide NCCs have been produced by an anti-solvent cocrystallization approach [97]. During the production of these cocrystals, the particle size was controlled by the concentration of stabilizers, solvent–anti-solvent ratio, stirring rate, sonication time, and temperature.

4.3.3 Supercritical Fluid-Based Methods

Supercritical technology has been effectively exploited for producing pharmaceutical cocrystals. This process can be designed by using the different properties of the supercritical carbon dioxide (CO_2), which can act as solvent, atomization enhancer, and anti-solvent (Figure 4.10). Some of the important supercritical CO_2-based methods are discussed below.

4.3.3.1 Cocrystallization with Supercritical Solvent

This method utilizes the power of supercritical CO_2 for dissolving the target drugs and coformers with the subsequent cocrystallization. In this method, the cocrystallization of the target molecule along with the coformer is controlled by the processing parameters including temperature, pressure, and concentration of the API and coformers. There are reports covering the production of cocrystals through this approach including sulfamethazine, carbamazepine, theophylline, and indomethacin [99, 100].

4.3.3.2 Rapid Expansion of Supercritical Solvents

This method is based on the principle of saturation of API and coformer(s) in supercritical CO_2 before depressurizing through a coaxial nozzle into the drying chamber. The critical process parameters such as solution concentration, flow

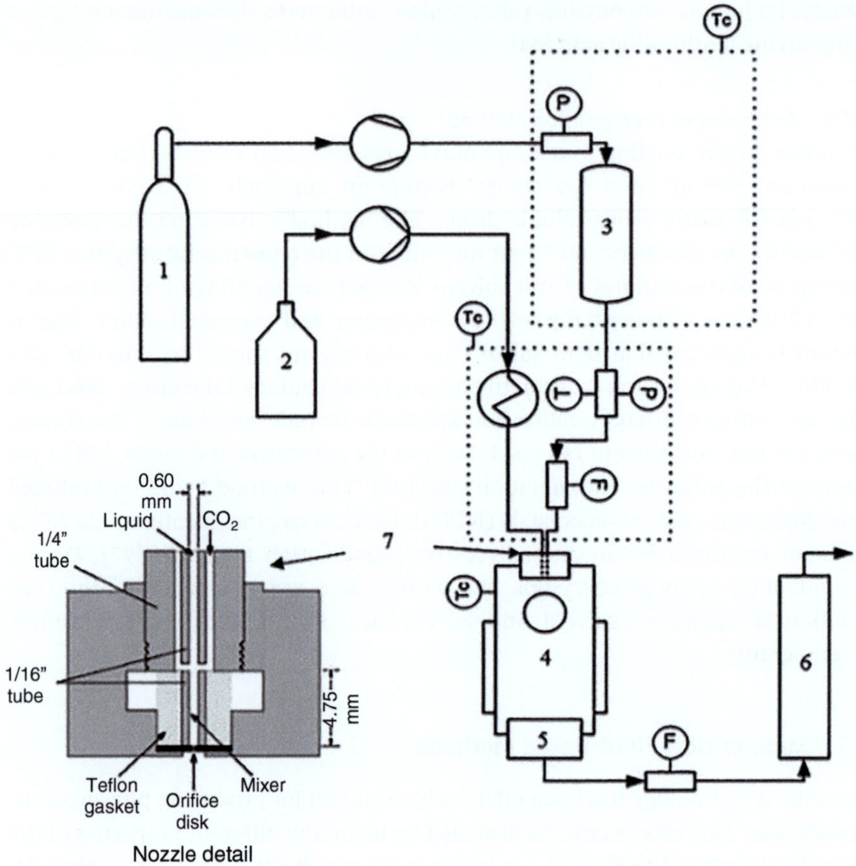

Figure 4.10 Schematic diagram of the SEA apparatus. (1) CO_2 cylinder; (2) liquid solution flask; (3) temperature-controlled CO_2 storage cylinder; (4) precipitator; (5) filter; (6) solvent trap; and (7) detail of the nozzle cap. *P*, *T*, *F*: instruments for pressure, temperature, and flow measurements, respectively. T_c is for temperature control and measurement. *Source:* Padrela et al. [98] / with permission of Elsevier.

rate, and pressure can significantly impact the particle size and morphology of the produced cocrystals. The challenge with this method is that both the API and coformer should demonstrate adequate solubility in the supercritical CO_2. Muller et al. reported the cocrystals of ibuprofen–nicotinamide through this method [101].

4.3.3.3 Supercritical Anti-solvent (SAS) Approach

The supercritical anti-solvent approach is based on the principle of anti-solvent crystallization, in which the solubility of the coformer(s) and target API is reduced in the supercritical CO_2 with the subsequent nucleation and crystallization [99]. The morphology and size of the cocrystals are tailored by concentration of the API and coformer in the organic solvents and supercritical CO_2, flow rate, and pressure of CO_2. The coformer and API are dissolved in the organic solvents, followed by spraying into the supercritical CO_2, which will lead to supersaturation and crystallization. There are numerous reports covering the production of cocrystals using this method such as indomethacin–saccharin, naproxen–nicotinamide, and diflunisal–nicotinamide [102, 103].

4.3.3.4 Supercritical Assisted Spray Drying (SASD)

The supercritical assisted spray drying (SASD) utilizes the supercritical CO_2 as an atomization enhancer for disrupting the liquid jet in which the drug and coformer(s) are dissolved. In this method, the drug and coformer solution along with the supercritical CO_2 are simultaneously sprayed through the coaxial nozzle into the drying chamber which has atmospheric pressure. Interestingly, this method can be used for producing both micro-cocrystals and NCCs by controlling the process conditions. The key process conditions for tailoring the particle size in this method include CO_2 pressure, solution flow rate, solution concentration, and CO_2 flow rate. Padrela and his group have reported several cocrystals through this method, including celecoxib, theophylline, and indomethacin cocrystals [104].

4.3.4 Miscellaneous Methods

4.3.4.1 Spray Drying

Spray drying is a single-step continuous process for producing cocrystals. In this method, the coformer and API solution is sprayed into the hot drying chamber, which leads to rapid evaporation of the solvent with the subsequent cocrystal formation [105]. The main advantage of this method is that high yield of the cocrystals is produced through a single step. Some of the examples of the cocrystals produced through this method include indomethacin–nicotinamide and carbamazepine–nicotinamide [82, 106].

4.3.4.2 Freeze Drying

This is a low-energy method for producing cocrystals that involves sublimation of frozen solution of API and coformer(s). This approach is mainly employed and useful for thermally sensitive materials, which can potentially make them highly porous cocrystals with a large surface area. After freeze drying, the residual moisture may be removed by a secondary drying process. In addition, this is a slow process, which will take a bit longer to get the dried powder. The solvent selection depends on the solubility of the compounds and compatibility with the equipment. Various solvents have been used to produce cocrystals through this method, including water, *t*-butanol, chloroform, acetonitrile, DMSO, and dioxane [107]. The examples of cocrystals produced through this method include, caffeine–citric, theophylline–oxalic acid, and ibuprofen–nicotinamide [2].

4.3.4.3 Ultrasound-Assisted Approach

This method has got a noticeable attraction for cocrystallization as a green technology, which uses high-frequency sound waves that are typically in the range of

20–100 kHz and induce controlled cocrystallization of the APIs and coformers. The underlying mechanism for the generation of cocrystals through this approach includes acoustic cavitation. This reduces the processing time with improved cocrystal purity and quality. Several cocrystals have been prepared through this method including carbamazepine–nicotinamide and caffeine–maleic acid [108, 109].

4.4 Characterization Techniques of Pharmaceutical Cocrystals

Analytical techniques used for pharmaceutical cocrystal characterization can be categorized into four main groups. The first set of techniques includes powder X-ray diffraction (PXRD) and single-crystal X-ray diffraction (XRD). Thermal analysis tools (differential scanning calorimetry [DSC] and thermal gravimetric analysis [TGA]) are the second category. Third, spectroscopic methods are applied for cocrystal investigation, which include Fourier transform infrared (FTIR), Raman spectroscopy, and solid-state nuclear magnetic resonance (SSNMR) spectroscopy. The final category consists of microscopy techniques such as scanning electron microscopy (SEM).

This section provides a brief description of pharmaceutical cocrystal characterization technologies along with different current examples of studies which applied these techniques.

4.4.1 Powder X-ray Diffraction (PXRD) and Single-Crystal X-ray Diffraction (XRD) Analysis

The X-ray crystallography analysis is based on the deflection of X-ray at specific angles (called Bragg angles) when it interacts with the recurrent arrangement of atoms in the crystal lattice of the material [110, 111]. The generated diffraction pattern is unique for each material's crystal structure; therefore, this technique can be used to detect cocrystal formation by comparing diffraction peaks of raw materials and the obtained product [112]. In addition, XRD can be applied to investigate the crystalline pattern of cocrystals, while PXRD can be used for quantification of cocrystal product. Moreover, they can be adopted as continuous monitoring tools during cocrystallization manufacturing, as they are able to confirm cocrystal formation, determine the percentage of cocrystal formation and quantify the remaining amount of the raw materials [113, 114].

Al-Hanafi et al. have monitored the formation of indomethacin: saccharine cocrystals by interfacial cocrystallization using the PXRD analysis at different time intervals during the process (Figure 4.11) [115].

4.4.2 Thermal Analysis

4.4.2.1 Differential Scanning Calorimetry (DSC)

DSC is a highly sensitive and precise thermal analysis tool applied to study the thermal phase transitions of the material by measuring the heat flow as a function

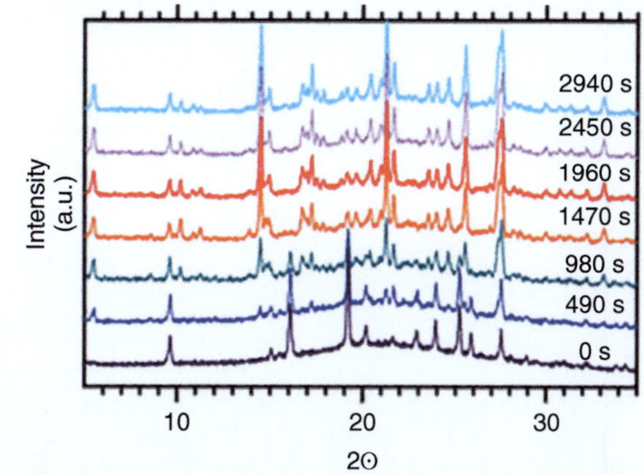

Figure 4.11 Consecutive XRD diffractograms of indomethacin oily phase mixed with saccharin taken at intervals of 490 seconds (T_1–T_7, from bottom to top) showing the change in diffraction pattern upon formation of cocrystals by time. *Source:* Al-Hanafi et al. [115] / with permission of American Chemical Society.

of temperature and time [116]. Upon heating or cooling the sample and reference material in DSC machine, valuable thermal properties of the material can be determined such as melting point, glass transition, and crystallization, which are observed as a change in heat flow. Data from DSC analysis are plotted as peaks of heat flow versus temperature or time, with endothermic peaks representing heat-absorbing processes such as melting and exothermic peaks indicating heat-releasing processes such as crystallization [116].

For pharmaceutical cocrystals, DSC has been used as a rapid, efficient, green technique that utilizes only negligible amount of sample for characterization [117]. Application of DSC for cocrystal screening relies on the fact that there is a remarkable change in endothermic as well as exothermic peaks of cocrystalline product in comparison to the parent pure components [118]. There is an assumption that DSC peak of cocrystal lies between DSC peaks of the raw materials used for their formation; however, some of the reported literature detects cocrystal endothermic melting peak beyond the range of the drug and coformer melting points, as stated for salicylic acid and caffeine cocrystals [119].

DSC was used as "real time" analytical technique for detection of indomethacin: saccharin cocrystal formation (endothermic melting point peak at 184 °C) [120]. In another study performed by Patrycja Saganowska and Marek Wesolowski, DSC was used for cocrystal screening of eight benzodiazepines and nine coformers at 1:1 molar ratio [121]. According to DSC thermograms data, they identified 15 physical mixtures of the studied compounds that showed potential for cocrystals formation and required further investigation. An example of their DSC data is presented in Figure 4.12.

4.4.2.2 Thermogravimetric Analysis (TGA)

TGA belongs to thermal analytical techniques for characterization of pharmaceutical cocrystals. The analysis is based on the measurement of the change in the mass of the sample upon heating over a certain period [116]. The recorded mass transitions can be used to detect thermal stability, composition, and degradation behavior of pharmaceutical cocrystals. In addition, TGA is applied to determine the moisture content of the pharmaceutical cocrystals by monitoring weight loss at a certain temperature range, which normally corresponds to boiling point of

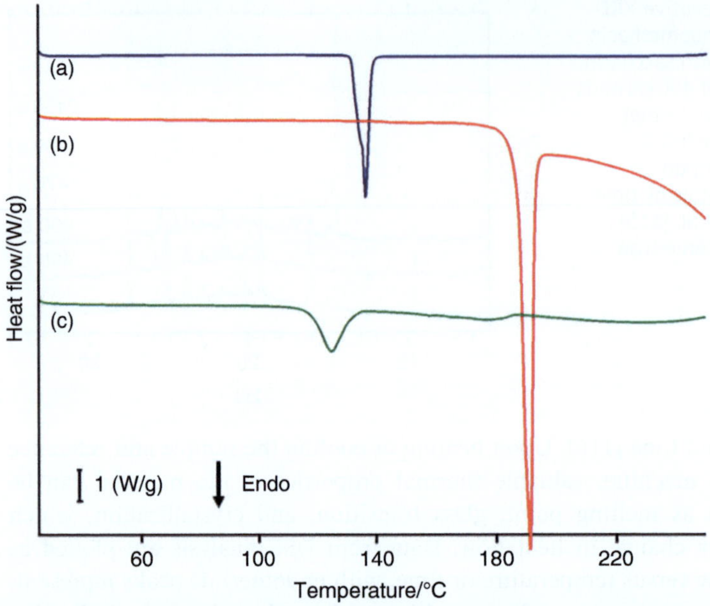

Figure 4.12 DSC thermograms of (a) diazepam, (b) succinic acid, and (c) their cocrystals. *Source:* Saganowska and Wesolowski [121] / Springer Nature / CC BY 4.0.

water. TGA is commonly used as complementary analysis with other cocrystal characterization techniques such as PXRD and DSC to provide comprehensive data of cocrystal properties.

TGA has been applied to study the thermal stability of dimethyl fumarate: nicotinamide cocrystals by comparing the onset temperature of decomposition for cocrystals, parent components, and their physical mixture. Results showed that the cocrystallization provided appropriate thermal stability to dimethyl fumarate with 62 °C higher temperature of decomposition, as presented in Figure 4.13 [122].

4.4.2.3 Hot Stage Microscopy (HSM)

It is a common thermal analysis method for screening and characterization of cocrystals, especially to detect some tiny thermal transitions of the material that may be missed by DSC and TGA. The technique is a combination of conventional optical microscopy and a temperature-adjusted platform where the sample is placed. It is used for real-time measurement of thermal properties of the sample such as phase transition, melting, and crystallization. Hot stage microscopy (HSM) can be coupled with other techniques such as FTIR, Raman, DSC, and SEM to generate more valuable characterization data [123–126].

In arena of pharmaceutical cocrystals, HSM is used as quick and green thermal analysis method which requires small quantities of API and coformer for cocrystallization screening using Kofler mixed fusion method. During the analysis, the physical mixture of cocrystal components is placed on a glass slide and allowed to melt, which is then mixed with a coverslip. If cocrystallization occurs, the cocrystals

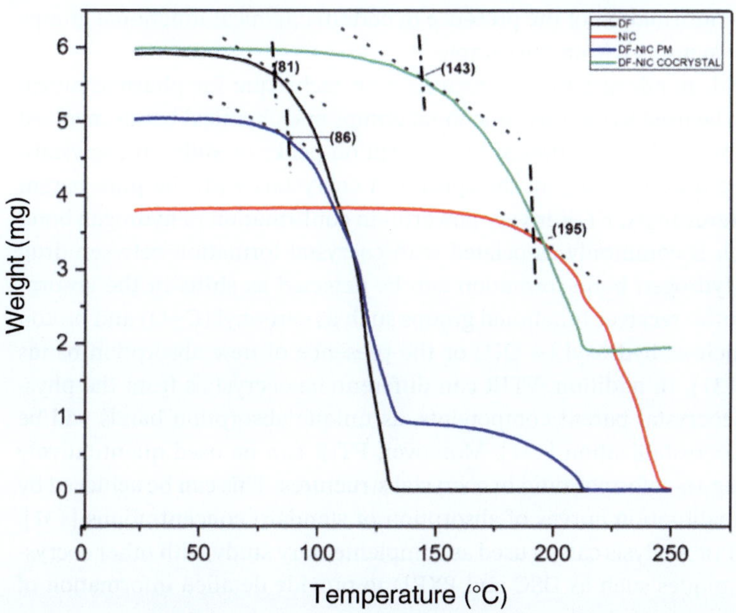

Figure 4.13 TGA thermograms of DF (dimethyl fumarate; drug), NIC (nicotinamide; coformer), DF-NIC PM (physical mixture), and DF-NIC cocrystals. *Source:* Alam et al. [122] / with permission of American Chemical Society.

will be detected at the interface between the raw materials' melts [127, 128]. For instance, HSM has been applied for cocrystal screening of the coformer nicotinamide with seven APIs using Kofler mixed fusion method [129]. More details and examples of the application of HSM in cocrystals screening and characterization can be found in the review by Kumar et al. [130].

Thermal analysis techniques are commonly applied for characterization of pharmaceutical cocrystals; however, they can only be used for thermostable APIs and coformers.

4.4.3 Spectroscopic Techniques

These techniques can be classified into two main categories: vibrational spectroscopy (FTIR and Raman) and nuclear magnetic resonance (NMR). FTIR and Raman spectroscopy are based on the absorption and scattering of light, respectively, caused by certain functional groups within the molecule structure, while NMR provides comprehensive structural elucidation of the multicomponent systems such as cocrystals by analyzing atomic environments and intermolecular interactions.

4.4.3.1 Fourier Transform Infrared (FTIR) Spectroscopy

FTIR spectroscopy analysis is based on the interactions between chemical bonds within the material and infrared light when it passes through the sample. The molecules of the sample absorb or emit infrared radiation at specific wavelengths or frequencies, triggering them to vibrate. These vibration spectra act as fingerprints for confirmation of the presence of certain chemical functional groups and molecular structures within the sample.

FTIR is a rapid, nondestructive characterization technique for pharmaceutical cocrystals. It can be used to identify functional groups and chemical bonds involved in cocrystal formation by detection of the absorption peaks or shifts in cocrystals' spectra. It is also used to compare the spectra of cocrystals with the pure parent components. Interestingly, FTIR is very powerful in confirmation of hydrogen bond formation, which is commonly associated with cocrystal formation between drug and coformer. Hydrogen bond formation can be detected as shifts in the absorption bands of proton receiver functional groups such as carbonyl (C=O) and proton donor groups such as hydroxyl (—OH) or the presence of new absorption bands in the spectra [131]. In addition, FTIR can differentiate cocrystals from the physical mixture of cocrystal parent components, as unique absorption bands will be detected upon cocrystallization [131]. Moreover, FTIR can be used quantitatively to determine drug-to-coformer ratio in cocrystal structures. This can be achieved by construction of calibration curves of absorption of standard concentrations [131]. Furthermore, FTIR analysis can be used as complementary study with other cocrystal analysis techniques such as DSC and PXRD to provide detailed information of cocrystal structures.

FTIR has been applied to investigate the formation of theophylline and citric acid cocrystals by comparing vibrational spectra of cocrystalline product, physical and ground mixture, and pure components (Figure 4.14) [132].

4.4.3.2 Raman Spectroscopy

Raman spectroscopy is a real-time technique for confirmation of cocrystallization between API and coformer/s [133]. The analysis is based on exposing the sample to monochromatic light source, usually laser. Upon passing through the material, part of that light exhibits inelastic scattering, also known as Stokes–Raman scattering, due to molecular vibrations causing a change in both energy and wavelength of the scattered light [134]. Raman spectra are characteristic for each crystal structure and are used to distinguish drugs, coformers, and their cocrystals. In addition, Raman peaks can provide valuable information related to the chemical bond type, strength, and length. The position and intensity of Raman bands can reveal the type of interaction within cocrystal structure, especially hydrogen bonding [131]. Furthermore, Raman spectroscopy is more accurate, sensitive, and precise than FTIR.

Ketkar et al. have used Raman spectroscopy to detect the formation of caffeic acid phenethyl ester (CAPE) cocrystals with three coformers (caffeine, nicotinamide, and iso-nicotinamide) [135]. The results from this report demonstrated that CAPE: nicotinamide cocrystals increased the aqueous solubility of the parent drug by 2.76-fold. Figure 4.15 shows Raman spectra of pure components and their cocrystals.

4.4.3.3 Solid-State Nuclear Magnetic Resonance (SSNMR)

SSNMR is a powerful tool to provide extensive information on pharmaceutical cocrystal structures [136]. Structure elucidation is made via identification of

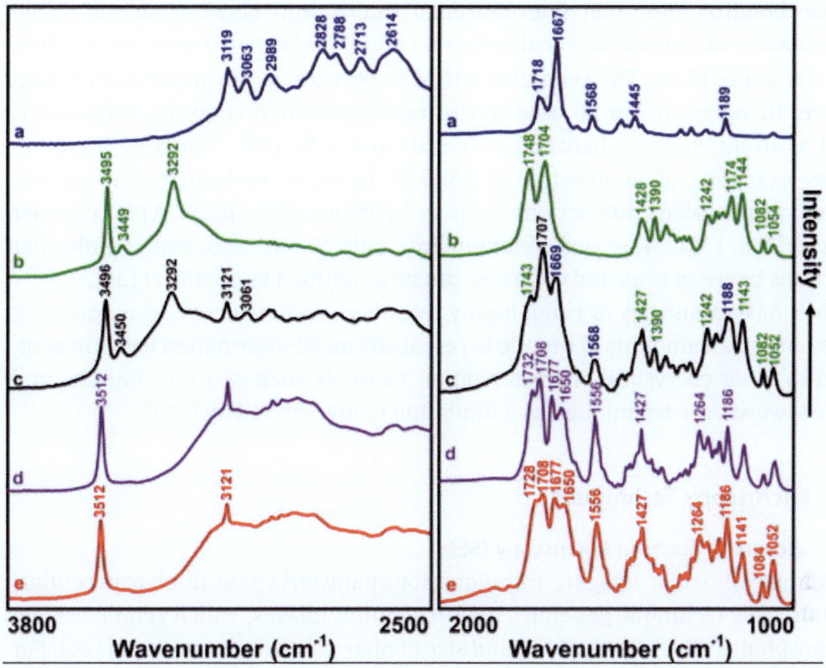

Figure 4.14 FTIR spectra of theophylline (a), citric acid (b), physical mixture (c), ground mixture (d), and intact cocrystal (e) of theophylline and citric acid showing characteristic FTIR bands of each material. *Source:* Lin et al. [132] / with permission of Elsevier.

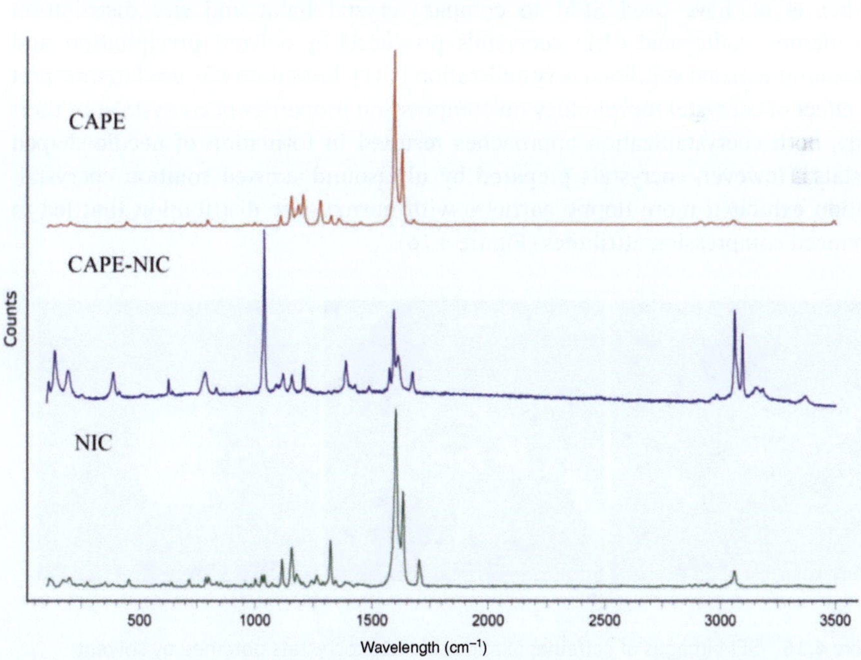

Figure 4.15 Raman spectra of caffeic acid phenethyl ester (CAPE), caffeic acid phenethyl ester: nicotinamide cocrystals (CAPE-NIC), and nicotinamide (NIC). *Source:* Ketkar et al. [135] / with permission of American Chemical Society.

hydrogen bonding and other intermolecular interactions between atomic nuclei which manifest as distinct chemical shifts that reveal molecular arrangement within the crystal lattice [137]. The generated SSNMR spectrum is unique for each crystal structure; therefore, it can be used to distinguish different cocrystal polymorphs as well as to differentiate between cocrystals and salts [24]. The cited literature includes examples of application of SSNMR to study molecular associations, especially noncovalent interactions such as hydrogen bonding in API: coformer cocrystals [138, 139]. Moreover, stoichiometric ratio as well as potential molecular interactions between drug and coformer can be quantified by SSNMR [136].

SSNMR has advantages of being highly selective, nondestructive technique that utilizes only small amounts of sample to reveal advanced information levels in comparison to other cocrystals' characterization methods such as FTIR, Raman, and PXRD. However, the technique has a limitation of low sensitivity [139].

4.4.4 Microscopy Techniques

4.4.4.1 Scanning Electron Microscopy (SEM)

SEM is a very efficient imaging technique for characterization of pharmaceutical cocrystals. This technique generates high-resolution images, which can be used to study morphology, surface topology, and size of pharmaceutical cocrystals [140]. For SEM analysis, samples are coated with gold or carbon and mounted on an aluminum stub using bilateral gold tape or self-adhesive carbon mounts. Samples are then placed in vacuum to enhance their conductivity before scanning by electron beam.

Aher et al. have used SEM to compare crystal habit and size distribution of caffeine: oxalic acid (2:1) cocrystals produced by solvent precipitation and ultrasound-assisted solution cocrystallization [141]. SEM data was used to interpret the effect of cocrystal morphology on compression properties of cocrystals. In their study, both cocrystallization approaches resulted in formation of needle-shaped crystals. However, cocrystals prepared by ultrasound-assisted solution cocrystallization exhibited more floppy particles with narrow size distribution that led to enhanced compression attributes (Figure 4.16).

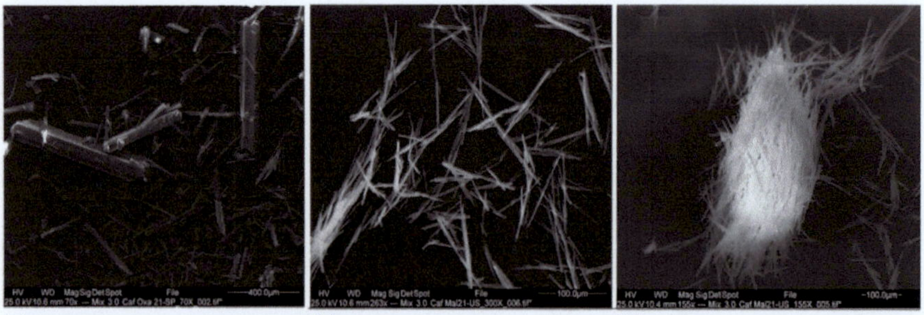

(a) (b) (c)

Figure 4.16 SEM images of caffeine: oxalic acid (2:1) cocrystals obtained by solvent precipitation (a) and ultrasound-assisted solution cocrystallization (b and c). *Source:* Aher et al. [141] / with permission of Taylor & Francis.

4.5 Solubility Aspects of Pharmaceutical Cocrystals

4.5.1 Thermodynamic Principles of Solubility

Solubility of the drug can be defined quantitatively as the concentration of the drug in a saturated solution (i.e. a solution in which there is a dynamic equilibrium between soluble material and its intact form) at specified temperature. It can be described qualitatively as the capability of the drug to dissolve in a solvent to form a homogeneous dispersion. Solubility of drugs can be expressed using different terms such as grams of drug per litre of the solvent and molarity units [142]. However, US Pharmacopeia and National Formulary record drug solubility as "the volume of solvent required to dissolve 1 gram of the drug at a specified temperature." For the drugs whose exact solubility has not been identified, general solubility terms are used, as described in Remington's Pharmaceutical Sciences and presented in Table 4.1 [143].

The solubility of a drug can be influenced by several factors such as temperature, chemistry of drug molecule, and solvent [142]. Adequate water solubility is crucial for drugs, as only solubilized drug is absorbed into bloodstream. USFDA provides the Biopharmaceutics Classification System (BCS), which classifies drugs into categories according to their aqueous solubility and intestinal permeability. BCS Class II and IV drugs have poor water solubility; therefore, enhancing their aqueous solubility is important product development aspect. Cocrystallization of drugs with suitable coformer/s can be used to improve solubility of drugs with poor aqueous solubility [144].

The saturation solubility of a cocrystal is directly proportional to the solubility of its components, particularly that of coformer [1]. It was reported that a solubility difference of 10 between API and coformer is essential to improve cocrystals' solubility [145]. An investigation conducted by the same group showed 152 times enhancement in carbamazepine solubility when cocrystallized with nicotinamide.

Table 4.1 Approximate solubility expression terms.

Solubility expression term	Parts of solvent required for 1 part of solute
Very soluble	<1 part
Freely soluble	1–10 parts
Soluble	10–30 parts
Sparingly soluble	30–100 parts
Slightly soluble	100–1000 parts
Very slightly soluble	1000–10,000 parts
Practically insoluble, or insoluble	>10,000 parts

Source: Adapted from Remington [143].

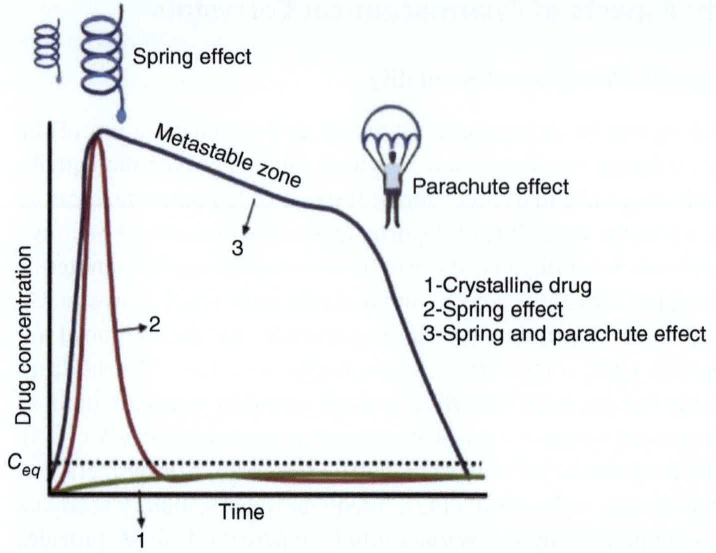

Figure 4.17 Spring and parachute effect for generation of higher apparent solubility for aqueous insoluble drugs. *Source:* Kara and Rathnanand [147] / MDPI / CC BY 4.0.

Another study of solubility increase upon cocrystallization was conducted by Oberoi et al., and their results showed more than 7.5 times increase in ibuprofen: nicotinamide cocrystal solubility when compared to the parent drug [146].

According to the Nernst equation, there is an increase in cocrystal dissolution rate due to an increase in apparent solubility that is explained by the spring and parachute concept (Figure 4.17). When cocrystals are placed in a dissolution medium, the organized periodic array of constituents in the crystal lattice is lost. The randomly oriented molecules form supramolecular aggregates similar to the amorphous phase of the drug, resulting in significant increase in solubility (spring effect). This high solubility is maintained (the parachute effect) over a sufficient period (typically 4–6 hours) until these amorphous aggregates slowly transform into a metastable polymorphic form of the drug and ultimately into a stable thermodynamic crystal form according to Ostwald's step rule. Ideally, most of the drug will have been absorbed before this transformation occurs. This slow transformation process is inhibited by polymers and excipients present in the gastrointestinal tract alongside the drug [1].

For the determination of cocrystal solubility, various dissolution media are used such as simulated gastric fluid or intestinal fluid to find an in vivo–in vitro correlation. Cocrystal particle size was reported to influence its dissolution behavior. Carbamazepine: saccharin cocrystal dissolution rate was found to be inversely related to cocrystal particle size [148]. It is important to address these issues during cocrystal development.

Studies on the effect of cocrystallization of API on intrinsic solubility are limited. McNamara et al. reported 18 times increase in intrinsic solubility of glutaric acid 2-[4-(4-chloro-2-flouorofainoxy)-phenyl] pyrimidine-4-carboxamide cocrystals compared to the parent drug [128].

Cocrystals of ethenzamide–gentisic acid and curcumin with resorcinol and pyrogallol have shown improved intrinsic dissolution rate and solubility, respectively [74, 149].

4.6 Pharmaceutical Cocrystals: Physicochemical Properties and Applications

This section covers the application of pharmaceutical cocrystals for formulation, pharmacokinetics, and therapeutic advantages based on the enhanced physicochemical attributes, with examples provided when applicable.

4.6.1 Application of Pharmaceutical Cocrystals for Formulation Advantages

Cocrystallization is a powerful crystal engineering approach to improve physical, chemical, mechanical, and optical properties of drugs to achieve improvement of parent pharmaceutical formulation.

4.6.1.1 Cocrystals and Physical Stability

Physical stability of solids can be defined as the resistance to change in the physical properties of materials such as melting point, hygroscopicity, solubility, and hardness. Cocrystallization has been reported to enhance physical properties of drugs as well as to preserve their physical stability during manufacturing and storage [150]. In this subsection, only melting point and hygroscopicity will be discussed, and some other physical properties such as solubility will be highlighted in the subsequent subsections.

In comparison to the liquid dosage forms, solid dosage forms are preferred by both pharmaceutical manufacturers and patients. Processing solid drug forms is more appropriate from a manufacturing perspective, considering isolation/purification, identification, transport, and storage. The use of drug in a solid form is also more acceptable for most patients with respect to administration and handling. Nevertheless, some of the drugs, due to their low melting points, are liquids at room temperature. For low-melting-point drugs, cocrystallization with an appropriate coformer can modify their melting temperatures. Propofol is an anesthetic and sedative agent, which is commonly formulated as an emulsion owing to its low melting point (18 °C). This liquid formula was associated with instability as well as patient discomfort issues. Studies have shown that propofol can be converted to crystalline solids upon introducing coformers such as isonicotinamide, bipyridine, and phenazine into its crystal lattice, leading to increment in the melting point when compared to the pure drug [151, 152].

Determination of drug's hygroscopicity is very crucial in pre-formulation study, as it may impact other physicochemical properties of the molecule such as solubility, stability, dissolution rate, and mechanical properties. For instance, the anhydrous form of the drug dasatinib showed higher aqueous solubility compared to its monohydrate form [153]. Numerous approaches have been used to maintain the stability of nonhydrated drugs during drug development, involving addition of excipients, control of humidity in the manufacturing areas, application of proper packaging systems, and coating the drug product with certain polymers [150]. It is important to highlight here that absorption of atmospheric moisture can occur due to the presence of hygroscopic excipients such as citric acid, which is an essential ingredient for effervescent formulations. Many studies have investigated the application of cocrystallization approach to hinder moisture absorption by drugs and excipients. It is speculated that the formed hydrogen bonds between API and coformer prevent water molecules from attaching to the hydrogen donor and acceptor groups within API chemical structure, therefore enhancing the hygroscopic stability of the drug [150, 154, 155]. Lithium chloride is an antipsychotic drug used for the treatment of certain mental illnesses. It is highly hygroscopic even at very low humidity. Upon cocrystallization with glucose, lithium chloride was reported to have improved hygroscopic stability while retaining most of the pharmacokinetic properties of the pure drug [150]. Citric acid is an essential excipient for effervescent dosage forms, which interacts with sodium bicarbonate in a fizzy reaction upon contact with water. Citric acid can absorb moisture easily from the surrounding atmosphere, leading to initiation of the mentioned reaction during product processing or storage. Accordingly, several costly moisture-free adaptation techniques have been used during the manufacturing of effervescent products and for packaging system as well. Pagire et al. have prepared citric acid: nicotinamide cocrystals to block the hygroscopic sites of citric acid, therefore, enhancing its hygroscopic stability and ultimately reducing production and packaging costs [156]. Figure 4.18 shows a comparison between pure citric acid and citric acid: nicotinamide-containing effervescent tablets following manufacturing and during stability study.

4.6.1.2 Cocrystals and Chemical Stability

Chemical stability can be defined as the ability of drug or drug dosage form to maintain its molecular integrity and labeled potency within specified limits over time under certain conditions of use, handling, and storage. Several factors can cause chemical degradation of API or drug products such as temperature, light, humidity, and formulation excipients [157]. Chemical instability can occur through various reactions e.g. hydrolysis, oxidation, and photolysis, leading to the formation of less potent or even toxic pharmaceutical products. Therefore,

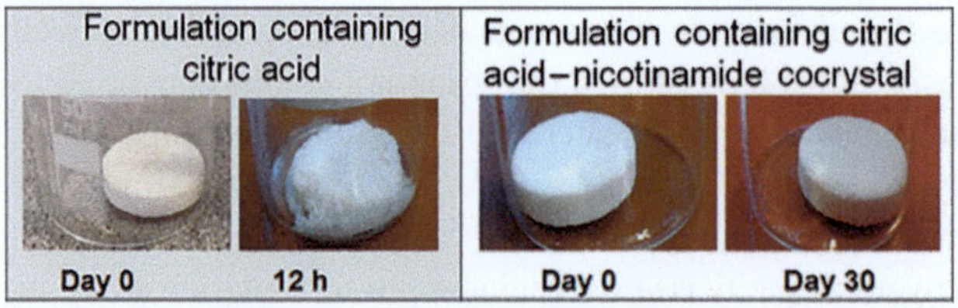

Figure 4.18 Stability comparison between free citric acid and citric acid: nicotinamide effervescent tablets. *Source:* Pagire et al. [156] / with permission of American Chemical Society.

maintaining chemical stability during the stated shelf life is crucial to ensure preparation of safe and efficient drug products. Pharmaceutical cocrystallization strategy has been applied to ensure chemical stability of solid drugs [158, 159]. The prevention of chemical degradation by cocrystallization is proposed through the modification of crystal arrangement of API through five main mechanisms, namely blocking and increasing the distance of the reaction sites, reduction of reaction cavity's size, enhancement of π–π stacking, and alteration of the local molecular environment [160].

Within solid crystal lattice, the space adjacent to the reaction groups is designated as reaction cavity. It has been reported that the large reaction cavity increases the molecular motion as well as the possibility of chemical reactions within the crystalline solids [161]. Photodegradation through isomerization is observed for Epalrestat (anti-neuropathic agent) from (*E*, *Z*) isomer to (*Z*, *Z*) isomer (Figure 4.19) as a change in the drug color from orange to pale yellow [162–164]. The degradation reaction adversely impacts both drug manufacturing and pharmacological efficacy. This issue has been mitigated by cocrystallization of Epalrestat with betaine, in which the drug–cocrystal complexes exhibited enhanced photostability (i.e. no color change) upon exposure to 6000 lux illumination for one day (Figure 4.20) [162]. The improved stability was attributed to rigid drug–coformer intermolecular

Figure 4.19 Isomerization of Epalrestat upon exposure to light. *Source:* Putra et al. [162] / with permission of American Chemical Society.

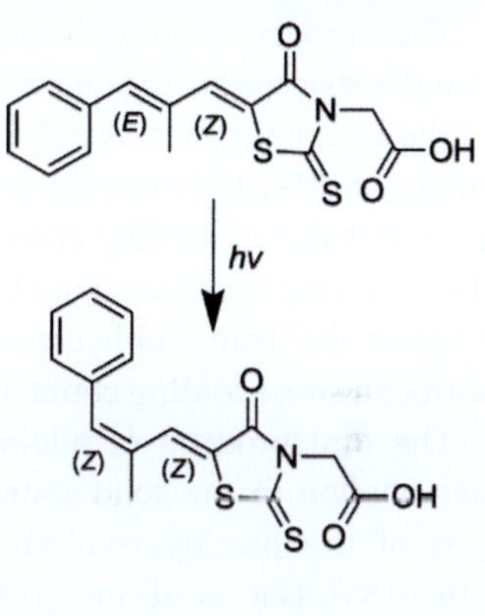

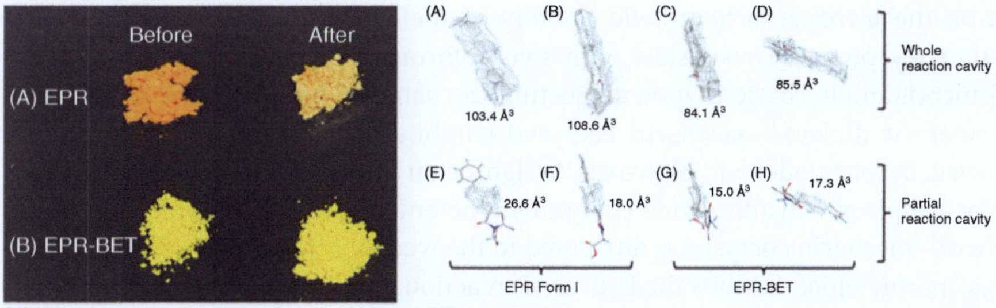

(a) (b)

Figure 4.20 (a) Optical manifestation of Epalrestat and Epalrestat: betaine cocrystals before (A) and after (B) exposure to illumination. (b) Whole (A–D) and partial (E–H) reaction cavity within Epalrestat and Epalrestat: betaine cocrystal lattices. *Source:* Putra et al. [162] / with permission of American Chemical Society.

bonds and a reduced reaction cavity that restricts molecular motion and inhibits isomerization.

The distance between the reactive sites is crucial for chemical stability of solid-state crystals. Cocrystallization approaches have been applied to enhance drug's chemical stability by increasing the distance between the reactive sites, as reported for multidrug cocrystals of vitamin D2 and vitamin D3 [165]. Nicorandil, a drug used for the treatment of cardiovascular disorders, is prone to chemical degradation by high moisture, temperature, and mechanical stress (Figure 4.21a) [158]. In nicorandil crystal lattice, the distance between the lone pair electrons on pyridines (N1) and C8 of another nearby molecule is 3.367 Å (Figure 4.21b). This distance is shorter than the combined van der Waals lengths of carbon and nitrogen (3.484 Å), facilitating the degradation via stabilization of carbocations ($C8^+$) of the degradation intermediate (Ia) (Figure 4.21a). Researchers developed cocrystals of nicorandil with various acids (1-hydroxy-2-naphthoic acid [1-HNA], salicylic acid (SA), 3-hydroxybenzoic acid (3-HBA), and 2,5-dihydroxybenzoic acid (2,5-DHBA)), which exhibited superior hydrothermal stability over nicorandil (Figure 4.21c) [158]. N1—C8 distances within the crystal structures of the prepared cocrystals were extended, surpassing the combined van der Waals lengths of carbon and nitrogen atoms (Figure 4.21d). Therefore, preventing the stabilization of the carbocation ($C8^+$) by N1 and terminates the catalyzed decomposition of nicorandil [158].

The nitrogen atom of the piperazine ring within brexpiprazole (an antipsychotic drug) is susceptible to react with peroxide functional group which is present in some of the excipients such as polyvinyl pyrrolidone [166]. The reaction results in the formation of N-oxide impurities with poor photostability. The brexpiprazole: catechol cocrystal shows stability at 40 °C and 75% RH for one month [166]. This stability is due to strong hydrogen bond formation between the nitrogen atom in the piperazine rings of the drug and hydroxyl groups of the coformer, which blocks the reactive nitrogen site, avoiding contact with peroxides.

The oral prodrug of adefovir, known as adefovir dipivoxil, is susceptible to degradation in the solid state due to two main degradation mechanisms: hydrolysis of the pivaloyloxymethyl group and dimerization of the adenine moiety [167–169]. Gao et al. investigated the impact of cocrystallization coformer selection on the increase in hydrolytic stability of adefovir dipivoxil in its solid state [169]. They prepared cocrystals with two coformers: saccharin (weakly acidic) and nicotinamide (basic). Upon subjecting the samples to 30-day stress stability, the adefovir dipivoxil–saccharin cocrystal exhibited the minimum degradation, followed by pure adefovir dipivoxil. A significant degradation was observed for adefovir dipivoxil–nicotinamide cocrystals. The enhanced stability of the adefovir dipivoxil–saccharin cocrystal is attributed to the weakly acidic conditions provided by saccharin, which inhibits the hydrolysis reaction. On the contrary, nicotinamide generates a basic environment that accelerates hydrolysis degradation.

A [2 + 2] photodimerization degradation can occur in menadione (vitamin K) in the crystalline solid state (Figure 4.22a). Upon exposure to sunlight, menadione molecules undergo topochemical reactions to form cyclobutane photodimers [170, 171]. To address this photostability issue, Zhu et al. cocrystallized menadione

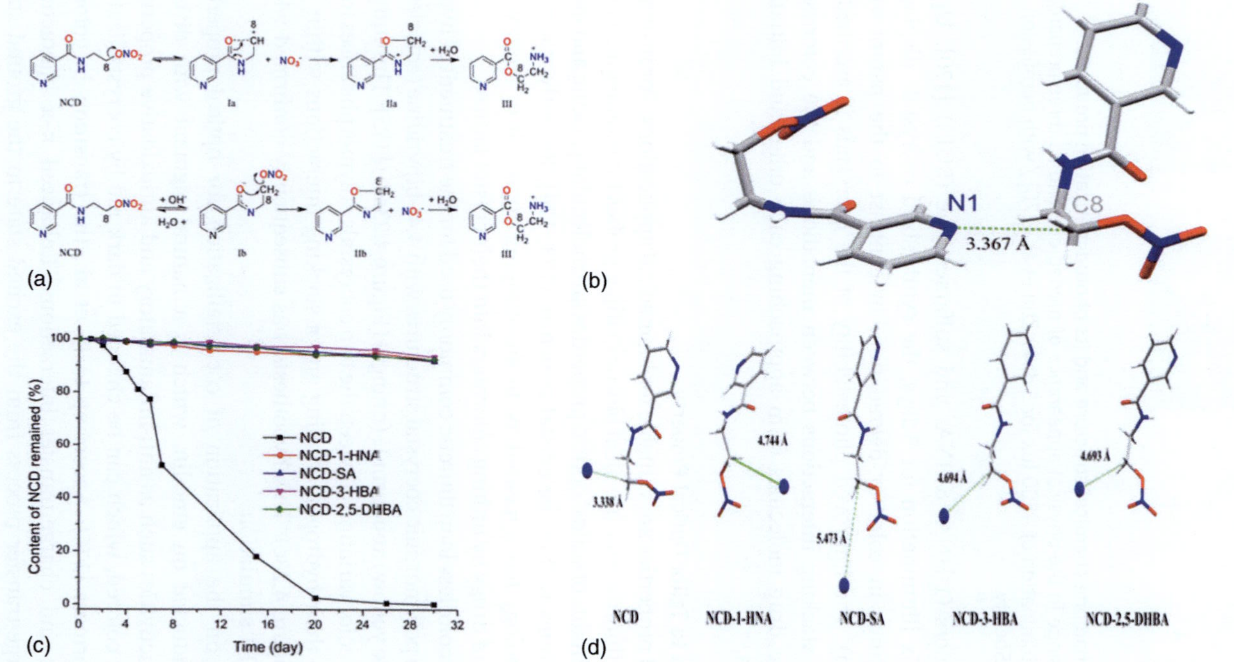

Figure 4.21 (a) Chemical degradation mechanism of nicorandil upon exposure to different environmental stresses. (b) The distance between N1 of pyridine ring and C8 of ethyl group of nearby nicorandil molecules in the drug crystal lattice. (c) The remaining amount of nicorandil after hydrothermal stability study of parent drug and the various formed drug cocrystals. (d) Outline and comparison of N1–C8 distance in the crystal structures of the prepared cocrystals. *Source:* Guo et al. [158] / with permission of American Chemical Society.

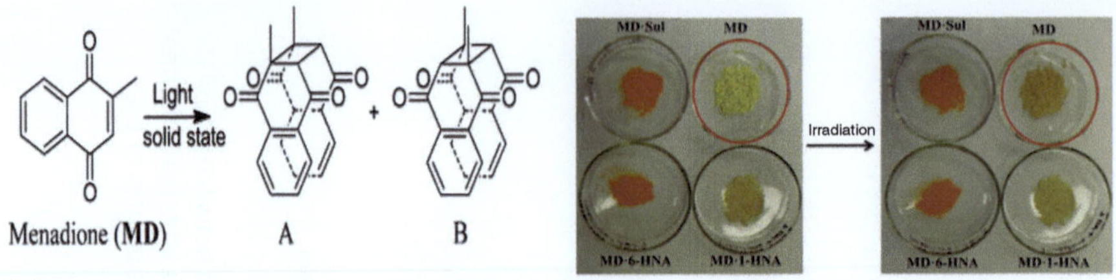

(a) (b)

Figure 4.22 (a) Menadione chemical structure and its photodimerization products. (b) Review of the change in the physical appearance of menadione and its three prepared cocrystals with an illumination of 4500 lux for 5 days. Zhu et al. [170] / with permission of American Chemical Society.

with three coformers (1-HNA, 6-HNA, and sulfamerazine (SUL)) [170]. Upon exposure to 4500 lx illumination for 5 days, the synthesized cocrystals displayed no significant change in color or degradation, in contrast to the parent vitamin (Figure 4.22b). The improved photostability of the cocrystals is ascribed to strengthened π–π stacking interactions between menadione and the coformers, which prevent the drug molecules from approaching each other and initiating photodimerization.

4.6.1.3 Cocrystals to Tailor Optical Property

Drugs with optical properties are valuable for biomedical applications. For instance, drugs with strong fluorescence are ideal biocompatible probes for bioimaging [172]. The major determinants of solids' optical properties are molecular packing and intermolecular interactions within the crystal structure [173, 174]. Recently, cocrystal engineering technology has appeared as an encouraging approach to modify the optical properties of drugs by uniting coformers into their crystal lattices.

Furosemide is a colorless loop diuretic commonly used for the treatment of hypertension. It has two polymorphic cocrystal structures with 4,4′-bipyridine (colorless), named form I (pale yellow) and form II (orange) (Figure 4.23a,b) [175]. The authors suggested that the color variation between the two cocrystals polymorphs arises from the differences in their hydrogen bonding and π-stacking interactions within the crystal lattices (Figure 4.23c). This hypothesis was subsequently confirmed when they performed DFT simulation.

Another study on the application of cocrystallization for optical properties tailoring was conducted on emodin, which is a natural pigment with diverse pharmacological activities such as anti-inflammatory and antioxidative properties. Emodin is yellow colored, which can be changed to dark red by cocrystallization with various coformers [176]. According to Li et al., the transition in color is attributed to π–π and charge-transfer interactions. Enhanced π–π interactions facilitate the charge-transfer process from the excited state to the ground state, resulting in a redshift in the maximum absorption wavelength [176].

4.6.1.4 Cocrystals and Mechanical Properties

When considering development of solid dosage forms, it is essential to investigate the mechanical properties of the crystalline drugs, which may impact the

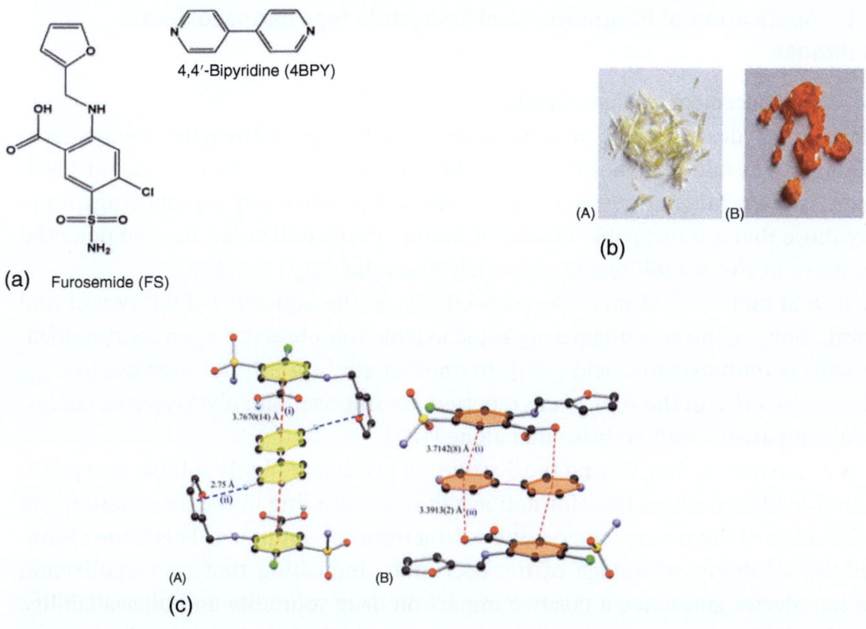

Figure 4.23 (a) Chemical structure of furosemide and 4,4′-bipyridine. (b) Physical appearance of furosemide: 4,4′-bypridine cocrystals form I and form II. (c) Crystal structure of furosemide: 4,4′-bypridine cocrystals form I and form II. *Source:* Sangtani et al. [175] / with permission of American Chemical Society.

manufacturing processes such as mixing, particle size reduction, granulation, and tableting. The mechanical deformation mechanisms of solids comprise elastic, plastic, viscoelastic, and fragmentation processes. Usually, materials demonstrating superior plasticity predictably display improved compressibility, undergoing permanent plastic deformation upon the removal of stress. However, many organic compounds exhibit suboptimal mechanical properties, posing challenges in the development of tablet dosage forms. These poor mechanical parameters can be effectively modified by cocrystallization with appropriate coformer/s [177–180].

Tabletability is the capability of the material to convert into a tablet of specified tensile strength under a given compaction pressure. For organic solids, good tabletability relies on having more plastic distortion and fewer elastic recovery. The presence of slip planes within the crystal structure would ease plastic deformation and ultimately enhance the compression behavior [181]. Cocrystallization approach has been applied to enhance the tabletability of caffeine by complexing it with 3-nitrobenozic acid [179].

One of the common problems during tablet manufacturing is sticking of the tablets to tablet press punches during tableting. Drugs with elevated plasticity show significant punch sticking. An example of drug with serious punch sticking is celecoxib due to the presence of slip planes in its crystal lattice; this issue has been reduced by cocrystallizing it with L-proline, which strengthened the crystal structure and deactivated the slip planes [182].

4.6.2 Application of Pharmaceutical Cocrystals for Pharmacokinetic Advantages

4.6.2.1 Enhancement of Bioavailability

Bioavailability describes the amount of the intact drug that reaches the systemic circulation. The failure of many drug candidates to pass the preclinical stage of development is frequently due to low bioavailability. For poorly water-soluble drugs, especially those that are incapable of salt formation, cocrystallization has the potential to improve *in vivo* solubility and ultimately bioavailability [183–185].

A 10-fold and 6.7-fold increase, respectively, in the aqueous solubility and oral bioavailability of the antifungal drug ketoconazole was observed upon cocrystallization with *p*-aminobenzoic acid [186]. In another study, a 2.7-fold increase in C_{max} and a twofold rise in the AUC were reported for ambrisentan: glycylglycine cocrystals in comparison with ambrisentan alone [187].

It is worth noting that, during the dissolution process of highly soluble cocrystals, solution-mediated phase transformation can occur, leading to the dissociation and precipitation of the poorly water-soluble drug from the solutions. This issue diminishes the solubility advantage of the cocrystals, indicating that cocrystallization does not always guarantee a positive impact on drug solubility and bioavailability. Accordingly, delaying or preventing the process of solution-mediated phase transformation is critical to extending supersaturation, thus improving the bioavailability of drugs with low aqueous solubility [184, 188]. The oral bioavailability of danazol: vanillin cocrystals was increased 10-fold when dissolved in the presence of 1% vitamin E-TPGS (solubilizing agent) and 2% Klucel (crystallization inhibitor). In the absence of these additives, danazol: vanillin cocrystals showed only 1.7-fold improvement in the AUC compared to the parent drug [189]. Other examples include carbamazepine: succinic acid cocrystals with crystallization inhibitors (2% polyvinyl caprolactam–polyvinyl acetate polyethylene glycol graft copolymer® and 1% hydroxypropylmethylcellulose acetate succinate [HPMCAS]) as well as carbamazepine–saccharin cocrystals with the surfactant sodium lauryl sulfate acting as a solubilizing agent and supersaturation extender [190].

4.6.2.2 Modulation of Drug Release Profiles

In comparison to immediate-release dosage forms, sustained-release formulations possess superior advantages in terms of reducing dosing frequency, improved patient compliance, and minimizing side effects owing to upholding stable-state blood levels with fewer plasma fluctuations. Cocrystallization can be used to achieve a sustained-release profile of drugs [191–193].

The first-line antitubercular drug isoniazid has a larger dose and a prolonged treatment course, causing reduced patient compliance, dangerous hepatotoxicity, and the development of drug resistance. A sustained-released form of isoniazid has been developed by cocrystallization with curcumin (hydrophobic coformer) (Figure 4.24a) [86]. In their investigation, dissolution studies of free isoniazid, isoniazid–curcumin physical mixture, and isoniazid–curcumin cocrystals were performed. Complete release of isoniazid from the physical mixture occurred after

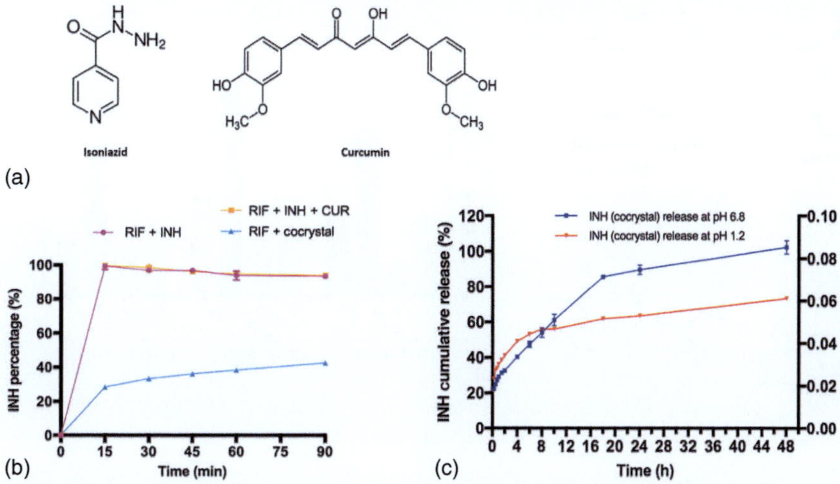

Figure 4.24 (a) Chemical structure of isoniazid and curcumin. (b) Dissolution profile of isoniazid from different sample groups in pH 1.2 buffer solution. (c) Dissolution profile of isoniazid form 2:1 isoniazid: curcumin cocrystals in the presence of rifampicin in pH 1.2 and 6.8 buffer solutions ($n = 3$). *Source:* Xuan et al. [86] / with permission of American Chemical Society.

15 minutes at pH 1.2. On the other hand, only 28% drug release was observed after 15 minutes for cocrystals and 73% isoniazid release was detected after 48 hours (Figure 4.24b and c). The researchers ascribed the reduction of drug release to blockage of the solvation site in the crystal lattice of isoniazid by curcumin and as well as the inhibition of dissolution due to the precipitation of curcumin form III on the cocrystal's surface.

A sustained-release profile of drugs by cocrystallization can be attained by reduction of intrinsic dissolution rate as reported for piperazine ferulate (drug salt) when cocrystallized with the coformer pyrazinamide [192]. In addition, in another investigation, the retardation of the drug release by cocrystallization was attributed to an increase in the crystal lattice energy and intermolecular interactions strength caused by incorporation of coformer that ultimately reduces the solubility and dissolution rate of the parent drug. This mechanism has been proposed for pirfenidone: fumaric acid and pirfenidone: trimesic acid cocrystals [194].

4.6.3 Application of Pharmaceutical Cocrystals for Therapeutic Advantages

The beneficial and intended outcome of the treatment is referred to as a therapeutic effect, which is often affected by the physicochemical properties of the drug. The therapeutic effect of BCS Class II and IV drugs is constrained by low aqueous solubility or permeability. This restricted therapeutic outcome has been addressed by cocrystallization [195, 196]. When the anticancer drug tegafur was cocrystallized with the nutraceutical syringic acid (Figure 4.25a,b), an enhancement in solubility, the dissolution rate, and permeability was observed [197]. This improvement in

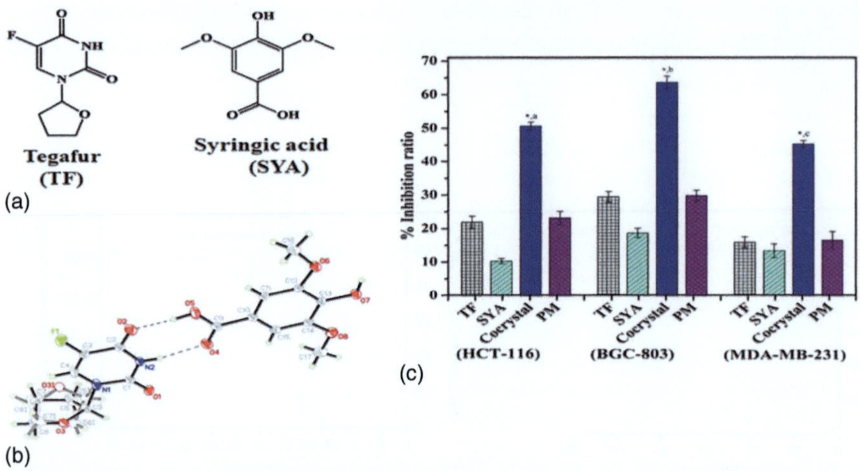

Figure 4.25 (a) Chemical structures of tegafur and syringic acid. (b) Crystal structure of tegafur–syringic acid cocrystal. (c) Cytotoxicity evaluation of individual materials, cocrystal, and physical mixture against the three tumor cell strains HCT-116 (human colon cancer), MDAMB-231 (breast cancer), and MGC-803 (human gastric carcinoma) by sulforhodamine beta (SRB) assays. *Note:* (a–c) represent the CI values of the cocrystal of 0.79, 0.71, and 0.67 for HCT-116, MGC-803, and MDA-MB-231 tumor cell lines, respectively, $^{*}P < 0.05$. *Source:* Yu et al. [197] / with permission of The Royal Society of Chemistry.

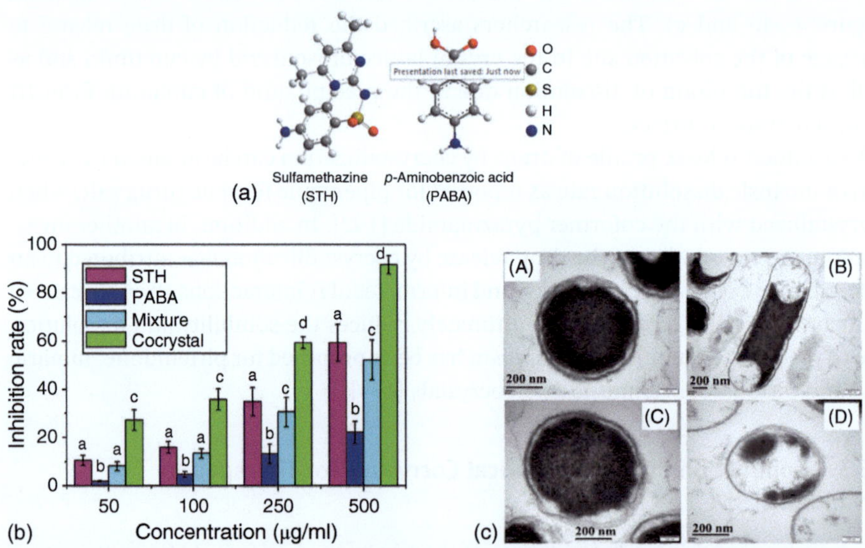

Figure 4.26 (a) Chemical structure of sulfamethazine and *p*-aminobenzoic acid. (b) Antibacterial effect of sulfamethazine and *p*-aminobenzoic acid, physical mixture, and their cocrystal on *Escherichia coli*. Different letters within figure denote a significant difference ($p < 0.05$). (c) Transmission electron microscopy images of *E. coli* before (A) and after treatment with 500 μg/ml sulfamethazine (B), *p*-aminobenzoic acid (C), and cocrystal (D) for 12 hours. *Source:* Pan et al. [198] / with permission of American Chemical Society.

the drug's physicochemical properties leads to enhanced *in vitro* cytotoxicity due to higher drug concentration near tumor cells and enhanced permeability across tumor cell membranes (Figure 4.25c).

A twofold increase in antibacterial activity was recorded for sulfamethazine: *p*-aminobenzoic acid cocrystals when compared to the parent drug (Figure 4.26a,b) [198]. The authors attributed this improved antibacterial activity to the increased solubility of cocrystals and significant cellular leakage when applying cocrystals formula (Figure 4.26c).

More examples of application of cocrystallization approach for therapeutic advantage can be found in the cited review [160].

4.7 Regulatory Considerations and Approval Pathways

The regulatory guidelines for pharmaceutical cocrystals and their formulations strongly influence the methodologies applied in development and quality control. The first regulatory agency to provide a classification for pharmaceutical cocrystals was the USFDA, which defines them as "solids which are crystalline materials composed of two or more molecules in the same crystal lattice" [8]. According to the USFDA definition, pharmaceutical cocrystals are considered as drug product intermediates with the expectation of enhancing the physicochemical attributes of the parent drug. Consequently, the USFDA guidelines listed the data required to support the submission and implementation of new drug applications and abbreviated new drug applications based on some evidence on the subsequent facts:

- Presence of drug and coformer in the same crystal lattice.
- pK_a difference of API and coformer/s of <1.
- The extent of proton transfer.
- Dissociation of the coformer from the drug preceding to drug's site of therapeutic activity.

Cocrystals have recently been categorized as "new active substances" and defined as "The different salts, esters, ethers, isomers, mixtures of isomers, complexes, or derivatives of an active substance shall be considered to be the same active substance unless they differ significantly in properties concerning safety and/or efficacy" [199]. From a scientific perspective, solvates, including hydrates, are regarded as a subclass of cocrystals, although regulatory interpretations may vary. The EMA does not classify multi-phase materials prepared via co-precipitation or physical mixing as cocrystals [200].

EMA considers cocrystals as a "new active substance" under the condition that they are safe and efficacious [9]. In addition, EMA states that the documentation procedure for salts can be adopted for cocrystals due to their conceptual similarities. However, if complex coformers are used, additional scientific documentation may be necessary.

A comparison of cocrystals' regulatory parameters between USFDA and EMA can be checked in the cited review [201].

Table 4.2 Examples of cocrystals-based products in the market and in clinical trials.

Name	Description	Uses	Developer	References
Seglentis®	A tramadol and celecoxib cocrystals approved by USFDA in 2021	Treatment of acute pain	Kowa Pharmaceuticals, USA	[203]
Entresto®	Cocrystal of sacubitril and valsartan, approved by USFDA in July 2015 and EMA in November 2015 Commercially available as film-coated tablets in variety of strengths	Treatment of symptomatic chronic heart failure	Novartis	[202]
Suglat®	Drug–nutraceutical cocrystals of ipragliflozin-L-proline, available as film-coated tablets	Treatment of type 2 diabetes in Japan	Astellas Pharma and Kotobuki Pharmaceuticals	[204]
TAK-020 gentisic acid cocrystals	Cocrystal product in phase I clinical trial	Treatment of rheumatoid arthritis	Takeda Pharmaceuticals	[202]
Lexapro®	Escitalopram oxalate cocrystal	Treatment of depression	Lundbeck Australia Pty. Ltd.	[18]
E58425 (tramadol/celecoxib) cocrystals	Ionic cocrystals of tramadol and celecoxib in ongoing clinical trials (Phase III)	Treatment of pain	ESTEVE Incorporation	[205]
Ertugliflozin-L-pyroglutamic acid cocrystal	Cocrystals of Ertugliflozin-L-pyroglutamic acid in phase III clinical trials	Prevention of hyperglycemia in type 2 diabetes	Pfizer	[206]
Lamivudine-zidovudine Teva®	Drug–drug cocrystals approved in 2011 by EMA	Combination therapy in treatment of human immunodeficiency virus (HIV) infection	Teva Pharmaceuticals	[207]

4.8 Commercially Available Cocrystals-Based Pharmaceuticals

Over the last 25 years, significant research has been conducted in cocrystallization, and many patents have been granted. This flourishing exploration led to the commercialization of Entresto® (sacubitril: valsartan), which was approved by the USFDA for the treatment of heart failure in 2015. This was followed by another approval of Seglentis® (tramadol–celecoxib cocrystals) in 2021 [202]. In addition, the successful ongoing clinical trials of cocrystals such as E-58425 (racemic tramadol: celecoxib, ionic cocrystals) for the treatment of acute pain further support the increase in the demand for generation of such crystalline complexes [202]. Table 4.2 shows examples of cocrystals-based products in the market and in clinical trials along with their uses and the developers.

4.9 Challenges and Future Perspectives

4.9.1 Challenges in Cocrystals-Based Product Development

Despite their valuable advantages, cocrystal product development encounters many challenges, which hurdle their broad pharmaceutical application. Tackling such challenges is necessary to allow their full utilization in pharmaceutical formulations. One of the main obstacles is the lack of systematic approach in cocrystal development, particularly selection of the potential coformer. In recent years, some effective screening methods have been applied to identify possible coformers such as pK_a, Cambridge Structural Database, HSP, and supramolecular synthon approach. However, the need to develop more efficient tools remains [208]. In addition, sometimes it is difficult to determine whether a cocrystal or a salt is formed. Based on USFDA guidance, the pK_a difference of ≥ 1 with the proton transfer between API and coformer is associated with salt formation, and that <1 without proton transfer is related to cocrystal formation. However, the proton transfer process can be affected by temperature and crystallization environment [209]. It was reported that escitalopram oxalate has the ability to cocrystallize with oxalic acid and may behave as a salt or as a cocrystal [18]. Scalability is a major concern for cocrystal synthesis. A limited number of techniques are available for continuous cocrystal formation. Application of such manufacturing methods for large-scale production of cocrystals necessitates in-depth knowledge of the critical process parameters.

Furthermore, the cocrystal formation process has a high degree of variability, and many critical parameters (e.g. purity of solid entities, amorphous content of cocrystals, and polymorphism) are encountered. These parameters affect the performance and the yield of the production technique and the cocrystals' purity. To produce cocrystal-based products of reasonable quality, the adjustment of these parameters before manufacturing is crucial. Another significant challenge for cocrystals is maintaining their structural integrity and physicochemical properties during the formulation and development of dosage forms. Instability of cocrystals in the presence of commonly applied formulation excipients such as the diluent calcium phosphate anhydrous, the lubricant calcium stearate, and the emulsifier sodium citrate has been reported [72]. Another obstacle observed for highly soluble cocrystals is that they tend to release the parent, less soluble drug during dissolution. This issue can be mitigated by the inclusion of supersaturation inhibitors such as polymers and surfactants. Therefore, in the formulation of cocrystals-based product, careful selection of coformers must be performed to ensure cocrystal stability [210]. It is important to note that there is missing data related to the evaluation of physicochemical and biopharmaceutical characteristics in most of the published literature [211–213].

Additionally, cocrystals usually have a fixed stoichiometric ratio between their components, which can complicate dose adjustment because this ratio may not correspond to the therapeutic dose required for clinical efficacy.

4.9.2 Multidrug Cocrystals

Multidrug cocrystals (MDCs) are molecular complexes of two or more therapeutically active solids in a stochiometric ratio held together by nonionic interactions and rarely via a combination of ionic and nonionic interactions in the same crystal lattice with or without the presence of solvate molecules [214]. MDCs surpass the traditional fixed-dose combinations in terms of stability, solubility, and compatibility aspects. These therapeutic hybrids can reduce a drug's adverse effects via reduction of the required dose, decreasing the incidence of drug resistance, improving patient compliance, and enabling cost-effective manufacturing. Moreover, MDCs are reported to have synergistic and additive effects [215–219]. These advantages are necessary for the effective management of complex disorders such as cardiovascular diseases, diabetes, cancer, infectious diseases, and HIV/AIDS. However, the development of MDCs may present development hurdles, including the selection of an appropriate pharmaceutically acceptable hybrid taking into consideration dose adjustment, drug–drug interactions, expected therapeutic benefits, stability, compatibility, and solubility.

4.9.3 Drug–Nutraceutical Cocrystals

"Nutraceutical" is a term initially introduced by Stephen Felice in 1989 and ascribes to naturally occurring compounds present in food or food products with ability to aid in disease treatment or even prevention [220]. Polyphenols represent a common class of these compounds with disease-preventing potential and have been used as antioxidants, radical scavengers, and in cosmetics [220, 221]. These compounds could be attractive coformers for pharmaceutical cocrystallization as they have robust supramolecular synthons i.e. functional groups with cocrystal-forming ability, and that also makes them attractive target for IP. They can offer several advantages including synergistic effects when cocrystallized with API and improving API's physicochemical properties such as solubility. They have antioxidant attributes, which can be applied to stabilize oxygen-labile drugs or to reduce the incidence of diseases associated with oxidative stress. Drug–nutraceutical cocrystals are easier to develop and have the ability to overcome challenges associated with drug–drug cocrystal formation [214]. Successful cocrystal formation as acetonitrile solvate between sinapic acid (nutraceutical) and the anti-hyperthyroidism drug, 6-propyl-2-thiouracil (PTU) was reported [220]. This opened the door to new pharmaceutical formulations of this API. However, this hybrid exhibited only one-week stability at 40 °C, 75% RH (accelerated stability conditions, ICH) with desolvation. Despite their conspicuous health benefits, some nutraceuticals suffer from poor physicochemical properties i.e. poor stability, solubility, and bioavailability [221]. Cocrystallization of these nutraceuticals with soluble drugs offers the opportunity to improve their physicochemical and biopharmaceutical properties. Gallic acid-(*S*)-oxiracetam cocrystals showed improved physical stability over time and relative humidity. Detailed examples of such an approach have been interestingly addressed in a review titled "cocrystallization of nutraceuticals" [221]. A new range of powerful therapeutic hybrids will emerge upon cocrystallization of nutraceuticals with APIs with synergistic benefits and minimized adverse effects and may find great applications in multiple disorders such as Alzheimer's disease, arthritis, diabetes, cancer, and cardiovascular disease. Recently, ionic cocrystals of lithium and racetam were reported [222–225]. It is worth noting that formation of ionic cocrystals of inorganic systems, particularly lithium and magnesium salts, with nutraceuticals could offer new treatment avenues for psychiatric disorders and neuropathic pain accordingly [226, 227].

4.9.4 Nano-cocrystals

NCCs present numerous advantages such as enhanced dissolution rate and bioavailability, rapid onset of action, and prolonged circulation time linked to their size in nano range. There is limited research published on NCCs. Recently, a new hope was given to this area upon successful development of caffeine–oxalic acid, caffeine–glutaric acid, and itraconazole–carboxylic acid NCCs [228, 229]. The first two cocrystals were prepared by an innovative continuous spray flash evaporation process; this technique can be further explored for NCCs formation. Undoubtedly, the arena of nanotechnology is an emerging area for cocrystal research, especially for disorders such as cancer, in which drug–drug hybrids can address bioavailability and dissolution issues of the parent drugs. However, significant effort and investigations are required to identify robust synthons in existing drugs to form MDCs. Owing to their complex nature, NCCs require sophisticated analytical techniques to identify the process conditions that are critical to achieving acceptable limits for key attributes in their production processes.

4.9.5 Process Analytical Technology and Cocrystals

The key ingredient to make a process innovation in pharmaceutical industry is to have deep understanding of the process. Hence, USFDA encourages the implementation of quality by design and process analytical technology (PAT) approaches in the pharmaceutical operations [230]. PAT is a tool to measure the critical quality attributes that have potential influence on the final product quality. Hence, it enables planning, analyzing, and controlling of pharmaceutical manufacturing processes. PAT aims to improve the final product quality and production efficiency and to reduce the manufacturing costs. Various PAT tools are applied to crystallization processes [231]. Recently, near-infrared (NIR) has been extensively used as a fast, nondestructive, and cost-effective PAT tool. The use of NIR to monitor cocrystal formation by HME and to predict cocrystal concentration in a powder mixture was reported in the literature [232, 233].

4.9.6 Cocrystal Polymorphism

The polymorphism of cocrystals has gathered substantial interest due to its potential to improve commercially available formulations by providing greater flexibility in modifying various attributes. This polymorphism can lead to improvement in solubility, stability, and bioavailability of drugs. In addition, the ability to produce multiple crystal structures of a cocrystal offers deliberate advantages in intellectual property protection (patent life cycle management), as each distinct polymorph can be independently patented.

4.10 Conclusion

For numerous crystalline drugs, their therapeutic efficacy and market availability are frequently restrained by poor physicochemical attributes such as low aqueous solubility, limited bioavailability, and instability, which can prevent their effective delivery and intended treatment outcomes. Such properties can be modified by tailoring drug's crystal structure through noncovalent interactions with appropriate coformer/s, i.e. cocrystallization. Pharmaceutical cocrystallization emerges as a potentially exuberant method for enhancing solubility, dissolution rate, and consequent bioavailability of poorly soluble drugs, while preserving appropriate physicochemical stability. Modern advances in coformer screening, supported by computational approaches, have boosted the process of cocrystal development, while innovative preparation techniques permit greater control over the generation of cocrystals and stable products.

Consequently, and owing to massive research, several cocrystal-based products have been successfully marketed, fulfilling regulatory body's requirements and demonstrating advanced characteristics over conventional preparations. However, the development of cocrystals-based pharmaceutical products faces many challenges related to cocrystal formation, stability, and scale-up manufacturing.

International multidisciplinary collaboration between crystallographers, chemists, analysts, pharmaceutical scientists, and high-tech experts in *in silico* screening approaches such as computational chemistry tools and ML methods is required to take these therapeutic hybrids to the apex of commercial success. In-depth understanding of supramolecular interactions in cocrystals and their constituents' association and dissociation is critical to ensure that they provide their intended pharmaceutical outcomes. To achieve this, robust prediction models need to be developed. Drug–coformer interactions must be considered during selection of the combination and possible cocrystal design/manufacturing.

One can conclude that cocrystals offer an interesting platform for new patents in the pharmaceutical industry owing to their novelty, utility, and nonobviousness, and many generic products are expected to emerge in the near future.

List of Abbreviations

APIs	active pharmaceutical ingredients
BCS	Biopharmaceutics Classification System
CCDC	Cambridge Crystallographic Data Centre
COSMO-RS	conductor-like screening model for real solvents
CSP	crystal structure prediction
COD	crystallography open database
DFT	density functional theory
DSC	differential scanning calorimetry
EMA	European Medicines Agency
FTIR	Fourier transform infrared spectroscopy
GRAS	generally recognized as safe
HSPs	Hansen solubility parameters
HME	hot melt extrusion
HSM	hot stage microscopy
ML	machine learning
MDs	molecular dynamics simulation
MDCs	multidrug cocrystals
NCCs	nanococrystals
NMR	nuclear magnetic resonance
PXRD	powder X-ray diffraction
PAT	process analytical technology
QSAR	quantitative structure–activity relationship
QM	quantum mechanical
SEM	scanning electron microscopy
SCXRD	single-crystal X-ray diffraction
SSNMR	solid-state nuclear magnetic resonance
TGA	thermogravimetric analysis
USFDA	U.S. Food and Drug Administration

References

1 Karagianni, A., Malamatari, M., and Kachrimanis, K. (2018). Pharmaceutical cocrystals: new solid phase modification approaches for the formulation of APIs. *Pharmaceutics* 10 (1): 18.

2 Kumar Bandaru, R., Rout, S.R., Kenguva, G. et al. (2021). Recent advances in pharmaceutical cocrystals: from bench to market. *Front. Pharmacol.* 12: 780582.

3 Gadade, D.D. and Pekamwar, S.S. (2016). Pharmaceutical cocrystals: regulatory and strategic aspects, design and development. *Adv. Pharm. Bull.* 6 (4): 479.

4 Savale, A., Mogal, R., Talele, S. et al. (2023). Pharmaceutical cocrystals: a novel systematic approach for the administration of existing drugs in new crystalline form. *Biosci. Biotechnol. Res. Asia* 20 (4): 1195–1210.

5 Stahly, G.P. (2009). A survey of cocrystals reported prior to 2000. *Cryst. Growth Des.* 9 (10): 4212–4229.

6 Nangia, A. (2010). Supramolecular chemistry and crystal engineering. *J. Chem. Sci.* 122 (3): 295–310.

7 Swiss Patent F. Hoffmann-La Roche & Co, Basel (Switzerland) (1937). Verfahren zur Darstellung von 2,4-Dioxo- 3,3-diäthyl-tetrahydropyridin. CH 187826.

8 Food and Drug Administration (2013). Regulatory classification of pharmaceutical co-crystals, guidance for industry. https://www.fda.gov/downloads/Drugs/Guidances/UCM281764.pdf (accessed 25 July 2025).

9 EMA (2014). Reflection paper on the use of cocrystals and other solid state forms of active substances in medicinal products. European Medicines Agency.

10 Schultheiss, N. and Newman, A. (2009). Pharmaceutical cocrystals and their physicochemical properties. *Cryst. Growth Des.* 9 (6): 2950–2967.

11 Kalepu, S. and Nekkanti, V. (2015). Insoluble drug delivery strategies: review of recent advances and business prospects. *Acta Pharm. Sin. B* 5 (5): 442–453.

12 Xie, B., Liu, Y., Li, X. et al. (2024). Solubilization techniques used for poorly water-soluble drugs. *Acta Pharm. Sin. B* 14: 4683–4716.

13 Blagden, N., de Matas, M., Gavan, P.T., and York, P. (2007). Crystal engineering of active pharmaceutical ingredients to improve solubility and dissolution rates. *Adv. Drug Deliv. Rev.* 59 (7): 617–630.

14 Pandi, P., Bulusu, R., Kommineni, N. et al. (2020). Amorphous solid dispersions: An update for preparation, characterization, mechanism on bioavailability, stability, regulatory considerations and marketed products. *Int. J. Pharm.* 586: 119560.

15 Gavezzotti, A., Colombo, V., and Lo Presti, L. (2016). Facts and factors in the formation and stability of binary crystals. *Cryst. Growth Des.* 16 (10): 6095–6104.

16 Food and Drug Administration (2015). Guidances (drugs).

17 U.S. Food and Drug Administration (2018). *Regulatory classification of pharmaceutical co-crystals guidance for industry*. Silver Spring, MD: FDA. Available at: https://www.fda.gov/files/drugs/published/Regulatory-Classification-of-Pharmaceutical-Co-Crystals.pdf (accessed 25 July 2025).

18 Harrison, W.T.A., Yathirajan, H.S., Bindya, S., and Anilkumar, H.G. (2007). Escitalopram oxalate: co-existence of oxalate dianions and oxalic acid molecules in the same crystal. *Acta Crystallogr. Sect. C: Cryst. Struct. Commun.* 63 (2): o129–o131.

19 Ngilirabanga, J.B. and Samsodien, H. (2021). Pharmaceutical co-crystal: an alternative strategy for enhanced physicochemical properties and drug synergy. *Nano Select* 2 (3): 512–526.

20 Sakhiya, D.C. and Borkhataria, C.H. (2024). A review on advancement of cocrystallization approach and a brief on screening, formulation and characterization of the same. *Heliyon* 10 (7): e29057.

21 Desiraju, G.R. (2010). Crystal engineering: a brief overview. *J. Chem. Sci.* 122: 667–675.

22 Banerjee, M., Nimkar, K., Naik, S., and Patravale, V. (2022). Unlocking the potential of drug-drug cocrystals–a comprehensive review. *J. Control. Release* 348: 456–469.

23 Fukte, S.R., Wagh, M.P., and Rawat, S. (2014). Coformer selection: an important tool in cocrystal formation. *Int J Pharm Pharm Sci* 6 (7): 9–14.

24 Kumar, S. and Nanda, A. (2017). Pharmaceutical cocrystals: an overview. *Indian J. Pharm. Sci.* 79 (6): 858–871.

25 Rehder, S., Klukkert, M., Löbmann, K.A. et al. (2011). Investigation of the formation process of two piracetam cocrystals during grinding. *Pharmaceutics* 3 (4): 706–722.

26 Jung, S., Choi, I., and Kim, I.W. (2015). Liquid-assisted grinding to prepare a cocrystal of adefovir dipivoxil thermodynamically less stable than its neat phase. *Crystals* 5 (4): 583–591.

27 von Essen, C. and Luedeker, D. (2023). In silico co-crystal design: assessment of the latest advances. *Drug Discov. Today* 28 (11): 103763.

28 Issa, N., Karamertzanis, P.G., Welch, G.W., and Price, S.L. (2009). Can the formation of pharmaceutical cocrystals be computationally predicted? I. Comparison of lattice energies. *Cryst. Growth Des.* 9 (1): 442–453.

29 Kumar, A. and Nanda, A. (2021). In-silico methods of cocrystal screening: a review on tools for rational design of pharmaceutical cocrystals. *J. Drug Delivery Sci. Technol.* 63: 102527.

30 Lemli, B., Pál, S., and Széchenyi, A. (2024). Prioritizing computational cocrystal prediction methods for experimental researchers: a review to find efficient, cost-effective, and user-friendly approaches. *Int. J. Mol. Sci.* 25 (22): 12045.

31 Bu, F.-Z., Yu, Y.-M., Shen, Y.-L. et al. (2023). Cocrystallization with nutrient ferulic acid towards reducing the dissolubility behaviors of antifungal drug 5-fluorocytosine: an integrated theoretical and experimental case research. *J. Mol. Struct.* 1275: 134601.

32 Bofill, L., Prohens, R., Barbas, R., and Frontera, A. (2020). DFT analysis of uncommon π··· H-bond array interaction in a new pterostilbene/theophylline cocrystal. *Cryst. Growth Des.* 20 (10): 6691–6698.

33 Van Mourik, T., Bühl, M., and Gaigeot, M.-P. (2014). *Density Functional Theory Across Chemistry, Physics and Biology*, 20120488. The Royal Society Publishing.

34 Du, J.J., Lai, F., Váradi, L. et al. (2018). Monoclinic paracetamol vs. paracetamol-4, 4′-bipyridine co-crystal; what is the difference? a charge density study. *Crystals* 8 (1): 46.

35 Ahmadi, S., Ghanavati, M.A., and Rohani, S. (2024). Machine learning-guided prediction of cocrystals using point cloud-based molecular representation. *Chem. Mater.* 36 (3): 1153–1161.

36 Gaussian, Inc. (2025). *Gaussian 16: electronic structure modeling software*. Available at: https://gaussian.com/gaussian16/ (accessed 25 March 2026).

37 Dassault Systèmes BIOVIA (2025). *BIOVIA Materials Studio: an integrated, multi-scale modeling environment*. Available at: https://www.3ds.com/products/biovia/materials-studio (accessed 25 March 2026).

38 Quantum ESPRESSO Foundation (2025). *Quantum ESPRESSO: integrated open-source suite for electronic-structure calculations and materials modelling*. Available at: https://www.quantum-espresso.org/ (accessed 25 March 2026).

39 CASTEP Developers Group (2025). *CASTEP: first-principles materials modelling using density functional theory*. Available at: https://www.castep.org/ (accessed 25 March 2026).

40 VASP Software GmbH (2025). *VASP: Vienna Ab initio Simulation Package—first-principles electronic-structure and molecular-dynamics simulations*. Available at: https://www.vasp.at/ (accessed 25 March 2026).

41 Meng, X.-Y., Zhang, H.-X., Mezei, M., and Cui, M. (2011). Molecular docking: a powerful approach for structure-based drug discovery. *Curr. Comput. Aided Drug Des.* 7 (2): 146–157.

42 Wieczorek, G. and Niedzialek, D. (2020). Molecular dynamics. In: *Encyclopedia of Life Sciences (eLS); International Editorial Advisory Board*, 1–18. Wiley-Blackwell.

43 Dhibar, M., Chakraborty, S., Basak, S. et al. (2023). Critical analysis and optimization of stoichiometric ratio of drug-coformer on cocrystal design: molecular docking, in vitro and in vivo assessment. *Pharmaceuticals* 16 (2): 284.

44 Pagadala, N.S., Syed, K., and Tuszynski, J. (2017). Software for molecular docking: a review. *Biophys. Rev.* 9 (2): 91–102.

45 Chaudhari, K.R., Savjani, J.K., Savjani, K.T., and Shah, H. (2021). Improved pharmaceutical properties of ritonavir through co-crystallization approach with liquid-assisted grinding method. *Drug Dev. Ind. Pharm.* 47 (10): 1633–1642.

46 AUTODOCK (2025). AutoDock: automated molecular docking software https://autodock.scripps.edu/ (accessed 25 March 2026).

47 Vina, A. (2025). AutoDock Vina: molecular docking and virtual screening software. https://vina.scripps.edu/ (accessed 25 March 2026).

48 GOLD C (2025). GOLD: Protein–ligand docking software. https://www.ccdc.cam.ac.uk/solutions/software/gold/ (accessed 25 March 2026).

49 CCDC (2025). CCDC Software Suites. https://www.ccdc.cam.ac.uk/suites/ (accessed 25 March 2026).

50 Schrödinger, Inc (2025). *Desmond: high-performance molecular dynamics engine*. Available at: https://www.schrodinger.com/platform/products/ms-desmond/ (accessed 25 March 2026).

51 LAMMPS Developers (2025). *LAMMPS: Large-scale Atomic/Molecular Massively Parallel Simulator*. Available at: https://www.lammps.org/ (accessed 25 March 2026).

52 Desiraju, G.R. (2002). Cryptic crystallography. *Nat. Mater.* 1 (2): 77–79.

53 CALYPSO Developers. *CALYPSO: an efficient structure prediction method and computer software*. Available at: http://www.calypso.cn/home/ (accessed 25 March 2026).

54 University College London (UCL), CPOSS Group. *DMACRYS: Computational database for organic crystal structure prediction*. Available at: https://www.chem.ucl.ac.uk/cposs/dmacrys/index.html (accessed 25 March 2026).

55 Cambridge Crystallographic Data Centre (CCDC). *CSP Blind Test*. Available at: https://www.ccdc.cam.ac.uk/community/ccdc-for-the-community/partnerships-and-initiatives/csp-blind-test/ (accessed 25 March 2026).

56 Klamt, A. (2018). The COSMO and COSMO-RS solvation models. *Wiley Interdiscip. Rev. Comput. Mol. Sci.* 8 (1): e1338.

57 Abramov, Y.A., Loschen, C., and Klamt, A. (2012). Rational coformer or solvent selection for pharmaceutical cocrystallization or desolvation. *J. Pharm. Sci.* 101 (10): 3687–3697.

58 Mohammad, M.A., Alhalaweh, A., and Velaga, S.P. (2011). Hansen solubility parameter as a tool to predict cocrystal formation. *Int. J. Pharm.* 407 (1-2): 63–71.

59 HSPiP. *Hansen solubility parameters in practice*. Available at: https://www.hansen-solubility.com/HSPiP/ (accessed 25 March 2026).

60 Ho, T.K. (1998). The random subspace method for constructing decision forests. *IEEE Trans. Pattern Anal. Mach. Intell.* 20 (8): 832–844.

61 scikit-learn Developers. *scikit-learn: machine learning in Python*. Available at: https://scikit-learn.org/stable/index.html (accessed 25 March 2026).

62 Roy, K., Kar, S., and Das, R.N. (2015). *A Primer on QSAR/QSPR Modeling: Fundamental Concepts*. Springer.

63 KNIME *KNIME Analytics Platform*. Available at: https://www.knime.com/knime-analytics-platform (accessed 25 March 2026).

64 Molajafari F, Li T, Abbasichaleshtori M, ZD MH, Cozzolino AF, Fandrick DR, et al. Computational screening for prediction of co-crystals: method comparison and experimental validation. *CrystEngComm* 2024;26(11):1620-36.

65 CCDC. *ConQuest*. Available at: https://www.ccdc.cam.ac.uk/ (accessed 25 March 2026).

66 Cambridge Crystallographic Data Centre (CCDC). *Mercury*. Available at: https://www.ccdc.cam.ac.uk/solutions/software/mercury/ (accessed 25 March 2026).

67 CCDC. *CSD Python API*. Available at: https://www.ccdc.cam.ac.uk/solutions/software/csd-python/ (accessed 25 March 2026).

68 Crystallography Open Database (COD). *Crystallography Open Database*. Available at: https://www.crystallography.net/cod/ (accessed 25 March 2026).

69 DrugBank. *DrugBank Online*. Available at: https://www.drugbank.com/ (accessed 25 March 2026).

70 Korde, S., Pagire, S., Pan, H. et al. (2018). Continuous manufacturing of cocrystals using solid state shear milling technology. *Cryst. Growth Des.* 18 (4): 2297–2304.

71 Maheshwari, C., Jayasankar, A., Khan, N.A. et al. (2009). Factors that influence the spontaneous formation of pharmaceutical cocrystals by simply mixing solid reactants. *CrystEngComm* 11 (3): 493–500.

72 Duggirala, N.K., Vyas, A., Krzyzaniak, J.F. et al. (2017). Mechanistic insight into caffeine–oxalic cocrystal dissociation in formulations: role of excipients. *Mol. Pharm.* 14 (11): 3879–3887.

73 Friščić, T., Childs, S.L., Rizvi, S.A., and Jones, W. (2009). The role of solvent in mechanochemical and sonochemical cocrystal formation: a solubility-based approach for predicting cocrystallisation outcome. *CrystEngComm* 11 (3): 418–426.

74 Sanphui, P., Goud, N.R., Khandavilli, U.B.R., and Nangia, A. (2011). Fast dissolving curcumin cocrystals. *Cryst. Growth Des.* 11 (9): 4135–4145.

75 Daurio, D., Medina, C., Saw, R. et al. (2011). Application of twin screw extrusion in the manufacture of cocrystals, part I: four case studies. *Pharmaceutics* 3 (3): 582–600.

76 Wood, C.J. (2016). Investigation of a solvent-free continuous process to produce pharmaceutical co-crystals. Understanding and developing solvent-free continuous cocrystallisation (SFCC) through study of co-crystal formation under the application of heat, model shear and twin screw extrusion, including development of a near infrared spectroscopy partial least squares quantification method. University of Bradford.

77 Medina, C., Daurio, D., Nagapudi, K., and Alvarez-Nunez, F. (2010). Manufacture of pharmaceutical co-crystals using twin screw extrusion: a solvent-less and scalable process. *J. Pharm. Sci.* 99 (4): 1693–1696.

78 Dhumal, R.S., Kelly, A.L., York, P. et al. (2010). Cocrystalization and simultaneous agglomeration using hot melt extrusion. *Pharm. Res.* 27: 2725–2733.

79 Karimi-Jafari, M., Padrela, L., Walker, G.M., and Croker, D.M. (2018). Creating cocrystals: a review of pharmaceutical cocrystal preparation routes and applications. *Cryst. Growth Des.* 18 (10): 6370–6387.

80 Boksa, K., Otte, A., and Pinal, R. (2014). Matrix-assisted cocrystallization (MAC) simultaneous production and formulation of pharmaceutical cocrystals by hot-melt extrusion. *J. Pharm. Sci.* 103 (9): 2904–2910.

81 Narala, S., Nyavanandi, D., Alzahrani, A. et al. (2022). Creation of hydrochlorothiazide pharmaceutical cocrystals via hot-melt extrusion for enhanced solubility and permeability. *AAPS PharmSciTech* 23 (1): 56.

82 Patil, S.P., Modi, S.R., and Bansal, A.K. (2014). Generation of 1: 1 carbamazepine: nicotinamide cocrystals by spray drying. *Eur. J. Pharm. Sci.* 62: 251–257.

83 do Amaral, L.H., do Carmo, F.A., Amaro, M.I. et al. (2018). Development and characterization of dapsone cocrystal prepared by scalable production methods. *AAPS PharmSciTech* 19: 2687–2699.

84 Kaviani, R., Jouyban, A., and Shayanfar, A. (2025). Chiral resolution of RS-ofloxacin through a novel diastereomeric cocrystal formation with

L-glutamic acid by evaporative crystallization and quantification with capillary electrophoresis. *Sep. Purif. Technol.* 354: 129040.

85 Ito, F., Mochiduki, Y., Fujimoto, Y. et al. (2024). Effect of α-substitution on dibenzoylmethanatoboron difluoride on evaporative crystallization. *J. Phys. Chem. C* 128 (3): 1469–1476.

86 Xuan, B., Wong, S.N., Zhang, Y. et al. (2020). Extended release of highly water soluble isoniazid attained through cocrystallization with curcumin. *Cryst. Growth Des.* 20 (3): 1951–1960.

87 Dutt, B., Choudhary, M., and Budhwar, V. (2020). Preparation, characterization and evaluation of aspirin: benzoic acid cocrystals with enhanced pharmaceutical properties. *Futur. J. Pharm. Sci.* 6: 1–16.

88 Bhatia, M. and Devi, S. (2024). Co-crystallization: a green approach for the solubility enhancement of poorly soluble drugs. *CrystEngComm* 26 (3): 293–311.

89 Sathisaran, I., Bhatia, D.D., and Dalvi, S.V. (2020). New curcumin-trimesic acid cocrystal and anti-invasion activity of curcumin multicomponent solids against 3D tumor models. *Int. J. Pharm.* 587: 119667.

90 Chun, N.-H., Lee, M.-J., Song, G.-H. et al. (2014). Combined anti-solvent and cooling method of manufacturing indomethacin–saccharin (IMC–SAC) co-crystal powders. *J. Cryst. Growth* 408: 112–118.

91 Yu, Z.Q., Chow, P.S., and Tan, R.B. (2010). Operating regions in cooling cocrystallization of caffeine and glutaric acid in acetonitrile. *Cryst. Growth Des.* 10 (5): 2382–2387.

92 Zhang, G.G., Henry, R.F., Borchardt, T.B., and Lou, X. (2007). Efficient co-crystal screening using solution-mediated phase transformation. *J. Pharm. Sci.* 96 (5): 990–995.

93 Pawar, N., Saha, A., Nandan, N., and Parambil, J.V. (2021). Solution cocrystallization: a scalable approach for cocrystal production. *Crystals* 11 (3): 303.

94 Ahuja, D., Ramisetty, K.A., Sumanth, P.K. et al. (2020). Microwave assisted slurry conversion crystallization for manufacturing of new co-crystals of sulfamethazine and sulfamerazine. *CrystEngComm* 22 (8): 1381–1394.

95 Wang, I.-C., Lee, M.-J., Sim, S.-J. et al. (2013). Anti-solvent co-crystallization of carbamazepine and saccharin. *Int. J. Pharm.* 450 (1-2): 311–322.

96 Kudo, S. and Takiyama, H. (2018). Solubility determination for carbamazepine and saccharin in methanol/water mixed solvent: basic data for design of cocrystal production by antisolvent crystallization. *J. Chem. Eng. Data* 63 (2): 451–458.

97 Thakor, P., Yadav, B., Modani, S., and Shastri, N.R. (2020). Preparation and optimization of nano-sized cocrystals using a quality by design approach. *CrystEngComm* 22 (13): 2304–2314.

98 Padrela, L., Rodrigues, M.A., Tiago, J. et al. (2014). Tuning physicochemical properties of theophylline by cocrystallization using the supercritical fluid enhanced atomization technique. *J. Supercrit. Fluids* 86: 129–136.

99 Neurohr, C., Marchivie, M., Lecomte, S. et al. (2015). Naproxen–nicotinamide cocrystals: racemic and conglomerate structures generated by CO_2 antisolvent crystallization. *Cryst. Growth Des.* 15 (9): 4616–4626.

100 Pando, C., Cabañas, A., and Cuadra, I.A. (2016). Preparation of pharmaceutical co-crystals through sustainable processes using supercritical carbon dioxide: a review. *RSC Adv.* 6 (75): 71134–71150.

101 Müllers, K.C., Paisana, M., and Wahl, M.A. (2015). Simultaneous formation and micronization of pharmaceutical cocrystals by rapid expansion of supercritical solutions (RESS). *Pharm. Res.* 32: 702–713.

102 Padrela, L., Rodrigues, M.A., Velaga, S.P. et al. (2009). Formation of indomethacin–saccharin cocrystals using supercritical fluid technology. *Eur. J. Pharm. Sci.* 38 (1): 9–17.

103 Neurohr, C., Erriguible, A., Laugier, S., and Subra-Paternault, P. (2016). Challenge of the supercritical antisolvent technique SAS to prepare cocrystal-pure powders of naproxen-nicotinamide. *Chem. Eng. J.* 303: 238–251.

104 Padrela, L., Rodrigues, M.A., Velaga, S.P. et al. (2010). Screening for pharmaceutical cocrystals using the supercritical fluid enhanced atomization process. *J. Supercrit. Fluids* 53 (1-3): 156–164.

105 Mehta, B. (2016). The formation of pharmaceutical co-crystals by spray drying. An investigation into the chemical and physical factors affecting the production of pharmaceutical co-crystals by fast evaporation and spray drying. University of Bradford.

106 Tawfeek, H.M., Chavan, T., and Kunda, N.K. (2020). Effect of spray drying on amorphization of indomethacin nicotinamide cocrystals; optimization, characterization, and stability study. *AAPS PharmSciTech* 21: 1–13.

107 Eddleston, M.D., Patel, B., Day, G.M., and Jones, W. (2013). Cocrystallization by freeze-drying: preparation of novel multicomponent crystal forms. *Cryst. Growth Des.* 13 (10): 4599–4606.

108 Alwati, A.A. (2017). Ultrasound assisted processing of solid state pharmaceuticals. The application of ultrasonic energy in novel solid state pharmaceutical applications, including solvent free co-crystallisation (SFCC) and enhanced compressibility. University of Bradford.

109 Apshingekar, P.P., Aher, S., Kelly, A.L. et al. (2017). Synthesis of caffeine/maleic acid co-crystal by ultrasound-assisted slurry co-crystallization. *J. Pharm. Sci.* 106 (1): 66–70.

110 Cullity, B.D. and Smoluchowski, R. (1957). Elements of X-ray Diffraction. *Phys. Today* 10 (3): 50.

111 Bragg, W.H. (1913). The reflection of X-rays by crystals (II.). *Proc. R. Soc. Lond. Ser. A Contain. Pap. Math. Phys. Character* 89 (610): 246–248.

112 Lapidus, S.H., Stephens, P.W., Arora, K.K. et al. (2010). A comparison of cocrystal structure solutions from powder and single crystal techniques. *Cryst. Growth Des.* 10 (10): 4630–4637.

113 Chandel, N., Gupta, V., Pandey, A., and Saxena, S. (2011). Co-crystalization of aceclofenac and paracetamol and their characterization. *Int. J. Pharm. Life Sci.* 2 (8).

114 Bhattacharyya, S. and Manjunath, A. (2023). Pharmaceutical cocrystal-a deft technique for solubility enhancement. *Pharm. Sci. Asia.* 50 (4): 361–370.

115 Al-Hanafi, M.A.S., Alsirawan, M.B., Seaton, C.C. et al. (2022). Interfacial cocrystallization using oily phase via liquid–liquid phase separation. *Cryst. Growth Des.* 22 (10): 5845–5851.

116 Gaisford, S., Kett, V., and Haines, P. (2019). *Principles of Thermal Analysis and Calorimetry*. Royal Society of Chemistry.

117 Park, B., Yoon, W., Yun, J. et al. (2019). Emodin-nicotinamide (1: 2) cocrystal identified by thermal screening to improve emodin solubility. *Int. J. Pharm.* 557: 26–35.

118 Bruni, G., Maggi, L., Mustarelli, P. et al. (2019). Enhancing the pharmaceutical behavior of nateglinide by cocrystallization: physicochemical assessment of cocrystal formation and informed use of differential scanning calorimetry for its quantitative characterization. *J. Pharm. Sci.* 108 (4): 1529–1539.

119 Lu, E., Rodríguez-Hornedo, N., and Suryanarayanan, R. (2008). A rapid thermal method for cocrystal screening. *CrystEngComm* 10 (6): 665–668.

120 Lin, H.-L., Zhang, G.-C., and Lin, S.-Y. (2015). Real-time co-crystal screening and formation between indomethacin and saccharin via DSC analytical technique or DSC–FTIR microspectroscopy. *J. Therm. Anal. Calorim.* 120: 679–687.

121 Saganowska, P. and Wesolowski, M. (2018). DSC as a screening tool for rapid co-crystal detection in binary mixtures of benzodiazepines with co-formers. *J. Therm. Anal. Calorim.* 133: 785–795.

122 Alam, Q., Ganeshpurkar, A., Singh, S.K., and Krishnamurthy, S. (2023). Novel gastroprotective and thermostable cocrystal of dimethyl fumarate: its preparation, characterization, and in vitro and in vivo evaluation. *ACS Omega* 8 (29): 26218–26230.

123 Šimek, M., Grünwaldová, V., and Kratochvíl, B. (2014). Hot-stage microscopy for determination of API particles in a formulated tablet. *Biomed. Res. Int.* 2014 (1): 832452.

124 Abu Bakar, M., Nagy, Z., and Rielly, C. (2010). A combined approach of differential scanning calorimetry and hot-stage microscopy with image analysis in the investigation of sulfathiazole polymorphism. *J. Therm. Anal. Calorim.* 99 (2): 609–619.

125 Lin, S.Y. and Cheng, W.T. (2012). The use of hot-stage microscopy and thermal micro-Raman spectroscopy in the study of phase transformation of metoclopramide HCl monohydrate. *J. Raman Spectrosc.* 43 (8): 1166–1170.

126 Verma, S., Raynaud, G., and Rapp, R. (1981). Hot-stage scanning electron microscope for high-temperature in-situ oxidation studies. *Oxid. Met.* 15 (5): 471–483.

127 Kofler, L. and Kofler, A. (1952). *Thermal Micromethods for the Study of Organic Compounds and Their Mixtures*. Wagner.

128 McNamara, D.P., Childs, S.L., Giordano, J. et al. (2006). Use of a glutaric acid cocrystal to improve oral bioavailability of a low solubility API. *Pharm. Res.* 23: 1888–1897.

129 Berry, D.J., Seaton, C.C., Clegg, W. et al. (2008). Applying hot-stage microscopy to co-crystal screening: a study of nicotinamide with seven active pharmaceutical ingredients. *Cryst. Growth Des.* 8 (5): 1697–1712.

130 Kumar, A., Singh, P., and Nanda, A. (2020). Hot stage microscopy and its applications in pharmaceutical characterization. *Appl. Microsc.* 50 (1): 12.

131 Garbacz, P. and Wesolowski, M. (2018). DSC, FTIR and Raman spectroscopy coupled with multivariate analysis in a study of co-crystals of pharmaceutical interest. *Molecules* 23 (9): 2136.

132 Lin, H.-L., Hsu, P.-C., and Lin, S.-Y. (2013). Theophylline–citric acid co-crystals easily induced by DSC–FTIR microspectroscopy or different storage conditions. *Asian J. Pharm. Sci.* 8 (1): 19–27.

133 Tanaka, R., Hattori, Y., Ashizawa, K., and Otsuka, M. (2019). Kinetics study of cocrystal formation between indomethacin and saccharin using high-shear granulation with in situ Raman spectroscopy. *J. Pharm. Sci.* 108 (10): 3201–3208.

134 Larkin, P. (2017). *Infrared and Raman Spectroscopy: Principles and Spectral Interpretation*. Elsevier.

135 Ketkar, S., Pagire, S.K., Goud, N.R. et al. (2016). Tracing the architecture of caffeic acid phenethyl ester cocrystals: studies on crystal structure, solubility, and bioavailability implications. *Cryst. Growth Des.* 16 (10): 5710–5716.

136 Vogt, F.G., Clawson, J.S., Strohmeier, M. et al. (2009). Solid-state NMR analysis of organic cocrystals and complexes. *Cryst. Growth Des.* 9 (2): 921–937.

137 L-x, L.I.A.N.G. and HOU Guang-jin, D.E.N.G.F. (2019). Quantitative cross polarization magic-angle spinning NMR spectroscopy in solids [J]. *Chinese J. Magn. Reson.* 37 (1): 1–15.

138 Peach, A.A., Hirsh, D.A., Holmes, S.T., and Schurko, R.W. (2018). Mechanochemical syntheses and 35 Cl solid-state NMR characterization of fluoxetine HCl cocrystals. *CrystEngComm* 20 (20): 2780–2792.

139 Zhao, L., Hanrahan, M.P., Chakravarty, P. et al. (2018). Characterization of pharmaceutical cocrystals and salts by dynamic nuclear polarization-enhanced solid-state NMR spectroscopy. *Cryst. Growth Des.* 18 (4): 2588–2601.

140 Jayram, P. and Sudheer, P. (2020). Pharmaceutical co-crystals: a systematic review. *Int. J. Pharm. Invest.* 10 (3).

141 Aher, S., Dhumal, R., Mahadik, K. et al. (2013). Effect of cocrystallization techniques on compressional properties of caffeine/oxalic acid 2: 1 cocrystal. *Pharm. Dev. Technol.* 18 (1): 55–60.

142 Savjani, K.T., Gajjar, A.K., and Savjani, J.K. (2012). Drug solubility: importance and enhancement techniques. *Int. Sch. Res. Notices* 2012 (1): 195727.

143 Remington, J.P. (2006). *Remington: The Science and Practice of Pharmacy*. Lippincott Williams & Wilkins.

144 Kuminek, G., Rodríguez-Hornedo, N., Siedler, S. et al. (2016). How cocrystals of weakly basic drugs and acidic coformers might modulate solubility and stability. *Chem. Commun.* 52 (34): 5832–5835.

145 Good, D.J. and Rodríguez-Hornedo, N. (2009). Solubility advantage of pharmaceutical cocrystals. *Cryst. Growth Des.* 9 (5): 2252–2264.

146 Oberoi, L.M., Alexander, K.S., and Riga, A.T. (2005). Study of interaction between ibuprofen and nicotinamide using differential scanning calorimetry, spectroscopy, and microscopy and formulation of a fast-acting and possibly better ibuprofen suspension for osteoarthritis patients. *J. Pharm. Sci.* 94 (1): 93–101.

147 Kara, D.D. and Rathnanand, M. (2022). Cocrystals and drug–drug cocrystals of anticancer drugs: a perception towards screening techniques, preparation, and enhancement of drug properties. *Crystals* 12 (10): 1337.

148 Hickey, M.B., Peterson, M.L., Scoppettuolo, L.A. et al. (2007). Performance comparison of a co-crystal of carbamazepine with marketed product. *Eur. J. Pharm. Biopharm.* 67 (1): 112–119.

149 Aitipamula, S., Chow, P.S., and Tan, R.B.H. (2009). Trimorphs of a pharmaceutical cocrystal involving two active pharmaceutical ingredients: potential relevance to combination drugs. *CrystEngComm* 11 (9): 1823–1827.

150 Duggirala, N.K., Smith, A.J., Wojtas, Ł. et al. (2014). Physical stability enhancement and pharmacokinetics of a lithium ionic cocrystal with glucose. *Cryst. Growth Des.* 14 (11): 6135–6142.

151 McKellar, S.C., Kennedy, A.R., McCloy, N.C. et al. (2014). Formulation of liquid propofol as a cocrystalline solid. *Cryst. Growth Des.* 14 (5): 2422–2430.

152 Bacchi, A., Capucci, D., Giannetto, M. et al. (2016). Turning liquid propofol into solid (without freezing it): thermodynamic characterization of pharmaceutical cocrystals built with a liquid drug. *Cryst. Growth Des.* 16 (11): 6547–6555.

153 Roy, S., Quiñones, R., and Matzger, A.J. (2012). Structural and physicochemical aspects of dasatinib hydrate and anhydrate phases. *Cryst. Growth Des.* 12 (4): 2122–2126.

154 Deng, J.-H., Lu, T.-B., Sun, C.C., and Chen, J.-M. (2017). Dapagliflozin-citric acid cocrystal showing better solid state properties than dapagliflozin. *Eur. J. Pharm. Sci.* 104: 255–261.

155 Huang, S., Xue, Q., Xu, J. et al. (2019). Simultaneously improving the physicochemical properties, dissolution performance, and bioavailability of apigenin and daidzein by co-crystallization with theophylline. *J. Pharm. Sci.* 108 (9): 2982–2993.

156 Pagire, S.K., Seaton, C.C., and Paradkar, A. (2020). Improving stability of effervescent products by co-crystal formation: a novel application of crystal engineered citric acid. *Cryst. Growth Des.* 20 (7): 4839–4844.

157 Velagaleti, R. (2008). Stability and shelf life of pharmaceutical products. In: *Pharmaceutical Sciences Encyclopedia: Drug Discovery, Development, and Manufacturing* (ed. S.C. Gad), 559. Wiley. doi: 10.1002/9780470571224.pse390.

158 Guo, C., Zhang, Q., Zhu, B. et al. (2020). Pharmaceutical cocrystals of nicorandil with enhanced chemical stability and sustained release. *Cryst. Growth Des.* 20 (10): 6995–7005.

159 Bofill, L., de Sande, D., Barbas, R., and Prohens, R. (2020). New cocrystal of ubiquinol with high stability to oxidation. *Cryst. Growth Des.* 20 (8): 5583–5588.

160 Guo, M., Sun, X., Chen, J., and Cai, T. (2021). Pharmaceutical cocrystals: a review of preparations, physicochemical properties and applications. *Acta Pharm. Sin. B* 11 (8): 2537–2564.

161 Ohashi, Y., Yanagi, K., Kurihara, T. et al. (1981). Crystalline state reaction of cobaloxime complexes by x-ray exposure. 1. Direct observation of cobalt-carbon bond cleavage in [(R)-1-cyanoethyl][(S)-(-)-. alpha.-methylbenzylamine] bis (dimethylglyoximato) cobalt (III). *J. Am. Chem. Soc.* 103 (19): 5805–5812.

162 Putra, O.D., Umeda, D., Nugraha, Y.P. et al. (2018). Simultaneous improvement of epalrestat photostability and solubility via cocrystallization: a case study. *Cryst. Growth Des.* 18 (1): 373–379.

163 Ramirez, M.A. and Borja, N.L. (2008). Epalrestat: an aldose reductase inhibitor for the treatment of diabetic neuropathy. *Pharmacother. J. Hum. Pharm. Drug Therapy* 28 (5): 646–655.

164 Steele, J.W., Faulds, D., and Goa, K.L. (1993). Epalrestat: a review of its pharmacology, and therapeutic potential in late-onset complications of diabetes mellitus. *Drugs Aging* 3: 532–555.

165 Wang, J.-R., Yu, Q., Dai, W., and Mei, X. (2016). Drug–drug co-crystallization presents a new opportunity for the development of stable vitamins. *Chem. Commun.* 52 (17): 3572–3575.

166 Arabiani, M.R., Bhunia, S., Teja, P.K. et al. (2019). Brexpiprazole–catechol cocrystal: structure elucidation, excipient compatibility and stability. *CrystEngComm* 21 (44): 6703–6708.

167 Yuan, L.-C., Dahl, T.C., and Oliyai, R. (2000). Effect of carbonate salts on the kinetics of acid-catalyzed dimerization of adefovir dipivoxil. *Pharm. Res.* 17: 1098–1103.

168 Lin, R.-Z., Sun, P.-J., Tao, Q. et al. (2016). Mechanism study on stability enhancement of adefovir dipivoxil by cocrystallization: degradation kinetics and structure-stability correlation. *Eur. J. Pharm. Sci.* 85: 141–148.

169 Gao, Y., Gao, J., Liu, Z. et al. (2012). Coformer selection based on degradation pathway of drugs: a case study of adefovir dipivoxil–saccharin and adefovir dipivoxil–nicotinamide cocrystals. *Int. J. Pharm.* 438 (1-2): 327–335.

170 Zhu, B., Wang, J.-R., Zhang, Q., and Mei, X. (2016). Improving dissolution and photostability of vitamin K3 via cocrystallization with naphthoic acids and sulfamerazine. *Cryst. Growth Des.* 16 (1): 483–492.

171 Werbin, H. (1968). Photochemistry of electron-transport quinones. I. Model studies with 2-methyl-1, 4-naphthoquinone (vitamin K3). *J. Am. Chem. Soc.* 90 (26): 7296–7301.

172 Nguyen, H.N., Ha, P.T., Sao Nguyen, A. et al. (2016). Curcumin as fluorescent probe for directly monitoring in vitro uptake of curcumin combined paclitaxel loaded PLA-TPGS nanoparticles. *Adv. Nat. Sci. Nanosci. Nanotechnol.* 7 (2): 025001.

173 Yan, D. and Evans, D.G. (2014). Molecular crystalline materials with tunable luminescent properties: from polymorphs to multi-component solids. *Mater. Horiz.* 1 (1): 46–57.

174 Yan, D., Delori, A., Lloyd, G.O. et al. (2011). A cocrystal strategy to tune the luminescent properties of stilbene-type organic solid-state materials. *Angew. Chem. Int. Ed.* 50 (52): 12483–12486.

175 Sangtani, E., Sahu, S.K., Thorat, S.H. et al. (2015). Furosemide cocrystals with pyridines: an interesting case of color cocrystal polymorphism. *Cryst. Growth Des.* 15 (12): 5858–5872.

176 Li, M., Li, Z., Zhang, Q. et al. (2018). Fine-tuning the colors of natural pigment emodin with superior stability through cocrystal engineering. *Cryst. Growth Des.* 18 (10): 6123–6132.

177 Gawade, A., Kuchekar, A., and Wagh, P. (2021). Application of cocrystals technology in pharmaceuticals. *Indo Am. J. Pharm. Res.* 11 (1): 62–65.

178 Bhatt, J.A., Bahl, D., Morris, K. et al. (2020). Structure-mechanics and improved tableting performance of the drug-drug cocrystal metformin: salicylic acid. *Eur. J. Pharm. Biopharm.* 153: 23–35.

179 Singaraju, A.B., Bahl, D., Wang, C. et al. (2019). Molecular interpretation of the compaction performance and mechanical properties of caffeine cocrystals: a polymorphic study. *Mol. Pharm.* 17 (1): 21–31.

180 Karki, S., Friščić, T., Fábián, L. et al. (2009). Improving mechanical properties of crystalline solids by cocrystal formation: new compressible forms of paracetamol. *Adv. Mater.* 21 (38-39): 3905–3909.

181 Sun, C.C. and Hou, H. (2008). Improving mechanical properties of caffeine and methyl gallate crystals by cocrystallization. *Cryst. Growth Des.* 8 (5): 1575–1579.

182 Wang, C., Paul, S., Sun, D.J. et al. (2020). Mitigating punch sticking propensity of celecoxib by cocrystallization: an integrated computational and experimental approach. *Cryst. Growth Des.* 20 (7): 4217–4223.

183 Smith, A.J., Kavuru, P., Wojtas, L. et al. (2011). Cocrystals of quercetin with improved solubility and oral bioavailability. *Mol. Pharm.* 8 (5): 1867–1876.

184 Childs, S.L., Kandi, P., and Lingireddy, S.R. (2013). Formulation of a danazol cocrystal with controlled supersaturation plays an essential role in improving bioavailability. *Mol. Pharm.* 10 (8): 3112–3127.

185 Huang, Y., Zhang, B., Gao, Y. et al. (2014). Baicalein–nicotinamide cocrystal with enhanced solubility, dissolution, and oral bioavailability. *J. Pharm. Sci.* 103 (8): 2330–2337.

186 Martin, F., Pop, M., Kacso, I. et al. (2020). Ketoconazole-*p*-aminobenzoic acid cocrystal: revival of an old drug by crystal engineering. *Mol. Pharm.* 17 (3): 919–932.

187 Haneef, J., Markad, D., and Chadha, R. (2020). Interaction map driven cocrystallization of ambrisentan: structural and biopharmaceutical evaluation. *Cryst. Growth Des.* 20 (7): 4612–4620.

188 Wang, C., Tong, Q., Hou, X. et al. (2016). Enhancing bioavailability of dihydromyricetin through inhibiting precipitation of soluble cocrystals by a crystallization inhibitor. *Cryst. Growth Des.* 16 (9): 5030–5039.

189 Ullah, M., Raza Shah, M., Hassan Bin Asad, M.H. et al. (2017). Improved in vitro and in vivo performance of carbamazepine enabled by using a succinic acid cocrystal in a stable suspension formulation. *Pak. J. Pharm. Sci.* 30 (6).

190 Alhalaweh, A., Ali, H.R.H., and Velaga, S.P. (2014). Effects of polymer and surfactant on the dissolution and transformation profiles of cocrystals in aqueous media. *Cryst. Growth Des.* 14 (2): 643–648.

191 Aitipamula, S., Cadden, J., and Chow, P.S. (2018). Cocrystals of zonisamide: physicochemical characterization and sustained release solid forms. *CrystEngComm* 20 (21): 2923–2931.

192 Yu, X.-Z., Wang, L.-Y., Liu, F. et al. (2020). Sustained-release dual-drug ternary salt cocrystal of piperazine ferulate with pyrazinamide: synthesis, structure, and Hirshfeld surface analysis. *Cryst. Growth Des.* 20 (3): 2064–2073.

193 Chen, J.-M., Li, S., and Lu, T.-B. (2014). Pharmaceutical cocrystals of ribavirin with reduced release rates. *Cryst. Growth Des.* 14 (12): 6399–6408.

194 Kumari, N., Bhattacharya, B., Roy, P. et al. (2019). Enhancing the pharmaceutical properties of pirfenidone by mechanochemical cocrystallization. *Cryst. Growth Des.* 19 (11): 6482–6492.

195 Liu, F., Wang, L.-Y., Li, Y.-T. et al. (2018). Protective effects of quercetin against pyrazinamide induced hepatotoxicity via a cocrystallization strategy of complementary advantages. *Cryst. Growth Des.* 18 (7): 3729–3733.

196 Gautam, M.K., Besan, M., Pandit, D. et al. (2019). Cocrystal of 5-fluorouracil: characterization and evaluation of biopharmaceutical parameters. *AAPS PharmSciTech* 20: 1–17.

197 Yu, Y.-M., Yu, M.-C., Wang, L.-Y. et al. (2020). A supramolecular adduct of tegafur and syringic acid: the first tegafur-nutraceutical cocrystal with perfected in vitro and in vivo characteristics as well as synergized anticancer activities. *New J. Chem.* 44 (37): 15994–16005.

198 Pan, X., Zheng, Y., Chen, R. et al. (2019). Cocrystal of sulfamethazine and p-aminobenzoic acid: structural establishment and enhanced antibacterial properties. *Cryst. Growth Des.* 19 (4): 2455–2460.

199 Sekhon, B.S. (2009). Pharmaceutical co-crystals-a review.

200 U.S. Food and Drug Administration (USFDA). *Guidances | Drugs*. **2025**. Available at: https://www.fda.gov/drugs/guidance-compliance-regulatory-information/guidances-drugs (accessed 25 March 2026).

201 Thayyil, A.R., Juturu, T., Nayak, S., and Kamath, S. (2020). Pharmaceutical co-crystallization: regulatory aspects, design, characterization, and applications. *Adv. Pharm. Bull.* 10 (2): 203.

202 Chavan, R.B., Thipparaboina, R., Yadav, B., and Shastri, N.R. (2018). Continuous manufacturing of co-crystals: challenges and prospects. *Drug Deliv. Transl. Res.* 1–14.

203 U.S. Food and Drug Administration (USFDA). *Drugs@FDA: FDA-approved drugs*. Available at: https://www.accessdata.fda.gov/scripts/cder/daf/index.cfm?event=BasicSearch.process (accessed 25 March 2026).

204 Poole, R.M. and Dungo, R.T. (2014). Ipragliflozin: first global approval. *Drugs* 74 (5): 611–617.

205 Almansa, C., Mercè, R., Tesson, N. et al. (2017). Co-crystal of tramadol hydrochloride–celecoxib (ctc): a novel API–API co-crystal for the treatment of pain. *Cryst. Growth Des.* 17 (4): 1884–1892.

206 Bernhardson, D., Brandt, T.A., Hulford, C.A. et al. (2014). Development of an early-phase bulk enabling route to sodium-dependent glucose cotransporter 2 inhibitor ertugliflozin. *Org. Process Res. Dev.* 18 (1): 57–65.

207 Ltd TPI. (2020) *Using co-crystals to produce alternative generic products.* https://www.teva-api.com/knowledge-center/using-co-crystals-to-produce-alternative-generic-products/ (accessed 25 March 2026).

208 Kumar, S. and Nanda, A. (2018). Approaches to design of pharmaceutical cocrystals: a review. *Mol. Cryst. Liq. Cryst.* 667 (1): 54–77.

209 Seliger, J. and Žagar, V. (2014). Hydrogen bonding and proton transfer in cocrystals of 4, 4′-bipyridyl and organic acids studied using nuclear quadrupole resonance. *Phys. Chem. Chem. Phys.* 16 (34): 18141–18147.

210 Kale, D.P., Zode, S.S., and Bansal, A.K. (2017). Challenges in translational development of pharmaceutical cocrystals. *J. Pharm. Sci.* 106 (2): 457–470.

211 Delori, A., Galek, P.T., Pidcock, E. et al. (2013). Knowledge-based hydrogen bond prediction and the synthesis of salts and cocrystals of the anti-malarial drug pyrimethamine with various drug and GRAS molecules. *CrystEngComm* 15 (15): 2916–2928.

212 Pathak, C., Savjani, K., Gajjar, A.K., and Savjani, J.K. (2013). Cocrystal formation of paracetamol with indomethacin and mefenamic acid: An efficient approach to enhance solubility. *Int J Pharm Pharm Sci* 5 (4): 414–419.

213 Zaitu, S., Miwa, Y., and Taga, T. (1995). A 2: 1 molecular complex of theophylline and 5-fluorouracil as the monohydrate. *Cryst. Struct. Commun.* 51 (9): 1857–1859.

214 Thipparaboina, R., Kumar, D., Chavan, R.B., and Shastri, N.R. (2016). Multidrug co-crystals: towards the development of effective therapeutic hybrids. *Drug Discov. Today* 21 (3): 481–490.

215 Lu, J. and Rohani, S. (2010). Synthesis and preliminary characterization of sulfamethazine-theophylline co-crystal. *J. Pharm. Sci.* 99 (9): 4042–4047.

216 Cosgrove, S.D., Jonaitis, D., Sutch, T., and Derrick, J.C. Novel ticagrelor co-crystal. Astrazeneca UK Limited. WO 2012/164286, AstraZeneca UK Limited, 6 December 2012.

217 Dandela, R., Reddy, J.S., Viswanadha, G.S. et al. (2012) Novel cocrystals/molecular salts of mesalamine to be used as improved anti-inflammatory drug. Nutracryst Therapeutics Private Limited. WO 2012/090224, Nutracryst Therapeutics Private Limited.

218 Reddy, J.S., Dandela, R., Viswanadha, G.S. et al. (2012) Novel cocrystals/molecular salts of metformin with oleoylethanolamide as an effective antidiabetic + anti-obesity agent. Nutracryst Therapeutics Private Limited. WO 2012/090225.

219 Reddy, J.S., Dandela, R., Viswanadha, G.S. et al. Novel cocrystals/molecular salts of metformin with oleoylethanolamide as an effective antidiabetic+anti-obesity agent. Nutracryst Therapeutics Private Limited. WO Patent 2,012,090,225.

220 Sinha, A.S., Khandavilli, U.B.R., O, Connor, E.L. et al. (2015). Novel co-crystals of the nutraceutical sinapic acid. *CrystEngComm* 17 (26): 4832–4841.

221 Sinha, A.S., Maguire, A.R., and Lawrence, S.E. (2015). Cocrystallization of nutraceuticals. *Cryst. Growth Des.* 15 (2): 984–1009.

222 Ong, T.T., Kavuru, P., Nguyen, T. et al. (2011). 2: 1 Cocrystals of homochiral and achiral amino acid zwitterions with Li+ salts: water-stable zeolitic and diamondoid metal–organic materials. *J. Am. Chem. Soc.* 133 (24): 9224–9227.

223 Smith, A.J., Kim, S.-H., Duggirala, N.K. et al. (2013). Improving lithium therapeutics by crystal engineering of novel ionic cocrystals. *Mol. Pharm.* 10 (12): 4728–4738.

224 Grepioni, F., Wouters, J., Braga, D. et al. (2014). Ionic co-crystals of racetams: solid-state properties enhancement of neutral active pharmaceutical ingredients via addition of Mg^{2+} and Ca^{2+} chlorides. *CrystEngComm* 16 (26): 5887–5896.

225 Wouters, J., Grepioni, F., Braga, D. et al. (2013). Novel pharmaceutical compositions through co-crystallization of racetams and Li^{+} salts. *CrystEngComm* 15 (44): 8898–8902.

226 Alda, M. (2015). Lithium in the treatment of bipolar disorder: pharmacology and pharmacogenetics. *Mol. Psychiatry* 20 (6): 661.

227 Crosby, V., Wilcock, A., Mrcp, D., and Corcoran, R. (2000). The safety and efficacy of a single dose (500 mg or 1 g) of intravenous magnesium sulfate in neuropathic pain poorly responsive to strong opioid analgesics in patients with cancer. *J. Pain Symptom Manag.* 19 (1): 35–39.

228 Spitzer, D., Risse, B., Schnell, F. et al. (2014). Continuous engineering of nano-cocrystals for medical and energetic applications. *Sci. Rep.* 4: 6575.

229 De Smet, L., Saerens, L., De Beer, T. et al. (2014). Formulation of itraconazole nanococrystals and evaluation of their bioavailability in dogs. *Eur. J. Pharm. Biopharm.* 87 (1): 107–113.

230 Munson, J., Freeman Stanfield, C., and Gujral, B. (2006). A review of process analytical technology (PAT) in the US pharmaceutical industry. *Curr. Pharm. Anal.* 2 (4): 405–414.

231 Lawrence, X.Y., Lionberger, R.A., Raw, A.S. et al. (2004). Applications of process analytical technology to crystallization processes. *Adv. Drug Deliv. Rev.* 56 (3): 349–369.

232 Kelly, A.L., Gough, T., Dhumal, R.S. et al. (2012). Monitoring ibuprofen–nicotinamide cocrystal formation during solvent free continuous cocrystallization (SFCC) using near infrared spectroscopy as a PAT tool. *Int. J. Pharm.* 426 (1-2): 15–20.

233 Wood, C., Alwati, A., Halsey, S. et al. (2016). Near infra red spectroscopy as a multivariate process analytical tool for predicting pharmaceutical co-crystal concentration. *J. Pharm. Biomed. Anal.* 129: 172–181.

5

Self-Microemulsifying and Self-Emulsifying Drug Delivery Systems to Improve Solubility and Bioavailability

Dipak B. Bari[1], Rakhee Kapadia[2], Raju O. Sonawane[1], Kamalesh D. Mali[1,3], Vivek P. Rajkule[1], Mahesh G. Umbarkar[1], and Chandrakantsing V. Pardeshi[1]

[1] *R. C. Patel Institute of Pharmaceutical Education and Research, Industrial Pharmacy Laboratory, Department of Pharmaceutics, Karwand Naka, Shirpur, Maharashtra 425405, India*
[2] *Sanjeev Agarwal Global Educational University (SAGE) University, SIRT-Pharmacy, Ayodhya Bypass Road, Bhopal, Madhya Pradesh 462041, India*
[3] *R. C. Patel Institute of Pharmacy, Department of Pharmaceutics, Karwand Naka, Shirpur, Maharashtra, 425405, India*

5.1 Introduction

The oral route is a popular and convenient mode of drug delivery because of its simplicity of administration, excellent patient acceptability, and affordability [1–3]. However, sometimes it suffers several drawbacks *viz.* inadequate solubility, limited absorption from the gastrointestinal (GI) tract, poor stability, a large hepatic first-pass effect [4, 5], and degradation in the GI environment [6]. Nanotechnology may assist in addressing these issues by enhancing drug absorption via the oral route [7]. In recent years, self-emulsifying drug delivery systems (SEDDS) have extended their potential as drug carriers for oral administration [8]. Emulsions find a prominent role as pharmaceutical carriers in enhancing oral bioavailability owing to high absorption characteristics [9]. There is limited oral bioavailability and an uneven absorption pattern for Biopharmaceutical Classification System (BCS) Class II and IV drugs [10, 11]. Considering certain digestive system restrictions, hydrophilic macromolecular therapeutics such as polysaccharides, therapeutic peptides, and DNA-based formulations have poor oral bioavailability. To combat this, several strategies have been explored *viz.* structural drug modifications, adjunct substances, and the development of self-nano (SNEDDS) and self-microemulsifying drug delivery systems (SMEDDS) [12, 13].

Recently, there has been increased interest in lipid-based transport systems to improve the absorption of less soluble drugs in the GI tract [14, 15]. Emulsions [16], microemulsions [17], solutions, suspensions [18], dry emulsions [19], and SEDDS [6] are a few well-explored examples of these systems. Numerous lipid-based products *viz.* lipoplexes [20], solid lipid-based nanoparticles [21], macroemulsions [22], microemulsions, nanoemulsions, and SMEDDS have been studied by various research groups [23].

The primary objective of the present book chapter is to comprehensively discuss the fundamental concepts relevant to SEDDS and SMEDDS, their merits and demerits, formulation components, mechanism of self-emulsification, challenges in the development of SEDDS/SMEDDS, and applications in solubility and bioavailability enhancement of poorly soluble therapeutics.

5.2 Self-Emulsifying Drug Delivery Systems

Isotropic mixes of either synthetic or natural oils, liquid and solid surfactants, or, alternately, a combination of aqueous solvents and co-solvents/surfactants are known as self-emulsifying drug delivery systems. Under modest agitation and dilution in aqueous media like GI fluids, such structures can produce microemulsions or tiny oil-in-water (o/w) emulsions [24].

5.2.1 Self-Microemulsifying Drug Delivery System

SMEDDSs are isotropic mixtures composed of solid or liquid surfactants, synthetic or natural oils/lipids, or hydrophilic solvents and co-solvents/co-surfactants [25].

5.2.1.1 Formulation Development

Scientists must evaluate and pay close attention while selecting the additives to stability, permeability, and effectiveness as critical factors for the best possible outcomes [26]. Figure 5.1 illustrates the components of a SEDDS/SMEDDS formulation.

5.2.2 Self-Nanoemulsifying Drug Delivery System

As a combination of therapeutic drug, oil, surfactant, and co-surfactant, SNEDDS creates a nanoemulsion when it meets water, with droplet sizes usually between 10 and 200 nm. This improves drug administration and absorption [27].

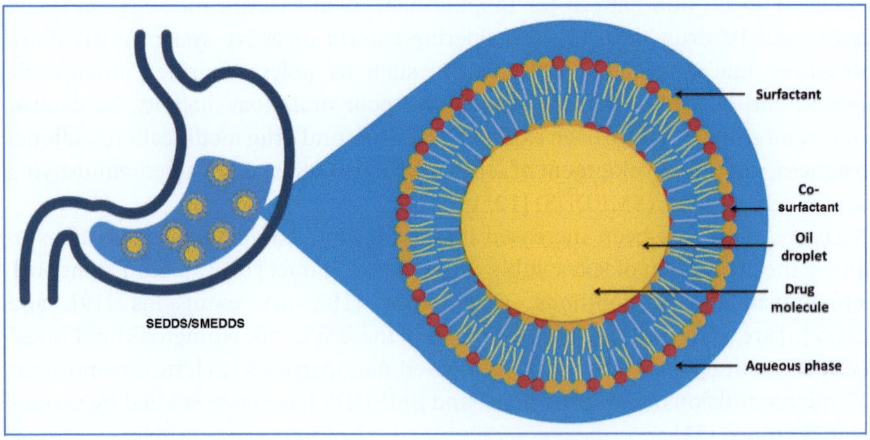

Figure 5.1 Schematic illustrating the components of SEDDS.

Table 5.1 Differences between SEDDS, SMEDDS, and SNEDDS.

Attributes	SEDDS	SMEDDS	SNEDDS
Globule size	~300 nm	<250 nm	<100 nm
The appearance of the system	Turbid	Clear	Clear
HLB value of the surfactant	>12	<12	<12
Classification as per LFCS	Type II	Type III-B	Type III-B
Oil phase	40–80%	>20%	>20%
Surfactant's concentration	30–40%	40–80%	40–80%
Phases	Biphasic	Monophasic	Monophasic
Stability	Unstable	Stable	Stable
Viscosity	Low	High	High
Interfacial tension	High	Ultra-low	Ultra-low
Energy requirement	High	Low	Low
Dosage forms	Solution	Solution, tablet, capsule, pellet, etc.	Solution, tablet, capsule, pellet, etc.

HLB: hydrophilic–lipophilic balance, LFCS: lipid formulation classification system.
Source: Rajpoot et al. [28] and Ango et al. [29].

The significant attributes differentiating SEDDS, SMEDDS, and SNEDDS are presented in Table 5.1.

5.3 Advantages and Disadvantages of SEDDS/SMEDDS

The SEDDS/SMEDDS offer several advantages *viz.* improved oral bioavailability of poorly soluble APIs and targeted drug delivery. However, GI irritation, drug precipitation following dilution in gut, and hurdles in maintaining formulation stability are the major challenges in scaling up the SEDDS/SMEDDS to the commercial market.

The advantages and disadvantages offered by SEDDS/SMEDDS are listed in Tables 5.2 and 5.3, respectively.

5.4 Formulation of SEDDS and SMEDDS

SMEDDSs are isotropic mixes of precisely the right amounts of hydrophilic solvents, natural or synthetic oils/lipids, surfactants, and co-solvents/co-surfactants (Figure 5.2). To assist formulators in choosing the best component composition for SMEDDS formulation while considering the issues of stability, permeability, and efficacy, Pouton [40] created a lipid formulation classification system (LFCS). This tool facilitates the selection of excipients from a wide variety, guaranteeing the effective formulation of SMEDDS. Based on composition, twofold impact of dilution, and avoidance of drug precipitation *via* digestion, the LFCS divides lipid

Table 5.2 Advantages of SEDDS/SMEDDS.

Aspects	Advantages
Boost bioavailability	SEDDS and SMEDDS can increase a drug's oral bioavailability. SEDDS may additionally assist molecules flow more easily via the intestinal lymphatic system, preventing the initial hepatic metabolism [30].
Increases solubility	SEDDS and SMEDDS have the ability to increase the hydrophobicity of drugs [30].
Wide range of applications	SEDDS/SMEDDS is appropriate for delivering a range of therapeutic categories *viz.* antineoplastics, antiepileptics, antivirals, etc. [6].
Storage	SMEDDS and similar emulsions promote solubility of hydrophobic drugs and are thermodynamically stable, making them easier to maintain than macroemulsions that cream over time [31].
Stability	Compared to micro/nanoemulsions, self-emulsification also increases the stability of several drugs like carvedilol [32].
Compliance	Most SMEDDS preparations are in tablet or capsule forms, are convenient to administer, and thus enhance patients' acceptability [26].
Palatability	SMEDDS composition is easily loaded into capsules, eliminating the palatability concerns related to lipid preparations [26].
Reduces variability	Food does not affect drug's absorption from SMEDDS formulations, thereby reducing intersubject variability [33].
Quick onset of action	SMEDDS, in tablet or capsule forms, promotes quick oral absorption and boosts oral bioavailability of several drugs *viz.* vitamin A [34].
Ease of production and scale up	SMEDDS can be easily produced at large scale since it demands inexpensive manufacturing facilities [35].

formulation systems into four categories *viz.* oils (Class I), SEDDS (Class II), SEDDS and SMEDDS (Class IIIA and IIIB), and lipid-free (Class IV) [40].

Triglyceride-soluble drugs, mixed glycerides, or an o/w emulsion with minimal emulsifiers are classified as type I lipid formulations [41]. In addition to their low aqueous dispersion, these formulations must be broken down by enzymes to release products of amphiphilic lipid degradation and speed up drug delivery. Nonetheless, this technique works well for easily absorbed formulations, since it facilitates drug solubilization and dispersion. When introduced in aqueous medium, type II lipid plans, often referred to as SEDDS, are isotropic mixtures of lipids and lipophilic surfactants that self-emulsify to create fine o/w emulsions [42, 43]. Self-emulsification occurs at surfactant concentrations above 25%, but may break down at higher concentrations. SEDDS offers a solution for poorly aqueous-phase soluble drugs, and the final formulation can be encapsulated in capsules. Type II-LBDDS overcome challenges like slow dissolution from the solid dosage forms by generating a larger effective surface area [44]. On the other hand, Type IV-LBDDS, which contains

Table 5.3 Disadvantages of SEDDS/SMEDDS.

Aspects	Disadvantages
Drug precipitation	As the solubilization capacity of the formulation falls, like once it is diluted in GI fluid, precipitation of drugs can occur. This may lead to a decrease in effective drug concentration and/or reduced efficacy [36].
Handling	It might be challenging to handle and store liquid SMEDDS [26].
in vitro models	There are only a few in vitro models available for estimation of efficacy of SMEDDS preparations ([37]; [38]).
Amount of surfactants	The high concentration of surfactants (about 30–70%) in SEDDS preparations has the potential to irritate the GI tract or create chemical instability in the drug material [39].
Volatile co-solvents	In SMEDDS products, volatile co-solvents have the potential to diffuse into the soft or hard gelatin capsule shells, resulting in the precipitation of lipophilic medicines [30].
Multicomponent formulation	It may be more challenging to validate SEDDS/SMEDDS formulas with multiple ingredients [28].
Oxidation	Oxidation of unsaturated fatty acid, component of SEDDS/SMEDDS, is possible with SMEDDS [28].

Figure 5.2 Components of SEDDS/SMEDDS formulations.

water-soluble surfactants, co-solvents, low-hydrophilic–lipophilic balance (HLB) surfactants at concentrations of 0–20%, and high-HLB surfactants at concentrations of 30–80%, creates fine dispersion and improves drug absorption and solubility [15].

5.4.1 Oils/Lipids

Lipids are a complex combination of organic molecules found in plants, animals, and microbes. They are necessary for all living organisms and comprise one of the three main nutritional groups, along with proteins and carbs. In addition to acting as permeability enhancers and assisting in the creation of controlled-releasing products, colloidal targeting systems, and dispersion/solubilization matrices, lipids and lipophilic excipients have a substantial influence on the absorption and circulation of lipophilic drugs. A natural phospholipid made from soy is used as vehicle-forming component for nutrition and pharmaceutical drugs. These lipids are non-inflammatory and immune-suppressive, making them a safe and biocompatible delivery approach. Phospholipids, such as lecithin, phosphatidylglycerol, phosphatidylinositol, phosphatidylethanolamine, and phosphatidylserine, are the key amphiphilic components of cell membranes and are used in nano-vectors due to their ability to self-assemble in an aqueous environment. The lipids/oils can be divided into uni-, oligo-, and multilamellar categories and allow the inclusion of hydrophilic medicines in the aqueous core and lipophilic medicines in the lipid bilayer. Phosphatidylserine (PS), phosphatidic acid (PA), phosphatidylglycerol (PG), phosphatidylcholine (PC), phosphatidylinositol (PI), and phosphatidyl ethanolamine (PE) are the most commonly used phospholipids from animal and plant sources. These lipids play a crucial role in the delivery of nutrients and drugs into target cells [45].

SMEDDS have used a variety of lipids, including triglycerides, partial glycerides, fatty acids, steroids, and waxes. Mixtures of mono-, di-, and triglycerides of fatty acids with varying chain lengths and degrees of unsaturation make up oils and fats [46, 47]. The factors such as physicochemical structure, drug solubility, solid/liquid lipid miscibility, and physiological tolerance play a crucial role in the use of lipids for drug administration. At room temperature, lipids are *generally recognized as safe* (GRAS) and are considered to be safe for clinical use. Solubility of the drug in the oil phase is vital for formulation development and manufacturing purposes. Low drug entrapment and loading can occur if the drug is not solubilized in the lipid core and instead binds to the surfaces of the particles or forms micelles in the aqueous phase. To verify the macroscopic lipid-phase homogeneity and separation below the fat melting point, miscibility tests are required [48]. Various lipids used in the SEDDS/SMEDDS formulations have been tabulated in Table 5.4.

5.4.2 Surfactants

One or several surfactants with contents ranging from 1.5% to 5% can stabilize SMEDDS/SNEDDS. Surfactant types and concentration have significant effects on their fabrications. It is common practice to combine lipophilic and hydrophilic surfactants to enhance stability and functional qualities. Biosurfactants have

Table 5.4 A summary of the various lipids used in SMEDDS/SNEDDS formulations.

Lipid	Outcome	References
Phosphatidylcholine	Improved the solubility of telmisartan	[49]
Phosphatidylcholine	Improved the solubility of resveratrol	[50]
Cardiolipin and phosphatidylserine	Inhibit glycoprotein permeability	[51]
1,2-Dipalmitoyl-sn-glycero-3-phosphatidic acid disodium salt	Reversed charge of SMEDDS	[52]
Phosphatidylcholine (HSPC), dimyristoylphosphatidylcholine (DMPC), dimyristoyl phosphatidylglycerol (DMPG), distearoyl phosphatidylcholine (DSPC)	Improved oral bioavailability of nisoldipine	[53]
Monoacyl phosphatidylcholine	Nanoemulsion formulation with medium-chain glyceride	[54]
Capryol 90	Improved solubility and bioavailability of lovastatin	[55]

been developed recently for use in biocompatible drug delivery systems [48]. For SNEDDS to be formulated, a surfactant with a high HLB is essential because it can create o/w nanoemulsions that efficiently absorb drugs and stop precipitation in the GI lumen. The intended HLB can be obtained by blending different surfactants, but nonideal mixing effects may lead to synergy. The composition and molecular structure of blended surfactants also influence the generation of nanoemulsions. On the other hand, there aren't many published studies on SNEDDS made using mixed polysorbates that take the mixture's HLB into account. It is crucial to look at the effects of various surfactant mixtures [56]. Through the drug dissolution, improved permeability, tight junction permeability, and reduced efflux, surfactants in SMEDDS/SNEDDS improve bioavailability. Large doses, however, may irritate the GI tract or result in reversible changes in intestinal wall permeability, necessitating formulation and surfactant concentration research [55]. Various surfactants and co-surfactants used in SEDDS/SMEDDS formulations have been enlisted in Table 5.5.

5.4.3 Co-surfactant

Though alcohols and volatile co-solvents are rarely utilized because they tend to permeate the capsule shell, alter mechanical stability, and increase the risk of drug precipitation, co-surfactants permit more excellent drug and hydrophilic surfactant solubility in lipids [61]. It is unusual to get transient negative interfacial tension with just one surfactant, so co-surfactants are frequently needed. To create a fluid interfacial film, a co-surfactant is used. A hard film develops without a co-surfactant, resulting in microemulsion within a restricted concentration range. Medium chain–length alcohols, for example, are co-surfactants that increase system entropy and interface fluidity while lowering interfacial tension [62].

Table 5.5 Different surfactants and co-surfactants used in SMEDDS/SNEDDS formulations.

Surfactant	Co-surfactant	Outcome	References
Mixture of Tween® 20, Tween® 80, Cremophor® RH40, or Cremophor® EL, Span® 20 or Span® 80	1,2-Octanediol	Produced SNEDDS with nano-sized emulsions after dispersion. Improved the solubility of nifedipine.	[56]
Tween 20	Polyethylene glycol 600	Produced SMEDDS after aqueous dispersion. Improved oral absorption of nevirapine.	[57]
Kolliphor EL	Lauroglycol FCC	Solubility of dolutegravir sodium enhanced after the development of SNEDDS.	[58]
Spans 20 and Tweens 80	1,2-octanediol	The dissolution rate of ibuprofen is enhanced by fabrication in SNEDDS.	[59]
Cremophor EL/Solutol HS-15	Akoline-MCM	Improved bioavailability of cefpodoxime proxetil.	[60]

The effects of oil, surfactant, and co-surfactant on self-microemulsification were investigated by Agubata et al. Artemether's solubility in Peceol® oil, Labrasol® surfactant, and Transcutol® co-surfactant and their combinations were examined, as well as pseudoternary-phase diagrams built with the water titration technique to measure the water and surfactant effectiveness, and the ability to solubilize was investigated. Artemether SEDDS were prepared by dissolving the APIs in the oil before adding surfactant and co-surfactant combinations that are evaluated for the tests such as phase stability, self-emulsification, pH, viscosity, freeze–thaw cycling, drug precipitation, centrifugation, and drug release and distribution. In the pseudoternary-phase diagrams, the SEDDS made with a 3:1 ratio of surfactant–co-surfactant (Smix) combination exhibited the biggest zone of microemulsion and the maximum surfactant efficiency. Higher Labrasol® concentration compositions demonstrated quicker self-emulsification, while a 1:2 Peceol/Smix ratio and a 3:1 Labrasol/Transcutol® ratio produced the best and quickest artemether release and dispersion from capsule-filled SEDDSs.

Figure 5.3a–e displays the resulting pseudoternary-phase diagrams. According to the study, SEDDS made with a 3:1 ratio of Labrasol®–Transcutol® surfactant combination had the greatest microemulsion zone. The microemulsions of surfactant–co-surfactant mixtures in 3:1 and 4:1 ratios persisted even after diluting oil/Smix pre-concentrates or in infinite water titration. Formulations including 4:1 Smix exhibited a more extensive microemulsion zone at lower oil content and greater surfactant concentration, but a smaller zone at higher Peceol® oil content. The proper concentrations of water, surfactant, co-surfactant, and oil inside the microemulsion area can be chosen to create microemulsions. The right

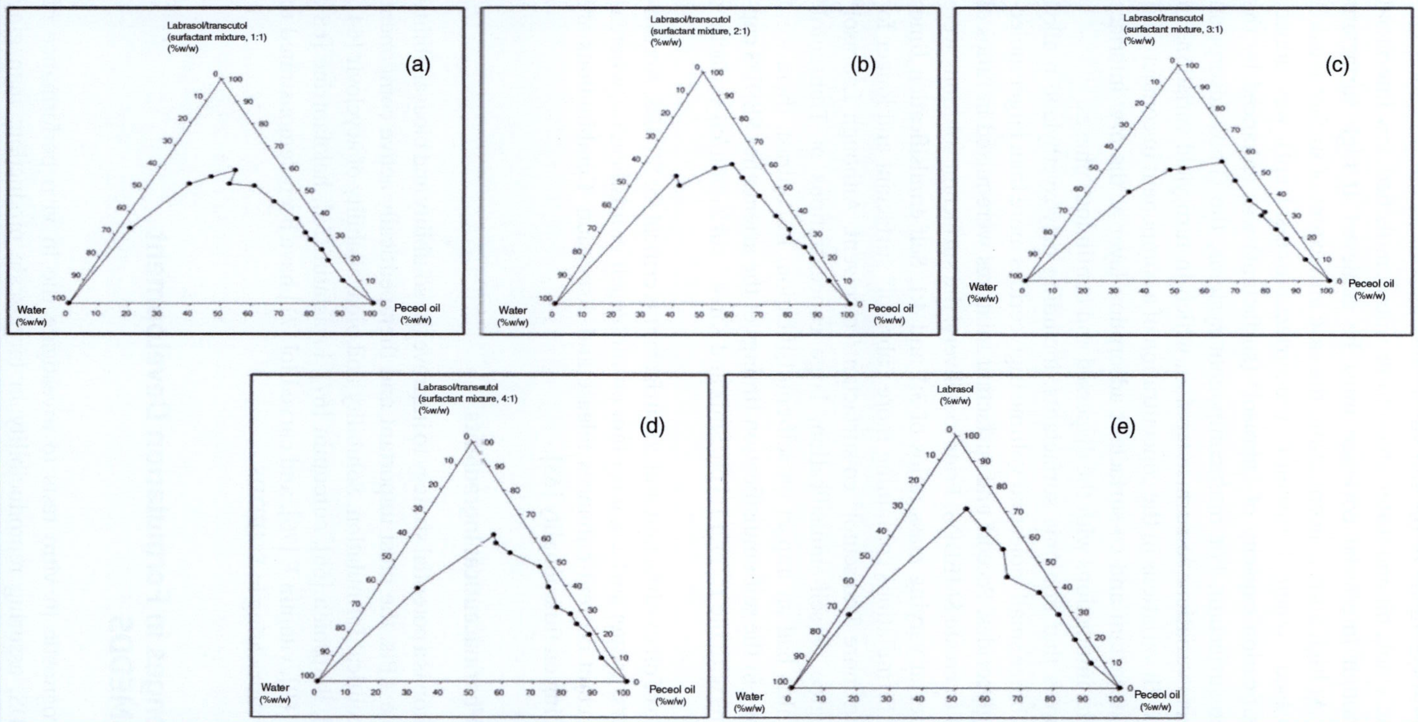

Figure 5.3 (a) Pseudoternary-phase diagram of Pecerol® and Labrasol®/Transcutol® (1:1); (b) Pecerol®, Labrasol®/Transcutol® (2:1), and water; (c) Pecerol®, Labrasol®/Transcutol® (3:1), and water; (d) Pecerol®, Labrasol®/Transcutol® (4:1), and water; and (e) Peceol®, Labrasol®/Transcutol®, and water. *Source:* Agubata et al. [63] / with permission of Austin Publishing Group.

proportions of oil, surfactant, and co-surfactant were chosen for this investigation using pre-concentrates made of oil and surfactants as well as the pseudoternary diagram. For the whole oil and water phase to be microemulsified, low interfacial tension and enough interfacial coverage must be achieved at high surfactant concentrations. At high Smix concentration, the ease and extent of surface tension reduction increased. When Transcutol® (co-surfactant/co-solvent) was added, the microemulsification capacity of Labrasol® (surfactant) was enhanced by the addition of a co-surfactant. For multicomponent systems, the Gibbs adsorption isotherm connects variations in surface tension, which in turn yield variations in surface energy, with variations in the concentration of a component in contact with a surface. The surfactant and co-surfactant adsorption layer at the o/w interface may influence the interactions with the dispersed and continuous phases.

The study shows that nonionic surfactants, including polyoxyethylene n-alkyl ethers, may solubilize small molecular volume triglycerides more than larger molecular volume triglycerides. Peceol® and surfactant mixtures were mixed in ratios of 4:6, 1:2, and 1:3 to create SMEDDS. Prior to delivery, the surfactant solutions were mixed with water at varying mass ratios of 3:1 and 4:1. Self-emulsification times were longer for formulations containing more Labrasol® surfactant and longer for those containing more Transcutol® co-surfactant/co-solvent. Although Labrasol® has a high degree of self-emulsification, large concentrations of Transcutol® decreased it, which had an impact on self-emulsification. In contrast, there were notable variations in the self-emulsification timings of the artemether SEDDS capsule formulations with the 1:2 and 1:3 oil/Smix and Km 4:1 surfactant/co-surfactant combination.

The ideal ratio of oil to surfactant and co-surfactant is critical for SEDDS. According to researchers, greater surfactant content promotes self-emulsification, whereas greater co-surfactant content enhances release and dispersion. Combinations are selected depending on functionality [63].

5.4.4 Active Pharmaceutical Ingredients

Self-emulsification is a potential strategy to improve the solubility and bioavailability of poorly soluble APIs, the most important and therapeutically active component of any pharmaceutical formulation. Solubility and bioavailability of acyclovir [64], fenofibrate [65], ibuprofen [66], curcumin [67], lovastatin [55], halofantrine [68], coenzyme Q10 [69], vitamin E [70], and carvedilol [71] have been demonstrated to be increased by fabricating in SMEDDS.

5.5 Challenges in Formulation Development of SEDDS/SMEDDS

The lack of prognostic in vitro tests to investigate the in vivo performance of SEDDS/SMEDDS, assuring reproducibility in large-scale production; managing potential taste and odor issues; choosing the right oils, surfactants, and co-surfactants; assuring compatibility with the drug; balancing safety, efficacy, and regulatory approval status; maintaining sufficient drug solubility; optimizing emulsification efficiency; maintaining stability during storage; controlling droplet size; and maximizing drug loading capacity are some of the key challenges [72, 73] (Figure 5.4).

5.5.1 Drug Precipitation

For the SMEDDs formulation, drug precipitation upon *in vivo administration* is an undesirable phenomenon in which drug solute precipitates as a result of reduced solubilization capacity. Dilution with body fluids, variation in pH, or breakdown of emulsifiers are the major reasons behind drug precipitation. Drug effectiveness is delayed or diminished when precipitation decreases drug concentration [72]. According to the study, decreased solubility of drug in oils/Smix causes precipitation and hence increases the particle size of SMEDDS, which induces globule aggregation and system instability [74]. Considering partition kinetics = favoring the drugs residing in the lipid droplets, there is less chance of precipitation upon diluting in the GI tract if the drug is solubilized in an oily layer [75]. In SMEDDS formulation, the drug must remain partitioned within the o/w emulsion droplets, following dilution of SMEDDS formulation with aqueous media in the GI tract [72]. It has been reported that when 60% w/w oil was used to formulate the SMEDDS, a small amount of globule aggregation occurs due to drug precipitation [26]. Precipitation of drugs involves three steps: supersaturation, nucleation, and crystal formation. A drug's precipitation from supersaturated solutions is influenced by supersaturation solubility, presence of impurities, temperature, and solution viscosity [61]. Supersaturation is caused by various processes, including solvent evaporation, cooling, chemical reactions, and the addition of compounds with higher solubility [76]. At constant supersaturation, increasing the solubility of SMEDDS increases the probability of intermolecular collision, which thereby increases nucleation rate [72]. Impurities promote nucleation, lowering the energy barrier to nuclei formation and resulting in crystal formation. Precipitation inhibitors, like hydroxy propyl methyl cellulose (HPMC), are used to prevent drug precipitation in the GI tract [77]. Precipitation takes place through three mechanisms: reticulation formation, hydrogen bonding, and the spring and parachute mechanism. Reticulate formation entails forming a network of swelling clusters containing hydrophobic substituents that can prevent nucleation and crystal development. Approximately 60% of the hydroxyl groups in HPMC are not substituted. HPMC or the hydrophilic polymers can thus form both intramolecular and intermolecular hydrogen bonds with drug, which is likely to retard drug precipitation. In a supersaturated system, the drug solution is in thermodynamically unstable system and tends to return to the equilibrium state by drug precipitation. However, when delivered in vivo, these formulations can cause precipitation, necessitating their optimization for maximum performance. Supersaturation can be controlled by excipients such as cellulosic polymers, surfactants, and cyclodextrins [72].

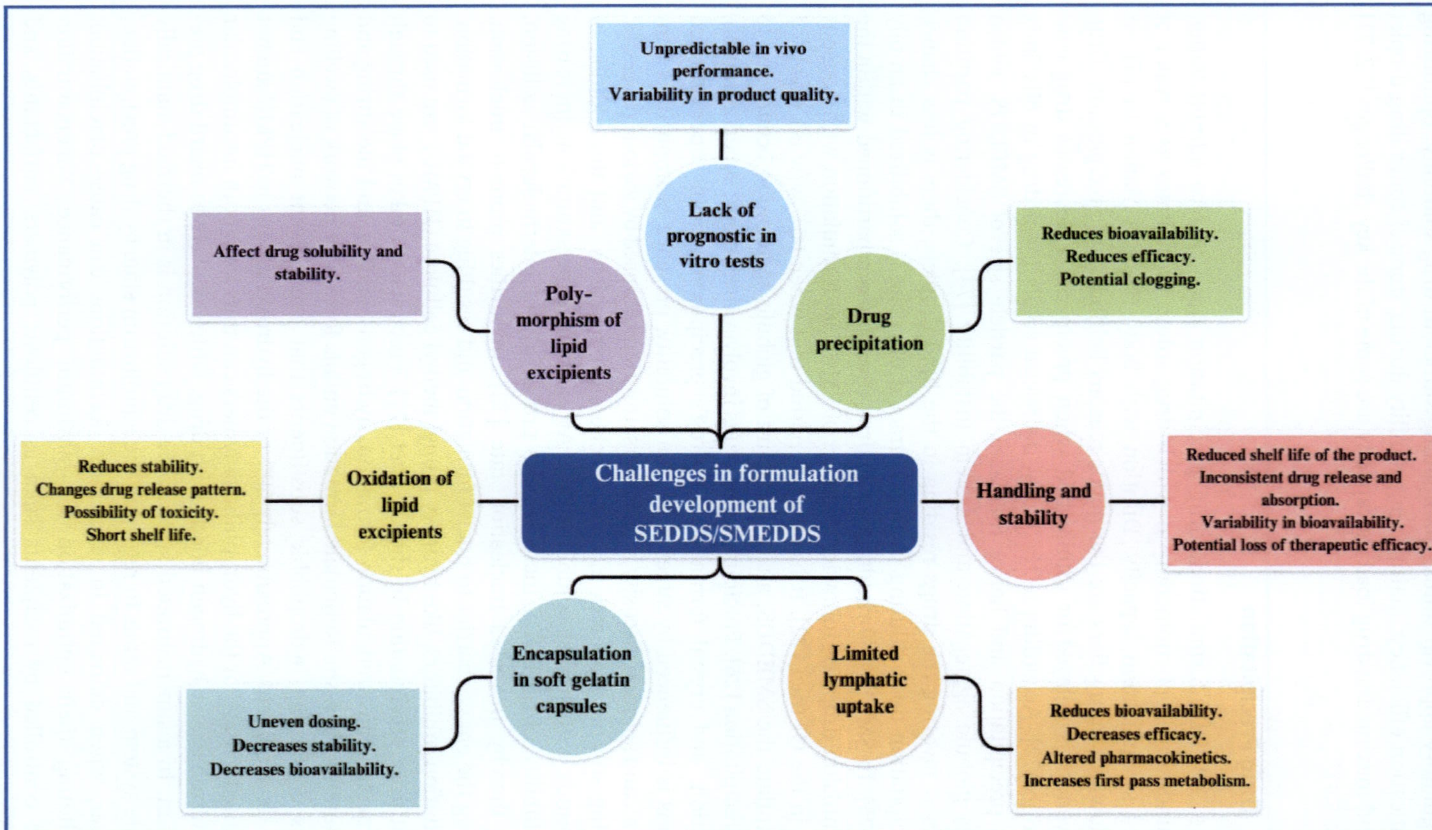

Figure 5.4 Challenges in SMEDDS formulation and their effects on SMEDDS. *Source:* Modified after Dokania and Joshi [72].

5.5.2 Handling and Stability

The temperature and light sensitivity of SMEDDS necessitates controlled storage at optimum temperature since formulation comprises lipids [78]. SMEDDS ought not to be exposed to light during manufacturing or stored to halt their degradation storage and prevent any possible degradation [79]. Long-chain glycerides are susceptible to oxidation, which leads to instability of SMEDDS; hence, medium-chain triglycerides should be preferred for development of SMEDDS as they are resistant to oxidation [80].

5.5.3 Limited Lymphatic Uptake

The lymphatic uptake of SEDDS/SMEDDS presents a challenge as it is important to the absorption of lipid-based formulations and lipophilic drugs [81]. Lymphatic absorption can be impacted by emulsion droplet size, drug's characteristics, and lipid content [82]. Adding certain excipients, changing the drug molecule, and optimizing the composition of lipids are some strategies to increase lymphatic absorption [57]. in vitro techniques, in vitro cell culture models employing lymphatic endothelial cells, and in vivo research utilizing lymph duct-cannulated animal models are among the methods used to assess lymphatic absorption [83]. Liao et al. [84] studied the oral absorption and lymphatic transport of the baicalein–phospholipid complex SMEDDS (BAPC-SMEDDS) and compared it with the conventional baicalein-SMEDDS (CBA-SMEDDS). Authors observed the improved oral bioavailability and lymphatic transport of baicalein [84].

5.5.4 Lack of Prognostic In Vitro Tests

Pharmaceutical research is severely hampered by the lack of predictive in vitro tests for SEDDS [85]. Because of the complicated physiological circumstances in the GI system, the absence of standardized procedures, and the poor connection between in vitro and in vivo data, it is challenging to predict their efficacy in vivo [86]. The drawbacks of current in vitro testing include their inability to capture the dynamic nature of digestion adequately, their lack of physiological relevance in terms of pH variations and enzyme concentrations, and their failure to take into consideration the effects of digestion on drug release and uptake [87]. Utilizing improved imaging methods, creating in vitro–in vivo correlation models, and establishing relevant media for predictive modeling are some interesting possibilities for advancement [88].

5.5.5 Encapsulation in Soft Gelatin Capsules

SEDDS in soft gelatin capsules experiences a few hurdles, particularly capsule shell compatibility, instability caused by lipid oxidation, and fabrication difficulties [89]. The traditional liquid SEDDS has disadvantages such as low drug loading capacity, drug leakage, limited dosage forms, incompatibility of some excipients with capsules, stability issues, and the likelihood of irreversible drug or excipient precipitation [90]. The liquid form raises the probability of leakage and drug precipitation [91]. Scaling up from laboratory-scale preparations to industrial manufacturing can be difficult [92]. Consequently, careful formulation design, excipient choice, and manufacturing process optimization are crucial for developing stable and effective SEDDS in soft gelatin capsules [93]. Huynh et al. [94] developed softgel capsules of cyclosporin A (CsA)-encapsulated pine oil-based SMEDDS system with an objective to improve the oral bioavailability of CsA [94].

5.5.6 Oxidation of Lipid Excipients

For SEDDS, lipid oxidation presents an important challenge since it can lead to reduced stability, altered drug release patterns, possible toxicity, and shortened shelf life [95, 96]. When lipid excipients undergo oxidation, they produce secondary oxidation products like aldehydes and ketones as a result of the breakdown of lipid hydroperoxides, which are the primary oxidation products; these by-products can significantly impact the quality and stability of SEDDS formulations [97]. Antioxidants, appropriate packaging, regulated storage, stable lipid excipient selection, and chelating substances to bind metallic ions are some strategies to reduce the oxidation of lipid excipients in SEDDS formulations [89]. When assessing the oxidation level of a lipid sample, "peroxide value" measures the primary oxidation products (hydroperoxides) through iodometric titration; "thiobarbituric acid reactive compounds (TBARS) assay" detects secondary oxidation products like malondialdehyde (MDA) using a colorimetric reaction with thiobarbituric acid, while "gas chromatography for volatile oxidation products" analyzes the specific volatile compounds formed during oxidation, providing a more detailed picture of the degradation process [98].

5.5.7 Polymorphism of Lipid Excipients

In SEDDS, polymorphism of the lipid excipients poses an imminent threat since it affects drug solubility and stability [61]. The solubility of distinct polymorphs can vary, influencing the bioavailability and release patterns of drugs. Stability problems may arise during preparation or storage, which might cause the formulation to become unstable [99]. Unintended polymorphic transitions might present manufacturing challenges. Variations in polymorphism can lead to discrepancies in performance, impacting droplet size, emulsification time, and overall biological performance of the SEDDS formulation [100]. Thorough characterization, polymorph screening, stabilization methods, cautious lipid excipient selection, process optimization, monitoring, and quality control are some strategies to deal with polymorphism issues [99, 101]. To create stable, efficient, and reliable SEDDS formulations, an integrated strategy combining analytical chemistry, formulation science, and process engineering is essential. Developing stable, efficient, and reliable SEDDS formulations requires a comprehension of and response to polymorphism in lipid excipients [6].

5.6 Mechanism of Self-Emulsification

The free energy of emulsion is determined by the following formula:

$$\Delta G = \sum N\Pi r2\sigma$$

where ΔG is the free energy, N is the number of droplets, r is the radius of droplets, and σ is the interfacial energy.

Reiss stated that, according to this equation, free energy decreases as interfacial energy decreases. When the energy required for forming droplets is less than the energy involved in dispersion, self-emulsification takes place [102]. Conventional emulsions have high free energy because it takes an enormous amount of energy to create a new surface between two immiscible phases, such as water and oil. On the other hand, because of the flexible interface and low or even negative free energy, emulsion formation occurs instantly in SMEDDS. A dispersed liquid crystalline phase is produced due to greater water penetration, depending on the surfactant concentration—an interface formed when water is mixed with an oil and surfactant/co-surfactant combination. Along the contact, the aqueous phase enters and becomes dissolved in the oil phase. The system can achieve equilibrium because microemulsions are thermodynamically stable [31].

5.7 Self-Emulsification and Uptake of Drugs In Vivo

Lipid-rich chyme stimulates the release of the intestinal hormone cholecystokinin into the duodenum. Cholecystokinin relaxes the hepatopancreatic sphincter, allowing bile salts, phospholipids, and cholesterol into the small intestine.

Additionally, endogenous GI fluids produced by pancreatic lipase/co-lipase are released into the intestines and can solubilize ingested lipids. By creating exogenous lipid carriers, lipid-based formulations take advantage of this innate process of lipid digestion and create the perfect environment for lipophilic drugs to partition. The drug and other by-products of the lipid substrate's digestion are transported to the intestinal wall's enterocytes for absorption. The formation of a concentration gradient for lipid absorption, as well as the solubilization of medications that are poorly soluble in water, is accomplished by the bile–phospholipid mixed micelles. It is commonly recognized that combining drugs with a fatty meal may cause the "food effect." Lipid-based formulations can improve drug transport to the intestinal lymphatic and systemic circulations and improve drug uptake, efflux, and disposal via altering enterocyte-based transport. Considering it can shield lipophilic drugs from enzymatic hydrolysis in the GI tract wall and avoid first-pass metabolism, the lymphatic system is an intriguing route for drug absorption. This increases the

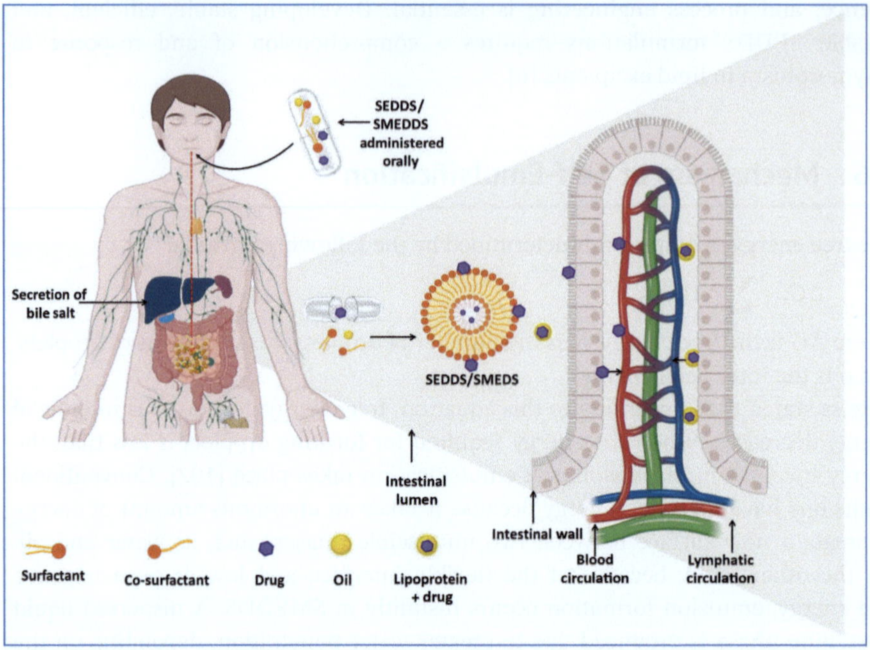

Figure 5.5 Mechanism of self-emulsification and uptake of lipophilic drugs in vivo.

bioavailability of the drug [92]. Self-emulsification and uptake of lipophilic drug through blood and lymphatic vessels are shown in Figure 5.5.

5.8 Applications of SEDDS

5.8.1 SEDDS/SMEDDS for Nasal Drug Delivery

Leichner et al. developed and described a SEDDS for delivering dimenhydrinate using the nasal route, which was selected based on drug solubility and stability research. The size distribution of the droplets was 60–220 nm. The permeability and toxicity of tissue were examined on the bovine nasal mucosa. The penetration was found to be 2.8 times higher than that of pure dimenhydrinate. There was no discernible severe tissue injury. The possibility of irritation caused by SEDDS was also assessed. According to the study, adding dimenhydrinate to SEDDS may be good and excellent way to deliver antiemetics through the nose. In homogenous mixes of surfactant (Labrasol, caprylocaproyl polyoxyl-8 glycerides HLB 12) and oily phase (Transcutol HP, diethylene glycol monoethyl ether HLB 4); phase diagrams were made to determine the region of emulsion formation. The mixtures were made with volume ratios ranging from 9:1 to 1:9, and water was gradually added while stirring constantly at 25 °C. The visual assessment showed that there were continuous stages that appeared as dazzling white, bluish white, less clear, and faintly blueish. Triplot version 4.1.2 was the program used to map the emulsion area. Phase diagrams were made for both drug-loaded formulations with 75 mg/ml of dimenhydrinate and blank formulations. The components are shown as a percentage of volume.

In the pseudoternary-phase diagram (Figure 5.6), the gray areas show the amount of water, oil, and surfactant needed for developing o/w emulsions. The emulsions develop in less than 2 minutes and range in appearance from faintly bluish to dazzling white. The system's emulsifying capabilities improve with increasing emulsion area. F1 and F2's pseudoternary diagrams were comparable, apart from F1's larger emulsion area. When Labrasol and Cremophor EL are combined, the self-emulsifying area in F3 can expand to a more significant lipid/surfactant ratio. These results are consistent with particle sizes: F3 shows the best emulsifying action with the smallest droplet size, while F1 is slightly smaller than F2. At decreasing ratios of lipid to surfactant, the emulsion area shrank due to the addition of dimenhydrinate [103].

Table 5.6 lists the compositions of different SEDDS/SMEDDS/SNEDDS formulations investigated for nasal delivery.

5.8.2 SEDDS/SMEDDS for Oral Drug Delivery

Wang et al. [59] developed the SNEDDS formulation with an objective to improve the dissolution rate of ibuprofen. A range of oils and surfactants was evaluated for SNEDDS according to droplet size. Several formulations were implemented for characteristics examination. The optimal SNEDDS formulation delivered over 95% of the ibuprofen within 30 minutes, having a mean nanoemulsion droplet diameter of 58 nm in phosphate buffer (pH 6.8). Authors utilized methyl decanoate (MD), methyl oleate (MO), and isopropyl myristate (IPM) as vehicles to dissolve ibuprofen. For SNEDDS, tweens and spans are used as surfactants, and 1.2-octanediol is used as a co-surfactant. These new SNEDDS carriers seem to be effective in managing the release rates of water-insoluble ibuprofen [59].

In this investigation, homogeneous mixtures of oils and emulsifiers were blended and homogenized for three days at 25 °C, and three samples of the chosen concentration were prepared. The materials were combined using a magnetic stirrer after being pre-equilibrated and diluted 100 times with water. Using an argon laser and dynamic light scattering at a 173° scattering angle, the droplet size of the resultant emulsion was measured.

Figure 5.7a shows the effect of emulsifier concentration on droplet size. It demonstrates droplet size drops with emulsifier concentrations up to 50 wt%, which is explained by increasing emulsifier adsorption along the o/w interface and a reduction in interfacial tension inside the system. This encourages the production of smaller droplet nanoemulsions. Droplet size does, however, grow over 50 wt% of emulsifier concentration. For SNEDDS containing IPM as an oil phase, an extremely viscous liquid crystalline phase appears, as shown by the pseudoternary-phase diagram (Figure 5.7b). This results in an increase in the droplet size of nanoemulsion due to the increased interfacial stiffness.

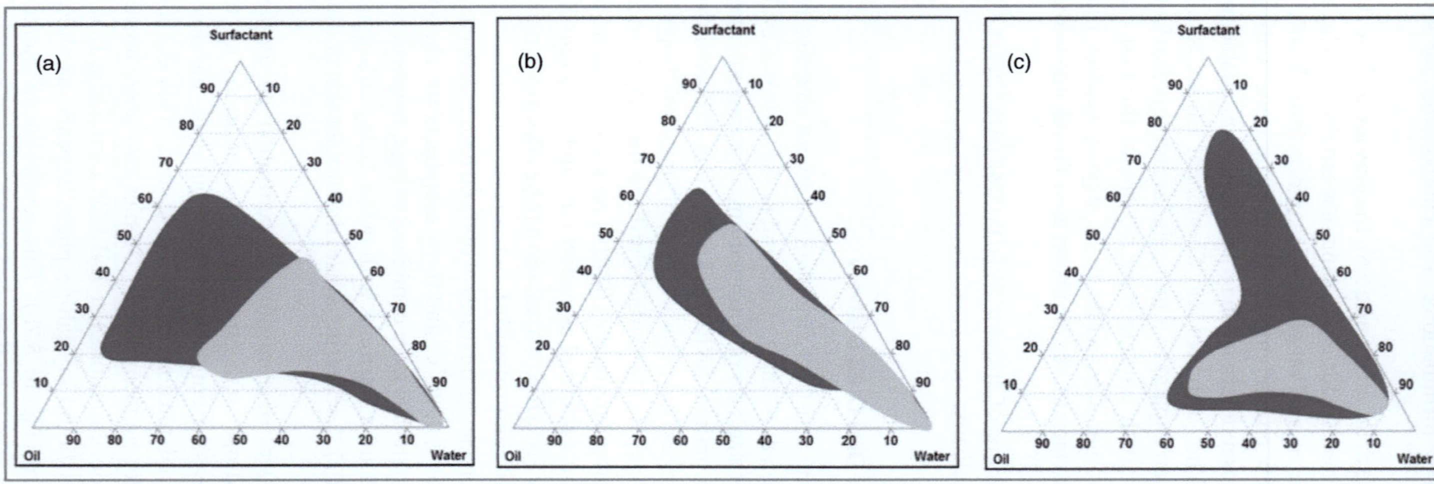

Figure 5.6 Pseudoternary-phase diagrams of formulation F1 (a), F2 (b), and F3 (c). Components are represented in volume percent, and the shaded areas depict the regions of emulsion. (a) Surfactant phase (Cremophor EL/PEG 200/PG [1.25/1/1]), oil phase (Transcutol HP/Capmul MCM [1/1.3]); (b) surfactant phase (Cremophor EL/PEG 400/PG [1.5/1/1]), oil phase (Transcutol HP/Capmul PG-8 [1/1]); and (c) surfactant phase (Cremophor EL/PEG 400/PG [2/1/1]), oil phase (Transcutol HP/Capmul MCM [2/1]). Excipients were mixed at 25 °C. Dark gray area, blank formulation; light gray area, dimenhydrinate loaded formulation. *Source:* Leichner et al. [103] / Springer Nature / CC BY 4.0.

Table 5.6 Compositions of different types of SEDDSs investigated for nasal delivery.

Types	Oil/lipid	Surfactant/ co-surfactant/ emulsifier	Drug	References
SNEDDS	Butylidene phthalide	Kolliphor® HS 15 (K-HS15), PEG-400	Butylidenephthalide	[104]
SMEDDS	Miglyol®	Kolliphor® RH40 and Transcutol HP	Perampanel	[105]
SNEDDS	Captex® 8000	Kolliphor® RH 40, Capmul® MCM NF	Insulin	[106]
SNEDDS,	Cremophor® RH 40 and Cremophor® EL	Lauroglycol™ FCC, Lauroglycol™ 90 and Transcutol® P	Morin	[107]

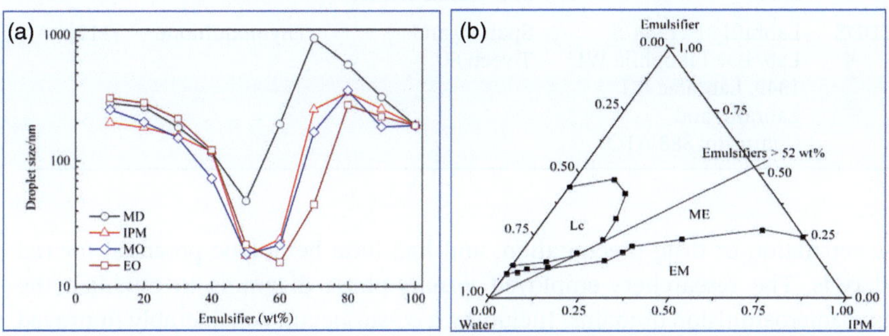

Figure 5.7 (a) Droplet size of emulsions (1 wt%) as a function of emulsifier concentration in oil/emulsifier (Tween 80–Span 20–1,2-octanediol)/water system (polydispersity index = 0.25 ± 0.05); (b) pseudoternary-phase diagram in the system of water/emulsifier (T80:S20:1,2-octanediol = 49:21:30 (w/w))/IPM. *Source:* Wang et al. [59] / with permission of Elsevier.

Table 5.7 lists the compositions of different types of SEDDS formulations investigated for oral drug delivery.

5.8.3 SEDDS/SMEDDS for Parenteral Delivery

Tacrolimus (FK 506), an immunosuppressant, is presently available in a non-aqueous formulation, including hydrogenated castor oil derivatives, linked to acute anaphylactic responses. Borhade et al. [113] produced a novel parenterally acceptable formulation employing SMEDDS, which uses ethyl oleate as the oily phase, Solutol HS 15 as the surfactant, and glycofurol as a co-surfactant. SMEDDS dissolved 0.8% FK 506, resulting in a spontaneous microemulsion with globule size smaller than 30 nm. The resultant microemulsion was iso-osmotic, did not display

Table 5.7 Compositions of different types of SEDDS investigated for oral delivery.

Types	Oil/lipid	Surfactant/ co-surfactant/ emulsifier	Drug	References
SNEDDS	Coconut oil, walnut oil, and almond oil	Tween 80, Tween 40, Tween 20, DMSO, Transcutol, and PEG	Citral	[108]
SNEDDS	Captex® 300	Kolliphor®RH 40 and Transcutol®HP	Pemigatinib	[109]
SNEDDS	Capmul MCM EP, Captex® 300	Kolliphor® RH40, Rhamnolipid	—	[110]
SNEDDS	Eucalyptus oil, almond oil, sunflower oil, ethyl oleate	Lauroglycol 90, Poloxamer 188, and Transcutol HP	Cephalexin	[111]
SNEDDS	Labrafil M 2125 CS, Labrafac Lipophile WL 1349, Labrafac PG, Labrafil, and Compritol 888 ATO	Span 80 and Tween 80	Thymoquinone	[112]

phase separation or drug precipitation, and had little hemolytic potential for red blood cells. The researchers employed ternary-phase diagrams to establish the limits of microemulsion domains. Including a co-surfactant considerably improved the area of microemulsion production at all K_m values when compared to Solutol HS 15. The most significant area was achieved at $K_m = 3.0$, which allowed the most ethyl oleate to be incorporated. The increase in the microemulsion region was more concentrated along the o/w axis. The study also discovered that Solutol HS 15 made a substantial contribution to microemulsion production. However, Solutol HS 15 alone might result in a very limited microemulsion area, necessitating the use of co-surfactants such as glycofurol. The study discovered that both high and low total surfactant concentrations resulted in gel-like structures and emulsion-type dispersion systems, independent of K_m (Figure 5.8).

The study effectively investigated the parenteral administration of the immunosuppressant FK 506, a weakly water-soluble drug, using SMEDDS. Without hydrogenated castor oil derivative, the proposed approach increased solubility and could be reconstituted as a microemulsion for intravenous administration. It proved to be dilutable, safe, and low in toxicity, which made it appropriate for a range of lipophilicity [113].

Table 5.8 lists the compositions of different types of SEDDS/SMEDDS/SNEDDS formulations investigated for oral drug delivery.

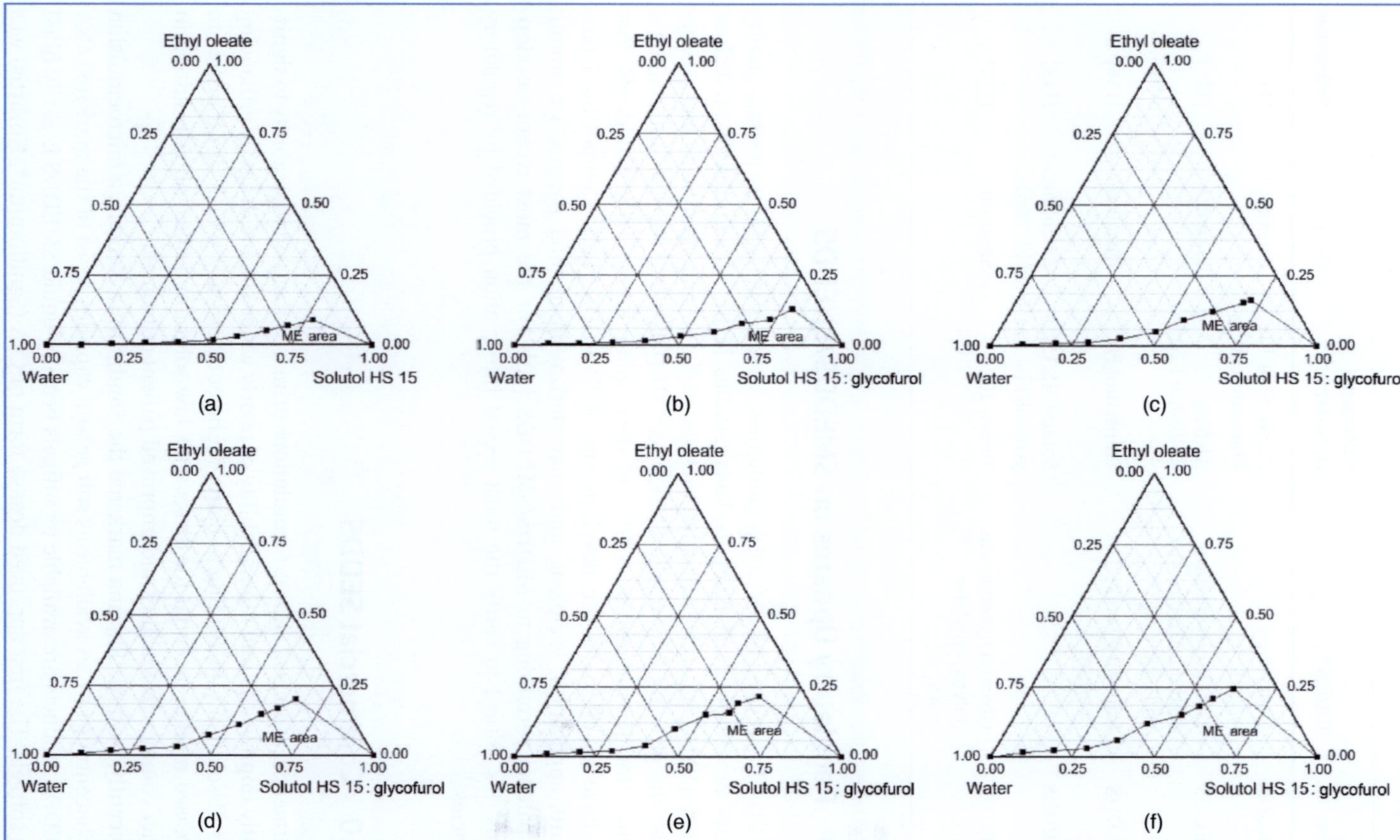

Figure 5.8 Pseudoternary-phase diagrams indicating o/w microemulsion existence region (ME) of ethyl oleate: Solutol HS 15: water system in the absence of glycofurol (a) and in the presence of glycofurol at various Km values 0.5 (b), 1.0 (c), 1.5 (d), 2.0 (e), and 3.0 (f). *Source:* Borhade et al. [113] / with permission of Taylor & Francis.

Table 5.8 Compositions of different types of SEDDS/SNEDDS investigated for parenteral delivery.

Types	Oil/lipid	Surfactant/ co-surfactant	Drug	References
SNEDDS	Capryol 90	Cremophor EL, Transcutol HP	Cilostazol	[114]
SNEDDS	*Zanthoxylum rhetsa* seed oil, Miglyol 810, and castor oil	KHS15, Kolliphor HS15	Propofol	[115]
SNEDDS	Capryol 90 and Labrasol	Transcutol HP	Docetaxel	[116]
SMEDDS		Solutol HS 15, glycofurol	Tacrolimus (FK 506)	[113]
SNEDDS	Glyceryl trioctanoate, polyoxyethylene castor oil	Tween 20	Sesamin	[117]

5.9 Regulatory Updates on SMEDDS/SNEDDS

The FDA has revised its SMEDDS guidelines, which cover stability testing, quality control procedures, and bioequivalence studies [118]. The European Medicines Agency (EMA) has released guidelines on pediatric and nanomedicines, recognizing the growing importance of SEDDS/SMEDDS [92, 119]. The International Council for Harmonization (ICH) is developing harmonizing standards that will have an effect on the development of SMEDDS [120]. Composition, procedure, application, excipient, and characterization method patents are among the patents pertaining to SEDDS/SMEDDS [30]. For the most recent developments, it's critical to verify the most recent information provided by regulatory agencies.

5.10 Commercial SEDDS

Commercial aspects of SEDDS formulations primarily focus on their ability to significantly improve the oral bioavailability of poorly water-soluble drugs by enhancing their dissolution rate, leading to better therapeutic efficacy, which can translate to increased market potential for drugs with low solubility issues, often resulting in higher commercial viability and improved patient outcomes.

Recently, Novartis Pharma marketed the Sandimmun Neoral®, a microemulsion preconcentrate of CsA available as soft gelatin capsules. Most of the marketed CsA SMEDDS products are available as softgels because liquid SMEDDS is easily filled into softgels. The first approved dosage form of CsA (Sandimmun®, Novartis) was designed as preconcentrate of o/w emulsion, which was based on corn oil, linoleoyl macroglyceride, and ethanol. A new dosage form, Neoral® (Novartis), which is considered advancement of Sandimmun®, forms microemulsions in aqueous media and provides more advanced and consistent bioavailability (www.novartis.com). Neoral cyclosporine formulation in which corn oil, mono-, di-, and triglycerides were used as lipid phase, cremophor RH40 as a surfactant, and propylene glycol and ethanol as co-solvent along with α-tocopherol as an antioxidant. *Neoral* is a thermodynamically stable dispersion with droplet size below 100 nm when introduced into an aqueous medium.

5.11 Patents on SEDDS

There are a few patents available on SEDDS formulations and here we present an updated account of patents on SEDDS formulation compositions (Table 5.9).

5.12 Clinical Status of SEDDS Formulations

The encouraging findings at preclinical level bought the SEDDS formulations into the clinical trials. A detailed account of clinical trials conducted on SEDDS formulations is tabulated in Table 5.10.

5.13 Conclusion

SEDDS primarily deals with selection of appropriate combination of lipids, surfactants, and solvents with high solubilization capabilities in order to enhance the solubility and bioavailability of poorly soluble APIs.

Here, authors have highlighted the major aspects of the SEDDS formulations *viz.* components of the SEDDS systems, challenges in their development, and applications. SEDDS systems provide a practical solution to the solubility issues associated with the poorly soluble drugs. The in situ emulsion-forming system could be used as an efficient approach in achieving the high drug stability in harsh GI conditions.

In conclusion, SEDDSs are poised to play a crucial role in modern pharmaceutical development, particularly for BCS Class II and IV drugs that face solubility challenges. Their ability to enhance bioavailability while maintaining formulation stability makes them a promising avenue for future research and application in drug delivery strategies. As advancements continue in formulation techniques and characterization methods, SEDDS will likely become integral to the development of effective therapeutic agents. Further developments in this area could lead to the emergence of novel drug delivery technology with improved physical stability and high commercial impact.

Table 5.9 A summary of patents on SEDDS formulations.

Title of the invention	Patent number	Inventors	Applicants	Year
Omega-3 fatty acid self-emulsifying composition	EP3248599B1	Hiromitsu Ito Hirosato Fujii Motoo Yamagata Daichi TANAKA	Mochida Pharmaceutical Co Ltd	2016
Solid self-emulsifying cannabinoid compositions	US20210212946A1	Doron Friedman	ICD Pharma Ltd	2019
An improved oral dosage form of tetrahydrocannabinol and a method of avoiding and/or suppressing hepatic first–pass metabolism via targeted chylomicron/lipoprotein delivery	WO2012033478A1	Ram B. Murty Santos B. Murty	Murty Pharmaceuticals Inc.	2010
Process for preparing a finely self-emulsifiable pharmaceutical composition	IL164163A0	—	Pharmacia Corporation	2003
Celecoxib self-emulsifying oral liquid and preparation method thereof	CN113750043A	Zhang Daizhou Li Dawei Yang Xuehua Wang Changbin Liu Zhengping Zhao Renyongwuli Hua Xue Song Mao Kaifan Wang Lili	Shandong Academy of Pharmaceutical Sciences	2021
Celecoxib-loaded solid oral compositions using self-nanoemulsifying drug delivery system and methods for preparing the same	KR102635930B1	Jin Seong-gyu Choi Han-gon Kim Jong-oh Kim Jeong-seok Woo Mi-ran Cheon Seung-hyeon		2021

Table 5.10 A summary of clinical trials on SEDDS formulations.

Study title	Trial ID	Number of enrolled subjects	Phase of study	Study type	Sponsor/ investigator
Tipranavir Dose-response Study Using a Prototype Self-emulsifying Drug Delivery System (SEDDS) Formulation in Treatment-naive HIV-1 Infected Patients Report on the Post-study Option, a Treatment of Triple Therapy with Delavirdine, ZDV and 3TC Without Tipranavir	NCT02249130	18	II	Interventional	Boehringer Ingelheim
Bioavailability of EPA + DHA in a SMEDS Formulation	NCT03443076	24	—	Interventional	Midwest Center for Metabolic and Cardiovascular Research
Clinical Assessment of Itraconazole Self Nano Emulsifying Drug Delivery System Intermediate Gel	NCT04110834	30	II	Interventional	Sara Botros, Minia University
Clinical Assessment of Voriconazole Self Nano Emulsifying Drug Delivery System Intermediate Gel	NCT04110860	30	II	Interventional	Sara Botros, Minia University

References

1 Alqahtani, M.S., Kazi, M., Alsenaidy, M.A., and Ahmad, M.Z. (2021). Advances in oral drug delivery. *Front. Pharmacol.* 12: 618411. doi: 10.3389/fphar.2021.618411.

2 Kim, J. and De Jesus, O. (2024). Medication routes of administration. In: *StatPearls [Internet]*. Treasure Island (FL): StatPearls Publishing. https://www.ncbi.nlm.nih.gov/books/NBK568677/.

3 Singh, S., Singh, G., Sehrawat, S. et al. (2021). Chapter 25 - Conclusion and future considerations of dendrimers. In: *Dendrimer-Based Nanotherapeutics* (ed. P. Kesharwani), 449–458. Academic Press. https://doi.org/10.1016/B978-0-12-821250-9.00005-6.

4 Azman, M., Sabri, A.H., Anjani, Q.K. et al. (2022). Intestinal absorption study: challenges and absorption enhancement strategies in improving oral drug delivery. *Pharmaceuticals (Basel)* 15 (8): 975. doi: 10.3390/ph15080975.

5 Claassen, V. (1994). Oral drug administration. *Techn. Behav. Neural Sci.* 12: 59–115. doi: 10.1016/B978-0-444-81871-3.50011-4.

6 Salawi, A. (2022). Self-emulsifying drug delivery systems: a novel approach to deliver drugs. *Drug Deliv.* 29 (1): 1811–1823. doi: 10.10.1080/10717544.2022.2083724.

7 Date, A.A., Hanes, J., and Ensign, L.M. (2016). Nanoparticles for oral delivery: design, evaluation and state-of-the-art. *J. Control. Release* 240: 504–526. doi: 10.1016/j.jconrel.2016.06.016.

8 Leonaviciute, G. and Bernkop-Schnürch, A. (2015). Self-emulsifying drug delivery systems in oral (poly) peptide drug delivery. *Expert Opin. Drug Deliv.* 12 (11): 1703–1716. doi: 10.1517/17425247.2015.1068287.

9 Zhu, Y., Ye, J., and Zhang, Q. (2020). Self-emulsifying drug delivery system improve oral bioavailability: role of excipients and physico-chemical characterization. *Pharm. Nanotechnol.* 8 (4): 290–301. doi: 10.2174/2211738508666200811104240.

10 Markovic, M., Zur, M., Ragatsky, I. et al. (2020). BCS class IV oral drugs and absorption windows: regional-dependent intestinal permeability of furosemide. *Pharmaceutics* 12 (12): 1175. doi: 10.3390/pharmaceutics12121175.

11 Schittny, A., Huwyler, J., and Puchkov, M. (2020). Mechanisms of increased bioavailability through amorphous solid dispersions: a review. *Drug Deliv.* 27 (1): 110–127. doi: 10.1080/10717544.2019.1704940.

12 Mahmood, A. and Bernkop-Schnürch, A. (2019). SEDDS: a game changing approach for the oral administration of hydrophilic macromolecular drugs. *Adv. Drug Deliv. Rev.* 142: doi: 10.1016/j.addr.2018.07.001.

13 Page, S. and Szepes, A. (2021). Impact on drug development and drug product processing. In: *Solid State Development and Processing of Pharmaceutical Molecules: Salts, Cocrystals, and Polymorphism*, vol. 79 (ed. M. Gruss), 325–364. Wiley. doi: 10.1002/9783527823048.ch6.

14 Plaza-Oliver, M., Santander-Ortega, M.J., and Lozano, M.V. (2021). Current approaches in lipid-based nanocarriers for oral drug delivery. *Drug Deliv. Transl. Res.* 11 (2): 471–497. doi: 10.1007/s13346-021-00908-7.

15 Kesharwani, R., Jaiswal, P., Patel, D.K., and Yadav, P.K. (2023). Lipid-based drug delivery system (LBDDS): an emerging paradigm to enhance oral bioavailability of poorly soluble drugs. *Biomed. Mater. Devices* 1 (2): 648–663. doi: 10.1007/s44174-022-00041-0.

16 McClements, D.J. (2018). Enhanced delivery of lipophilic bioactives using emulsions: a review of major factors affecting vitamin, nutraceutical, and lipid bioaccessibility. *Food Funct.* 9 (1): 22–41. doi: 10.1039/c7fo01515a.

17 He, C.X., He, Z.G., and Gao, J.Q. (2010). Microemulsions as drug delivery systems to improve the solubility and the bioavailability of poorly water-soluble drugs. *Expert Opin. Drug Deliv.* 7 (4): 445–460. doi: 10.1517/17425241003596337.

18 Patel, V.R. and Agrawal, Y.K. (2011). Nanosuspension: an approach to enhance solubility of drugs. *J. Adv. Pharm. Technol. Res.* 2 (2): 81–87. doi: 10.4103/2231-4040.82950.

19 Pohlen, M., Zdovc, J.A., Trontelj, J. et al. (2022). Relative bioavailability enhancement of simvastatin via dry emulsion systems: comparison of spray drying and fluid bed layering technology. *Eur. J. Pharm. Biopharm.* 172: 228–239. doi: 10.1016/j.ejpb.2021.12.004.

20 Nour, S., Bolandi, B., and Imani, R. (2020). Chapter 4 - Nanotechnology in gene therapy for musculoskeletal regeneration. In: *Nanoengineering in Musculoskeletal Regeneration* (ed. M. Razavi), 105–136. Academic Press. doi: 10.1016/B978-0-12-820262-3.00004-9.

21 Ghasemiyeh, P. and Mohammadi-Samani, S. (2018). Solid lipid nanoparticles and nanostructured lipid carriers as novel drug delivery systems: applications, advantages and disadvantages. *Res. Pharm. Sci.* 13 (4): 288–303. doi: 10.4103/1735-5362.235156.

22 Yadav, K.S., Soni, G., Choudhary, D. et al. (2023). Microemulsions for enhancing drug delivery of hydrophilic drugs: exploring various routes of administration. *Med. Drug Discov.* 20: doi: 10.1016/j.medidd.2023.100162.

23 Gupta, M., Sharma, V., and Chauhan, N.S. (2017). Chapter 16: Nanotechnology for oral delivery of anticancer drugs: an insight potential. In: *In Micro and Nano Technologies, Nanostructures for Oral Medicine* (ed. E. Andronescu and A.M. Grumezescu), 467–510. Elsevier. doi: 10.1016/B978-0-323-47720-8.00017-1.

24 Alothaid, H., Aldughaim, M.S., Yusuf, A.O. et al. (2021). A comprehensive study of the basic formulation of supersaturated self-nanoemulsifying drug delivery systems (SNEDDS) of albendazolum. *Drug Deliv.* 28 (1): 2119–2126. doi: 10.1080/10717544.2021.

25 Anand, S., Gupta, R., and Sk, P. (2016). Self microemulsifying drug delivery system. *Asian J. Pharm. Clin. Res.* 9 (8): 33–38. doi: 10.22159/ajpcr.2016.v9s2.13180.

26 Jaiswal, P., Aggarwal, G., Harikumar, S.L., and Singh, K. (2014 Oct). Development of self-microemulsifying drug delivery system and solid-self-microemulsifying drug delivery system of telmisartan. *Int. J. Pharm. Investig.* 4 (4): 195–206. doi: 10.4103/2230-973X.143123.

27 Sambhakar, P.S., Malik, R., Bhatia, S. et al. (2023). Nanoemulsion: an emerging novel technology for improving the bioavailability of drugs. *Scientifica (Cairo)* 2023: 6640103. doi: 10.1155/2023/6640103.

28 Rajpoot, R., Tekade, M., Pandey, V. et al. (2020). Self-microemulsifying drug-delivery system: ongoing challenges and future ahead. In: *Advances in Pharmaceutical Product Development and Research: Drug Delivery Systems* (ed. R.K. Tekade), 393–454. Academic Press. doi: 10.1016/B978-0-12-814487-9.00009-0.

29 Ango, B., Nobel, K., Johnson, R. et al. (2019). Self – micro emulsifying drug delivery system: a potential solution to the challenges of oraldelivery of poorly water – soluble drugs. *Int. J. Pharm. Sci. Rev. Res.* 58 (1): 34–44.

30 Meirinho, S., Rodrigues, M., Santos, A.O. et al. (2022). Self-emulsifying drug delivery systems: an alternative approach to improve brain bioavailability of poorly water-soluble drugs through intranasal administration. *Pharmaceutics* 14 (7): 1487. doi: 10.3390/pharmaceutics14071487.

31 Akula, S., Gurram, A.K., and Devireddy, S.R. (2014 Dec 8). Self-microemulsifying drug delivery systems: an attractive strategy for enhanced therapeutic profile. *Int. Sch. Res. Notices* 2014: 964051. doi: 10.1155/2014/964051.

32 Mahmoud, E.A., Bendas, E.R., and Mohamed, M.I. (2009). Preparation and evaluation of self-nanoemulsifying tablets of carvedilol. *AAPS PharmSciTech* 10 (1): 183–192. doi: 10.1208/s12249-009-9192-7.

33 Silberstein, S., Spierings, E.L.H., and Kunkel, T. (2023). Celecoxib oral solution and the benefits of self-microemulsifying drug delivery systems (SMEDDS) technology: a narrative review. *Pain Ther.* 12: 1109–1119. doi: 10.1007/s40122-023-00529-7.

34 Taha, E., Ghorab, D., and Zaghloul, A.A. (2007). Bioavailability assessment of vitamin A self-nanoemulsified drug delivery systems in rats: a comparative study. *Med. Princ. Pract.* 16 (5): 355–359. https://doi.org/10.1159/000104808.

35 Shah, M.G. and Agrawal, A. (2020). Self-microemulsifying system. In: *Colloid Science in Pharmaceutical Nanotechnology* (ed. S. Karakuş). IntechOpen. doi: 10.5772/intechopen.88603.

36 Rangaraj, N., Sampathi, S., Junnuthula, V. et al. (2022). Fast-fed variability: insights into drug delivery, molecular manifestations, and regulatory aspects. *Pharmaceutics* 14 (9): 1807. doi: 10.3390/pharmaceutics14091807.

37 Tang, B., Cheng, G., Gu, J.C., and Xu, C.H. (2008). Development of solid self-emulsifying drug delivery systems: preparation techniques and dosage forms. *Drug Discov. Today* 13 (13-14): 606–612. doi: 10.1016/j.drudis.2008.04.006.

38 Zhang, P., Liu, Y., Feng, N., and Xu, J. (2008). Preparation and evaluation of self-microemulsifying drug delivery system of oridonin. *Int. J. Pharm.* 355 (1–2): 269–276. doi: 10.1016/j.ijpharm.2007.12.026.

39 Lee, Y.Z., Seow, E.K., Lim, S.C. et al. (2021). Formulation and *in vivo* evaluation of a solid self-emulsifying drug delivery system using oily liquid tocotrienols as model active substance. *Pharmaceutics* 13 (11): 1777. doi: 10.3390/pharmaceutics13111777.

40 Pouton, C.W. (2000). Lipid formulations for oral administration of drugs: nonemulsifying, self-emulsifying and 'self microemulsifying' drug delivery systems. *Eur. J. Pharm. Sci.* 11 (Suppl. 2): S93–S98. doi: 10.1016/j.addr.2007.10.010.

41 Shah, A., Thool, P., Sorathiya, K. et al. (2018). Effect of different polysorbates on development of self-microemulsifying drug delivery systems using medium chain lipids. *Drug Dev. Ind. Pharm.* 44 (2): 215–223. doi: 10.1080/03639045.2017.1386202.

42 Bakala N'Goma, J.C., Amara, S., Dridi, K. et al. (2012). Understanding the lipid-digestion processes in the GI tract before designing lipid-based drug-delivery systems. *Ther. Deliv.* 3 (1): 105–124. doi: 10.4155/tde.11.138.

43 Boyd, B.J. and Clulow, A.J. (2021). The influence of lipid digestion on the fate of orally administered drug delivery vehicles. *Biochem. Soc. Trans.* 49 (4): 1749–1761. doi: 10.1042/BST20210168.

44 Nair, A.B., Singh, B., Shah, J. et al. (2022). Formulation and evaluation of self-nanoemulsifying drug delivery system derived tablet containing sertraline. *Pharmaceutics* 14 (2): 336. doi: 10.3390/pharmaceutics14020336.

45 Bari, D.B. and Pardeshi, C.V. (2024). Nanocochleates as controlled and novel drug delivery systems. In: *Advances in Controlled and Novel Drug Delivery System* (ed. N.K. Jain), 902–930. CBS Publishers & Distributors.

46 Hauss, D.J. (2007). Oral lipid-based formulations. *Adv. Drug Deliv. Rev.* 59 (7): 667–676. doi: 10.1016/j.addr.2007.05.006.

47 Souto, E.B. and Müller, R.H. (2010). Lipid nanoparticles: effect on bioavailability and pharmacokinetic changes. *Handb. Exp. Pharmacol.* 197: 115–141. doi: 10.1007/978-3-642-00477-3_4.

48 Elmowafy, M. and Al-Sanea, M.M. (2021). Nanostructured lipid carriers (NLCs) as drug delivery platform: advances in formulation and delivery strategies. *Saudi Pharm. J.* 29 (9): 999–1012. doi: 10.1016/j.jsps.2021.07.015.

49 Son, H.Y., Chae, B.R., Choi, J.Y. et al. (2018). Optimization of self-microemulsifying drug delivery system for phospholipid complex of telmisartan using D-optimal mixture design. *PLoS One* 13 (12): e0208339. doi: 10.1371/journal.pone.0208339.

50 Luo, X., Wang, D., Wang, M. et al. (2021). Development of phospholipid complex loaded self-microemulsifying drug delivery system to improve the oral bioavailability of resveratrol. *Nanomedicine (London)* 16 (9): 721–739. doi: 10.2217/nnm-2020-0422.

51 Gohel, M.C., Parikh, R.K., Modi, S.C. et al. (2011). Novel drug delivery approaches to bypass P-glycoprotein efflux pump. *Pharma. Rev.* 9 (1).

52 Akkuş-Dağdeviren, Z.B., Wolf, J.D., Kurpiers, M. et al. (2021 May 1). Charge reversal self-emulsifying drug delivery systems: a comparative study among various phosphorylated surfactants. *J. Colloid Interface Sci.* 589: 532–544. doi: 10.1016/j.jcis.2021.01.025.

53 Nekkanti, V., Rueda, J., Wang, Z., and Betageri, G.V. (2016). Comparative evaluation of proliposomes and self-micro-emulsifying drug delivery system for improved oral bioavailability of nisoldipine. *Int. J. Pharm.* 505 (1-2): 79–88. doi: 10.1016/j.ijpharm.2016.03.065.

54 Tran, T., Xi, X., Rades, T., and Müllertz, A. (2016). Formulation and characterization of self-nanoemulsifying drug delivery systems containing monoacyl phosphatidylcholine. *Int. J. Pharm.* 502 (1-2): 151–160. doi: 10.1016/j.ijpharm.2016.02.026.

55 Goyal, U., Arora, R., and Aggarwal, G. (2012). Formulation design and evaluation of a self-microemulsifying drug delivery system of lovastatin. *Acta Pharma.* 62 (3): 357–370. doi: 10.2478/v10007-012-0022-1.

56 Weerapol, Y., Limmatvapirat, S., Nunthanid, J., and Sriamornsak, P. (2014). Self-nanoemulsifying drug delivery system of nifedipine: impact of hydrophilic-lipophilic balance and molecular structure of mixed surfactants. *AAPS PharmSciTech* 15 (2): 456–464. doi: 10.1208/s12249-014-0078-y.

57 Chintalapudi, R., Murthy, T.E., Lakshmi, K.R., and Manohar, G.G. (2015). Formulation, optimization, and evaluation of self-emulsifying drug delivery systems of nevirapine. *Int. J. Pharm. Investig.* 5 (4): 205–213. doi: 10.4103/2230-973X.167676.

58 Giddam, R., Hepsiba, S., and Mukeri, I.H. (2022). Formulation of self emulsifying drug delivery system of Dolutegravir sodium. *J. Xi'an Shiyou Univ. Nat. Sci. Ed.* 18 (12): 195–209.

59 Wang, L., Dong, J., Chen, J. et al. (2009). Design and optimization of a new self-nanoemulsifying drug delivery system. *J. Colloid Interface Sci.* 330 (2): 443–448. doi: 10.1016/j.jcis.2008.10.077.

60 Date, A.A. and Nagarsenker, M.S. (2007). Design and evaluation of self-nanoemulsifying drug delivery systems (SNEDDS) for cefpodoxime proxetil. *Int. J. Pharm.* 329 (1-2): 166–172. doi: 10.1016/j.ijpharm.2006.08.038.

61 Park, H., Ha, E.S., and Kim, M.S. (2020 Apr 16). Current status of supersaturable self-emulsifying drug delivery systems. *Pharmaceutics* 12 (4): 365. doi: 10.3390/pharmaceutics12040365.

62 Lawrence, M.J. and Rees, G.D. (2000). Microemulsion-based media as novel drug delivery systems. *Adv. Drug Deliv. Rev.* 45 (1): 89–121. doi: 10.1016/s0169-409x(00)00103-4.

63 Agubata, C.O., Nzekwe, I.T., Obitte, N.C. et al. (2014). Effect of oil, surfactant and co-surfactant concentrations on the phase behavior, physicochemical properties and drug release from self-emulsifying drug delivery systems. *J. Drug Discov. Dev. Deliv.* 1: 7.

64 Djekic, L., Janković, J., Rašković, A., and Primorac, M. (2018 Aug 30). Semisolid self-microemulsifying drug delivery systems (SMEDDSs): effects on pharmacokinetics of acyclovir in rats. *Eur. J. Pharm. Sci.* 121: 287–292. doi: 10.1016/j.ejps.2018.06.005.

65 Mohsin, K., Alamri, R., Ahmad, A. et al. (2016 Jun 14). Development of self-nanoemulsifying drug delivery systems for the enhancement of solubility and oral bioavailability of fenofibrate, a poorly water-soluble drug. *Int. J. Nanomed.* 11: 2829–2838. doi: 10.2147/IJN.S104187.

66 Penjuri, S.C.B., Damineni, S., Ravouru, N., and Poreddy, S.R. (2017). Self-emulsifying drug delivery system (SEDDS) of Ibuprofen: formulation, *in vitro* and *in vivo* evaluation. *Ceska Slov. Farm.* 66 (1): 23–34. English.

67 Kamal, M.M., Salawi, A., Lam, M. et al. (2020). Development and characterization of curcumin-loaded solid self-emulsifying drug delivery system (SEDDS) by spray drying using Soluplus® as solid carrier. *Powder Technol.* 369: 137–145. doi: 10.1016/j.powtec.2020.05.023.

68 Khoo, S.M., Humberstone, A.J., Porter, C.J.H. et al. (1998). Formulation design and bioavailability assessment of lipidic self-emulsifying formulations of halofantrine. *Int. J. Pharm.* 167 (1–2): 155–164. doi: 10.1016/S0378-5173(98)00054-4.

69 Kommuru, T.R., Gurley, B., Khan, M.A., and Reddy, I.K. (2001). Self-emulsifying drug delivery systems (SEDDS) of coenzyme Q10: formulation development and bioavailability assessment. *Int. J. Pharm.* 212 (2): 233–246. doi: 10.1016/s0378-5173(00)00614-1.

70 Alqahtani, S., Alayoubi, A., Nazzal, S. et al. (2014). Enhanced solubility and oral bioavailability of γ-tocotrienol using a self-emulsifying drug delivery system (SEDDS). *Lipids* 49 (8): 819–829. doi: 10.1007/s11745-014-3923-6.

71 Salimi, A., Sharif Makhmal Zadeh, B., Hemati, A.A., and Akbari, B.S. (2014). Design and evaluation of self-emulsifying drug delivery system (SEDDS) of carvedilol to improve the oral absorption. *Jundishapur J. Nat. Pharm. Prod.* 9 (3): e16125. doi: 10.17795/jjnpp-16125.

72 Dokania, S. and Joshi, A.K. (2015). Self-microemulsifying drug delivery system (SMEDDS) – challenges and road ahead. *Drug Deliv.* 22 (6): 675–690. doi: 10.3109/10717544.2014.896058.

73 Kallakunta, V.R., Dudhipala, N., Nyavanandi, D. et al. (2023). Formulation and processing of solid self-emulsifying drug delivery systems (HME S-SEDDS): a single-step manufacturing process via hot-melt extrusion technology through response surface methodology. *Int. J. Pharm.* 642. doi: 10.1016/j.ijpharm.2023.123055.

74 Dixit, A.R., Rajput, S.J., and Patel, S.G. (2010). Preparation and bioavailability assessment of SMEDDS containing valsartan. *AAPS PharmSciTech* 11 (1): 314–321. doi: 10.1208/s12249-010-9385-0.

75 Jaiswal, P., Aggarwal, G., Harikumar, S.L., and Singh, K. (2013). Bioavailability enhancement of poorly soluble drugs by SMEDDS: a review. *J. Drug Deliv. Ther.* 3 (1): 98–109.

76 Gan, Y., Baak, J.P.A., Chen, T. et al. (2023). Supersaturation and precipitation applicated in drug delivery systems: development strategies and evaluation approaches. *Molecules* 28 (5): 2212. doi: 10.3390/molecules28052212.

77 Zhang, J., Guo, M., Luo, M., and Cai, T. (2023). Advances in the development of amorphous solid dispersions: the role of polymeric carriers. *Asian J. Pharm. Sci.* 18: 4. doi: 10.1016/j.ajps.2023.100834.

78 Muhammed, S.A. and Al-Kinani, K.K. (2023 May 30). Formulation and *in vitro* evaluation of meloxicam as a self-microemulsifying drug delivery system. *F1000Res* 12: 315. doi: 10.12688/f1000research.130749.2.

79 Kim, D.S., Cho, J.H., Park, J.H. et al. (2019). Self-microemulsifying drug delivery system (SMEDDS) for improved oral delivery and photostability of methotrexate. *Int. J. Nanomed.* 5 (14): 4949–4960. doi: 10.2147/IJN.S211014.

80 Gurram, A.K., Deshpande, P.B., Kar, S.S. et al. (2015 May-Jun). Role of components in the formation of self-microemulsifying drug delivery systems. *Indian J. Pharm. Sci.* 77 (3): 249–257. doi: 10.4103/0250-474x.

81 Li, F., Hu, R., Wang, B. et al. (2017 May). Self-microemulsifying drug delivery system for improving the bioavailability of huperzine A by lymphatic uptake. *Acta Pharm. Sin. B* 7 (3): 353–360. doi: 10.1016/j.apsb.2017.02.002.

82 Yáñez, J.A., Wang, S.W., Knemeyer, I.W. et al. (2011 Sep 10). Intestinal lymphatic transport for drug delivery. *Adv. Drug Deliv. Rev.* 63 (10–11): 923–942. doi: 10.1016/j.addr.2011.05.019.

83 Bekisz, S., Baudin, L., Buntinx, F. et al. (2022). *in vitro*, *in vivo*, and *in silico* models of lymphangiogenesis in solid malignancies. *Cancers (Basel)* 14 (6): 1525. doi: 10.3390/cancers14061525.

84 Liao, H., Gao, Y., Lian, C. et al. (2019). Oral absorption and lymphatic transport of baicalein following drug-phospholipid complex incorporation in self-microemulsifying drug delivery systems. *Int. J. Nanomedicine* 14: 7291–7306. doi: 10.2147/IJN.S214883.

85 Hedge, O.J., Höök, F., Joyce, P., and Bergström, C.A.S. (2021). Investigation of self-emulsifying drug-delivery system interaction with a biomimetic membrane under conditions relevant to the small intestine. *Langmuir* 37 (33): 10200–10213. doi: 10.1021/acs.langmuir.1c01689.

86 Sensoy, I. (2021 Apr). A review on the food digestion in the digestive tract and the used *in vitro* models. *Curr. Res. Food Sci.* 14 (4): 308–319. doi: 10.1016/j.crfs.2021.04.004.

87 Huang, Y., Yu, Q., Chen, Z. et al. (2021 Aug). *in vitro* and *in vivo* correlation for lipid-based formulations: current status and future perspectives. *Acta Pharm. Sin. B* 11 (8): 2469–2487. doi: 10.1016/j.apsb.2021.03.025.

88 Lu, Y., Kim, S., and Park, K. (2011 Oct 10). *in vitro-in vivo* correlation: perspectives on model development. *Int. J. Pharm.* 418 (1): 142–148. doi: 10.1016/j.ijpharm.2011.01.010.

89 Musakhanian, J., Rodier, J.D., and Dave, M. (2022 May 20). Oxidative stability in lipid formulations: a review of the mechanisms, drivers, and inhibitors of oxidation. *AAPS PharmSciTech* 23 (5): 151. doi: 10.1208/s12249-022-02282-0.

90 Mahajan, N., Mujtaba, M.A., Fule, R. et al. (2024 Feb 3). Self-emulsifying drug delivery system for enhanced oral delivery of tenofovir: formulation, physicochemical characterization, and bioavailability assessment. *ACS Omega* 9 (7): 8139–8150. doi: 10.1021/acsomega.3c08565.

91 Czajkowska-Kośnik, A., Szekalska, M., Amelian, A. et al. (2015). Development and evaluation of liquid and solid self-emulsifying drug delivery systems for atorvastatin. *Molecules* 20 (12): 21010–21022. doi: 10.3390/molecules201219745.

92 Seo, E.B., du Plessis, L.H., and Viljoen, J.M. (2022). Solidification of self-emulsifying drug delivery systems as a novel approach to the management of uncomplicated malaria. *Pharmaceuticals* 15 (2): 120. doi: 10.3390/ph15020120.

93 Gupta, S., Kesarla, R., and Omri, A. (2013). Formulation strategies to improve the bioavailability of poorly absorbed drugs with special emphasis on self-emulsifying systems. *ISRN Pharm.* 2013: 848043. doi: 10.1155/2013/848043.

94 Huynh, N.A., Do, T.H., Le, X.L. et al. (2022). Development of softgel capsules containing cyclosporine a encapsulated pine essential oil based self-microemulsifying drug delivery system. *J. Drug Delivery Sci. Technol.* 68: 103115. doi: 10.1016/j.jddst.2022.103115. https://www.novartis.com/sg-en/sites/novartis_sg/files/SandimmunNeoral-Jul2020.SIN-app091120.pdf (accessed 24 February 2025).

95 van Staden, D., Haynes, R.K., and Viljoen, J.M. (2023). The science of selecting excipients for dermal self-emulsifying drug delivery systems. *Pharmaceutics* 15 (4): 1293. doi: 10.3390/pharmaceutics15041293.

96 Wang, D., Xiao, H., Lyu, X. et al. (2023). Lipid oxidation in food science and nutritional health: a comprehensive review. *Oil Crop Sci.* 8 (1): 35–44. doi: 10.1016/j.ocsci.2023.02.002.

97 Shahidi, F. and Hossain, A. (2022). Role of lipids in food flavor generation. *Molecules* 27 (15): 5014. doi: 10.3390/molecules27155014.

98 Abeyrathne, E.D.N.S., Nam, K., and Ahn, D.U. (2021). Analytical methods for lipid oxidation and antioxidant capacity in food systems. *Antioxidants (Basel)* 10 (10): 1587. doi: 10.3390/antiox10101587.

99 Censi, R. and Di Martino, P. (2015). Polymorph impact on the bioavailability and stability of poorly soluble drugs. *Molecules* 20 (10): 18759–18776. doi: 10.3390/molecules201018759.

100 Shah, R.R., Taccone, M., Monaci, E. et al. (2019 Aug 8). The droplet size of emulsion adjuvants has significant impact on their potency, due to differences in immune cell-recruitment and -activation. *Sci. Rep.* 9 (1): 11520. doi: 10.1038/s41598-019-47885-z.

101 Chistyakov, D. and Sergeev, G. (2020 Jan 1). The polymorphism of drugs: new approaches to the synthesis of nanostructured polymorphs. *Pharmaceutics* 12 (1): 34. doi: 10.3390/pharmaceutics12010034.

102 Reiss, H. (1975). Entropy-induced dispersion of bulk liquids. *J. Colloid Interface Sci.* 53 (1): 61–70. doi: 10.1016/0021-9797(75)90035-1.

103 Leichner, C., Baus, R.A., Jelkmann, M. et al. (2019). *in vitro* evaluation of a self-emulsifying drug delivery system (SEDDS) for nasal administration of dimenhydrinate. *Drug Deliv. Transl. Res.* 9 (5): 945–955. doi: 10.1007/s13346-019-00634-1.

104 Chen, Y.S., Chiu, Y.H., Li, Y.S. et al. (2019). Integration of PEG 400 into a self-nanoemulsifying drug delivery system improves drug loading capacity and nasal mucosa permeability and prolongs the survival of rats with malignant brain tumors. *Int. J. Nanomedicine* 14: 3601–3613. doi: 10.3390/pharmaceutics14071487.

105 Meirinho, S., Rodrigues, M., Santos, A.O. et al. (2023). Nose-to-brain delivery of perampanel formulated in a self-microemulsifying drug delivery system improves anticonvulsant and anxiolytic activity in mice. *Int. J. Pharm.* 25 (642): 123145. doi: 10.1016/j.ijpharm.2023.123145.

106 Shah, D., Guo, Y., Ban, I., and Shao, J. (2022 Mar 25). Intranasal delivery of insulin by self-emulsified nanoemulsion system: *in vitro* and *in vivo* studies. *Int. J. Pharm.* 616: 121565. https://doi.org/10.1016/j.ijpharm.2022.121565.

107 Zhang, J., Peng, Q., Shi, S. et al. (2011). Preparation, characterization, and *in vivo* evaluation of a self-nanoemulsifying drug delivery system (SNEDDS) loaded with morin-phospholipid complex. *Int. J. Nanomedicine* 2011: 3405–3414. doi: 10.2147/IJN.S25824.

108 Mohd Izham, M.N., Hussin, Y., Aziz, M.N.M. et al. (2019). Preparation and characterization of self nano-emulsifying drug delivery system loaded with citraland its antiproliferative effect on colorectal cells *in vitro*. *Nanomaterials (Basel)* 9 (7): 1028. doi: 10.3390/nano9071028.

109 Reddy, M.R. and Gubbiyappa, K.S. (2022). Formulation development, optimization and characterization of Pemigatinib-loaded supersaturable self-nanoemulsifying drug delivery systems. *Futur. J. Pharm. Sci.* 8: 45. doi: 10.1186/s43094-022-00434-4.

110 Khan, M.A., Ullah, K., Rahman, N.U. et al. (2022). Formulation, characterization and in-vitro evaluation of self-nanoemulsifying drug delivery system containing rhamnolipid biosurfactant. *J. Drug Delivery Sci. Technol.* 75: doi: 10.1016/j.jddst.2022.103673.

111 Zafar, A., Yasir, M., Alruwaili, N.K. et al. (2022). Formulation of self-nanoemulsifying drug delivery system of cephalexin: physiochemical characterization and antibacterial evaluation. *Polymers* 14 (5): 1055. doi: 10.3390/polym14051055.

112 Rathore, C., Hemrajani, C., Sharma, A.K. et al. (2023). Self-nanoemulsifying drug delivery system (SNEDDS) mediated improved oral bioavailability of thymoquinone: optimization, characterization, pharmacokinetic, and hepatotoxicity studies. *Drug Deliv. Transl. Res.* 13 (1): 292–307. doi: 10.1007/s13346-022-01193-8.

113 Borhade, V.B., Nair, H.A., and Hegde, D.D. (2009). Development and characterization of self-microemulsifying drug delivery system of tacrolimus for intravenous administration. *Drug Dev. Ind. Pharm.* 35 (5): 619–630. doi: 10.1080/03639040802498856.

114 Mahmoud, D.B., Shukr, M.H., and Bendas, E.R. (2014). *in vitro* and *in vivo* evaluation of self-nanoemulsifying drug delivery systems of cilostazol for oral and parenteral administration. *Int. J. Pharm.* 476 (1–2): 60–69. doi: 10.1016/j.ijpharm.2014.09.045.

115 Kazi, M., Alqahtani, A., Alharbi, M. et al. (2023). The development and optimization of lipid-based self-nanoemulsifying drug delivery systems for the intravenous delivery of propofol. *Molecules* 28 (3): 1492. doi: 10.3390/molecules28031492.

116 Seo, Y.G., Kim, D.H., Ramasamy, T. et al. (2013). Development of docetaxel-loaded solid self-nanoemulsifying drug delivery system (SNEDDS) for enhanced chemotherapeutic effect. *Int. J. Pharm.* 452 (1–2): 412–420. doi: 10.1016/j.ijpharm.2013.05.034.

117 Wang, C.Y., Yen, C.C., Hsu, M.C., and Wu, Y.T. (2020). Self-nanoemulsifying drug delivery systems for enhancing solubility, permeability, and bioavailability of sesamin. *Molecules* 25 (14): 3119. doi: 10.3390/molecules25143119.

118 Buya, A.B., Beloqui, A., Memvanga, P.B., and Préat, V. (2020 Dec 9). Self-nano-emulsifying drug-delivery systems: from the development to the current applications and challenges in oral drug delivery. *Pharmaceutics* 12 (12): 1194. doi: 10.3390/pharmaceutics12121194.

119 Abouhussein, D.M.N., Bahaa El, D., Mahmoud, D., and Mohammad, F.E. (2019). Design of a liquid nano-sized drug delivery system with enhanced solubility of rivaroxaban for venous thromboembolism management in paediatric patients and emergency cases. *J. Liposome Res.* 29 (4): 399–412. doi: 10.1080/08982104.2019.1576732.

120 Šahinović, M., Hassan, A., Kristó, K. et al. (2023). Quality by design-based development of solid self-emulsifying drug delivery system (SEDDS) as a potential carrier for oral delivery of lysozyme. *Pharmaceutics* 15 (3): 995. doi: 10.3390/pharmaceutics15030995.

6

Liposomes and Mixed Micelles for Improvement of Drug Solubility

Chukwuebuka E. Umeyor[1], Diksha Narula[2], Ankita Anure[2], Andrew Ezegbe[3,4], Prafull Kolekar[2], and Vandana B. Patravale[2]

[1]*Nnamdi Azikiwe University, Department of Pharmaceutics and Pharmaceutical Technology, Faculty of Pharmaceutical Sciences, Agulu Campus Awka-Ekwulobia Road, Awka, Anambra 422001, Nigeria*
[2]*Institute of Chemical Technology, Department of Pharmaceutical Sciences and Technology, Nathalal Parekh Marg, Matunga, Mumbai, Maharashtra 400019, India*
[3]*University of Nigeria, Department of Pharmaceutical Technology and Industrial Pharmacy, Faculty of Pharmaceutical Sciences, Nsukka, Enugu 410001, Nigeria*
[4]*ABC Federal University, Human and Natural Science Center, Avenida dos Estados, 5001, Sao Paulo 09280-560, Brazil*

6.1 Introduction

The development of lipid-based systems (LBS) is an important avenue for effective delivery of drugs and small molecules and to improve drug disposition in clinical settings. In the last decade, many lipid formulations have been approved and translated into tangible products for improved management of various diseases with decreased toxicity reactions [1]. LBS are usually developed using a blend of oils or lipids, surfactants, and co-surfactants, which creates the lipid milieu that accommodates any pharmacologically active entity loaded into the system. Effective loading and encapsulation of a drug in LBS enhances its solubility, absorption, and bioavailability in order to produce the desired effect [2]. The growing utilization of LBS in drug delivery is due to their versatility and the biocompatibility of the excipients used in the formulations with physiological tissues. Hence, several lipid formulations like solid lipid nanoparticles, lipid nanoemulsions, micelles, and liposomes have been used to explore the possibilities of enhancing the solubility and bioavailability of both hydrophilic and lipophilic molecules [3].

Through the innovative and flexible dispositions of LBS, drug delivery through the oral, parenteral, vaginal, transdermal, ocular, rectal, nasal routes, etc., has resulted in enhanced in vivo performance of entrapped biomolecules. Liposomes have been widely reported as versatile and feasible LBS for targeted delivery of encapsulated drugs and small molecules. It consists of an aqueous core surrounded by a lipid bilayer, which facilitates its cellular uptake and targeted delivery of a payload. The lipid constituent of liposomes is biodegradable, biocompatible, and amphiphatic, giving it flexible biophysical characteristics, encourages ease of manufacture and the potential to encapsulate both hydrophilic and lipophilic molecules, and is suitable for clinical applications. Liposomes could be categorized as charged, conventional, stimuli-sensitive, gas-entrapped, and targeted liposomes [4, 5]. Furthermore, mixed micelles are formulated when lipid bilayers are solubilized in surfactants, leading to marked variations in the morphological properties of the bilayer, and the structural changes could be evaluated by microscopy, e.g. cryo-TEM (cryotransmission electron microscope), which visualizes the topography of the micellar system in a three-dimensional format [6]. Mixed micelles are developed as an innovative approach to optimize disease therapy and enhance the solubility and stability of encapsulated drugs. To guarantee effective solubilization of poorly water-soluble drugs using mixed micelles, surfactant type selected for the formulation and the amount to be incorporated are critical because their influence on the development of a stable micellar structure. Therefore, comprehensive and extensive studies on the solubilization capacity of the chosen surfactant are required [7].

This chapter will provide an overview of the various approaches used in the manufacture of the different types of liposomes and mixed micelles. It will discuss the concept of critical micelle concentration (CMC) and consider the importance of in vivo safety of liposomes and mixed micelles. In addition, it will consider regulatory guidelines in the use of liposomes and mixed micelles, their clinical applications, and enumerate some patented liposomes and mixed micelles.

6.2 Drug Carrier Solubility Enhances Drug Bioavailability

It is well-established that a drug's solubility, dissolution rate, and permeability are very critical parameters that control its rates of absorption and bioavailability. Traditional drug carrier systems like tablets, capsules, injectables, and powders are not able to increase the solubility of poorly soluble drugs which lowers the uptake, bioavailability, and targeting efficiency of loaded drugs. Recent drug delivery systems have changed this unpleasant scenario as they have led to higher solubilization of poorly soluble drugs. In addition to that, they present more advantages like enhanced drug dissolution, uptake, absorption, bioavailability, and targeting to desired sites for optimum therapeutic or prophylactic effects [8]. From the foregoing, a drug carrier system is very essential to facilitate effective drug delivery of therapeutic agents. Certain factors should be considered in the design and development of a carrier system including the type of material to be used and its physicochemical properties, interaction with drug molecules, desired route of administration, fabrication technique, stability profile, ability of the carrier system to cross barriers, dock at target sites, and interaction of the system with biological tissues [9, 10]. With advances in nanotechnology, novel drug delivery systems are formulated using biomaterials, e.g. lipids that are biocompatible and biodegradable and facilitate the production of nanocarriers (LBS) for drug encapsulation. These systems are prepared using state-of-the-art technologies but at affordable cost and are scalable [11].

Lipid-based nanocarriers like liposomes and mixed micelles are biodegradable, biocompatible, and non-immunogenic. If needed, these carriers could be functionalized to improve their targeting potential, modulate the efficiencies of drug encapsulation and release, enhance transport across membranes, and extend their systemic circulation [12]. The secretion of bile and lipases in the gastrointestinal tract (GIT) is stimulated by lipids. Lipid products catalyzed by lipases combine with bile and lipids to produce micelles in the intestinal milieu. When lipophilic drugs are encapsulated in lipid nanocarriers, they are solubilized in the lipid colloidal system and are deposited by transcellular and paracellular transport systems into the hepatic portal vein, where they are metabolized for delivery into the systemic circulation [13, 14]. Furthermore, following uptake by the microvilli cells, the lipid nanocarriers are trafficked to the lymphocytes from where they are delivered to the lymphatic system. Thus, drug-laden lipid nanoparticles are delivered to the systemic circulation through the hepatic portal vein and the lymphatic system. Also, reports have shown that when lipid nanoparticles are modified, they produce better therapeutic outcomes in vivo than non-functionalized systems [15, 16]. Since oral drug delivery is the most popular, convenient, flexible, and appealing among patients, drug-loaded LBS for oral delivery must possess very high solubilization potential because solubility of the carrier system is the rate-limiting parameter required to obtain high drug bioavailability and the expected pharmacological response. Formulating the LBS as a nanocarrier such as liposomes, nanostructured lipid carriers (NLC), lipid nanoemulsions, self-nanoemulsifying formulations, and mixed micelles enables it to overcome the challenge of poor bioavailability linked to low solubility and permeability, which is common in conventional drug carriers [17].

The low solubility of LBS is a common barrier in its use for drug delivery because it limits drug absorption, bioavailability, produces unpredictable therapeutic drug levels, and increases the risk of toxicity reactions. As an innovative approach, nanotechnology-based systems promise to address these problems and improve drug safety and efficacy. These nanocarriers produce consistent absorption and correct the erratic plasma concentrations, thereby emphasizing the critical role of lipid nanocarriers for enhanced drug solubility and bioavailability, and creating a balance between the desired pharmacological outcome and adverse reactions [18, 19].

6.3 Process Development of Liposomes

Determining an appropriate method for the development of liposomes could be a challenging task, but the challenges are surmountable. Efficient liposomal formulation is aimed at developing a monodisperse system with desirable properties like prolonged stability, efficient drug encapsulation and release, and minimum toxicity. Some of the factors which are considered during liposome development are presented below.

6.3.1 Liposomes Composition

Liposomes (shown in Figure 6.1) are composed of phospholipids with hydrophobic tail and hydrophilic head bilayers, which self-seal in aqueous media due to its

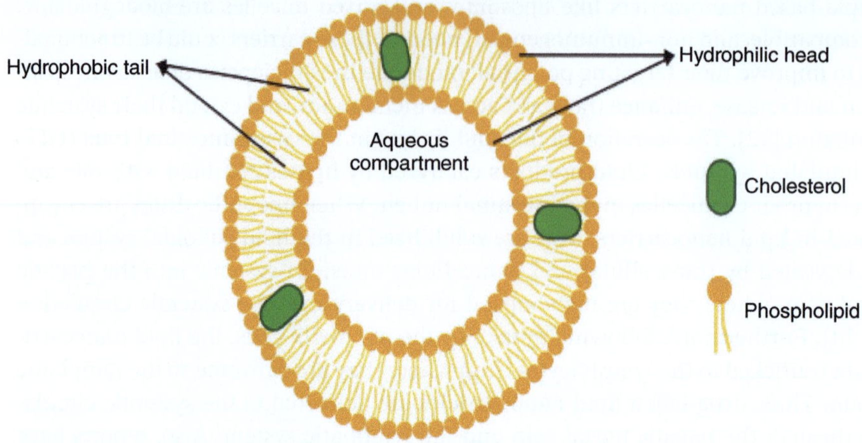

Figure 6.1 Structure of a liposome showing its constituents.

amphiphilic nature. Liposomes are composed of one or two bilayers with equal number of aqueous compartments. Liposomes can be used to deliver amphipathic, hydrophilic, and hydrophobic drugs. It is used to encapsulate the drug and also protect the drug from the toxic external environment. They are biocompatible, nontoxic, and non-immunogenic [20]. The phospholipids can be categorized into glycerophospholipids and sphingomyelins. Glycerophospholipids are the main phospholipids in eukaryotic cells, with glycerol as the backbone, whereas the sphingomyelins are the main phospholipids in animal cells with sphingosine as the backbone. Phospholipids can be either naturally or synthetically derived. Phosphatidylserine (PS), phosphatidylcholine (PC), and phosphatidylethanolamine (PE) are derived from natural sources, whereas synthetic phospholipids include dioleoyl PC, disteroyl PC, and dioleoyl PE [21, 22].

Liposomes can be effectively used as drug delivery systems due to its biocompatibility, non-toxicity, and ability to encapsulate the drug. Another component of liposome includes cholesterol that modulates membrane permeability and the fluidity of the membrane [23]. Hydroxyl groups position themselves toward aqueous core and hydrophobic group of cholesterol positions toward hydrophobic core during the liposome formation. Cholesterol reduces permeability of hydrophilic compounds through the liposome membrane. If cholesterol is not used in the formulation of liposomes, they tend to be less rigid and have a fluid bilayer, resulting in drug carrier destabilization [24].

Based on bilayer membranes and size, liposomes are categorized into [25]:

- Small unilamellar vesicles (SUV): 20–100 nm
- Large unilamellar vesicles (LUV): >100 nm
- Giant unilamellar vesicles (GUV): >1000 nm
- Oligolamellar vesicles (OLV): 100–500 nm
- Multilamellar vesicles (MLV): >500 nm
- Multivesicular vesicles (MVV): >1000 nm

Small unilamellar vesicles (SUV) possess a single bilayer and can be prepared by sonication or extrusion technique or can be prepared by reducing the size of LUV and MLV. Large unilamellar vesicles (LUV) have abundant space for loading biomolecules. They possess single bilayer and have large aqueous volume to lipid ratio. Giant unilamellar vesicles (GUV) have a size range of 1–200 μm and can be prepared by electroformation, natural swelling, etc. Multivesicular vesicles (MVV) have high drug-loading capacity and consist of several nonconcentric vesicles enclosed within a single bilayer [26].

6.3.2 Formulation Methods

Liposomes can be prepared by using different approaches depending on the type of liposome to be prepared and the phospholipid type. The conventional methods of liposomal formulation include the following:

6.3.2.1 Thin Film Hydration Method

Using this method, organic solvent is used to dissolve lipids and then dried to form a thin lipid film at the bottom of the flask (by evaporating the solvent under vacuum). Using an aqueous medium, the obtained lipid film is hydrated to produce liposomal dispersion. Liposomes with different structural organization can be created depending on the hydration conditions. Vigorous hydration can lead to production of heterogeneous MLVs, while gentle hydration of lipid film produces GUVs. The main advantage of this method includes use of common organic solvents like chloroform and methanol for lipid dissolution, and by extrusion of the hydrated lipid phase, homogeneous unilamellar vesicles are formed. Scalability, formation of heterogeneous lamellar, and poor encapsulation are the challenges of this method [24].

6.3.2.2 Reverse Phase Evaporation

It is an alternative to thin film hydration method. This method is based on creation of inverted micelles. In this method, a water-in-oil emulsion is formed, wherein lipids dissolved in organic solvents are mixed with aqueous buffer that contains a hydrophilic drug. Using a rotary evaporator at reduced pressure, organic solvents are evaporated, leading to formation of aqueous solution dispersed lipid vesicles. There is a slow depletion of organic solvents, which leads to gel-like consistency formation that collapses to form liposomes at critical point. The size and polydispersity of the vesicles are reduced by extrusion. With this method, encapsulation efficiency of around 65% can be achieved. The main drawback of this technique is that proteins can possibly be denatured, if intended for encapsulation [23, 27].

6.3.2.3 Detergent Removal Method

In this method, solutions containing detergents at their CMC are used for lipid hydration. Triton X-100, sodium cholate, and sodium deoxycholate are the commonly used detergents with a high CMC. On removal of detergents, the phospholipids self-assemble into bilayers and are free from the reaction medium. Detergent removal can be obtained through gel permeation chromatography or dialysis bag. After solution concentration (detergent removal), mixed micelles become lipid-rich, and LUV vesicles can be obtained [23]. The simplest method of detergent removal is dilution (10–100-fold) using buffer. The drawback of this method is low entrapment efficiency and low concentration of liposomes. Detergent removal by dialysis method leads to the formation of homogeneous-sized liposomes. It leads to vesicle formation with no degradation of relevant biological activity [28].

6.3.2.4 Solvent Injection Method

In this method, lipids dissolved in organic solvent (ethanol or ether) are injected into the aqueous solution, leading to liposome formation. Depending on whether the organic solvent is water-miscible or not, the experiment is conducted at room temperature or at a higher temperature (such as 60 °C). Since ethanol forms an azeotrope mixture with water, the liposomes made by the solvent injection technique are primarily polydispersed and heavily polluted by organic solvents. Solvent injection has a number of disadvantages, such as the therapeutic agents' ongoing exposure to high temperatures and organic solvents, which may compromise the stability and safety of the liposomes.

Ethanol Injection Method The lipids, dissolved in ethanol, are injected into an aqueous medium, resulting in liposome formation. When ethanol is diluted into water below a critical concentration, it favors self-assembly of lipids dissolved in the aqueous phase. Precipitation of lipid molecules followed by successive formation of bilayer planar fragments, encapsulating the aqueous phase, is favored in rapid ethanol dilution. With depletion of ethanol, fusion of lipid fragments occurs and closed unilamellar vesicles are formed. A heterogeneous population of MLVs can be formed when ethanol is injected rapidly and if below 7.5% of the whole formulation volume, homogeneous SUVs are formed. Using a rotary evaporator, ethanol can be removed [28].

Ether Injection Method Lipids dissolved in ether are slowly injected into an aqueous phase, which contains components that are to be encapsulated, and are heated to a temperature of 55–65 °C. LUVs are generated by successive removal of organic solvent. An advantage of this method in comparison to ethanol injection method is more efficient organic solvent removal from final product, favoring concentrated liposome solutions with high entrapment efficiencies, whereas the main disadvantage is production of highly polydisperse liposomes [28].

Microfluidic Channel Method

The method uses microscopic channels in size range of 5–500 μm. Organic solvents are used to dissolve lipids, and the resulting solution is forced either perpendicularly or in opposite direction with respect to aqueous medium within microchannels. The aqueous and organic solutions form liposomes due to phospholipid diffusion in aqueous phase, encouraging self-assembly when continuous axial mixing is done. Homogeneous SUV and LUV of size ranging from 40 to 140 nm are formed.

Microfluidic droplets and pulsed jet flow methods are the other microfluidic methods. Giant liposomes of size 4–20 μm are formed by dissolving phospholipids in hexane in microfluidic droplet method, whereas in pulsed jet flow method, there is modification in conventional film hydration method by drying lipid solution in microtubes. The lipid film, when hydrated further in microtubes through perfusion, produces larger vesicles with enhanced encapsulation efficiency. These methods are flexible and versatile and produce vesicles with desired size. Disadvantages include use of organic solvent and drastic agitation [29].

Supercritical Fluidic Method

This method uses supercritical fluids such as carbon dioxide (CO_2) maintained under supercritical temperature and pressure. It produces LUV (100–1200 nm) with higher encapsulation efficiency than the equivalent conventional reverse phase evaporation method. It is an organic solvent-free method and uses CO_2 as a harmless solvent. This method can possibly control particle size and large-scale production in industry. High-pressure usage, low yield, and high cost are the main disadvantages of the method [29].

6.3.3 Drug Loading Improvement in Liposomes

Drug loading in liposomes shown in Figure 6.2 can be achieved through two methods, namely passive and active loading methods.

Passive loading is a method used for encapsulation of drugs during liposome preparation. Hydrophobic drugs are located in the bilayer of the liposomes, whereas the hydrophilic drugs are dispersed in the aqueous phase, which is inside and outside of the liposomes. Liposomes, while being formed, capture the aqueous volume which contains the hydrophilic drug. Low encapsulation efficiency is seen in this method, resulting in the leakage of large amount of nonencapsulated drug [26]. The entrapment efficiency of a drug varies due to drug solubility, lipid concentration, and preparation procedure of the liposomes. The entrapped drug does not retain stability as there is a weak association between the drug and the liposome, which results in poor drug retention and storage stability.

Active loading, also called remote loading, is a method involving creating transmembrane pH or ion gradient that effectively drives the drug through the lipid bilayer. Drug is solubilized in the aqueous medium, outside the liposomes. A gradient is created between the inside and outside of the liposomes. This method has more encapsulation efficiency and retention time of drug compared to the passive loading method [26]. In remote loading method, the drug can diffuse into intravesicular spaces through lipid bilayer, which results in chemical modification of drug which further prevents drug removal from the membrane, resulting in accumulation of drug [24]. Active loading method is typically based on two phenomena: the lipophilic molecule gains a charge upon entering the liposomes by diffusing through the lipid bilayer, and when the proton molecule cannot cross the bilayer, resulting in the accumulation of the encapsulating agent.

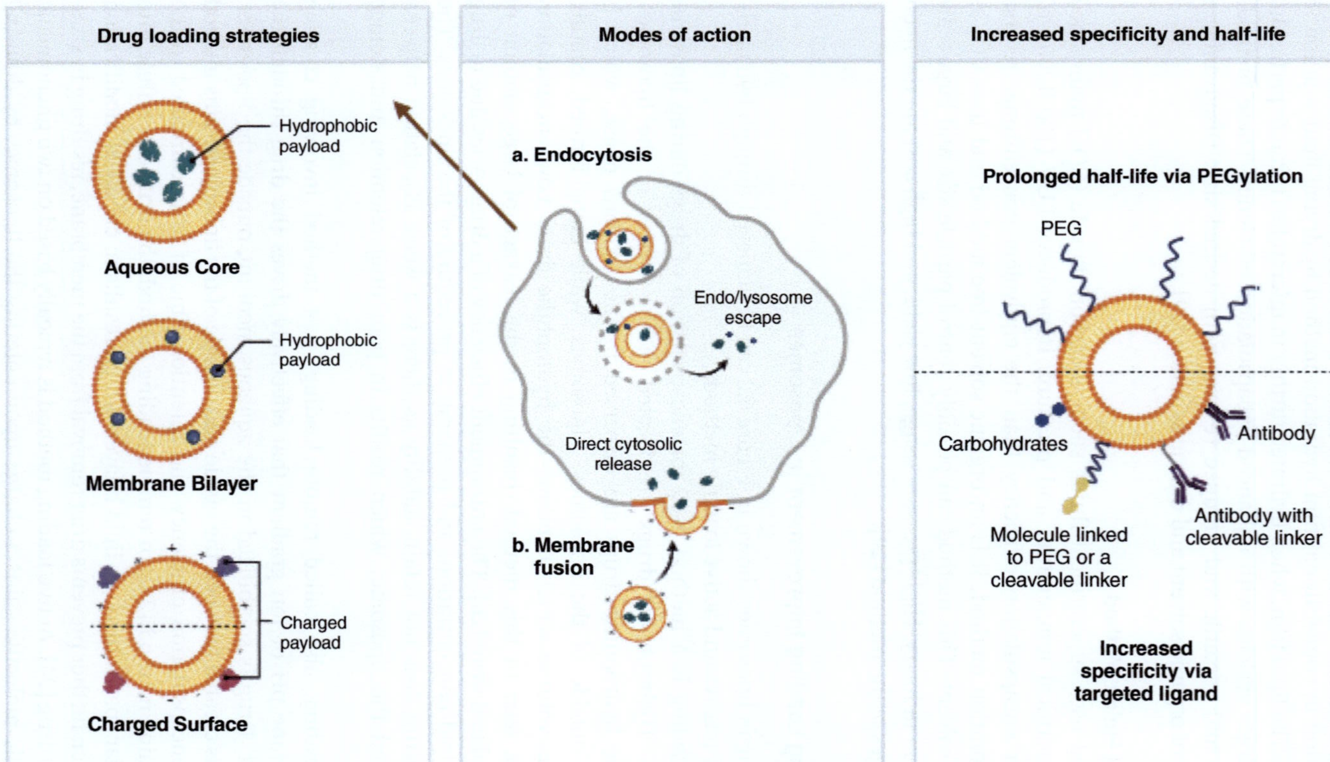

Figure 6.2 Drug-loading techniques in liposomes and their benefits in drug delivery. Adapted from Mosleh-Shirazi et al. [30].

Some of the frequently used active drug loading improvement strategies include:

- **The pH gradient method:** Drug loading in this method is driven by transmembrane pH gradient. Presence of ionizable functional groups in the drug structure and the aqueous solubility of the encapsulated drug are necessary for efficient loading. In extra liposomal aqueous phase pH, the drug exists in unionized form and migrates across the liposomal bilayer. After translocating the core, the drug changes to ionized form with the pH change. For amphiphilic drugs that are weak acids or bases, using pH gradient as driving force, drugs can be translocated and retained in the liposomes [30, 31].
- **Ammonium sulfate gradient method:** This method, also known as transmembrane sulfate gradient method, was introduced to encapsulate amines. Ammonium ion gradient is generated by its counterion sulfate that causes aggregation leading to stabilization of anthracycline sulfate salt. Liposomes prepared by thin layer method are dialyzed in phosphate buffer solution (PBS) to form gradient. After a gradient is formed, liposomes are incubated with drug solution for encapsulation. The drug, including the amine group, is able to penetrate the bilayer, leading to precipitation of drug as sulfate salt. Doxorubicin long-circulating liposomes were first produced by this method [31, 32].
- **Calcium acetate method:** This technique is based on permeability coefficient differences of acetic acid and calcium ions. Liposomes are prepared in calcium acetate solution. The calcium ions stay in liposomal aqueous core, and proton shuffles are acetic acid molecules, creating a pH gradient with the pH value higher inside the liposomes, resulting in weak amphiphilic acid molecules' entrapment inside liposomes.
- **Ionophore-generated pH gradient method:** Deamer et al. observed that when nigericin is added to SUVs with potassium salt, it leads to pH gradient generation of ~2 units. The researchers used this technique in ionophores like nigericin or A23187 to produce LUV, respectively, showing the gradients in transmembrane potential for K^+ or Mn^{2+}. Uptake of drug (present in external medium) is initiated by ionophore addition that couples outward transport of metal ion to H^+ inward movement. This produces an acidic pH gradient, which causes molecules with weak basic properties to be absorbed. When divalent cations are present, there is a requirement of ethylenediamine-tetra acetic acid (EDTA) as external chelating agent for cation binding that is released and to enhance uptake and also to avoid liposome aggregation [33].
- **EDTA ion gradient method:** This was designed to encapsulate the anticancer drug, idarubicin. Liposomes were prepared using thin lipid film method. The ion/pH gradient was generated by extravesicular liposomal solution exchange on Sephadex G-50 columns which were equilibrated with PBS buffers with different pH levels between 4.5 and 8.5 and were used for experimentation to study the effect of encapsulation of idarubicin on external pH. Liposomes with EDTA gradient were observed for idarubicin accumulation. Several liposome samples were prepared in EDTA disodium salt solution to check the effect of external pH on encapsulation of the drug. Higher external pH showed better drug

encapsulation, which was due to higher lipophilicity of drug. Idarubicin has very low solubility in EDTA salt solution at acidic pH. Using this method, rapid drug encapsulation with great retention capability was achieved [34].

- **Solvent-Assisted Active Loading Technology (SALT):** Traditional methods of drug loading lead to low encapsulation efficiency of poor water-soluble drugs. The SALT method was developed for loading poor water-soluble drugs into the aqueous liposomal core. In a study by Pauli et al., they hypothesized that adding a small amount of water-miscible solvent to the liposome and hydrophobic drug loading mixture could improve the efficiency of active loading encapsulation, increase drug penetration through the liposomal bilayer, and help solubilize the drug in the aqueous phase. The principle was that for a drug to successfully pass through the lipid bilayer, it must be dissolved in the free form in the outer phase of the liposome. For the molecule to be completely dissolved, a tiny amount of solvent is added to the liposomal suspension. The drug can interact with a trapping agent inside the liposomes to create insoluble precipitates for a stable "lock-in" effect after penetrating the liposomal core. Lastly, gel filtration or dialysis can be used to eliminate the solvent. An example where this technique was utilized includes Staurosporine, an anti-tumor agent, where the researchers hypothesized that using dimethyl sulfoxide (DMSO) in small amounts in the liposomal suspension could keep the drug soluble and help the drug to penetrate more effectively into the inner core of liposomes. Ammonium sulfate gradient method was used to trap the drug with help of sulfate–sodium thiosulfate (STS) nanoaggregates inside the core, resulting in 100% drug encapsulation along with high drug-to-lipid ratio. Further, gel filtration was utilized to remove the residual DMSO [35].
- **Chemical conjugation.** Functional groups of drugs can get attached to phospholipids bilayer of liposomes using several chemical modifications. Various amphiphilic molecules like PEGylated lipids, peptides, or suitable polymer could be used for covalent coupling. The molecules used should be able to interact with functional groups present on the phospholipid heads. Chemical linkages like disulfide, ester, and hydrazone are commonly used as they can release the drug in reducing conditions. For the release at low pH, esterase enzymes could be used. Conjugation step can be performed either during liposome synthesis or after liposome synthesis. When drug is incorporated during the synthesis, its presence can be seen inside and outside the liposomes, whereas when the conjugation is performed after synthesis, drugs are displayed only on the surface of the liposomes [36].

6.3.4 Analysis of Drug Encapsulation

For the quantification of encapsulated drug, the unencapsulated drug should first be removed from the carrier using ultrafiltration, ultracentrifugation, or dialysis. The unencapsulated drug can be studied with the measured or theoretical total drug amount, which will give the amount of drug encapsulated. This is described in Table 6.1.

Table 6.1 Methods for quantification of encapsulated drug.

Method	Technique	Example
RP-HPLC	Monolithic silica columns showed higher efficiency for separation and quantifying unencapsulated drug and drug-loaded particles [37]	Solid phase extraction (SPE) of Doxil from plasma, along with a C18 column for free or extracted doxorubicin analysis [38]
Capillary electrophoresis (CE)	Separation of samples based on charge-to-size ratio. This technique provides differences in analytes based on their electrophoretic mobility, providing information about liposome surface charge, size, and permeability of liposome membranes [37]	In a study by Ansar et al., CE conditions like voltage, pressure, buffer pH, and salt concentration were optimized, and the encapsulation efficiency of Doxil was determined. It was found that liposomes were stable during changes in voltage and injection pressures, and they also observed an enhanced drug release under high phosphate conditions, low sucrose content, and high pH [39]
Field-flow fractionation (FFF)	Separation of analytes is done according to their diffusion rate inside an open channel. It can be categorized as symmetrical and asymmetrical flow FFF (AF4) [37]	Fractionation of amphotericin B-loaded liposome formulation on the basis of their particle sizes using AF4 system was done by Van Haute et al. Quantification of drug and lipid contents was done using an offline HPLC–UV–CAD system [40]
^{1}H NMR	The principle of quantitative NMR includes integrated peak area of each signal, which tells the number of equivalent nuclei in that signal	In a study by Guimaraes et al., two different drug concentrations were determined using the method [41]

6.3.5 Stability of Liposomes

Liposomes are unstable colloidal system, and their stability can be evaluated using parameters such as chemical and physical stability, their sizes, and studying the influence of biological fluids on their properties. The main issues that influence liposomes include chemical and physical stability. To detect physical stability, size evaluation and visual inspection can be done. Chemical stability includes factors such as alterations in pH, electrolyte composition, and surface-active compound presence. Drug and phospholipid interactions can affect chemical stability. Degradation can cause interference in the permeability of the membrane.

Liposomes are not stable in an aqueous state, so to maintain their long-term stability, they can be stored in dry or powder form using the lyophilization method. This technique includes sublimation, wherein water is removed from the sample, followed by desorption under vacuum. Excipient choice for loading within liposomes is essential to maintain their size and avoid drug leakage. Sugars are the most commonly used excipients for stabilizing and protecting liposomal integrity during the process [26].

Liposomes formulated with unsaturated PC have low stability and high permeability, whereas saturated phospholipids create impermeable bilayer structure. Hydrophilic group charge on the lipids provides stability through electrostatic force. Cholesterol, when incorporated in lipid bilayer, improves stability of liposomes [23]. Approaches to enhance colloidal stability of liposomes include:

- Incorporating charged components such as anionic/cationic lipids. This will create a zeta potential among liposomes, avoiding aggregation effects. Negative charge added by negatively charged molecules evidently provides enhanced colloidal stability due to the generation of electrostatic repulsive forces [42]. A study showed that incorporating stearyl amine into PC forms positively charged liposomes from neutral liposomes, increasing the stability of liposomes. Incorporating diacetyl phosphate provides PC with a negative charge, which does not change the stability of liposomes, but if negatively charged sphingolipids are incorporated, they decrease the stability. Also, the incorporation of cholesterol increases the stability by 10–20%, which is due to changes in the physical properties of the membrane [43].
- Another approach involves incorporating polymers into the surface. PEGylation can be carried out by decorating liposomes with polyethylene glycol (PEG) molecules, which improves the colloidal stability of liposomes [42]. A study showed that an immunogenic response is produced due to the formation of anti-PEG antibodies. PEG is conjugated directly to the surface to enhance the stability and solubility of the drug, and it has a high solubility in aqueous media and is biocompatible. Due to the induced immunogenic response, new alternative polymers have been developed that mimic the PEG properties. Utilizing shorter hyperbranched polymers can mitigate antigenic response and also not compromise the stealth behavior of liposomes. For example, a new polymer, poly (oligo (ethylene glycol) methyl ether methacrylate) (POEGMA), had a 3D structure with numerous oligoethylene glycol-side moiety side chains, which showed a decrease in antigenicity. Polyvinylpyrrolidone (PVP) can be used as an alternative, providing high hydration, good biocompatibility, and lower immunogenicity [44].
- Functionalizing liposome surface with ligands like monoclonal antibodies, aptamers, and peptides allows specific liposome interaction and controlled drug release [42].

The chemical stability of liposomes is affected by microenvironmental pH. In a study on curcumin as a model drug, a novel approach was designed to enhance chemical stability of drug which was loaded in liposome. Delivery of curcumin is affected due to its insolubility and instability in fluids. They developed an approach and showed that the chemical stability and model efficacy of the drug could be increased if the microenvironment of liposome is made more acidic [45].

6.3.6 Types of Liposomes

The different types of liposomes that are categorized based on their physicochemical properties are discussed in this section.

6.3.6.1 Gas-Entrapped (Bubble) Liposomes

Gas-entrapped liposomes, also known as bubble liposomes, are a type of liposomal carrier that enhances drug solubility, increases bioavailability, and allows for targeted delivery of poorly water-soluble drugs. These liposomes are unique in that they encapsulate gas, usually inert gases such as perfluorocarbons or other biocompatible gases, within their lipid bilayer or aqueous core, resulting in stabilized bubbles. This structure not only provides drug-carrying capability but also introduces properties beneficial for advanced drug delivery applications, such as ultrasound-mediated drug release, increased solubility, and stability [46].

Structure of Gas-Entrapped (Bubble) Liposomes Gas-entrapped liposomes are made up of phospholipids (such as PC and cholesterol), which help to stabilize the bilayer. The lipids form a flexible yet stable bilayer that traps gas bubbles and incorporates hydrophobic drugs into the bilayer while transporting hydrophilic drugs through the aqueous regions. Biocompatible gases such as perfluorocarbon-based gases or nitrogen are trapped within the liposomal structure during the preparation process. These gases reduce the density of liposomes while increasing their stability, allowing for buoyancy and controlled delivery. Hydrophilic drugs are encapsulated in the aqueous core, whereas hydrophobic drugs are incorporated into the lipid bilayer. The presence of gas improves solubility by creating a stable interface that allows for drug dispersion in aqueous media [47].

Mechanism of Action of Gas-Entrapped (Bubble) Liposomes Gas-entrapped liposomes work using a combination of drug encapsulation, gas dynamics, and stimulus-responsive release mechanisms. The lipid bilayer contains hydrophobic drugs, whereas the aqueous core contains hydrophilic drugs. The gas trapped inside the liposomes or generated in situ forms an interface that improves the dispersion and solubility of poorly soluble drugs in aqueous environments. When exposed to external ultrasound stimuli, the gas bubbles inside the liposomes expand, oscillate, and collapse, resulting in acoustic cavitation. This causes the liposomal membrane to rupture or destabilize, releasing the encapsulated drug to the desired location. Ultrasound also improves cellular membrane permeability, allowing drugs to enter target cells more efficiently. Bubble liposomes can be internalized through endocytosis and cellular uptake. Surface-modified bubble liposomes containing targeting ligands (such as antibodies or peptides) selectively bind to receptors on diseased tissues (such as tumor cells), allowing for localized drug release. The gas bubbles in liposomes serve as contrast agents for ultrasound imaging, allowing for real-time monitoring of drug delivery and accumulation in target tissues. Bubble liposomes are ideal for precision medicine due to their dual function as both therapeutic and imaging agents [48].

Applications of Gas-Entrapped Liposomes

- **Cancer therapy:** Ultrasound-responsive bubble liposomes are used to deliver chemotherapeutics to tumor tissues, lowering systemic toxicity and increasing therapeutic efficacy.

- **Cardiovascular imaging and treatment:** Bubble liposomes, used as ultrasound contrast agents, can help visualize blood vessels while also delivering drugs to specific cardiovascular sites.
- **Gene delivery:** Gas-entrapped liposomes can be loaded with genetic material such as DNA or RNA, and ultrasound can facilitate their cellular uptake for gene therapy.
- Bubble liposomes enhance the solubility and dispersion of poorly soluble drugs in pulmonary systems, making them suitable for inhalation therapies [47, 49–51].

6.3.6.2 Charged Liposomes

Charged liposomes are LBS that improve drug solubility, stability, and targeted delivery by adding a surface charge to the liposome structure. Unlike neutral liposomes, charged liposomes have either a positive (cationic) or negative (anionic) charge on their surface, which improves their interaction with drugs, biological membranes, and target cells. The surface charge increases drug loading efficiency and cellular uptake and, in some cases, provides greater control over drug release. Charged liposomes are particularly effective for encapsulating and delivering poorly soluble drugs and macromolecules such as nucleic acids and proteins, making them useful in various therapeutic applications [4].

Composition of Charged Liposomes

- **Phospholipids:** As with conventional liposomes, the main lipids used to form the lipid bilayer are PC or PE.
- **Charged lipids:** To add surface charge, specific lipids or additives are used like cationic lipids (positive charge) such as 1,2-dioleoyl-3-trimethylammonium propane (DOTAP), stearylamine, and dimethyldioctadecylammonium bromide (DDAB). These lipids give liposomes a positive charge, which increases their ability to bind to negatively charged molecules such as nucleic acids or proteins, and anionic lipids (negative charge): Examples are PS, dicetylphosphateDCP, and phosphatidic acid. These carry a negative charge, which improves stability and controls aggregation.
- Cholesterol is commonly used to improve membrane rigidity, stability, and drug retention within the lipid bilayer.
- **PEGylation (polyethylene glycol):** Reduces immune system clearance and increases circulation time.
- **Targeting ligands:** The addition of antibodies, peptides, or carbohydrates enables charged liposomes to target specific sites.
- **Drug payload:** Hydrophobic drugs are integrated into the lipid bilayer, whereas hydrophilic drugs are encapsulated in the aqueous core. The charged surface enhances the electrostatic interaction between the drug and the liposome, which improves drug loading efficiency [52, 53].

Mechanism of Action of Charged Liposomes The charged surface of liposomes interacts electrostatically with negatively charged drug molecules, increasing drug loading efficiency. For example, cationic liposomes bind tightly to negatively charged drugs or macromolecules such as DNA, RNA, and proteins. Anionic liposomes can bind positively charged drugs, increasing encapsulation and solubility. Positively charged (cationic) liposomes interact with negatively charged cell membranes, facilitating endocytosis and increasing drug uptake. This property is particularly useful in gene delivery and cancer therapies, which require high intracellular concentrations. The charged surface reduces liposome aggregation through electrostatic repulsion, thereby improving liposome stability and dispersion in biological fluid. This increases the solubility and bioavailability of poorly water-soluble drugs. The surface charge influences membrane permeability, allowing for the gradual release of encapsulated drugs. Cationic liposomes, in particular, can destabilize endosomal membranes after cellular uptake, allowing for efficient drug release into the cytoplasm (endosomal escape). When charged liposomes are functionalized with targeting ligands such as antibodies or peptides, they can selectively bind to specific receptors on diseased cells, thereby improving drug accumulation at the target site. Multidrug resistance (MDR) can be reduced by using charged liposomes to bypass drug efflux pumps in cancer cells, thereby improving chemotherapeutic agent delivery and overcoming resistance [53, 54].

Applications of Charged Liposomes

- **Cancer therapy:** Cationic liposomes are used to deliver chemotherapeutic agents such as doxorubicin and paclitaxel to tumor cells, thereby improving drug uptake and reducing systemic toxicity, e.g. Doxil®/Caelyx®—these are PEGylated liposomal formulations of doxorubicin, approved for treating various cancers including breast and ovarian cancer [55].
- **Gene delivery:** Positively charged liposomes are ideal carriers for DNA, siRNA, and other genetic material, allowing for efficient transfection in gene therapy, e.g. cationic liposome-mediated E1A gene transfer. In a Phase I clinical trial, cationic liposomes were used to deliver the E1A gene in patients with breast and ovarian cancer, showing promising results in targeted gene therapy [56].
- **Vaccine delivery:** Charged liposomes are used to deliver vaccine antigens and adjuvants, which improves immunogenicity, e.g. CAF01 adjuvant system. This is a cationic liposome-based adjuvant containing dimethyldioctadecylammonium (DDA) and trehalose dibehenate (TDB). CAF01 is used in experimental vaccines to boost immunogenicity [57].
- Antimicrobial agents can be delivered via anionic or cationic liposomes to treat bacterial and viral infections, respectively.
- **Pulmonary drug delivery:** Charged liposomes improve drug deposition and retention in the lungs for the treatment of respiratory diseases, e.g. liposomal Beclometasone. This inhalable liposomal corticosteroid formulation has shown improved pulmonary retention and reduced systemic side effects in the treatment of asthma [58].
- **Overcoming MDR:** Charged liposomes deliver drugs that bypass efflux pumps in resistant cancer cells, improving therapeutic efficacy, e.g. liposomal formulations co-loaded with chemotherapeutic agents and efflux pump inhibitors have shown potential in restoring drug sensitivity in MDR cancer models [53].

6.3.6.3 Stimuli-Triggered Liposomes

Stimulus-triggered liposomes are advanced lipid-based drug delivery systems that release therapeutic payloads in response to specific internal or external stimuli. These liposomes increase drug solubility, improve bioavailability, and enable controlled and targeted drug release at the desired site, lowering systemic toxicity and improving therapeutic outcomes. Liposomes, which incorporate stimuli-responsive elements into their structure, can respond to physical, chemical, or biological signals, ensuring drug delivery is activated only under specific conditions [59].

Types of Stimuli-Triggered Liposomes Stimuli-triggered liposomes can be broadly classified as internal stimuli-responsive and external stimuli-responsive systems.

Internal stimuli-triggered liposomes react to changes in the microenvironment of the body such as pH, temperature, redox potential, and enzyme activity.

- Liposomes that respond to acidic environments such as tumor tissues (pH ~6.5) or endosomes/lysosomes (pH ~5–6). The composition contains pH-sensitive lipids such as dioleoyl phosphatidylethanolamine (DOPE) or acid-labile linkers that destabilize the liposomal membrane at low pH, allowing drug release. The acidic environment causes lipid protonation, which destabilizes the bilayer and releases the encapsulated drug. Targeted cancer therapy in which the acidic tumor microenvironment causes drug release.
- **Redox-responsive liposomes:** These liposomes are designed to release their payloads in response to elevated levels of intracellular glutathione (GSH), which are typically 10–100 times higher in tumor cells compared to normal cells. In the reductive environment of tumor cells, glutathione cleaves disulfide bonds present in the lipids or linkers of the liposome structure. This cleavage destabilizes the liposomal membrane, triggering targeted drug release. The mechanism allows for selective delivery of therapeutics in cancer treatment, exploiting the high concentrations of GSH within tumor cells.
- **Enzyme-responsive liposomes:** These liposomes are engineered to respond to elevated levels of specific enzymes—such as matrix metalloproteinases (MMPs) and phospholipases—commonly found in diseased tissues. Their membranes incorporate enzyme-sensitive peptides or lipid linkers that can be selectively cleaved by these overexpressed enzymes. Once cleaved, the liposomal structure becomes destabilized, triggering the release of the encapsulated drug. This targeted approach is particularly useful for treating conditions like cancer, inflammation, and other tissue-specific diseases.

External stimuli-triggered liposomes release drugs in response to physical stimuli like heat, light, ultrasound, or magnetic fields.

- **Temperature-responsive liposomes:** These liposomes are sensitive to slight temperature increases, such as hyperthermia (40–42 °C) near diseased tissues. Lipids such as DPPC (dipalmitoyl phosphatidylcholine) undergo phase transitions at high temperatures, destabilizing the membrane. Heating causes the lipid bilayer to transition from a gel to a liquid crystalline phase, increasing membrane permeability and releasing the drug. Cancer treatment using hyperthermia.

- **Light-responsive liposomes:** These liposomes contain photo-responsive lipids or photosensitive molecules such as azobenzene and photopolymerizable lipids. Light irradiation (UV or near-infrared) causes structural changes in lipids, disrupting the bilayer and allowing controlled drug release. Localized drug delivery with precise spatial and temporal control in cancer and photodynamic therapy.
- **Ultrasound-responsive liposomes:** When exposed to external ultrasound waves, these liposomes can disrupt the liposomal membrane via acoustic cavitation. The incorporation of gas-generating agents or microbubbles can increase ultrasound sensitivity. Ultrasound waves cause mechanical and thermal stress on liposomes, destabilizing the bilayer and releasing the drug. Noninvasive, targeted drug delivery for tumors and cardiovascular treatments.
- **Magnetic field-responsive liposomes:** These liposomes contain magnetic nanoparticles (such as iron oxide) embedded in the lipid bilayer or core. Exposure to an external magnetic field causes heat (magnetic hyperthermia) or mechanical force to disrupt the liposomal membrane, allowing drug release. Targeted drug delivery and real-time magnetic resonance imaging (MRI) [60, 61].

Mechanism of Action of Stimuli-Triggered Liposomes Internal or external stimuli cause structural changes in the lipid bilayer, resulting in increased permeability and drug release. Temperature or light causes a phase transition of specific lipids, disrupting the integrity of the bilayer and allowing drug release. Chemical linkers (pH-sensitive, redox-sensitive, enzyme-cleavable) are disrupted under specific conditions, causing drug release. Ultrasound or magnetic fields cause physical stress or heat, which ruptures the liposomal membrane. Stimuli (e.g. enzymes or heat) localized at the disease site permit precise and controlled release, improving therapeutic outcomes [59].

Applications of Stimuli-Triggered Liposomes

- In cancer treatment, temperature- and enzyme-sensitive liposomes enable targeted and controlled drug release at tumor sites. Additionally, light-responsive liposomes are utilized in photodynamic therapy to deliver photosensitizers specifically for cancer cell destruction.
- Inflammatory diseases are treated with enzyme-responsive liposomes, which release anti-inflammatory drugs into the affected tissues.
- **Noninvasive treatments:** liposomes that respond to ultrasound and magnetic fields enable targeted drug release without the need for invasive procedures.
- **Combination therapies:** Stimuli-triggered liposomes allow for the simultaneous delivery of multiple drugs or therapies (such as chemotherapy and hyperthermia) [62].

6.3.6.4 Conventional Liposomes

Conventional liposomes are the first generation of lipid-based vesicular systems intended to enhance the solubility, stability, and therapeutic efficacy of hydrophilic and hydrophobic drugs. Liposomes are spherical phospholipid bilayer vesicles with an aqueous core, making them excellent carriers for a wide range of drug molecules. Their composition and structural organization enable them to increase drug solubility, lower systemic toxicity, and achieve controlled release [63].

Composition of Conventional Liposomes Major components include natural or synthetic phosphatidylcholine (PC), phosphatidylethanolamine (PE), phosphatidylglycerol (PG), or phosphatidylserine (PS). These amphiphilic molecules form bilayers in aqueous environments because of their hydrophilic head and hydrophobic tail. Cholesterol is incorporated into the lipid bilayer to increase the membrane rigidity and stability of liposomes, thereby preventing premature drug leakage. Hydrophilic drugs are entrapped in the aqueous core of liposomes, whereas hydrophobic drugs are incorporated into the lipid bilayer. The addition of anionic (e.g. phosphatidic acid), cationic (e.g. DOTAP), or neutral lipids stabilizes the liposomal membrane and increases drug encapsulation efficiency [53].

Structure of Conventional Liposomes SUVs have a single phospholipid bilayer surrounding an aqueous core and its size ranges between 20 and 100 nm. It is ideal for encapsulating hydrophilic drugs and facilitating cellular uptake. LUVs have a structure similar to a single lipid bilayer but with a larger aqueous core. Its size ranges between 100 and 500 nm. It is suitable for encapsulating both hydrophilic and lipophilic drugs, and has a large loading capacity. MLVs are made up of multiple concentric lipid bilayers separated by aqueous spaces. MVVs are large vesicles with multiple smaller vesicles within. These liposomes find application for controlled and prolonged drug release [64].

Mechanism of Action of Conventional Liposomes Liposomes accumulate at tumor sites or inflamed tissues due to leaky vasculature and poor lymphatic drainage (enhanced permeability and retention [EPR] effect), allowing for passive drug targeting without surface modification. Liposomes are internalized by cells via endocytosis and then released within the endosome or cytoplasm, increasing intracellular drug availability. In some cases, the lipid bilayer of the liposome fuses with the cell membrane, allowing the drug to be released directly into the cytosol. Conventional liposomes protect encapsulated drugs by ensuring sustained release over time, thereby improving therapeutic outcomes [65].

Applications of Conventional Liposomes

- **Cancer therapy:** Using the EPR effect, encapsulation of chemotherapeutic agents such as doxorubicin reduces toxicity while increasing the tumor accumulation [66].
- **Antifungal treatments:** Amphotericin B-loaded liposomes are used to treat fungal infections with minimal nephrotoxicity [65].
- **Antimicrobial delivery:** Liposomes enhance the solubility and stability of antibiotics such as vancomycin and rifampicin [64].
- **Gene delivery:** Liposomes can transport DNA or RNA for gene therapy purposes [52].
- **Vaccines:** Liposomal formulations serve as adjuvants and carriers for antigens, which enhance immune responses [67].

6.3.6.5 Targeted Liposomes

Targeted liposomes are advanced drug delivery systems that selectively deliver therapeutic agents to specific tissues, cells, or organs, increasing treatment efficacy while reducing systemic toxicity. This targeted delivery is accomplished by functionalizing liposomes with specific ligands, antibodies, peptides, or other biomolecules that can bind to overexpressed receptors or markers at the target site [68]. Some targeted liposomes are shown in Table 6.2.

Types of Targeted Liposomes Targeted liposomes are broadly classified into active and passive targeting liposomes. Passive targeting liposomes have the following characteristics:

- Passive targeting takes advantage of the EPR effect, which is commonly seen in solid tumors.
- Liposomes accumulate preferentially at the tumor site due to leaky vasculature and inadequate lymphatic drainage.
- Liposomes coated with PEG, also known as stealth liposomes, evade immune system clearance, increasing circulation time and passive accumulation at the tumor site. Doxil® is the first FDA-approved liposomal formulation for doxorubicin (DOX). It has provided several benefits over free doxorubicin in the treatment of solid tumors. It has lowered the risk of cardiotoxicity, a major concern with DOX. The clinical approval of Doxil (1995) has proven a turning point for lipid-based drug delivery and cancer nanomedicine. These liposomes were conjugated with hydrophilic PEG molecules, creating a surface shell protecting from the immune system and prolonging half-life in the systemic circulation (doxorubicin-loaded stealth liposomes for cancer therapy) [74].

Active targeting entails modifying the liposomal surface with specific targeting ligands that bind to receptors or antigens found on target cells or tissues. These ligands include:

- **Antibodies (immunoliposomes):** Monoclonal antibodies or antibody fragments are attached to the liposomal surface to target specific receptors. Like Anti-HER2 antibodies are used to target cancer cells that are overexpressing HER2 [69].

Table 6.2 Representative targeted liposomes and their applications.

Drug delivery system	Target moiety	Application	References
HER2-targeted liposomes	Anti-HER2 Antibody	Breast cancer	[69, 70]
Folate-targeted liposomes	Folate	Folate receptor-positive tumors	[71]
RGD-modified liposomes	RGD peptide	Tumor angiogenesis	[72]
Mannose-modified liposomes	Mannose	Targeting macrophage	[73]

- **Peptides:** Peptide ligands identify overexpressed receptors (e.g. arginylglycylaspartic acid [RGD] peptides for integrins). For example, RGD-modified liposomes can target tumor vasculature [48].
- Aptamers are single-stranded oligonucleotides or peptide molecules that bind to their targets with high specificity. For example, cancer cells can be treated with aptamer-functionalized liposomes [53].
- **Small molecules:** Ligands, such as folate and transferrin, target overexpressed receptors. Folate-targeted liposomes, for example, are used to treat tumors that express the folate receptor.
- **Carbohydrates and glycoproteins:** Target lectins or specific carbohydrate receptors on cells. For example, liposomes functionalized with mannose to macrophages [74].

Mechanisms of Targeting Liposomes They use physical and physiological properties, such as the EPR effect, to accumulate liposomes in target tissues, particularly tumors. Liposome ligands interact with specific cell surface receptors or antigens, allowing for receptor-mediated endocytosis and intracellular drug delivery. Liposomes are designed to release their cargo in response to external or internal stimuli such as pH, temperature, enzymes, or light [74].

Applications of Targeted Liposomes

- **Cancer therapy:** Targeted liposomes are employed to deliver chemotherapeutic drugs directly to tumor tissues, such as HER2-targeted immunoliposomes used in the treatment of breast cancer.
- **Gene delivery: Targeted liposomes are used to deliver DNA, RNA, or siRNA to specific cells for gene therapy.**
- **Infectious diseases: Liposomes with targeting ligands deliver antibiotics or antifungal agents to infected tissues.**
- **Neurological disorders: Liposomes are being targeted across the blood–brain barrier (BBB) to treat diseases such as Alzheimer's and brain tumors.**
- **Inflammatory diseases: Targeted liposomes deliver anti-inflammatory agents to macrophages or inflamed tissues** [74].

6.4 Mixed Micelles for Enhanced Drug Solubility

One primary requirement associated with drug delivery systems is their ability to enhance the delivery of the active drug molecule to the particular site [75]. Some criteria must be put into consideration in the design of a drug delivery system [76]. The criteria include the rate of release of the medication to the site of action, toxicity of the drug, and the ability of the drug to penetrate cellular structures and tissues that ordinarily cannot be accessed from the overall general circulation [77]. The properties associated with the use of polymeric micelles in drug delivery include a small size, sustained delivery of active drug molecules, administration of low-molecular-weight drugs, and prevention of degradation of encapsulated materials. Polymeric micelles are defined as nanostructure core-shell materials formed in an aqueous solution composed of amphiphilic blocks. Formation of micelle in aqueous solution occurs at the area of CMC [78]. In the field of research and drug delivery, polymeric micelles have numerous applications, particularly when utilized as an intelligent drug delivery system [79]. Polymeric micelles can be utilized as carrier systems because they undergo dynamic physicochemical changes during drug entrapment and release in a molecular form [80]. Classification of copolymers based on their hydrophobic blocks includes poly(L-amino acid), poly(propylene oxide), and polyesters [81].

6.4.1 Types of Mixed Micelles

6.4.1.1 Copolymer-Mixed Micelles

The most commonly used commercially available polymer in the field of nanomedicine is pluronics [82]. They are classified under amphiphilic and temperature-responsive triblock copolymers, which consist of poly(propylene oxide) and poly(ethylene oxide) blocks. In the human body, pluronics cannot be degraded, but the human kidneys have a way of filtering these systems, which are subsequently removed via urination [83]. Some of the combinations used in the formation of mixed micelles include block copolymer/homopolymer mixtures, copolymer mixtures, polymer/biopolymer liposome, protein, and peptide mixtures [84]. A study reported by Rakshit et al. employed two polyglycerols which were grafted with β-cyclodextrin and lactobionic acid-based nanocarriers for targeted delivery of paclitaxel as the drug of choice for the treatment of carcinoma. The in vitro experiments showed a positive result indicated by the remission of the carcinoma, facilitated by inhibition of drug efflux transporters by the pluronic block copolymers, followed by enhanced transport of the therapeutic agent across endothelial cells and intestinal barrier cells in vitro [85, 86]. Another study conducted by Zana et al. [87] demonstrated the use of amphiphilic random and block copolymers, which consist of hydrophobic pyridyl disulfide (PDs) derivatives and hydrophilic PEGs. Development of polyion complex (PIC) micelles usually stems from the electrostatic interactions between two copolymers. The method of production of PIC micelles, which involves the elimination of residual organic solvents, enables them to be used as an excipient for sensitive cargo micelles. PIC micelles can also serve in encapsulating ionic substance [86, 88].

6.4.1.2 PEGylated Lipid Micelles

Some unique properties associated with the use of PEGylated micelles in drug delivery include their biocompatibility, non-toxicity, and encapsulation of drugs from enzymatic degradation [89, 90]. They are also known to increase the therapeutic index of encapsulating drugs and subsequently lower the toxicity associated with them [91–93]. It has been documented that PEGylation improves the overall performance of liposomes in terms of residence time. PEG has also been used in the surface modification of nanocarriers to increase their circulation half-lives [94–97].

Some of the PEGylated micelles and other nanocarriers cannot be taken up by the reticuloendothelial cells; therefore, there are higher chances that they will reach and deliver increased levels of therapeutics to target diseased organs when compared to non-PEGylated nanocarriers [98]. This is attributed to the fact that PEGylation has a stealth property on nanocarrier surfaces, which subsequently causes the PEGylated nanocarriers to experience a decreased uptake by the reticuloendothelial system [99].

The two moieties that constitute the PEG-phospholipid conjugates consist of hydrophilic PEG block and a hydrophobic phospholipid. The molecular weight of the PEG block is between 750 and 5000 Da. It has been used over the years in pre-clinical trials due to its associated properties such as neutrality, water solubility and non-toxicity [100]. PEG formation is associated with a highly hydrated chain, which primarily serves as a protective barrier and helps minimize interactions with proteins and metabolic enzymes. A study was conducted to examine the blood clearance of various polyethylene glycol-phosphatidylethanolamine (PEG-DSPE) micelles. The result obtained recorded that the different micelle formulations, which consist of PEG500, PEG2000-DSPE, and PEG750-DSPE, had different circulation half-lives of 2.1, 2.0, and 1.2 hours, respectively. The formulation with the highest molecular weight gave the highest circulation half-life due to the stearic protection phenomenon [101]. There are several reasons associated with the importance of the micelle size. It has an impact on the drug loaded within the micelle by preventing its evasion from the body. Renal clearance or extravasation is the route used in releasing particles that are less than 7 nm, while elimination route is used for particles that are larger than 10 nm. Particles within the size range of micelles (10–70 nm) can easily penetrate capillaries, thereby enhancing tissue distribution [102]. Another study conducted by Remsberg et al. dosed (10 mg/kg) rats intravenously with DSPE-PEG_{2000} micelle formulation of ridaforolimus. The results obtained showed a significant increase in the half-life of ridaforolimus by 170% and a decrease in clearance by 58% [103].

Some factors influence the shape of the micelle, which include the concentration of surfactants, hydrophilic segment, and hydrophilic segment size. There are several factors that influence the drug-loading capacity of micelles. They include hydrophobicity and the length of the core-forming block, the length of corona interactions between the drug and the core, and the preparation method [104, 105]. There are three central areas in a micelle where drugs can be localized, and these areas include the core, corona, and core–corona region [106]. For the diffusion of micelles into cell membrane, internalization mechanism is required [107].

A lot of approaches have been adopted in order to enhance the intracellular delivery of drugs loaded PEG–lipid micelles [108]. One commonly used approach involves the attachment of a targeted ligand or vector molecules that possess affinity toward ligands that are overexpressed at target sites to the surface of the micelles [109]. There will be an improvement in the micelles and incorporated drugs when the ligands are targeted to the hydrophilic corona. Targeted micelles have been made possible with the use of unique ligands such as transferrin, folate residues, and epidermal growth factors [110]. Antibodies been one of the specific ligands, have a great role to play in terms of diversity of targets and specificity of interaction. Monoclonal antibodies in combination with PEG-DSPE-based immunomicelles were prepared using activated *P*-nitrophenyl carbonyl (PNP) group. Excessive drug degradation occurs in lysosomes as a result of endocytosis-mediated capture of therapeutic micelles [110].

6.4.1.3 Stimuli-Sensitive Micelles

pH-Sensitive Polymeric Micelles The application of polymeric micelles in drug delivery can be enhanced when used in combination with polyacids and polybases. They tend to form polymeric micelles at a pH above the pK_a of the protonable group, leaving the hydrophobic group uncharged. Decrease in pH below the pK_a causes the polymers to experience electrostatic repulsions, which subsequently lead to destabilization of the micelles. There is great potential associated with the use of pH-sensitive polymeric micelles in systemic gene delivery. There is an interaction between the DNA molecule and the copolymers, leading to the formation of a complex that prevents enzymatic degradation [111].

Temperature-Sensitive Polymers They are helpful in the preparation of hydrophilic intelligent systems below their critical dissolution temperature. The behavior of the polymer changes with respect to changes in temperature. For example, the polymers become hydrophobic below the critical solution temperature. Temperature-sensitive micelles can also be obtained with the help of amphiphilic copolymers [112]. The release mechanism of the drug is activated above the critical temperature. For example, doxorubicin is released slowly at 37 °C, while above 42 °C, the release becomes faster [113]. When rightly channeled, the micelle could be adequately harnessed in determining the drug release rate [114].

Light-Sensitive Polymeric Micelles Light can be utilized as a triggering agent to enhance the release of drugs and minimize toxicity to adjacent tissues. Treatments involving the skin and mucous membrane can be enhanced using ultraviolet light. The mechanism of action of light in polymeric micelle is based on their ability to undergo reversible alterations of the hydrophilic–lipophilic balance. It has been reported that use of amphiphilic polymeric micelles allowed incorporation of coumarin chromophores. As an end result, the stabilization and de-stabilization of micelles was achieved by reversible photodimerization of coumarin groups and photocleavage of cyclobutene bridges [115].

Ultrasound-Sensitive Polymeric Micelles Ultrasound can be applied in pulsatile drug delivery into the tumors using polymeric micelles [114]. Polymeric micelles are known to accumulate in tumor cells. At the peak of the maximum micelle accumulation, an ultrasound should be deployed at the tumor site. Some factors can be manipulated to control the amount of drug released. These factors include frequency, power density, pulse strength, and inter-pulse intervals [115]. Over the years, the relevance of polymeric micelles in pharmaceutical industry cannot be overemphasized.

6.4.2 Method Used to Prepare Mixed Micelles

There are two major techniques used to prepare mixed micelles for drug encapsulation, and these methods are described as follows:

Polymer dispersion in an aqueous solvent: It involves the dispersion of the drug in either the polymer or an organic solvent. The organic solvent is added drop-wise to the copolymer dispersion and stirred. The organic solvent is then removed using the rotary evaporator. This technique is suitable for hydrophobic polymers, which may require heating of the aqueous dispersion to form micelles. Advantages associated with this technique include short preparation time (time-saving) and simplicity, easy scaling-up method, and high encapsulation efficiency [116]. One major limitation associated with this technique is that only a few drugs can be entrapped, thus obtaining high loading levels.

The second technique involves the use of organic solvents, such as DMSO, acetonitrile, methanol, and acetone, to dissolve the amphiphilic polymers along with the drug. The micelle is formed upon removal of the solvent [117]. For water-miscible organic solvents, deionized water can be used to dialyze the solution. Loading of doxorubicin in pH-sensitive Polyttis-PEG/Poly (L-lactic acid) (PLLA)–PEG mixed micelles with the application of this technique, is reported by Lee et al. [117]. The possibility of evaporating the organic phase can lead to the formation of a thin film that favors the polymer–drug interactions. Rehydration of the thin film with an aqueous phase is also called thin film method [118]. Emulsification method involves the use of water-immiscible organic solvents using the hydrophobic drugs, which could be physically entrapped. According to Hu et al. [119], the result obtained showed that there was an increase in the solubility of paclitaxel around sixfold when the drug was loaded in PLA–PEO–aryinylglycylaspartic acid (RGD/Pluronic F68) mixed micelles. Advantages associated with this technique include high drug loading levels and favorable polymer–drug interaction. The limitations associated with this technique are the use of organic solvents, poor scale-up feasibility and it is a multistep process.

6.4.3 Concept of Critical Micelle Concentration

The CMC is a technique used to evaluate the stability of different micellar systems [120]. Micelles are usually impacted by environmental variations such as pH change, ionic strength, media exposure, and dilution. Control of the micelle formation will help to prevent a rapid drug release and to ensure delivery to the site of action. In the preparation of micellar formulations, it is easy to determine the CMC of amphiphilic macromolecules. This could be achieved by measuring the surface tension, micellar size, and optical clarity of a solution [121]. Some of the techniques employed include fluorescence spectroscopy, surface tension measurement, UV spectroscopy, and dynamic light scattering (DLS) [122, 123]. Among these techniques, the use of fluorogenic dyes such as pyrene has been the most preferred due to its high sensitivity [124]. It works by partitioning of the pyrene on the hydrophobic core of the formed micelle, which then allows for the determination of the CMC [125]. UV–Vis spectroscopy could be used in determining CMC, and this is achieved by allowing the hydrophobic dye molecules to solubilize in the organic solvent. Due to the absence of micelles, no dye partition takes place [123]. Increase in the polymer concentration leads to a positive response in the absorbance reading. Another reliable technique that can be used to determine the CMC of samples is DLS. The working principle is based on changes in light intensity, while an increase in the scattering intensity leads to the formation of micelles [124]. Advantages associated with this technique include the contamination of the probe by dye molecules and small sample size.

6.4.4 Cloud Point of Micelles

A cloud point (CP) is defined as the formation of a cloud appearance by a nonionic surfactant solution at a given temperature. This results in the separation into two phases, which is caused by a decrease in the surfactant solubility in water. Increase in temperature increased the aggregation number of micelles. There is a direct relationship that exists between the CP of micellar dispersion and interactions between the micelles [126]. The binary system has one phase that contains a polymer-rich phase with little water, while the other phase that is water-rich contains little or no polymer [127]. Another important parameter that is considered is the CP of a mixed micellar system. This gives us an idea of the growth of the micelles and also helps to determine the possibility of administering via the intraperitoneal route. Visual observation is another precautionary measure that is considered in the formation of CP. It involves placing the glass vials in an oil bath at room temperature. Subsequently, the system is heated until the formation of a turbid color is observed [128].

6.5 In vivo Safety of Liposomes and Mixed Micelles Drug Carriers

Taking into account liposome fabrication techniques, "liposomes" are composed of natural and biodegradable lipids and are considered nontoxic and safe. As conventional drug delivery carriers are examined critically for clinical safety, liposomal formulations must be evaluated for their clinical safety and toxic side effects before being accepted as carriers [129].

As we look structurally, the composition of liposome-content vesicles is spherical, and these hydrophilic structures are entirely or partially covered with one or more lipidic layers via phospholipids. The amphiphilic nature of phospholipids provides closed boundary bilayer confrontation, which helps to lower the free energy surfaces. These properties make liposomes more stable in hydrophilic and hydrophobic domains with self-assembling characteristics. The modification in stability domains is altered via alteration in physicochemical parameters [130]. This unique property of liposomal carriers enables versatile drug delivery applications. To date, liposomes as drug delivery carriers have been explored in many disease indications, including cancer treatment, delivery of antibiotics, gene delivery, and delivery of drug combinations. The phospholipid in liposomal carriers plays a significant role in making it a better carrier for drug delivery: it has a similar structure to the cell membrane that can allow liposomes to pass the primary barriers of the cell. This passage provides access to the drug moiety and its distribution to a large extent [131].

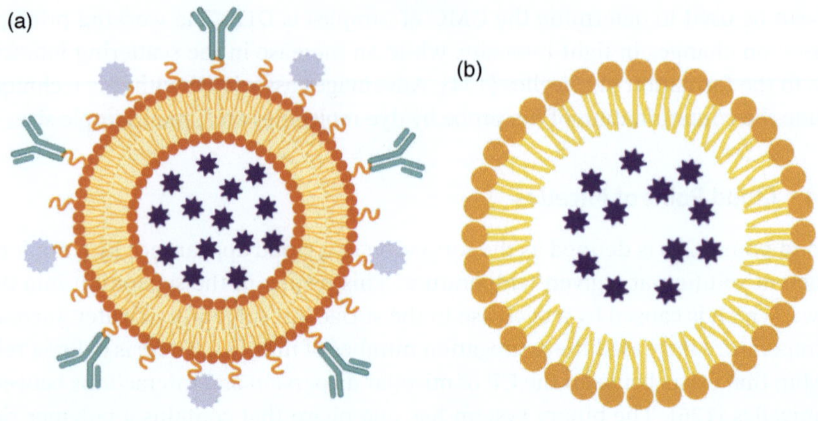

Representative structure of (a) liposomal drug delivery system (functionalized nanocarrier with drug) and (b) micelles.

6.6 Benefits and Limitations of Liposomes and Mixed Micelles

Liposomes and mixed micelles are beneficial for a range of drug delivery applications due to their ability to transport both hydrophilic and hydrophobic drugs. Liposomes are effectively used for treating a variety of terminal illnesses, including cancer. The lipid layer and the aqueous layer can both be enveloped by liposomes. Lipid components can be included in the carrier systems, and they enclose the target molecules, which are delivered to the site of action [132, 133]. Some of the benefits of liposomes and mixed micelles are as follows:

6.6.1 Targeted Drug Delivery

For the treatment of life-threatening diseases like cancer, most drugs have very narrow therapeutic index. In such conditions, the dosage regimen has to be critically monitored to avoid serious toxicity effects. To overcome this challenge, drug moieties are encapsulated into these carrier systems and delivered to the target cell. This delivery improves the therapeutic index without changing the activity. For example, the anticancer agent, doxorubicin, when administered orally, causes serious side effects, but when it is encapsulated in form of liposomes or mixed micelles, it shows better therapeutic activity without altering normal physiological functions [134].

In 1995, the US Food and Drug Administration (FDA) approved Doxil®, a PEGylated liposomal form of doxorubicin. By encasing the drugs in a PEG-coated stealth liposome, which greatly extends its bloodstream circulation period of time, it is intended to improve the therapeutic index of doxorubicin. Through the EPR effect, this alteration reduces exposure to healthy tissues while promoting enhanced accumulation in malignant tissues [1, 2].

On the other hand, Myocet® is a non-PEGylated liposomal version of doxorubicin that is approved in both Europe and Canada. It is mainly used to treat metastatic breast cancer and is frequently used in conjunction with cyclophosphamide. Myocet has distinct pharmacokinetics and a shorter circulation time than Doxil® because it does not have the PEG coating. Although Myocet's cardioprotective effect is marginally less effective than Doxil's, it also delivers reduced cardiotoxicity when compared to free doxorubicin [1, 2].

6.6.2 Cell Targeted Drug Delivery

The acceptance of liposomes and mixed micelles by cells when it administered via various routes is higher compared to the conventional drug delivery strategies. Most drugs in their pure forms are not taken up by cells because they are poorly soluble, but their cell uptake increases when encapsulated in these carrier systems [135, 136].

6.6.3 Controlled and Sustained Drug Delivery

The optimum plasma drug concentration needs to be maintained for a longer period for the desired therapeutic outcome to be achieved. Liposomes and mixed micelles could provide sustained and controlled delivery of targeted drug because of their chemical compositions and drug release patterns [137, 138].

6.6.4 Intraperitoneal Drug Delivery

Diseases like tumors in intraperitoneal cavity need special attention due to high drug clearance rate. Administration of liposomes and mixed micelles as drug carriers can lower drug release intraperitoneally and provide drug delivery in a prolonged manner. Maximum drug absorption and bioavailability could be achieved via liposomal and micellar encapsulation [139].

6.6.5 Delivery of Vaccines

The structural properties of liposomes make them an excellent vaccine delivery platform. The dual nature (lipophilic and hydrophilic) of liposomes makes them capable of accommodating different types of antigens. On the basis of the antigen, it can be incorporated into either aqueous or lipid bilayers. Antigen-incorporated liposomes produce a significantly stronger immunological response than the antigen alone. The first trial of vaccination using liposomal carriers was for diphtheria toxoid vaccine [58, 140].

6.6.6 Enhancement in Pharmacokinetic and Dynamic Properties of Actives

The dramatic improvement in the kinetic and dynamic properties of drugs can be achieved by using liposomes and mixed micelles. Therapeutic concentrations in the blood plasma are controlled through extending the circulation time of liposomes. And this circulation is dependent on the membrane permeation, fluidity, size of the carriers, and type of carriers. As an example, small-sized liposomes and neutrally charged liposomes stay longer in the systemic circulation. Liposomes with lower drug loading show lower side effects with sustained drug release [140, 141].

6.6.6.1 Limitations of Liposomes and Mixed Micelles as Drug Delivery Carriers

The clinical translation of liposome and mixed micelle drug delivery technologies has not advanced as rapidly as the abundance of encouraging data would suggest, despite extensive study over the past 50 years. Although liposomal and mixed micelles formulations have shown notable therapeutic benefits for a wide range of therapeutic applications, challenges linked to scale-up, regulatory compliance, and intellectual property have been identified as the need to be addressed. Other nano delivery methods encounter similar challenges to enter the clinical development stage [142]. Pharmaceutical development is mainly constrained by the cost of production and quality compliance. In line with the other nanocarrier systems, liposomes and mixed micelles are mainly impacted by (i) the process of manufacturing, (ii) batch-to-batch reproducibility, (iii) limited knowledge, (iv) denaturation or chemical instability of the labile APIs under the processing conditions, and (v) long-term stability [143].

Quality compliance encompasses issues related to the development and scale-up of the process. These days, one can use the now effective techniques for producing conventional liposomes on an industrial scale without the need for organic solvents or a number of multistep manufacturing (Figure 6.3).

The issues arise when the design and operation of the liposomal system become more intricate, such as when coating and/or ligands are introduced to alter the surface. It takes significant amount of chemical modifications and formulation optimizations to integrate different components into a nanocarrier, which invariably presents challenges for scale-up. Higher production costs have an impact on the advancement of these products.

Factors such as type of coating and ligand significantly impact the biocompatibility, distribution of drug and toxicological characteristics of liposomal formulations. Additional research is necessary to further comprehend how nanocarriers interact with biological tissues and cells. The synergy between academia and industry will be necessary to further enhance the regulatory framework for evaluating these formulations [144].

Patent protection for liposomal-based treatments can drive up complex issues and is likely to increase development expenses. This is linked to the type of encapsulation or entrapment methods used, as well as the ligands used to develop liposomes and mixed micelle carrier compositions. Any of these elements could hamper the intellectual property position of the formulation, making it less commercially appealing, which would affect the product's ability to advance in the drug discovery pipeline [144, 145].

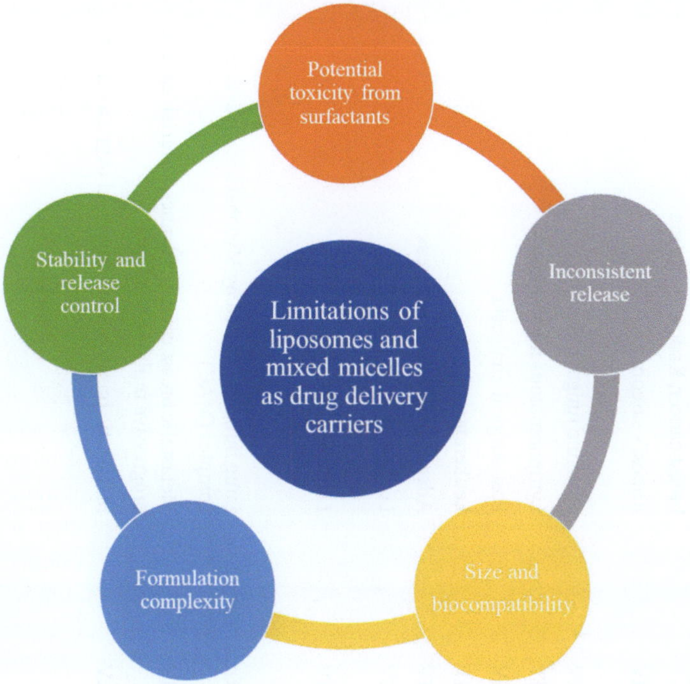

Figure 6.3 Constraints linked to the use of liposomes and mixed micelles as drug delivery vehicles. Adapted from Cerpnjak et al. [3].

6.7 Marketed Micellar and Liposomal Products

Nowadays, a substantial number of therapies based on liposomes and mixed micelles are accessible for clinical applications. Most liposomal formulations are suitable for intramuscular (IM) and intravenous (IV) administration. A concise overview of these products is summarized in Table 6.3.

6.8 Regulatory Guidelines for Liposomes and Mixed Micelles Use

Over the past few decades, the FDA and European Medicines Agency (EMA) have approved around 100 and 11 nanomedicines, respectively, while 48 nanomedicines are now being clinically tested in the European Union. The regulatory bodies around the globe including FDA, EMA, Ministry of Health, Labor and Welfare—Japan (MHLW), and Chinese National Medical Products Administration (NMPA) released a number of guidelines concerning nanomaterials and nanoproducts in light of the clinical application and exchange of regulatory knowledge to assess these formulations [146]. These guidelines cover a variety of nanodrug delivery formulations, including liposomes, nanoion colloidal materials, mixed micelles, and

Table 6.3 Liposomal and mixed micelles therapeutics approved by USFDA and European Medicines Agency (EMA).

Liposomal formulations				
Product and year of approval	**API**	**Dosage form**	**Route of administration**	**Clinical applications**
Doxil® (1995)	Doxorubicin	Suspension (doxorubicin liposome injection)	IV	Various cancer treatments including ovarian cancer, breast cancer, Kaposi's sarcoma
DaunoXome® (1996)	Daunorubicin	Suspension (daunorubicin citrate liposome injection)	IV	Kaposi's sarcoma
Ambisome® (1997)	Amphotericin	SU	IV	Systemic fungal infections
DepoCyt (1999)	Cytarabine	Suspension	IT	Lymphomatous meningitis
Exparel® (2011)	Bupivacaine	Suspension	IV	Post-surgical analgesia
Marqibo® (2012)	Vincristine sulfate	Suspension	IV	Leukemia
Onivyde® (2016)	Hydrochloride trihydrate	Suspension	IV	Adenocarcinoma of pancreas
Vyxeos® (2017)	daunorubicin, cytarabine	Lyophilized powder/suspension	IV	Leukemia
Shingrix® (2018)	Recombinant varicella-zoster virus glycoprotein E	Suspension	IV	Leukemia
Arikayce® (2020)	Amikacin sulfate	Suspension	Oral inhalation	Lung infection caused by Mycobacterium avium complex (MAC)
Zolsketil® (2022)	Doxorubicin	Concentrated dispersion form	IV	Metastatic breast cancer, Advanced ovarian cancer, progressive multiple myeloma
Celdoxome® (2022)	Doxorubicin	Concentrated dispersion form	IV	Metastatic breast cancer, advanced ovarian cancer

Mixed micelle formulations				
Apealea® (2018)	Paclitaxel	Suspension	IV	Epithelial ovarian cancer, fallopian tube cancer.
Genexol® (2007)	Paclitaxel	Lyophilized powder/suspension	IV	For suppressing hypersensitivity reactions
Neoral® (1994)	Cyclosporine A	Oral solution	Oral	Used to increase solubility and enhance absorption of cyclosporin in organ transplant patients.
Sandimmune® (1983)	Cyclosporine A	Oral capsule	Oral	Treatment of autoimmune disease
Emulsol® (1989)	Propofol	Intravenous emulsion	IV	Used in surgeries, intensive care units for procedural sedation
Abraxane® (2008)	Paclitaxel	Suspension	IV	Metastatic breast cancer

siRNA-loaded nanoformulations. The regulatory guidelines compiled for liposomes by the listed regulatory bodies originate from their widespread use and approval in clinical trials and markets [144, 146].

Despite the numerous advantages, there are frequent incidences of physical instabilities such as fusion or aggregation in liposomal formulations. These systems are thermodynamically unstable and sensitive to temperature fluctuations as they are multiphase systems.

Furthermore, the lipids used to develop liposomal formulations are prone to chemical and oxidative effects upon storage. Hence, careful selection of the grade of lipids and appropriate stability testing of these excipients is essential in developing any liposomal formulation [147]. The quality control (QC) methods used to evaluate liposomal formulations are outlined in Table 6.4.

The capacity of liposomes to cross biological barriers and clearance, are heavily influenced by their size and shape. Therefore, critical quality attributes like particle size and polydispersity index are parameters to be controlled. Transmission electron microscope (TEM) techniques such as negative staining, freeze-etching, and cryogenic microscopy, can directly visualize particle morphology and size. The disadvantage of negative-stain TEM is that it requires the material to be dried before staining, which may result in liposome shrinking, deformation, or aggregation. The other two techniques, however, do not require drying. Thus, cryo-TEM has the benefit of requiring little work for sample preparation. Scanning electron microscopy (SEM), although useful, is rarely employed for lipid nanoparticles, as it disrupts the material. In recent years, environmental scanning electron microscopy (ESEM) has been used to study liposomes in their hydrated condition and to check alterations in various environments. DLS, nanoparticle tracking analysis (NTA), and tenable resistive pulse sensing (TRPS) can properly estimate the PDI of dispersion and mean

Table 6.4 Quality control methods used for the analysis of liposomal products.

Quality control parameter	The method used for testing
pH	pH meter
Lipid concentrations (phospholipids)	Bartlett method
Osmolarity	Osmometer
Cholesterol concentration	An assay using HPLC/spectroscopic method
Drug content (assay and purity)	HPLC
Particle size	Dynamic light scattering
Surface charges	Electric fields/pH-sensitive probes
Drug release	in vitro drug release methods
Surface conjugation	Surface plasmon resonance
Sterility	Aerobic/anaerobic culture media method
Presence of ligands	Imaging methods like magnetic resonance imaging (MRI), and spectroscopic techniques like Raman spectroscopy

hydrodynamic radius (Rh). These single-particle sizing procedures are fast, simple to use, and inexpensive, making them ideal for rapid differentiation of polydispersed samples or quality control of materials with less dispersity. While DLS has historically been the most widely utilized technique, researches have shown that NTA and TRPS can deal with multifunctional nanoparticle mixtures more effectively than batch-mode DLS [144, 147].

Considering the immense potential of nanotechnology-based pharmaceuticals, regulatory bodies such as the USFDA and EMA have come up with the guidelines to meet the regulatory requirements of nano-formulation development. These guidelines help in the production of liposomes and their efficient quality check by defining the factors which must be considered during the development. The USFDA issued a revised draft guidance titled "Liposome Drug Products: Chemistry, Manufacturing, and Controls (CMC); Human Pharmacokinetics and Bioavailability; and Labelling Documentation" in April 2018, which includes comprehensive data on various liposomal drug product designs. Guidelines from the USFDA, MHLW, and Central Drugs Standard Control Organization (CDSCO) outline stability testing requirements for liposomal drug formulations. Regulatory bodies worldwide encourage adhering to the ICH criteria for stability testing, specifically Q1 and Q5C [148].

6.9 Clinical Application of Liposomes and Mixed Micelles

Liposomal and mixed micelles preparations are employed in medical applications as vehicles for physiologically active compounds. Because of their durability and advantageous toxicity profile over conventional therapies, liposomes have been the subject of much research in fields including gene therapy and medication delivery. Hydrophilic molecules or therapeutics can be encapsulated in liposomes to enhance their internalization through the lipid bilayers of cells [149 151].

6.9.1 Application in Drug Delivery

For the production of liposomes that target tumor cells selectively and deliver anticancer medications straight to tumor locations, a variety of approaches have been devised. Both passive and active targeting techniques are made possible by the surface functionalization of liposomes. Because tumor arteries leak, liposomal medication formulations freely move over capillary endothelium and into the interstitial fluid. Liposomes can target passively because the endothelial cells lining normal capillaries have space between them that range from 4 to 11 nm, while the endothelial cells lining tumor capillaries have voids between them that range from 95 to 780 nm. Therefore, if liposomes and mixed micelles are made in a size range suitable for extravasation into tumor tissues rather than normal tissues, optimal targeting can be achieved [152]. Various techniques are being used to create actively targeted liposomes in an effort to further reduce off-target negative effects.

Addressing ligands, which include aptamers, monoclonal antibodies, and peptides, are typically conjugated onto the liposome surface in order to prepare actively targeted liposomes. There are several methods for attaching ligands to liposomes, such as attaching directly to lipids or at the end of PEG chains. Incorporating ligands into the liposome formation process is another often-used strategy. These ligands can be used to take advantage of certain antigens and receptors that cancer cells produce. Targeting ligand-attached liposomes can connect with particular cell surface receptors to undergo endocytosis into cells. Furthermore, sufficient concentrations of targeting ligands must be added in order to achieve a higher affinity for cell surface receptors [153, 154].

6.9.2 Application in Clinical Diagnosis and Imaging

Clinical results are significantly influenced by the early diagnosis of diseases, especially malignancy. Consequently, a wide range of research has concentrated on developing novel diagnostic agents and methods as well as enhancing the functionality of already-existing ones. Despite the modality employed, diagnostic/imaging agents are typically required to obtain a sufficiently strong response from the target area in order to distinguish specific structural pattern from surrounding tissues. As a result, many particle systems have been studied as contrast agent carriers to help the imaging "contrast" agent accumulate in the necessary zone. Liposomes have garnered particular interest among these carriers due to their readily adjustable pharmacokinetics and physicochemical characteristics. Furthermore, the multifunctional nature of liposomes has made it possible to load a range of imaging agents and incorporate particular ligands for effective tissue targeting. Liposomes are commonly employed as carriers in numerous imaging modalities such as MRI, ultrasound, fluorescence, and nuclear imaging [140].

One of the most widely used diagnostic methods for visualizing gene expression, the location of biomolecules, and enzyme activity in live cells is fluorescence imaging. Enhancement in pharmacokinetic liposomes can aid in the targeted delivery of fluorescence imaging agents. The development of a hybrid nanocomposite for cancer imaging by encasing PEG-coated quantum dots (QDs) within the water-soluble phase of 1,2-dioleoyl-sn-glycero-3-phosphocholine (DOPC)-based liposomes was reported. These hybrid nanoparticles have a greater ability to target tissue from tumors while maintaining the fluorescence characteristics of the original QDs. Furthermore, as compared to unmodified QDs, these nanoparticles demonstrated improved tumor passage and retention characteristics in tumor spheroids and subcutaneous solid tumors in tumor model tests, increasing their diagnostic potential [155].

MRI is a noninvasive imaging method that visualizes organs and other internal structures by using radio electromagnetic pulses and a magnetic field. The utilization of paramagnetic contrast chemical compounds to visualize the tissue or organ being studied is essential to enhancing MRI performance. Liposome application in MRI was maximized by their capacity to deliver MRI contrast molecules to the location of interest and circulate in the blood for longer periods of time. In addition to improving tumor specificity localization, researchers created a multifunctional thermosensitive liposomal product that improves the release pattern and in vivo anticancer activity of the encapsulated medication by linking with a noninvasive MRI agent (gadopentetate dimeglumine, or GdDTPA) [140].

Oenography, often known as ultrasound imaging, is another popular noninvasive diagnostic imaging method that uses ultrasound to measure the echoes produced by tissues at various angles of reflection. Ultrasound imaging probes can be made from liposomes that include gas bubbles, which are effective sound reflectors. Preparing liposomes for sonography involves either creating the bubble directly inside the liposome through a chemical reaction, such as bicarbonate hydrolysis, producing carbon dioxide, or entrapping gas bubbles into the liposome. In animal experiments, gas bubbles trapped inside the phospholipid membrane demonstrated minimal toxicity and good efficacy of these contrast agents.

Computerized tomography (CT) is an imaging technique that produces fine-grained images of inside body regions using specialized X-ray equipment. CT contrast agents can be integrated into the liposome membrane or added to the inner aqueous core of the liposome. Liposome-based contrast agents present a viable, low-toxicity way to improve CT and ultrasound imaging, with enhanced efficiency in contrast resolution, targeted delivery of contrast agent, and improved patient experience. They have the capacity to incorporate a high amount of contrast agents, with good stability and generate better vision of internal body tissues and organs. The ability to functionalize liposomes-bearing contrast agents with ligands facilitates their use in clinical diagnosis for targeting diseased tissues. Iopromide, for instance, has been added to both plain and PEGylated liposomes and has shown promising biodistribution and imaging capabilities in rats and rabbits [156].

The establishment of theragnostics, which either integrates several imaging or therapeutic activities in a complex, has been a significant topic recently. Owing to adaptable encapsulation properties of liposomes, they can be further developed to co-deliver various therapeutic medications and imaging agents, enabling multimodal imaging and therapy impact assessment. Theragnostic liposomes can entrap nanosized diagnostic agents such as iron oxide nanoparticles, QDs, and gold nanoparticles, while therapeutic drugs can be enclosed in the core or embedded in the lipophilic bilayer shell. Advanced theragnostic liposomes are coupled with molecular biomarkers to provide a targeted effect. Stealth liposomes, also known as PEG-coated liposomes, were designed to be stable and have a prolonged half-life in the bloodstream, preventing immune system opsonization and rapid removal [149, 151, 156].

6.9.3 Application in Vaccine and Gene Delivery

The unique features of liposomes make them promising for disease treatment and immunization. Examples include antibiotics, protection from viruses, prostaglandins, steroid and non-steroid anti-inflammatory medications, insulin, and more. Recombinant DNA technologies and gene therapy research all rely on the in vitro and in vivo transportation of nucleic acids into cells. In vivo delivery requires a greater effort than in vitro methods, which can rely on various physical and chemical processes. DNA-carrier systems comprise a variety of colloidal particles. Cationic liposomes, for instance, have been shown to be able to combine with negatively charged DNA to create complexes. By causing the target cells to express the protein included in the DNA plasmid, these complexes have the ability to transfect cells in vitro. In vivo delivery is obviously favored for gene therapy, which treats disorders at the molecular level by turning genes on or off. Liposomes function as an adjuvant and co-adjuvant carrier, enhancing the immunological response to the vaccine antigen [157].

The susceptibilities of liposomes made from widely accessible lipids to enzymatic destruction, the low pH in the stomach, especially the bile salt breakdown in the intestine, have been a major obstacle to the production of liposomes for oral or mucosal administration of macromolecules. Therefore, common liposomes are unable to shield macromolecules from GIT enzymatic breakdown. Although liposomes that have been mechanically and sterically stabilized can withstand these circumstances, they are too stable in the colon to transport the medicine therein by the regular absorption process [4, 5].

6.10 Patent Landscape of Liposomes and Mixed Micelles

Over the past decades, several researchers and groups have submitted applications seeking to patent outstanding discoveries, developments, and outcomes of their studies in LBS. Documenting patent information in books or journal publications is not a popular practice, but it is essential for inspiring future research and leading the translation of promising research outcomes. Some of the patent applications have been officially granted, while some are still pending and Table 6.5 summarizes representative patents of lipid-based delivery systems.

6.11 Future Research and Conclusion

The field of LBS has the potential to revolutionize and advance effective and targeted drug delivery. Researchers are constantly developing newer approaches to improve drug disposition, and these emerging systems are the future of pharmaceutical drug development. Future projects utilizing LBS should focus on developing personalized medicines which are tailored to serve the needs of individual patients, taking into account patient, drug, and disease factors. Formulation of combination therapies containing multiple drugs should be explored using LBS to prevent repeated administration of single drugs, which might impede patient compliance. LBS should be integrated with nanotechnology to produce nanoformulations with desirable properties such as controlled release, targeted delivery, improved solubility, and bioavailability. They should be designed such that their scalability would be feasible and cost-effective. They have immense potential to address the problems associated with poor drug solubility and bioavailability. With continued innovations and advances, LBS have the ability to improve patient drug administration and care, with minimum toxicity reactions.

Table 6.5 Selected patents of liposomes and mixed micelles for various drug delivery applications.

Title of patent	Inventors/applicants	Patent number	Year of publication	References
Patents for liposomes				
Tailored liposomes for the treatment of bacterial infections	Draeger A, Babiychuk E	US10744089	2020	[20]
Co-loaded liposome and preparation method thereof	Chengdu Jinrui Found Biotechnology Co., Ltd	CN116270473B	2023	[158]
mRNA–liposome complex and application thereof	Zhejiang Zhida Pharmaceutical Co., Ltd.	CN112107680B	2019	[159]
Liposomes with ginsenoside as membrane material and preparations and use thereof	Li C, Wang Y, Zhan H	US10639276	2020	[20]
Patents for mixed micelles				
A kind of pH-sensitive mixed micelles and its preparation method and application	Yongjian W, Ao Y, Xinnan W, Dandan S, Xueming L	CN-107441043-A	2017	[160]
Aqueous suspension comprising mixed micelles loaded with a lipophilic physiologically active substance	Rooimans T, Tiofarma BV	EP3391909A1	2018	[161]
Quercetin-loaded pluronicâ® f127/chitosan myristic acid mixed micelles to improve the biopharmaceutical and hepatoprotective potential of quercetin	Ganorkar SB	551967	2024	[162]

References

1 Shrestha, H., Bala, R., and Arora, S. (2014). Lipid-based drug delivery systems. *J. Pharm.* 2014: 801820.

2 Sambhakar, P.S., Malik, R., Saurabh Bhatia, S. et al. (2024). Lipid horizons: recent advances and future prospects in LBDDS for oral administration of antihypertensive agents. *Int. J. Hypertens.* 2024: 2430147.

3 Cerpnjak, K., Zvonar, A., Gasperlin, M., and Vrecer, F. (2013). Lipid-based systems as a promising approach for enhancing the bioavailability of poorly water-soluble drugs. *Acta Pharma.* 63: 427–445.

4 Sercombe, L., Veerati, T., Moheimani, F. et al. (2015). Advances and challenges of liposome assisted drug delivery. *Front. Pharmacol.* 6: 286.

5 Deng, W., Chen, W., Clement, S. et al. (2018). Controlled gene and drug release from a liposomal delivery platform triggered by X-ray radiation. *Nat. Commun.* 9: 2713.

6 Almgren, M. (2000). Mixed micelles and other structures in the solubilization of bilayer lipid membranes by surfactants. *Biochim. Biophys. Acta* 1508: 146–163.

7 Feng, S., Catron, N.D., Zhu, A.D. et al. (2017). Predictive modeling of micellar solubilization by single and mixed nonionic surfactants. S0022-3549 (18): 30143–30146.

8 Losada-Barreiro, S., Celik, S., Sezgin-Bayindir, Z. et al. (2024). Carrier systems for advanced drug delivery: improving drug solubility/bioavailability and administration routes. *Pharmaceutics* 16: 852.

9 Patrick, M.G. and Vladimir, R.M. (2019). Pharmacokinetic and pharmacodynamic properties of drug delivery systems. *J. Pharmacol. Exp. Ther.* 370: 570.

10 Sadar, M.D. (2020). Discovery of drugs that directly target the intrinsically disordered region of the androgen receptor. *Expert Opin. Drug Discov.* 15: 551–560.

11 Ezike, T.C., Okpala, U.S., Onoja, U.L. et al. (2023). Advances in drug delivery systems, challenges and future directions. *Heliyon* 9 (6): e17488.

12 Xie, B., Liu, Y., Li, X. et al. (2024). Solubilization techniques used for poorly water-soluble drugs. *Acta Pharm. Sin. B* 14 (11): 4683–4716.

13 Porter, C.J.H., Pouton, C.W., Cuine, J.F., and Charman, W.N. (2008). Enhancing intestinal drug solubilisation using lipid-based delivery systems. *Adv. Drug Deliv. Rev.* 60: 673e91.

14 Huang, Y.P., Yu, Q., Chen, Z.J. et al. (2021). in vitro and in vivo correlation for lipid-based formulations: current status and future perspectives. *Acta Pharm. Sin. B* 11: 2469e87.

15 Clark, M.A., Jepson, M.A., and Hirst, B.H. (2001). Exploiting M cells for drug and vaccine delivery. *Adv. Drug Deliv. Rev.* 50: 81e106.

16 Wang, D., Sun, Y.T., Liu, Y.G. et al. (2018). Clinical translation of immunoliposomes for cancer therapy: recent perspectives. *Expert Opin. Drug Deliv.* 15: 893e903.

17 Bhalani, D.V., Nutan, B., Kumar, A., and Chandel, S.A.K. (2022). Bioavailability enhancement techniques for poorly aqueous soluble drugs and therapeutics. *Biomedicines* 10: 2055.

18 Zhuo, Y., Zhao, Y.G., and Zhang, Y. (2024). Enhancing drug solubility, bioavailability, and targeted therapeutic applications through magnetic nanoparticles. *Molecules* 29: 4854.

19 Liu, Y., Liang, Y., Yuhong, J. et al. (2024). Advances in nanotechnology for enhancing the solubility and bioavailability of poorly soluble drugs. *Drug Des. Devel. Ther.* 18: 1469–1495.

20 Andra, V.V.S.N.L., Pammi, S.V.N., Bhatraju, L.V.K.P., and Ruddaraju, L.K. (2022). A comprehensive review on novel liposomal methodologies, commercial formulations, clinical trials and patents. *BioNanoScience* 12 (1): 274–291.

21 Sharma, M., Joshi, J., Chouhan, N.K. et al. (2020, 2020). Liposome - a comprehensive approach for researchers. In: *Molecular Pharmacolgy* (ed. A. Catala and U. Ahmad). Intech Open eBooks. https://doi.org/10.5772/intechopen.93256.

22 Hamid, N.M.S.S., Hatwar, N.P.R., Bakal, N.R.L., and Kohale, N.N.B. (2024). A comprehensive review on liposomes: as a novel drug delivery system. *GSC Biol. Pharm. Sci.* 27 (1): 199–210.

23 Nsairat, H., Khater, D., Sayed, U. et al. (2022). Liposomes: structure, composition, types, and clinical applications. *Heliyon* 8 (5): e09394.

24 Pande, S. (2023). Liposomes for drug delivery: review of vesicular composition, factors affecting drug release and drug loading in liposomes. *Artif. Cells Nanomed. Biotechnol.* 51 (1): 428–440.

25 Laouini, A., Jaafar-Maalej, C., Limayem-Blouza, I. et al. (2012). Preparation, characterization and applications of liposomes: state of the art. *J. Colloid Sci. Biotechnol.* 1 (2): 147–168.

26 Guimarães, D., Cavaco-Paulo, A., and Nogueira, E. (2021). Design of liposomes as drug delivery system for therapeutic applications. *Int. J. Pharm.* 601: 120571.

27 Sinha, S., Ravi, P.R., Somvanshi, M., and Rashmi, S.R. (2024). Solid lipid nanoparticles for increased oral bioavailability of acalabrutinib in chronic lymphocytic leukaemia. *Discover Nano.* 19: 218.

28 Lombardo, D. and Kiselev, M.A. (2022). Methods of liposomes preparation: Formation and control factors of versatile nanocarriers for biomedical and nanomedicine application. *Pharmaceutics* 14 (3): 543.

29 Nkanga, C.I., Bapolisi, A.M., Okafor, N.I., and Krause, R.W.M. (2019). General perception of liposomes: formation, manufacturing and applications. In: *Liposomes – Advances and Perspectives* (ed. A. Catala). IntechOpen eBooks. https://doi.org/10.5772/intechopen.84255.

30 Mosleh-Shirazi, S., Abbasi, M., Moaddeli, M.R. et al. (2022). Nanotechnology advances in the detection and treatment of cancer: an overview. *Nanotheranostics* 6 (4): 400–423.

31 Abbasi, H., Kouchak, M., Mirveis, Z. et al. (2023). What we need to know about liposomes as drug nanocarriers: an updated review. *Adv. Pharm. Bull.* 13 (1): 7–23.

32 Haran, G., Cohen, R., Bar, L.K., and Barenholz, Y. (1993). Transmembrane ammonium sulfate gradients in liposomes produce efficient and stable entrapment of amphipathic weak bases. *Biochim. Biophys. Acta Biomembr.* 1151 (2): 201–215.

33 Gubernator, J. (2011). Active methods of drug loading into liposomes: recent strategies for stable drug entrapment and increasedin vivo activity. *Expert Opin. Drug Deliv.* 8 (5): 565–580.

34 Gubernator, J., Chwastek, G., Korycińska, M. et al. (2010). The encapsulation of idarubicin within liposomes using the novel EDTA ion gradient method ensures improved drug retention in vitro and in vivo. *J. Control. Release* 146 (1): 68–75.

35 Pauli, G., Tang, W.L., and Li, S.D. (2019). Development and characterization of the solvent-assisted active loading technology (SALT) for liposomal loading of poorly water-soluble compounds. *Pharmaceutics* 11 (9): 465.

36 Almeida, B., Nag, O.K., Rogers, K.E., and Delehanty, J.B. (2020). Recent progress in bioconjugation strategies for liposome-mediated drug delivery. *Molecules* 25 (23): 5672.

37 Fan, Y., Marioli, M., and Zhang, K. (2020). Analytical characterization of liposomes and other lipid nanoparticles for drug delivery. *J. Pharm. Biomed. Anal.* 192: 113642.

38 Yamamoto, E., Hyodo, K., Ohnishi, N. et al. (2011). Direct, simultaneous measurement of liposome-encapsulated and released drugs in plasma by on-line SPE-SPE-HPLC. *J. Chromatogr. B Analyt. Technol. Biomed. Life Sci.* 879 (30): 3620–3625.

39 Ansar, S.M., Jiang, W., and Mudalige, T. (2018). Direct quantification of unencapsulated doxorubicin in liposomal doxorubicin formulations using capillary electrophoresis. *Int. J. Pharm.* 549 (1–2): 109–114.

40 Van Haute, D., Jiang, W., and Mudalige, T. (2019). Evaluation of size-based distribution of drug and excipient in amphotericin B liposomal formulation. *Int. J. Pharm.* 569: 118603.

41 Guimarães, D., Noro, J., Loureiro, A. et al. (2019). Quantification of drugs encapsulated in liposomes by 1H NMR. *Colloids Surf. B Biointerfaces* 179: 414–420.

42 Franzé, S., Selmin, F., Samaritani, E. et al. (2018). Lyophilization of liposomal formulations: still necessary, still challenging. *Pharmaceutics* 10 (3): 139.

43 Hernandez-Caselles, T., Villalain, J., and Gomez-Fernandez, J.C. (1993). Influence of liposome charge and composition on their interaction with human blood serum proteins. *Mol. Cell. Biochem.* 120 (2): 119–126.

44 Thi, T.T.H., Pilkington, E.H., Nguyen, D.H. et al. (2020). The importance of poly(ethylene glycol) alternatives for overcoming PEG immunogenicity in drug delivery and bioconjugation. *Polymers* 12 (2): 298.

45 Shao, X.R., Wei, X.Q., Zhang, S. et al. (2017). Effects of micro-environmental pH of liposome on chemical stability of loaded drug. *Nanoscale Res. Lett.* 12 (1): 504.

46 Huang, S.L., McPherson, D.D., and Macdonald, R.C. (2008). A method to co-encapsulate gas and drugs in liposomes for ultrasound-controlled drug delivery. *Ultrasound Med. Biol.* 34 (8): 1272–1280.

47 Suzuki, R., Tanaka, K., Sawamura, K., and Takizawa, T. (2006). Development of the liposomes entrapped ultrasound imaging gas ("bubble liposomes") as novel gene delivery carriers. *AIP Conf. Proc.* 829: 568–573.

48 Nikolova, M.P., Kumar, E.M., and Chavali, M.S. (2022). Updates on responsive drug delivery based on liposome vehicles for cancer treatment. *Pharmaceutics* 14 (10): 2195.

49 Kheirolomoom, A., Dayton, P.A., Lum, A.F. et al. (2007). Acoustically-active microbubbles conjugated to liposomes: characterization of a proposed drug delivery vehicle. *J. Control. Release* 118 (3): 275–284.

50 Yoon, Y.I., Kwon, Y.S., Cho, H.S. et al. (2014). Ultrasound-mediated gene and drug delivery using a microbubble-liposome particle system. *Theranostics* 4 (11): 1133–1144.

51 Prabhakar, A. and Banerjee, R. (2019). Nanobubble liposome complexes for diagnostic imaging and ultrasound-triggered drug delivery in cancers: a theranostic approach. *ACS Omega* 4 (13): 15567–15580.

52 Juszkiewicz, K., Sikorski, A.F., and Czogalla, A. (2020). Building blocks to design liposomal delivery systems. *Int. J. Mol. Sci.* 21 (24): 9559.

53 Nsairat, H., Khater, D., Sayed, U. et al. (2022). Liposomes: structure, composition, types, and clinical applications. *Heliyon* 8: e09394.

54 Nisini, R., Poerio, N., Mariotti, S. et al. (2018). The multirole of liposomes in therapy and prevention of infectious diseases. *Front. Immunol.* 9: 155.

55 Majidinia, M., Mirza-Aghazadeh-Attari, M., Rahimi, M. et al. (2020). Overcoming multidrug resistance in cancer: recent progress in nanotechnology and new horizons. *IUBMB Life* 72 (5): 855–871.

56 Hortobagyi, G.N., Ueno, N.T., Xia, W. et al. (2001). Cationic liposome-mediated E1A gene transfer to human breast and ovarian cancer cells and its biologic effects: a phase I clinical trial. *J. Clin. Oncol.* 19 (14): 3422–3433.

57 Schwendener, R.A. (2014). Liposomes as vaccine delivery systems: a review of the recent advances. *Ther. Adv. Vaccines.* 2 (6): 159–182.

58 Daraee, H., Etemadi, A., Kouhi, M. et al. (2016). Application of liposomes in medicine and drug delivery. *Artif. Cells Nanomed. Biotechnol.* 44 (1): 381–391.

59 AlSawaftah, N., Pitt, W.G., and Husseini, G.A. (2021). Dual-targeting and stimuli-triggered liposomal drug delivery in cancer treatment. *ACS Pharmacol. Transl. Sci.* 4 (3): 1028–1049.

60 Lee, Y. and Thompson, D.H. (2017). Stimuli-responsive liposomes for drug delivery. *Wiley Interdiscip. Rev. Nanomed. Nanobiotechnol.* 9 (5): doi: 10.1002/wnan.1450.

61 Zhang, S., Xing, M., and Li, B. (2019). Recent advances in musculoskeletal local drug delivery. *Acta Biomater.* 93: 135–151.

62 Heidarli, E., Dadashzadeh, S., and Haeri, A. (2017). State of the art of stimuli-responsive liposomes for cancer therapy. *Iran J. Pharm. Res.* 16 (4): 1273–1304.

63 Tsermentseli, S.K., Kontogiannopoulos, K.N., Papageorgiou, V.P., and Assimopoulou, A.N. (2018). Comparative study of PEGylated and conventional liposomes as carriers for Shikonin. *Fluids* 3 (2): 36.

64 Vaishnav, S., Khairnar, K., and Kothavade, M. (2024). Liposomes: versatile nanocarriers in modern drug delivery systems - a comprehensive review. *Int. J. Pharm. Sci.* 2 (4): 723–734.

65 Liu, P., Chen, G., and Zhang, J. (2022). A review of liposomes as a drug delivery system: current status of approved products, regulatory environments, and future perspectives. *Molecules* 27 (4): 1372.

66 Cattel, L., Ceruti, M., and Dosio, F. (2004). From conventional to stealth liposomes: a new frontier in cancer chemotherapy. *J. Chemother.* 16 (Suppl 4): 94–97.

67 Navarro, G., Cabral, P., Cabrera, M. et al. (2011). 99mTc-labeling and biological evaluation of conventional liposomes. *Alasbimn J.* 13: 51.

68 Cattel, L., Ceruti, M., and Dosio, F. (2003). From conventional to stealth liposomes: a new frontier in cancer chemotherapy. *Tumori* 89 (3): 237–249.

69 Naik, H., Sonju, J.J., Singh, S. et al. (2021). Lipidated peptidomimetic ligand-functionalized HER2 targeted liposome as nano-carrier designed for doxorubicin delivery in cancer therapy. *Pharmaceuticals (Basel)* 14 (3): 221.

70 Espelin, C.W., Leonard, S.C., Geretti, E. et al. (2016). Dual HER2 targeting with trastuzumab and liposomal-encapsulated doxorubicin (MM-302) demonstrates synergistic antitumor activity in breast and gastric cancer. *Cancer Res.* 76 (6): 1517–1527.

71 Chitgupi, U., Qin, Y., Ghosh, S. et al. (2023). Folate-targeted nanoliposomal chemophototherapy. *Pharmaceutics* 15 (10): 2385.

72 Tie, Y., Zheng, H., He, Z. et al. (2020). Targeting folate receptor β positive tumor-associated macrophages in lung cancer with a folate-modified liposomal complex. *Signal Transduct. Target. Ther.* 5 (1): 6.

73 Yuba, E., Fukaya, Y., Yanagihara, S. et al. (2020). Development of mannose-modified carboxylated curdlan-coated liposomes for antigen presenting cell targeted antigen delivery. *Pharmaceutics* 12 (8): 754.

74 Wang, S., Chen, Y., Guo, J., and Huang, Q. (2023). Liposomes for tumor targeted therapy: a review. *Int. J. Mol. Sci.* 24 (3): 2643.

75 Lorenzo, C.A. and Concheiro, A. (2008). Intelligent drug delivery systems: polymeric micelles and hydrogels. *Mini-Rev. Med. Chem.* 8: 1065–1074.

76 Bawa, P., Pillay, V., and du Toit, L.C. (2009). Stimuli-responsive polymers and their applications in drug delivery. *Biomed. Mater.* 4 (2): 1035–1041.

77 Roy, D., Cambre, J.N., and Sumerlin, B.S. (2010). Future perspectives and recent advances in stimuli-responsive materials. *Prog. Polym. Sci.* 3278–3301.

78 Liu, F. and Marek, W. (2010). Recent advances and challenges in designing stimuli responsive polymers. *Urban Progr. Polym. Sci.* 35: 3–23.

79 Chi, A. and Clayton. (2013). Intelligent drug delivery devices based on micro- and nano-technologies. *Ther. Deliv.* 4 (1): 77–94.

80 Vila, A. and Sanchez, A. (2002). Design of biodegradable particles for protein delivery. *J. Control. Release* 78 (1–3): 15–24.

81 Yamanaka, Y.J. and Leong, K.W. (2008). Engineering strategies to enhance nanoparticle mediated oral delivery. *J. Biomater. Sci. Polym.* 19 (12): 1549–1570.

82 Alade, S.L., Brown, R.E., and Paquet, A. (1986). Polysorbate 80 and E-Ferol toxicity. *Pediatrics* 77: 593–597.

83 Kim, S.Y., Shin, I.G., Lee, Y.M. et al. (1998). Methoxy poly (ethylene glycol) and ε-caprolactone amphiphilic block co polymeric micelle containing indomethacin, II. Micelle formation and drug release behaviors. *J. Control. Release* 51: 13–22.

84 Kabanov, A.V. and Alakhov, V.Y. (2002). Pluronic block copolymers in drug delivery: from micellar nanocontainers to biological response modifiers. *Crit. Rev. Ther. Drug* 19: 1–73.

85 Rakshit, A.K. and Palepu, R.M. (2003). Additive effects on nonionic surfactant molecular assemblies: a review. *Recent Dev. Coll. Interface Res.* 1: 203–219.

86 Chu, B. (1991). *Laser Light Scattering. Basic Principles and Practice*, 2e. New York: Academic Press.

87 Zana, R. (2000). Fluorescence studies of amphiphilic block copolymers in solution. In: *Amphiphilic Block Copolymers: Self-Assembly and Applications* (ed. P. Alexandridis and B. Lindman), 191–220. Amsterdam: Elsevier.

88 Jada, A., Siffert, B., and Reiss, G. (1993). Adsorption at the solid-solution inter face and micelle formation in water of a PEO-PS-PEO triblock co polymer. *Colloids Surf. A Physicochem. Eng. Asp.* 75: 203–209.

89 Kumari, A., Singla, R., Guliani, A., and Yadav, S.K. (2014). Nanoencapsulation for drug delivery. *EXCLI J.* 13: 265–286.

90 Abu, L.A.S. and Ishida, T. (2017). Liposomal delivery Systems: design optimization and current applications. *Biol. Pharm. Bull.* 40 (1): 1–10.

91 Ando, H., Kobayashi, S., Abu, L.A.S. et al. (2015). Advanced therapeutic approach for the treatment of malignant pleural mesothelioma via the intrapleural administration of liposomal pemetrexed. *J. Control. Release* 220 (Pt A): 29–36.

92 Lu, L., Ding, Y., Zhang, Y. et al. (2018). Antibody-modified liposomes for tumor-targeting delivery of timosaponin AIII. *Int. J. Nanomedicine* 13: 1927–1944.

93 Abu, L.A.S., Kizuki, S., Doi, Y. et al. (2009). Oxaliplatin encapsulated in PEG-coated cationic liposomes induces significant tumor growth suppression via a dual-targeting approach in a murine solid tumor model. *J. Control. Release* 137 (1): 8–14.

94 Jevsevar, S., Kunstelj, M., and Porekar, V.G. (2010). PEGylation of therapeutic proteins. *Biotechnol. J.* 5 (1): 113–128.

95 Boissenot, T., Fattal, E., Bordat, A. et al. (2016). Paclitaxel-loaded PEGylated nanocapsules of perfluorooctyl bromide as theranostic agents. *Eur. J. Pharm. Biopharm.* 108: 136–144.

96 Pasut, G., Paolino, D., and Celia, C. (2015). Polyethylene glycol (PEG)-dendron phospholipids as innovative constructs for the preparation of super stealth liposomes for anticancer therapy. *J. Control. Release* 199: 106–113.

97 Suk, J.S., Xu, Q., and Kim, N. (2016). PEGylation as a strategy for improving nanoparticle-based drug and gene delivery. *Adv. Drug Deliv. Rev.* 99 (Pt A): 28–51.

98 Pisal, D.S., Kosloski, M.P., and Balu-Iyer, S.V. (2010). Delivery of therapeutic proteins. *J. Pharm. Sci.* 99 (6): 2557–2575.

99 Matthews, S.J. and McCoy, C. (2004). Peginterferon alfa-2a: a review of approved and investigational uses. *Clin. Ther.* 26 (7): 991–1025.

100 Veronese, F.M. and Pasut, G. (2005). PEGylation, successful approach to drug delivery. *Drug Discov. Today* 10: 1451–1458.

101 Croy, S.R. and Kwon, G.S. (2006). Polymeric micelles for drug delivery. *Curr. Pharm. Des.* 12: 4669–4684.

102 Wu, H., Zhu, L., and Torchilin, V.P. (2013). pH-sensitive poly(histidine)-PEG/DSPE-PEG co-polymer micelles for cytosolic drug delivery. *Biomaterials* 34: 1213–1222.

103 Remsberg, C.M., Zhao, Y., and Takemoto, J.K. (2012). Pharmacokinetic evaluation of a DSPE-PEG2000 micellar formulation of ridaforolimus in rat. *Pharmaceutics* 5: 81–93.

104 Ragi, C., Sedaghat-Herati, M.R., Ouameur, A.A., and Tajmir-Riahi, H.A. (2005). The effects of poly(ethylene glycol) on the solution structure of human serum albumin. *Biopolymers* 78: 231–236.

105 Liu, J., Lee, H., and Allen, C. (2006). Formulation of drugs in block copolymer micelles: drug loading and release. *Curr. Pharm. Des.* 12: 4685–4701.

106 Kataoka, K., Harada, A., and Nagasaki, Y. (2001). Block copolymer micelles for drug delivery: design, characterization and biological significance. *Adv. Drug Deliv. Rev.* 47: 113–131.

107 Sandvig, K., Pust, S., Skotland, T., and Van Deurs, B. (2011). Clathrin-independent endocytosis: mechanisms and function. *Curr. Opin. Cell Biol.* 23: 413–420.

108 Sawant, R.R. and Torchilin, V.P. (2010). Multifunctionality of lipid-core micelles for drug delivery and tumour targeting. *Mol. Membr. Biol.* 27: 232–246.

109 Wang, T., Petrenko, V.A., and Torchilin, V.P. (2010). Paclitaxel-loaded polymeric micelles modified with MCF-7 cell-specific phage protein: enhanced binding to target cancer cells and increased cytotoxicity. *Mol. Pharm.* 7: 1007–1014.

110 Lukyanov, A.N. and Torchilin, V.P. (2004). Micelles from lipid derivatives of water-soluble polymers as delivery systems for poorly soluble drugs. *Adv. Drug Deliv. Rev.* 56: 1273–1289.

111 Oerlemans, C., Bult, W., Bos, M. et al. (2010). Polymeric micelles in anti-cancer therapy: Targeting, imaging and triggered release. *Pharm. Res.* 27 (12): 2569–2589.

112 Aliabadi, H.M. and Lavasanifar, A. (2006). Polymeric micelles for drug delivery. *Expert Opin. Drug Deliv.* 3 (1): 139–162.

113 Chung, J.E., Yokoyama, M., Yamato, M. et al. (1999). Thermo-responsive drug delivery from polymeric micelles constructed using block copolymers of poly(N-isopropylacrylamide) and poly(butylmethacrylate). *J. Control. Release* 62 (1-2): 115–127.

114 Kanayama, N., Fukushima, S., Nishiyama, N. et al. (2006). A PEG-based biocompatible block catiomer with high buffering capacity for the construction of polyplex micelles showing efficient gene transfer toward primary cells. *Chem. Med. Chem.* 1 (4): 439–444.

115 Satturwar, P., Eddine, M.N., Ravenelle, F., and Leroux, J.C. (2007). pH-responsive polymeric micelles of poly (ethylene glycol) b-poly(alkyl(meth) acrylate-co-methacrylic acid): influence of the copolymer composition on self-assembling properties and release of candesartan cilexetil. *Eur. J. Pharm. Biopharm.* 65: 379–387.

116 Rijcken, C.J., Soga, O., Hennink, W.E., and van Nostrum, C.F. (2007). Triggered destabilisation of polymeric micelles and vesicles by changing polymers polarity: an attractive tool for drug delivery. *J. Control. Release* 120 (3): 131–148.

117 Lee, E.S., Na, K., and Bae, Y.H. (2003). Polymeric micelle for tumor pH and folate-mediated targeting. *J. Control. Release* 91: 103–113.

118 Zhang, H., Zhao, L., Chu, L. et al. (2014). Preparation, optimization, characterization and cytotoxicity in vitro of Baicalin-loaded mixed micelles. *J. Colloid Interface Sci.* 434: 40–47.

119 Hu, Z., Luo, F., Pan, Y. et al. (2008). Arg-Gly-Asp (RGD) peptide conjugated poly(lactic acid)-poly(ethylene oxide) micelle for targeted drug delivery. *J. Biomed. Mater. Res. A* 85 (3): 797–807.

120 Chiappetta, D.A., Facorro, G., Rubin de Celis, E., and Sosnik, A. (2011). Synergistic encapsulation of the anti-HIV agent efavirenz within mixed poloxamine/poloxamer polymeric micelles. *Nanomedicine* 7: 624–637.

121 Alexandridis, P. and Hatton, T.A. (1995). Poly (ethylene oxide)-poly (propylene oxide)-poly (ethylene oxide) block copolymer surfactants in aqueous solutions and at interfaces: thermodynamics, structure, dynamics, and modeling. *Colloids Surf. A Physicochem. Eng. Asp.* 96: 1–46.

122 Rub, M.A., Azuma, N., Kumar, D. et al. (2014). Micellization and microstructural studies between amphiphilic drug ibuprofen with non-ionic surfactant in aqueous urea solution. *J. Chem. Thermodyn.* 74: 91–102.

123 Moretton, M.A., Taira, C., Flor, S. et al. (2014). Novel nelfinavir mesylate loaded d-α-tocopheryl polyethylene glycol 1000 succinate micelles for enhanced pediatric anti-HIV therapy: in vitro characterization and in vivo evaluation. *Colloids Surf. B Biointerfaces* 123: 302–310.

124 Basu, R.G., Chakraborty, I., and Moulik, S.P. (2006). Pyrene absorption can be a convenient polarity. method for probing critical micellar concentration (cmc) and indexing micellar. *J. Colloid Interface Sci.* 294: 248–254.

125 Kalyanasundaram, K. and Thomas, J.K. (1977). Environmental effects on vibronic band intensities in pyrene monomer fluorescence and their application in studies of micellar systems. *J. Am. Chem. Soc.* 99: 2039–2044.

126 Lindman, B. and Wennerström, H. (1991). Nonionic micelles grow with increasing temperature. *J. Phys. Chem.* 95: 6053–6054.

127 Vora, S., George, A., Desai, H., and Bahadur, P. (1999). Mixed micelles of some anionic anionic, cationic-cationic, and ionic-nonionic surfactants in aqueous media. *J. Surfactant Deterg.* 2: 213–221.

128 Xiuli, L., Jian, X., Wanguo, H., and Dejun, S. (2004). Effect of additives on the cloud points of two tri-block copolymers in aqueous solution. *Colloids Surf. A Physicochem. Eng. Asp.* 237: 1–6.

129 Parnham, M.J. and Wetzig, H. (1993). Toxicity screening of liposomes. *Chem. Phys. Lipids* 64 (1–3): 263–274.

130 Hossen, S., Hossain, M.K., Basher, M.K. et al. (2019). Smart nanocarrier-based drug delivery systems for cancer therapy and toxicity studies: a review. *J. Adv. Res.* 15: 1–18.

131 Selim, A., Lila, A., and Ishida, T. (2017). Liposomal delivery systems: design optimization and current applications. *Biol. Pharm. Bull.* 40 (1): 1–10.

132 Wakaskar, R.R. (2018). General overview of lipid–polymer hybrid nanoparticles, dendrimers, micelles, liposomes, spongosomes and cubosomes. *J. Drug Target.* 26 (4): 311–318.

133 Gregoriadis, G. (2016). Liposomes in drug delivery: how it all happened. *Pharmaceutics* 8 (2): 19.

134 Forssen, E. and Willis, M. (1998). Ligand-targeted liposomes. *Adv. Drug Deliv. Rev.* 29 (3): 249–271.

135 Torchilin, V.P. (2012). Multifunctional nanocarriers. *Adv. Drug Deliv. Rev.* 64 (SUPPL): 302–315.

136 Lestini, B.J., Sagnella, S.M., Xu, Z. et al. (2002). Surface modification of liposomes for selective cell targeting in cardiovascular drug delivery. *J. Control. Release* 78 (1-3): 235–247.

137 Lian, T. and Ho, R.J.Y. (2001). Trends and developments in liposome drug delivery systems. *J. Pharm. Sci.* 90 (6): 667–680.

138 Zeng, L. and Wu, X. (2010). Modeling the sustained release of lipophilic drugs from liposomes. *Appl. Phys. Lett.* 97 (7): 073701.

139 Sadzuka, Y., Hirama, R., and Sonobe, T. (2002). Effects of intraperitoneal administration of liposomes and methods of preparing liposomes for local therapy. *Toxicol. Lett.* 126 (2): 83–90.

140 Kim, E.M. and Jeong, H.J. (2021). Liposomes: biomedical applications. *Chonnam Med. J.* 57 (1): 27.

141 Song, G., Wu, H., Yoshino, K., and Zamboni, W.C. (2012). Factors affecting the pharmacokinetics and pharmacodynamics of liposomal drugs. *J. Liposome Res.* 22 (3): 177–192.

142 Sercombe, L., Veerati, T., Moheimani, F. et al. (2015). Advances and challenges of liposome assisted drug delivery. *Front. Pharmacol.* 6: 163819.

143 Sawant, R.R. and Torchilin, V.P. (2012). Challenges in development of targeted liposomal therapeutics. *AAPS J.* 14 (2): 303–315.

144 Zylberberg, C. and Matosevic, S. (2016). Pharmaceutical liposomal drug delivery: a review of new delivery systems and a look at the regulatory landscape. *Drug Deliv.* 23 (9): 3319–3329.

145 Jensen, G.M. and Hodgson, D.F. (2020). Opportunities and challenges in commercial pharmaceutical liposome applications. *Adv. Drug Deliv. Rev.* 154–155: 2–12.

146 Wang, Y. and Grainger, D.W. (2022). Regulatory considerations specific to liposome drug development as complex drug products. *Front. Drug Deliv.* 2: 901281.

147 Schlich, M., Musazzi, U.M., Campani, V. et al. (2022). Design and development of topical liposomal formulations in a regulatory perspective. *Drug Deliv. Transl. Res.* 12 (8): 1811–1828.

148 International Conference on Harmonisation (ICH), *Stability Testing of New Drug Substances and Products Q1A(R2)*, ICH Harmonised Tripartite Guideline, 2003.

149 Gregoriadis, G. and Florence, A.T. (1993). Liposomes in drug delivery. Clinical, diagnostic and ophthalmic potential. *Drugs* 45 (1): 15–28.

150 Al-Jamal, W.T. and Kostarelos, K. (2011). Liposomes: from a clinically established drug delivery system to a nanoparticle platform for theranostic nanomedicine. *Acc. Chem. Res.* 44: 1094–1104.

151 Lamichhane, N., Udayakumar, T.S., D'Souza, W.D. et al. (2018). Liposomes: clinical applications and potential for image-guided drug delivery. *Molecules* 23 (2): 288.

152 Alavi, M. and Hamidi, M. (2019). Passive and active targeting in cancer therapy by liposomes and lipid nanoparticles. *Drug Metab. Pers. Ther.* 34 (1): 0032.

153 Pisani, S., Di Martino, D., Cerri, S. et al. (2023). Investigation and comparison of active and passive encapsulation methods for loading proteins into liposomes. *Int. J. Mol. Sci.* 24 (17): 13542.

154 Das, R.P., Gandhi, V.V., Singh, B.G., and Kunwar, A. (2019). Passive and active drug targeting: role of nanocarriers in rational design of anticancer formulations. *Curr. Pharm. Des.* 25 (28): 3034–3056.

155 Xia, Y., Xu, C., Zhang, X. et al. (2019). Liposome-based probes for molecular imaging: from basic research to the bedside. *Nanoscale [Internet]* 11 (13): 5822–5838.

156 Xing, H., Hwang, K., and Lu, Y. (2016). Recent developments of liposomes as nanocarriers for theranostic applications. *Theranostics* 6 (9): 1336–1352.

157 Li, M., Du, C., Guo, N. et al. (2019). Composition design and medical application of liposomes. *Eur. J. Med. Chem.* 164: 640 653.

158 Qiu, N., Kong, F., Ye, H., et al. (2023). Co-loaded liposome and preparation method thereof. Chinese Patent No. CN116270473B.

159 Smriti, A., Gao, X., and Huang, J. (2022). mRNA-liposome complex and application thereof. Chinese Patent No. CN112107680B.

160 Wang Yongjian, Yu Ao, Wang Xinnan, Sun Dandan, Lyu Xueming , (UNIV NANKAI) (2017) A kind of pH sensitive mixed micelles and its preparation method and application. https://pubchem.ncbi.nlm.nih.gov/patent/CN-107441043-A (accessed 20 January 2025).

161 Thijs ROOIMANS, (Tiofarma B.V) – 2018 Aqueous suspension comprising mixed micelles loaded with a lipophilic physiologically active substance. https://patentimages.storage.googleapis.com/6e/d1/c1/9557730f54a64d/EP3391909A1.pdf (accessed 15 January 2025).

162 Telange, D. R., Pethe, A. M., Telrandhe, U. B., Bobde, P. S., & Ganorkar, S. B. (2024). Quercetin-loaded pluronicâ®f127/chitosan myristic acid mixed micelles to improve the biopharmaceutical and hepatoprotective potential of quercetin. https://www.rcpatelpharmacy.co.in/files/patent/2024-25/patent1-070125.pdf (accessed 15 January 2025).

7

Improvement of Solubility and Bioavailability of Poorly Soluble Drugs Using Mesoporous Silica Particles

Jitendra Wankar[1,2,3] and Sonia Gera[4]

[1]*Division of Gastroenterology, Brigham and Women's Hospital, Harvard Medical School, 65 Landsdowne Street, Cambridge, Boston, MA 02139, USA*
[2]*The Laboratory for Translational Engineering, Broad Institute of MIT and Harvard, Stanley Building, 75 Ames St , Cambridge, Boston, MA 02142, USA*
[3]*UCD School of Veterinary Medicine, University College Dublin, Belfield, Dublin, D04 V1W8, Ireland*
[4]*Department of Materials, Imperial College London, South Kensington Campus, London SW7 2AZ, UK*

7.1 Introduction

To make an active pharmaceutical agent (API) into a commercially viable pharmaceutical product for human use, it needs to be solubilized in body fluids like gastric or intestinal fluid in its molecular form for drug diffusion across the gastrointestinal barriers. Several strategies have been utilized to improve the solubility profile of the API, for example, solubilization with cyclodextrin (CD), particle size reduction by micro-ionization, suspension, and solid dispersions [1]. The discovery of mesoporous silica has opened another door to improve the solubilization of pharmaceutical drugs. The unique feature of silica consists of a higher surface area, tuning its particle and pore size, and surface modification for targetability. Additionally, silica is biocompatible and biodegradable when presented in nanosized form. Chemically, silica is available as silicon dioxide (SiO_2), and its widely known physical forms are sand and quartz. At the molecular level, it exists in two forms—crystalline and amorphous forms. Historically, crystalline silica has been used to produce glass and concrete. In the modern world, it has been used to fabricate heat-resistant refractory materials and ceramics. Structurally, crystalline silica is arranged as a tetrahedral network consisting of a centrally located Si atom coordinating with four O atoms, and each O atom bonds with two Si atoms to form a Si—O—Si linkage representing the siloxane functional group. This SiO_4 tetrahedron is the main building block of silica, and its arrangement and connecting linkage (Si—O—Si) decide its characteristics [2]. The SiO_4 tetrahedron arrangement and its density, when packed in a periodic manner, exhibit glassy or quartz-like structures, and amorphous when irregularly arranged. The presence of +4 and −2 charges on silica and oxygen makes their structural forms overall neutral in nature. Therefore, a significant difference has been found in the structural feature of the tetrahedron pore size when compared to its bulk form. Surface groups on silica, siloxane (Si—O—Si), and silanol (SiOH) are primarily exposed, providing surface-modifiable functionalities that make them useful materials for pharmaceutical and biomedical engineering. Mesoporous silica nanoparticles (NPs) with pore sizes in the range of 2–50 nm are ideal for drug delivery and biomedical applications. The surface functionalization of mesoporous silica particles (MSNs) instilled additional attractive features to them, for example, targeted drug delivery when conjugated with targeting ligands, controlled release of drugs by gatekeeper functionalities, or making them stealthy by PEGylation to prevent opsonization [3]. The modification of the surface properties of silica is facilitated by its hydrophilic Si—OH groups, which helps improve the drug stability from enzymatic degradation in the gut and fabricate it into stimulus-responsive drug delivery system [4]. The safety of the silica particle remains debatable; however, its modification can improve its biosafety profile.

7.2 Different Techniques for Synthesis of Silica Particles

Various methods have been used to synthesize silica particles, which include top-down and bottom-up approaches. The top-down approach utilizes physical energy such as heat and mechanical energy to reduce particle size, such as pulverization and ball milling, while the bottom-up approach uses the atomic or molecular approach, for example, sol–gel, microemulsion, and flame synthesis. The sol–gel or Stöber method is the most widely used approach to form uniform, narrow-sized distribution particles with controlled morphology, using optimized parameters. The synthesis methods used for the particle preparation determine the structure, morphology, size, physicochemical properties, and functional performance of these particles.

7.2.1 Sol–Gel Method

The sol–gel method is a widely used method for the synthesis of silica nanoparticles, which involves hydrolysis and condensation of silica precursors such as silicon alkoxide, tetraethyl orthosilicate (TEOS), in a solution [5]. Catalytic conditions of acid or base facilitate the reaction of these precursors and convert its sol-based colloidal suspension into solid network of gel within a liquid. The typical sol–gel process involves the following steps: hydrolysis and condensation, aging, drying, and calcination. The hydrolysis of TEOS creates silanol bridges and condensation/polymerization between silanol or ethoxy groups, creating siloxane bridges (Si—O—Si) to form a silica structure [6]. The size and properties of the resulting silica nanoparticles can be controlled by adjusting factors such as reactant concentration, reaction temperature, solvent choice, pH, and drying techniques.

7.2.2 Microemulsion Method

Microemulsion methods provide invaluable tools to synthesize silica nanoparticles with controlled size and shape and act as a template for the sol–gel process. In a

Table 7.1 Types of templates and their examples reported in the literature.

Nonionic surfactants	• Poly (ethylene glycol)-block-poly (propylene glycol)-block-poly (ethylene glycol) (P123 copolymer) • Poly (ethylene oxide)-poly (propylene oxide)-poly (ethylene oxide) (Pluronic F-127)
Cationic surfactants	• Cetyltrimethylammonium bromide (CTABr) • Cetyltriethylammonium bromide (CTEABr)
Anionic surfactants	• Sodium dodecyl benzenesulfonate (SDBS) • Sodium dodecyl sulfate (SDS)

water-in-oil (W/O) microemulsion [7], the water droplets serve as confined reaction sites where silica precursors undergo hydrolysis and condensation. This method allows for the production of functionalized, core–shell, or hollow silica structures. Standard composition of this method includes cyclohexane as continuous phase/oil phase, alkylphenol ethoxylate as surfactant, water in droplet form, and TEOS as silica precursor. The sol–gel conversion process is induced by the addition of an aqueous ammonium hydroxide, which acts as base catalyst; in many cases, medium-chain alkyl alcohol (pentanol or hexanol) is added as cosurfactant [8].

7.2.3 Template-Assisted Synthesis

To create the MSNs, the key elements needed include a template and a silica precursor. The template and its removal from the solid structure of the silica particles direct the formation of pores inside them [9]. Several templates have been investigated. In literature, MSNs are synthesized by using different templates—surfactant (cationic, anionic, and nonionic) and ionic liquids (ILs), shown in Table 7.1.

These templates, when dissolved in water at certain concentrations, form micelles/nano-assembly. The added silica precursor (TEOS) in the above micelles, under controlled temperature conditions, undergoes condensation and forms a solid shell around the micelles arranged in spherical/cylindrical shape. To create pores in solid silica particles, the micelle-based template must be removed, thereby forming mesoporous structures in the silica matrix. Therefore, these micelles (template) are generally removed by conventional thermal treatment—calcination, where these particles are heated at high temperature, up to 550 °C under controlled air conditions. The synthesis scheme is shown in Figure 7.1.

This is the most effective method for complete removal of the template; however, it suffers from some limitations, such as partial destruction of ordered structure, shrinkage of the framework, reduced number of silanol groups on pores affecting post-modification, contaminant-generated high amounts of CO_2, organic amine waste, and carbon deposits. Moreover, it does not allow for recycling of the expensive organic template used in silica particle synthesis. Therefore, several novel methods have been investigated for the removal of the template from the solid silica matrix to overcome the obstacle in the conventional calcination process. In literature, these methods are categorized into two types: (i) physical method and (ii) chemical method [10], reported in Table 7.2. The physical methods include calcination at

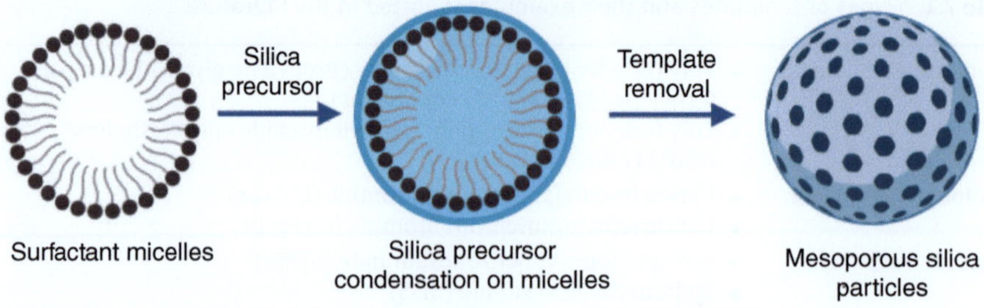

Figure 7.1 Schematic presentation of the synthesis of mesoporous silica particles.

Table 7.2 Summary of the methods used for template removal.

Physical method	Chemical method
Calcination	Solvent extraction
Supercritical SC–CO_2 extraction	Chemical oxidation
Microwave-assisted treatment	Ionic liquid treatment
Ultrasonic-assisted treatment	
Ozone treatment	
Plasma technology	

Source: Adapted from Ghaedi and Zhao [9].

high temperature (550 °C). Therefore, it suffers from challenges such as framework shrinkage, network collapse, carbon deposits, and the need for removal of the organic functionalities and recovery of the expensive organic templates. The chemical methods involve the utilization of alkaline or acidic solvents, which are effective but corrosive in nature. The choice of method depends on the desired material characteristics. Among them, methods such as microwave digestion and IL approach are quick and efficient approaches, which preserve the silica framework, minimize the silicon group loss, and are environmentally friendly. Moreover, these methods are economically viable as compared to conventional calcination methods.

7.3 Characterization of MSN

7.3.1 Particle, Pore Size, Their Distribution, and Morphology

The colloidal size distribution of these MSNs can be tested by dynamic light scattering (DLS) when they are in nanosized form with a narrow-sized distribution. The particles are slightly larger in size, around 200 nm, with wide size distribution that tend to settle in the measuring cuvette of the DLS. The obtained colloidal size measurement can be inconsistent. Therefore, precise size, shape, and morphological features of the MSNs can be tested by advanced microscopic techniques such as transmission electron microscopy (TEM) and cryo/field emission surface electron microscopy (SEM). The pore size and pore size distribution of these particles can be determined using nitrogen adsorption–desorption analysis, which provides results derived from models such as the BJH (Barrett–Joyner–Halenda), NLDFT (nonlocal density functional theory), and Bhuiyan (BET) theory for analyzing surface area, pore volume, and pore size distribution. The morphology of MSNs can control biological function, performance, and biodistribution of these particles. The synthesis method and conditions can influence the shape of silica particles from the most common sphere to rods, shells, hollow, oval shape, etc. [11]. The amount of the catalyst, the temperature for template removal, the surfactant, and the TEOS concentration can affect the shape. For example, tuning the sol–gel method for rod-shaped particles by adjusting the cosolvent ethanol and low surfactant concentration has been reported to affect morphology. Furthermore, controlling the aspect ratio of MSNs offers significant advantages for enhancing drug oral absorption. For instance, rod-shaped particles can control permeation across the mucous barrier, resulting in higher fenofibrate absorption, and they have been shown to affect cellular uptake and permeability coefficients and improve the oral bioavailability (BA) of drugs [12].

7.3.2 Crystallinity and Nature

The chemical characterization of silica is mainly conducted by X-ray diffraction (XRD) spectroscopy. The solid-state properties of silica such as pore size and its surface features impact the incorporation of the drug molecules. The molecular form of drugs presented in either crystalline or amorphous form in mesoporous silica can alter the dissolution and, subsequently, drug BA, and XRD is a powerful tool to detect these solid-state properties. Specifically, the XRD reveals the amorphous or crystalline nature of the silica and incorporated drug molecules, its mesoporous structural feature, and the presence of sharp diffraction peaks in the case of being doped with other inorganic nanoparticles. The broad hump at 15 and 30 θ indicates the amorphous silica, while sharp peaks at 3° 2θ values depict the crystalline nature. The peak position parameters determined from low-angle X-ray scattering (LAXS) can be used to calculate pore size, while the area under the LAXS profile shows a correlation with the size of MSNs. LAXS has several advantages over TEM for determining the size of MSNs in the presence of mesopores, which not only estimates their average particle size but also informs their pore size distribution [13].

7.4 Mesoporous Silica for Drug Dissolution and Dosage Form Development

Mesoporous silica is used as a carrier for active pharmaceutical ingredients to enhance their dissolution following oral administration. Along with that, it is widely used as glidant to enhance the flow properties of the powders, disintegrant, and stabilizer in conventional oral dosage forms such as tablets/capsules, as safe for human use and commercially available excipient approved by the Food and Drug

Administration (FDA) (Table 7.3) [24]. More precisely, it is approved as a stabilizer to develop a solid dispersion of itraconazole [25].

One promising strategy to enhance the solubility and stability of drugs is to develop them into amorphous formulations. The porous carriers, such as mesoporous silica, are the carriers that increase the solubility by adsorption of drugs into the pore structure [25]. Mesoporous silica improves the dissolution of poorly soluble drugs due to the presence of silanol groups for non-covalent interactions such as hydrogen bonding, large surface area, and porous architecture. The pore size, pore volume, and surface area primarily control drug release from these materials [26]. The drug molecules are incorporated within the silica porous structure in a noncrystalline/amorphous form, depending on the interaction between the drug molecule and the silica pore. This porous silica structure prevents the recrystallization of drugs, and the amorphous form of the drug has ideally higher solubility and dissolution rate [27]. The size and shape of the pores are critical to control the drug loading and drug release. Additionally, surface modification of silica can affect the solubility and dissolution of the drugs. Different manufacturing conditions, such as temperature and surfactant concentration, yield ordered and non-ordered silica materials. The ordered architecture of silica material has defined pore size, shape, and volume. Typical examples of highly ordered silica materials include MCM-41 (mobile crystalline material), which is obtained through hydrothermal synthesis and templating [28]. These have a defined, ordered structure at specific lengths, creating hexagonal channels identified by surface morphology and XRD patterns. It has higher thermal, hydrothermal, and hydrolytic properties. MCM-48 is the cubic form of MCM-41, with a space group that allows molecular accessibility, making it suitable for controlled-release formulations [29]. While the Santa Barbara Amorphous form SBA-15 has a larger pore size and facilitates the faster release of the drug [30].

7.5 Drug-Loading Methods in Mesoporous Silica Particles to Improve Drug Solubility

MSNs have gained attention as a drug delivery carrier in the pharmaceutical field. Spherical shape with enormous surface area and pores within the particles allows the molecular encapsulation of the drug, which ultimately improves the water solubility of hydrophobic drugs and improves their BA [31]. Several methods have been investigated for drug encapsulation; among them, solvent evaporation remains a widely used method as described below. The drug encapsulation process involves adsorption of the drug molecules in mesoporous silica inner/outer surfaces under conditions that favor drug–carrier interactions, such as van der Waals interactions, hydrogen bonding, electrostatic binding, or covalent bonding. An ideal drug-loading process should quickly load a large amount of the drug and then unload it in biorelevant media with the desired release profile. It is also desirable to minimize the fraction of the drug adsorbed on the external particle surface to avoid the possibility of its crystallization due to the lack of (pores) entrapment effects. The extent of drug loading depends on several parameters. The most significant factors are the surface area and the drug's affinity for the silica substrate.

Table 7.3 Use of silica nanoparticles in different dosage forms.

Dosage form	Types of silica and properties	Function of silica	Purpose and example
		Oral dosage form	
Tablets	Aerosil®, Syloid®, and MCM-41	Drug carrier, dissolution, and absorption of drugs/immediate or controlled release of drugs	Hydrophobic drug examples: Glimepiride [14] and celecoxib silica-loaded tablets [15]
Oral disintegrating tablets ODTs	Syloid® 244 FP, Syloid® XDP (silicon dioxide), Syloid® AL1 FP Porous (meso- and macropores), surface area 300–700 m^2/g	Enhanced/immediate dissolution in the mouth/saliva	Faster onset of action ODTs of celecoxib, itraconazole, and meloxicam loaded in Syloid 244 FP for improved solubility [16]
Capsules	MCM-41, SBA-15	Drug-loaded silica in the filled gelatin capsules	Poorly soluble drugs [17]
Suspensions/ solid dispersions	MCM-41, SBA-15, UVM-7 Ordered mesoporous (pore size 2–10 nm), high surface area (600–1000 m^2/g)	Stabilizes the hydrophobic drug in amorphous form	Carrier for amorphous solid dispersions Controls release and enhances dissolution May improve taste mask Improves the oral bioavailability of drugs
Powders/ granules	1. Aerosil® 200, Cab-O-Sil® M5 Nonporous, nanosized (~7–40 nm), surface area 200–400 m^2/g 2. Neusilin® US2, Neusilin® UFL2 Porous granular form	Excipient for stability, powder flow properties, compressibility, and absorption in liquid dosage	1. Dosage form development and reconstitution used in *captopril*, *cilnidipine*, *glimepiride* ODTs to improve blend flow and tablet hardness 2. Glimepiride, domperidone, ibuprofen ODTs prepared by melt adsorption on Neusilin US2

(Continued)

Table 7.3 (Continued)

Dosage form	Types of silica and properties	Function of silica	Purpose and example
		Injectable dosage form	
Hydrogels	Mesoporous silica (MSNs, SBA-15) 2–50 nm pores	Drug carrier in a polymer network	Sustained drug or peptide release [18, 19]
Suspensions	MCM-48 silica spheres, mesoporous silica colloidal/fumed silica 10–200 nm	Silica nanoparticles can improve solubility of hydrophobic drugs for parenteral administration	Solubility of anticancer drugs [20, 21]
		Topical/transdermal	
Creams/ gels/ ointments	Colloidal/amorphous silica	Solubility/ bioavailability, stability	Sunscreens, anti-inflammatory [22]
		Pulmonary/inhalation	
Dry powder inhalers	Colloidal/amorphous silica (Aerosil®, fumed silica, or surface-modified)	Stabilizes the drugs/or improves the aerosolization	Inhaled performance of corticosteroids and hydrophobic drugs [23]

7.5.1 Solvent-Based Method

In the solvent-based methods, the internal pore volume of the silica, polarity of the solvent, and concentration of the drug in the solution are critical for drug encapsulation [32]. The choice of drug-loading method affects the degree of drug loading, the distribution of the drug in silica, and the physicochemical properties of the loaded drug. Currently, the following methods have been investigated in the literature.

(i) **Adsorption method:** In this method, MSNs are incubated with concentrated drug–organic solvent mixture, and therefore the drug is adsorbed inside the pore walls and at the surface of the silica. The drug-adsorbed silica particles are further isolated by filtration or centrifugation, and finished particles are dried at optimum temperature condition (40 °C). This method is suitable for hydrophilic, hydrophobic drugs, and thermosensitive protein-based drug molecules. Critical parameter that can affect drug encapsulation using this method is solubility of the drug in organic solvent. Additionally, it suffers from some limitations, such as insufficient drug loading caused by blocking of the pores, time-consuming, variability in drug loading, and wastage of the drug during filtration and the centrifugation method. Also, the polarity of the solvent creates a competitive environment with the molecular hydrophilic environment of the MSNs, which can limit drug–silica interaction required for drug encapsulation [33].

(ii) **Incipient wetness impregnation method:** In this method, drug loading is driven by capillary diffusion forces within the elongated, capillary-sized pores of mesoporous silica. A precisely measured volume of highly concentrated drug solution is slowly added dropwise to the silica particles, allowing the drug to be absorbed effectively into the pores. Any excess drug is removed through a washing step, and the resulting powder is dried under optimized conditions to obtain the final drug-loaded silica particles. Critical parameters to be considered include the amount of concentrated drug solution, pore properties, and washing steps to eliminate excess drug, which affect the final drug-loading efficiency. This method requires only a small volume and is widely used in the catalysis field, but has been explored for loading expensive drugs into silica particles [34].

(iii) **Solvent evaporation:** It is one of the most frequently used methods for drug loading in MSNs, which involves rapid evaporation of the volatile organic solvent used to dissolve the drug and the silica particles, allowing the drug to adsorb within the particles' pores and on their surfaces. Evaporation is carried out under controlled temperature and pressure conditions using a rotary evaporator. Critical parameters to be considered are the type of volatile solvent and the rate of evaporation, which can affect the localization of the drug within pores and surface, its physical state (crystalline or amorphous), and finally its release profile from the particles [35].

(iv) **DiSupLo (Diffusion Supported Loading):** It is a recently developed technique for drug loading into MSNs, which involves the preparation of a homogenized physical mixture of silica and drug molecules or multiple components (drugs/silica) in a closed vessel containing an organic solvent (e.g. ethanol). The diffusing vapors of the organic solvent, upon condensation, dissolved the drugs, facilitating their transport within the pores of the silica particles. It is the easiest and fastest drug-loading method among reported techniques, which doesn't require specialized equipment or controlled conditions. Multiple drug loadings and the highest drug-loading efficiency have been achieved with minimum solvent, making it economical and environmentally friendly [36]

(v) **Supercritical fluid technology (SCF):** This technology can be used to load drugs into silica particles, offering several advantages over traditional methods. SCFs, particularly supercritical carbon dioxide ($SCCO_2$, 31.2 °C, 7.4 MPa), can dissolve drugs and be easily removed by depressurization, leaving the drugs incorporated into the silica matrix. This technique is beneficial for loading poorly water-soluble drugs into MSNs, improving their solubility and dissolution rate. Moreover, $SCCO_2$ is environmentally and cost-friendly due to its inertness, non-flammability, and non-toxicity. Additionally, it is superior to any other method for not using toxic organic cosolvent, and no residue will be left in the final drug-loaded silica product [37].

(vi) **One-pot drug-loading and synthesis procedure:** The conventional silica particle synthesis involves a sol–gel multistep method of using a micellar template and their removal by calcination to create the porous structure in silica. However, in this method, the drug, alone or in combination with surfactant or another polymer, forms the nano-assembly, which acts as a codirecting agent for the silica precursor (TEOS) to form a silica container by self-condensation around these nano-assembly. The obtained particles are washed to remove the unreacted silica precursor by ultracentrifugation. Therefore, it is one-step, one-pot drug loading while synthesizing silica particles with maximum drug loading possibility. Moreover, it doesn't involve expensive equipment or harsh experimental conditions, but it is time-consuming because it involves the slow process of self-condensation of the silica precursor over the drug–surfactant/polymer assembly [38].

(vii) **Covalent grafting:** This method involves covalent linking of the drug molecules to functional groups present on MSNs by click or conjugation chemistry of the amide, ester, disulfide, carbamate, and thiol. It is particularly used to alter drug release/stability, such as controlled release and protection of the drug from the harsh gut environment, making it stimulus-responsive drug delivery (enzyme/pH) or target-specific drug delivery. This method is associated with some disadvantages, such as the formation of inactive drug forms when bonded with silica, insufficient functional groups on the silica surface (synthetic method-based variability), and surface charge and steric hindrance that can negatively impact the covalent linking of the drug molecules [39].

7.5.2 Solvent-Free Methods

Compared to wet methods, solvent-free methods are less time-consuming and can achieve a high degree of drug loading. Moreover, the drug concentration in the mesoporous material used is readily predictable, as it is directly influenced by the API-to-MSN ratio. These methods are environmentally friendly and do not require monitoring the residual solvent in drug products. These methods can be categorized as "zero waste" chemistry methods, and each method is outlined as follows:

(i) **Melt method:** This solvent-free method involves heating the solid dispersion of the pharmaceutical drug and MSNs above the drug's melting point. It is the simplest method for efficiently encapsulating the drug in mesoporous silica. The significant limitations associated with this method are that only thermally stable drugs can be used. Moreover, drug should remain low viscous upon melting, as high viscosity can cause uneven drug loading and drug release profile [40].

(ii) **Microwave irradiation:** It is another solvent-free method for drug loading into MSNs, in which drug samples are heated under a controlled feedback system to monitor temperature during drug encapsulation in silica, while avoiding drug degradation from heat. It was found that the encapsulated drug showed an enhanced drug release profile when compared to other drug-loading methods. This may be due to the conversion of the crystalline drug to an amorphous form by microwave irradiation [41].

(iii) **Co-milling:** It is a known technique for reducing the particle size to the sub-micron level. A pharmaceutical drug is introduced as guest molecules into the silica particles during milling, leading to its encapsulation within the mesoporous silica. The milling process and the ratio of API to silica can be varied to obtain the drug-loaded silica particles. It is a simple technique without time-consuming steps and can be practically used at an industrial scale. However, mechanical stress introduced during the milling process can affect the drug-loading process, including the drug state, structure, and morphology of the drug-loaded particles. [42]

7.6 Mesoporous Silica Particles as a Potential Option to Improve Bioavailability (BA)

Orally delivered silica nanoparticles can improve the BA of hydrophobic and poorly soluble drugs by various mechanisms such as drug amorphization, increased dissolution, controlled release, and protection from degradative enzymes [43]. Researchers have used the following strategies to improve the oral BA of poorly soluble drugs.

7.6.1 Stealth Particle by PEGylation to Prevent Opsonization

A silica surface coating with polymers can overcome gastric barriers and facilitate release of the drug at the target site [44]. The coating using the polymers such as chitosan, polyethene glycol, polyethene amine, and poly-L-histidine can be used to coat MSNs. The choice of polymers depends on the applications such as sustained-release formulation, stimuli-specific or targeted drug delivery, or improving biocompatibility. MSNs have the ability to encapsulate a variety of molecules, including small molecules and large bioactives and biomolecules (proteins and growth factors), allowing their controlled or sustained zero-order release that can range from several weeks to months. PEG modification on silica surface can be achieved by covalent coupling of silane groups of the silica with PEG. The PEG functionalization is known to increase circulation time and permeability; for example, PEGylation improves the residence time of drugs by protecting them from harsh environments, particularly in conditions of inflammatory bowel disease (Schmid, 2023 #35). The PEGylated particles show more residence time and a lower clearance rate as compared to non-PEGylated silica nanoparticles. The stealth properties of PEG delay the uptake by the reticuloendothelial system to enhance stability, long circulation time, or sustained action of the drugs so as to increase their BA [45]. For example, the PEG-decorated silica particles (MSNs-SS-mPEG) attached by the disulfide SS linkage can be used to enhance the uptake of dyes and drugs inside the breast cancer cell lines. The trapped molecules can be released from the pores in response to glutathione (GSH)-rich cellular environment of the cancer cells. The presence of higher-concentration GSH in cancer cells than in normal cells causes the reduction of the -SS linkage and releases the drug intracellularly, specifically into the cancer cells. The higher the levels of GSH, the higher the release; such gatekeepers are a promising tool for cancer-specific drug delivery [46].

7.6.2 Cyclodextrin as a Gatekeeper on Silica for Controlled Release

The surface of MSNs can be derivatized using organic/inorganic molecules, as they act as gatekeepers to regulate the release of molecules based on external specific stimuli such as pH, temperature, alteration in redox potential, or photo/radiation. The controlled release is maintained by the gatekeepers' on/off mechanism. CD-gated silica nanoparticles are a promising approach to improve the oral BA of drugs and to provide controlled drug release [47]. The approach is beneficial and can overcome the shortcomings of solubility, dissolution, and permeability of poorly soluble drugs. CDs are oligosaccharides with a lipophilic inner cavity and a hydrophilic outer surface, capable of complexation or inclusion of a broad group of guest molecules [48]. CDs are approved as excipients for oral, transdermal, and parenteral routes of administration. The CDs have a ring size of 1.5 nm, which matches the pore diameter of silica particles. CDs serve as gatekeepers on silica particle pores, where drugs encapsulated within the pores are released in response to stimuli (pH, temperature, redox, or enzymes) through stimuli-responsive linkers that control pore opening. [49, 50]. For example, intracellular and cytosolic GSH levels in tumors have been used as redox-based signals to control drug release based on concentration differences [51]. Using CD-based gatekeepers by tethering the disulfide linker on the surface of silica nanoparticles can entrap and release guest molecules for a specific environment in response to GSH.

7.6.3 Targeted Drug Delivery by Silica Conjugates

The surface of mesoporous silica nanoparticles offers multifunctionality that can be used to conjugate ligands for specific targeting to cells/tissues or organs, thereby achieving maximum therapeutic efficiency and preventing off-target side effects. Ligands such as peptides can bind to specific receptors present on the target and facilitate uptake, thereby improving intracellular BA [52]. The stimulus-responsive, such as ultrasound/enzyme/temperature and pH-dependent drug release and delivery from silica particles, includes the engineered nanoparticles releasing the drug in response to a pH change in the environment around it. For example, in tumor tissue, pH is acidic compared with normal tissue, and a strategy can be used to design nanocarriers with pH-sensitive linkers attached to their surface, releasing the drug at a specific pH change and enabling targeted delivery. The targeted release ensures drug release specifically at the target site, improving efficiency and reducing side effects. pH-responsive silica particles are mainly obtained by polyelectrolyte supramolecular valves, pH-sensitive coating, and linkers or acid-labile materials. pH-responsive MSNs are primarily used in cancer therapy. Acid-labile linkages such as acetal, hydrazide, or ester bonds can undergo reactions to release the drug in response to a decrease in pH. These pH-sensitive linkers can be attached to the surface of silica particles, can cover the pore surface as nano gates controlling the drug release, or can be covalently attached in tunnels with drugs that undergo cleavage reaction under acidic conditions. One such example is functionalized MSNs for doxorubicin (DOX) release at the tumor site, using pseudo-rotaxanes to block drug release from the mesopores via pH-sensitive benzoic amine bonds that hydrolyze at acidic pH but remain stable at neutral pH [53]. The pseudo-rotaxanes contain the PEG and CD, where PEG serves as a guest for CD as a host. Under weak acidic conditions. The extracellular matrix of tumor tissue, at pH 6.8, breaks down and releases the DOX, while the amine group facilitates the drug's internalization. At a much lower endosomal pH of 5, the benzoic amine bond breaks, removing the poly pseudo-rotaxane caps and providing sustained release of the drug in the cytoplasm. Similarly, acetal linkages are used to cap gold nanoparticles on the surface of silica nanoparticles [54]. At neutral pH, this remains stable, whereas hydrolysis of the acetal at lower pH releases the cargo. Similarly, hydrazone bond linkers are used to conjugate DOX to silica nanoparticles that accumulate in the solid tumors of the liver. The drug's hydrolysis occurs due to the lower endosomal pH, leading to sustained release at the target site [55].

7.6.4 Improved Drug Stability by Protecting Drug from Gut Enzymatic Degradation

Potential barriers to the oral delivery of macromolecules, such as peptides and proteins, include pH, enzymes, and other factors that limit drug release and efficacy at the target site [56]. Silica nanoparticles can contribute to the drug stability by encapsulating the drug in pores, modifying the release environment, and protecting the degradation of the drug from enzymes like proteases, lipases, stomach acid, and premature drug release.

7.7 Commercial Success and Clinical Pipeline Using This Technology

There are numerous ongoing clinical trials using the silica particles for biomedical applications, including oral drug delivery, diagnostics, plasmon therapy, and photothermal ablation therapy, and this dosage form has shown good tolerability, safety, and improved pharmacokinetics after oral delivery [56]. The silica–lipid nanoparticles have demonstrated an increase in BA by a factor of 1.95 for a poorly soluble drug, ibuprofen, during first-in-human studies [57]. The lipoceramic hybrid silica improved the PK levels by a factor of 3.5-fold compared with the commercially available simvastatin formulation of Sandoz in a bioequivalence trial [58]. In another clinical study, in which 12 healthy adults participated, the oral delivery of mesoporous silica improved the BA of fenofibrate by 54% compared with the commercially available fenofibrate formulation (Lipanthy) [59]. The core–shell silica-gold conjugate (Aurolase and Auroshell) has shown success in clinical trials for the photothermal ablation of malignant cancer [60]. In one of the clinical trials, ferromagnetic core–shell gold silica particles significantly reduced coronary atherosclerosis when used as plasmon resonance therapy [61]. The other successful clinical products tested are outlined in Table 7.4.

Table 7.4 An overview of the clinical application of the silica-containing formulations.

Drug	Silica type	Applications	Clinical status	References
Itraconazole (Sporanox®, Onmel®)	Amorphous silica in solid dispersion: SBA-15, MCM-41	Improves solubility and oral BA	Approved	[62]
Fenofibrate tablets	Mesoporous silica MCM-41	Enhances dissolution and stability	Approved	[63]
Ibuprofen (silica–lipid hybrid)	MSN + lipid coating	~2× higher bioavailability vs standard ibuprofen	Phase I clinical trial	[57]
Simvastatin (silica–lipid hybrid)	MSN + lipid coating	~3.5× higher BA vs marketed simvastatin	Phase I clinical trial	[58]
C'Dots (Cornell dots)	Ultrasmall core–shell fluorescent silica NP	Tumor imaging/theranostics	Phase I clinical trial completed; follow-up theranostic trials ongoing	[64]
SiPore15	Oral mesoporous silica	Prediabetes/metabolic modulation	Pilot trials completed; pivotal SHINE trial ongoing	[65]
Sil-based tracers	Fluorescent silica nanoparticles	Lymph node mapping	First-in-human studies completed	[66]

7.8 Challenges Toward Clinical Translation of Mesoporous Silica Particles

Despite success in clinical trials, there are still many key hurdles for the translation of these drug delivery systems. The safety of this delivery system in terms of immunogenic response, clearance from the body, risk of accumulation in organs, and toxicity are some of the major concerns. The long-term safety of these particulate systems upon chronic exposure is a challenge, and toxicological concerns from different routes of exposure need to be explored further. The Critical Chemistry, Manufacturing, and Controls (CMC) challenges involve scalable manufacturing and maintaining batch-to-batch reproducibility in silica particle synthesis processes. These particles do not have a high cargo-loading capacity, especially for nucleic acid-based drugs. The mesoporous organosilica nanoparticles can load biologics such as proteins, peptides, and mRNA, though their biodistribution and long-term toxicity need to be critically evaluated in clinically relevant models to understand the potential risks and benefits of treatment. The regulatory hurdles, such as utility for parenteral administration and drug targeting, require strict preclinical and clinical requirements, Investigational New Drug (IND) filings, and Good Laboratory Practice (GLP) toxicology data. Regulatory requirements include stringent batch production and reproducibility, consistent drug loading, particle morphology, and stability under storage for plausible oral use. Additionally, manufacturing costs, clinical trials, and acceptability of these particular drug delivery systems as compared to conventional drugs are some of the other challenges in translation.

7.9 Future Prospects and Conclusions

The properties of high surface area and porosity of MSNs enhance the solubility and BA of poorly soluble drugs. The improved oral absorption and pharmacokinetics of drugs are attributed to their amorphous nature by trapping and confining the molecular form of the drug within the nanoporous structure. Additionally, it protects the drug from degradation in harsh environments such as gut enzymes and strong acidic pH. MSNs combined with lipids, polymers, and hybrid systems are being developed for targeted drug release and are functionalized with various ligands for improving tissue-specific or intracellular localized drug delivery. The surface functionalization with carboxyl, PEG, and amine groups improves loading, targeting, and the stability of drugs. The emerging applications of MSN include their use in imaging and diagnostics. Biodegradable and stimuli-responsive MSNs have also attracted significant attention for improving drug release and tailoring properties.

Abbreviations

API	active pharmaceutical agent
BA	bioavailability
BCS	Biopharmaceutical Classification System
C'Dots	Cornell dots
CD	cyclodextrin
CTAB	cetyltrimethylammonium bromide
DLS	dynamic light scattering
DOX	doxorubicin
FDA	Food and Drug Administration
GLP	Good Laboratory Practice
IL	ionic liquid
IND	Investigational New Drug
LAXS	low-angle X-ray spectroscopy
MCM-41	mobile crystalline material
MSN	mesoporous silica particles
NPs	nanoparticles
ODT	oral disintegrating tablet
PEG	polyethylene glycol

SBA-15	Santa Barbara Amorphous
SiO_2	silicon dioxide
TEOS	tetraethyl orthosilicate

References

1 Singh, A., Worku, Z.A., and Van den Mooter, G. (2011). Oral formulation strategies to improve solubility of poorly water-soluble drugs. *Expert Opin. Drug Deliv.* 8 (10): 1361–1378.

2 McLean, B. and Yarovsky, I. (2024). Structure, properties, and applications of silica nanoparticles: recent theoretical modeling advances, challenges, and future directions. *Small* 20 (51): e2405299.

3 Li, H., Chen, X., Shen, D. et al. (2021). Functionalized silica nanoparticles: classification, synthetic approaches and recent advances in adsorption applications. *Nanoscale* 13 (38): 15998–16016.

4 Xu, B., Li, S., Shi, R., and Liu, H. (2023). Multifunctional mesoporous silica nanoparticles for biomedical applications. *Signal Transduct. Target. Ther.* 8 (1): 435.

5 Bokov, D., Jalil, A.T., Chupradit, S. et al. Nanomaterial by sol-gel method: synthesis and application. *Adv. Mater. Sci. Eng.* 2021 (2021(1)): 5102014.

6 Rahman, I.A. and Padavettan, V. (2012). Synthesis of silica nanoparticles by sol-gel: size-dependent properties, surface modification, and applications in silica-polymer nanocomposites—a review. *J. Nanomater.* 2012 (1): 132424.

7 Aubert, T., Grasset, F., Mornet, S. et al. (2010). Functional silica nanoparticles synthesized by water-in-oil microemulsion processes. *J. Colloid Interface Sci.* 341 (2): 201–208.

8 Koźlecki, T., Polowczyk, I., Bastrzyk, A., and Sawiński, W. (2016). Improved synthesis of nanosized silica in water-in-oil microemulsions. *J. Nanopart.* 2016 (1): 8203260.

9 Ghaedi, H. and Zhao, M. (2022). Review on template removal techniques for synthesis of mesoporous silica materials. *Energy Fuel* 36 (5): 2424–2446.

10 Wu, S.-H., Mou, C.-Y., and Lin, H.-P. (2013). Synthesis of mesoporous silica nanoparticles. *Chem. Soc. Rev.* 42 (9): 3862–3875.

11 Pal, N., Lee, J.-H., and Cho, E.-B. (2020). Recent trends in morphology-controlled synthesis and application of mesoporous silica nanoparticles. *Nanomaterials* 10 (11): 2122.

12 Natarajan, S.K. and Selvaraj, S. (2014). Mesoporous silica nanoparticles: importance of surface modifications and its role in drug delivery. *RSC Adv.* 4 (28): 14328–14334.

13 Mohamed, F., Fathy, M., Fahmy, H.M., and Elshemey, W.M. (2021). Low-angle X-ray scattering for the determination of the size of mesoporous silica nanoparticles. *Radiat. Phys. Chem.* 179: 109235.

14 Tzankov, B., Voycheva, C., Aluani, D. et al. (2019). Improvement of dissolution of poorly soluble glimepiride by loading on two types of mesoporous silica carriers. *J. Solid State Chem.* 271: 253–259.

15 Kim, D., Koo, J., Yang, E.J. et al. (2023). Entrapment of celecoxib into mesoporous silica particles for tablets with improved dissolution through amorphization. *J. Drug Deliv. Sci. Technol.* 84: 104485.

16 Kapourani, A., Katopodis, K., Valkanioti, V. et al. (2024). Evaluation of suitable polymeric matrix/carriers during loading of poorly water soluble drugs onto mesoporous silica: physical stability and in vitro supersaturation. *Polymers* 16 (6): 802.

17 Abd-Elrahman, A.A., el Nabarawi, M.A., Hassan, D.H., and Taha, A.A. (2016). Ketoprofen mesoporous silica nanoparticles SBA-15 hard gelatin capsules: preparation and in vitro/in vivo characterization. *Drug Deliv.* 23 (9): 3387–3398.

18 Zhang, X., Xiao, Y., Guo, S. et al. (2025). pH-sensitive zinc oxide nanoparticle-controlled sodium alginate/silica hydrogel beads: for controlled curcumin release. *J. Mol. Struct.* 1324: 140967.

19 Yan, J., Siwakoti, P., Shaw, S. et al. (2024). Porous silicon and silica carriers for delivery of peptide therapeutics. *Drug Deliv. Transl. Res.* 14 (12): 3549–3567.

20 Benkő, F., Kristó, K., and Sovány, T. (2025). Mesoporous silica nanoparticles as drug delivery systems. *Pharmaceuticals* 18 (9): 1392.

21 Christoforidou, T., Giasafaki, D., Andriotis, E.G. et al. (2021). Oral drug delivery systems based on ordered mesoporous silica nanoparticles for modulating the release of aprepitant. *Int. J. Mol. Sci.* 22 (4): 1896.

22 Rahangdale, M., Solanki, S., Patil, P. et al. (2024). Fabrication and characterization of apremilast-loaded zinc oxide-mesoporous silica nanoparticles for psoriasis treatment. *Ther. Deliv.* 15 (6): 449–462.

23 van der Zwaan, I., Pilkington, G.A., Frenning, G. et al. (2024). Influence of particle diameter on aerosolization performance and release of budesonide loaded mesoporous silica particles. *Eur. J. Pharm. Sci.* 200: 106828.

24 Gao, Y., Zhang, Y., Hong, Y. et al. (2022). Multifunctional role of silica in pharmaceutical formulations. *AAPS PharmSciTech* 23 (4): 90.

25 McCarthy, C.A. et al. (2016). Mesoporous silica formulation strategies for drug dissolution enhancement: a review. *Expert Opin. Drug Deliv.* 13 (1): 93–108.

26 Valetti, S., Thomsen, H., Wankar, J. et al. (2021). Can mesoporous nanoparticles promote bioavailability of topical pharmaceutics? *Int. J. Pharm.* 602: 120609.

27 Qian, K.K. and Bogner, R.H. (2012). Application of mesoporous silicon dioxide and silicate in oral amorphous drug delivery systems. *J. Pharm. Sci.* 101 (2): 444–463.

28 Chen, C.-Y., Burkett, S.L., Li, H.X., and Davis, M.E. (1993). Studies on mesoporous materials II. Synthesis mechanism of MCM-41. *Microporous Mater.* 2 (1): 27–34.

29 Shah, P.V. and Rajput, S.J. (2017). A comparative in vitro release study of raloxifene encapsulated ordered MCM-41 and MCM-48 nanoparticles: a dissolution kinetics study in simulated and biorelevant media. *J. Drug Deliv. Sci. Technol.* 41: 31–44.

30 Quan, G., Wu, Q., Zhang, X. et al. (2016). Enhancing in vitro dissolution and in vivo bioavailability of fenofibrate by solid self-emulsifying matrix combined with SBA-15 mesoporous silica. *Colloids Surf. B: Biointerfaces* 141: 476–482.

31 Fatima, R., Katiyar, P., and Kushwaha, K. (2025). Recent advances in mesoporous silica nanoparticle: synthesis, drug loading, release mechanisms, and diverse applications. *Front. Nanotechnol.* 7: 1564188.

32 Kovtareva, S.Y., Kopishev, E.E., Zhang, H., and Filippov, S.K. (2025). Exploring the physicochemical interactions and loading strategies of mesoporous silicon dioxide nanoparticles for drug delivery. *Eur. J. Pharm. Biopharm.* 208: 114654.

33 Kumar, S., Malik, M., and Purohit, R. (2017). Synthesis methods of mesoporous silica materials. *Mater. Today Proc.* 4 (2): 350–357.

34 Vrålstad, T., Glomm, W.R., Øye, G. et al. (2005). Cobalt functionalization of mesoporous silica by incipient wetness impregnation and co-precipitation. *J. Dispers. Sci. Technol.* 26 (1): 87–94.

35 Limnell, T., Santos, H.A., Mäkilä, E. et al. (2011). Drug delivery formulations of ordered and nonordered mesoporous silica: comparison of three drug loading methods. *J. Pharm. Sci.* 100 (8): 3294–3306.

36 Trzeciak, K., Kaźmierski, S., Wielgus, E., and Potrzebowski, M.J. (2020). DiSupLo-New extremely easy and efficient method for loading of active pharmaceutical ingredients into the pores of MCM-41 mesoporous silica particles. *Microporous Mesoporous Mater.* 308: 110506.

37 Li-Hong, W., Xin, C., Hui, X. et al. (2013). A novel strategy to design sustained-release poorly water-soluble drug mesoporous silica microparticles based on supercritical fluid technique. *Int. J. Pharm.* 454 (1): 135–142.

38 Yong, W.Y.D., Zhang, Z., Cristobal, G., and Chin, W.S. (2014). One-pot synthesis of surface functionalized spherical silica particles. *Colloids Surf. A Physicochem. Eng. Asp.* 460: 151–157.

39 Giret, S., Wong Chi Man, M., and Carcel, C. (2015). Mesoporous-silica-functionalized nanoparticles for drug delivery. *Chem. Eur. J.* 21 (40): 13850–13865.

40 Zůza, D., Šoltys, M., Mužík, J. et al. (2019). Silica particles with three levels of porosity for efficient melt amorphisation of drugs. *Microporous Mesoporous Mater.* 274: 61–69.

41 Bahadur, N.M., Furusawa, T., Sato, M. et al. (2011). Fast and facile synthesis of silica coated silver nanoparticles by microwave irradiation. *J. Colloid Interface Sci.* 355 (2): 312–320.

42 Moutamenni, B., Tabary, N., Mussi, A. et al. (2023). Milling-assisted loading of drugs into mesoporous silica carriers: a green and simple method for obtaining tunable customized drug delivery. *Pharmaceutics* 15 (2): 390.

43 Florek, J., Caillard, R., and Kleitz, F. (2017). Evaluation of mesoporous silica nanoparticles for oral drug delivery–current status and perspective of MSNs drug carriers. *Nanoscale* 9 (40): 15252–15277.

44 Wang, Y., Zhao, Y., Cui, Y. et al. (2018). Overcoming multiple gastrointestinal barriers by bilayer modified hollow mesoporous silica nanocarriers. *Acta Biomater.* 65: 405–416.

45 Von Baeckmann, C., Kählig, H., Lindén, M., and Kleitz, F. (2021). On the importance of the linking chemistry for the PEGylation of mesoporous silica nanoparticles. *J. Colloid Interface Sci.* 589: 453–461.

46 Cui, Y., Dong, H., Cai, X. et al. (2012). Mesoporous silica nanoparticles capped with disulfide-linked PEG gatekeepers for glutathione-mediated controlled release. *ACS Appl. Mater. Interfaces* 4 (6): 3177–3183.

47 Yi, S., Zheng, J., Lv, P. et al. (2018). Controlled drug release from cyclodextrin-gated mesoporous silica nanoparticles based on switchable host–guest interactions. *Bioconjug. Chem.* 29 (9): 2884–2891.

48 Wankar, J., Kotla, N.G., Gera, S. et al. (2020). Recent advances in host–guest self-assembled cyclodextrin carriers: implications for responsive drug delivery and biomedical engineering. *Adv. Funct. Mater.* 30 (44): 1909049.

49 Rodrigo-Martínez, P., Barros, M., Terencio, M.C. et al. (2025). Mesoporous silica nanoparticles with an azobenzene gatekeeper as hypoxia-responsive nanocarriers for targeted doxorubicin delivery. *Drug Deliv. Transl. Res.* 1–12.

50 Wen, J., Yang, K., Liu, F. et al. (2017). Diverse gatekeepers for mesoporous silica nanoparticle based drug delivery systems. *Chem. Soc. Rev.* 46 (19): 6024–6045.

51 Luo, Z., Hu, Y., Cai, K. et al. (2014). Intracellular redox-activated anticancer drug delivery by functionalized hollow mesoporous silica nanoreservoirs with tumor specificity. *Biomaterials* 35 (27): 7951–7962.

52 Kankala, R.K., Han, Y.-H., Na, J. et al. (2020). Nanoarchitectured structure and surface biofunctionality of mesoporous silica nanoparticles. *Adv. Mater.* 32 (23): 1907035.

53 Garcia-Rio, L., Otero-Espinar, F.J., Luzardo-Alvarez, A., and Blanco-Méndez, J. (2014). Cyclodextrin based rotaxanes, polyrotaxanes and polypseudorotaxanes and their biomedical applications. *Curr. Top. Med. Chem.* 14 (4): 478–493.

54 Liu, R., Zhang, Y., Zhao, X. et al. (2010). pH-responsive nanogated ensemble based on gold-capped mesoporous silica through an acid-labile acetal linker. *J. Am. Chem. Soc.* 132 (5): 1500–1501.

55 Rahikkala, A., Pereira, S.A.P., Figueiredo, P. et al. (2018). Mesoporous silica nanoparticles for targeted and stimuli-responsive delivery of chemotherapeutics: a review. *Adv. Biosyst.* 2 (7): 1800020.

56 Bharti, C., Nagaich, U., Pal, A.K., and Gulati, N. (2015). Mesoporous silica nanoparticles in target drug delivery system: a review. *Int. J. Pharm. Invest.* 5 (3): 124.

57 Tan, A., Eskandar, N.G., Rao, S., and Prestidge, C.A. (2014). First in man bioavailability and tolerability studies of a silica–lipid hybrid (lipoceramic) formulation: a phase I study with ibuprofen. *Drug Deliv. Transl. Res.* 4 (3): 212–221.

58 Meola, T.R., Abuhelwa, A.Y., Joyce, P. et al. (2021). A safety, tolerability, and pharmacokinetic study of a novel simvastatin silica-lipid hybrid formulation in healthy male participants. *Drug Deliv. Transl. Res.* 11 (3): 1261–1272.

59 Bukara, K., Schueller, L., Rosier, J. et al. (2016). Ordered mesoporous silica to enhance the bioavailability of poorly water-soluble drugs: proof of concept in man. *Eur. J. Pharm. Biopharm.* 108: 220–225.

60 Riedel, R., Mahr, N., Yao, C. et al. (2020). Synthesis of gold–silica core–shell nanoparticles by pulsed laser ablation in liquid and their physico-chemical properties towards photothermal cancer therapy. *Nanoscale* 12 (5): 3007–3018.

61 Wu, M., Li, X., Guo, Q. et al. (2021). Magnetic mesoporous silica nanoparticles-aided dual MR/NIRF imaging to identify macrophage enrichment in atherosclerotic plaques. *Nanomedicine* 32: 102330.

62 Mellaerts, R., Mols, R., Jammaer, J.A.G. et al. (2008). Increasing the oral bioavailability of the poorly water soluble drug itraconazole with ordered mesoporous silica. *Eur. J. Pharm. Biopharm.* 69 (1): 223–230.

63 Zhao, Z., Gao, Y., Wu, C. et al. (2016). Development of novel core-shell dual-mesoporous silica nanoparticles for the production of high bioavailable controlled-release fenofibrate tablets. *Drug Dev. Ind. Pharm.* 42 (2): 199–208.

64 Phillips, E., Penate-Medina, O., Zanzonico, P.B. et al. (2014). Clinical translation of an ultrasmall inorganic optical-PET imaging nanoparticle probe. *Sci. Transl.Med.* 6 (260): 260ra149–260ra149.

65 Baek, J., Robert-Nicoud, G., Hidalgo, C.H. et al. (2022). Engineered mesoporous silica reduces long-term blood glucose, HbA1c, and improves metabolic parameters in prediabetics. *Nanomedicine* 17 (1): 9–22.

66 Zanoni, D.K., Stambuk, H.E., Madajewski, B. et al. (2021). Use of ultrasmall core-shell fluorescent silica nanoparticles for image-guided sentinel lymph node biopsy in head and neck melanoma: a nonrandomized clinical trial. *JAMA Netw. Open* 4 (3): e211936–e211936.

8

Gastrointestinal Dissolution and Absorption of BCS Class II Drugs

Almog Eliyahu Dahan and Arik Dahan

Ben-Gurion University of the Negev, School of Pharmacy, Faculty of Health Sciences, Department of Clinical Pharmacology, Ben-Gurion Blvd Beer-Sheva 8410501, Israel

8.1 Introduction

The Biopharmaceutical Classification System (BCS) is an advanced tool used for classifying medications based on dissolution, intestinal permeability, and solubility [1]. It also acts as a regulatory tool by allowing biowaivers that replace certain bioequivalence (BE) studies, particularly for immediate-release (IR) oral dosage forms of BCS Class I and, under specific conditions, Class III drugs. The BCS thus enables manufacturers to save time, reduce financial investment, and avoid unnecessary drug exposure to a significant number of healthy volunteers [2]. Since gastrointestinal (GI) dissolution is an essential step for oral absorption, low-solubility drugs often exhibit poor oral bioavailability (BA). These low-solubility drug candidates present a significant challenge for scientists, who must develop effective strategies to optimize their BA. In this context, the BCS has become an important framework for understanding and addressing the limitations of these compounds [3].

The aim of this chapter is to provide an overview of the principles related to the absorption of low-solubility, high-permeability drug substances, with particular attention to BCS Class II molecules. We will also explore the various factors influencing their intestinal absorption and provide insights into regulatory aspects related to this category of drug substances.

8.2 Drug Absorption and the BCS

Following oral administration, drug absorption involves a cascade of complex processes, with numerous factors determining both its rate and extent. These factors include physicochemical properties (e.g. pK_a, solubility, stability [chemical and physical], diffusivity, lipophilicity, surface area, particle size, crystal form), physiological parameters (e.g. GI pH, blood flow, gastric emptying, segmental transit time, absorption mechanisms, and disease state), and dosage form variables (e.g. tablets, capsules, solutions, suspensions, emulsions, gels) [4, 5].

When applying Fick' first law to a membrane, drug absorption across the GI mucosa under sink conditions can be written as:

$$J_W = P_W \times C_W = \frac{dM}{dt} \times \frac{1}{A} \tag{8.1}$$

where

J_W = mass transport across the GI wall (mass/area/time).
P_W = effective permeability.
C_W = concentration of the drug at the membrane.

As described by Amidon et al. [6–8], analysis of this equation shows that the key factors determining oral drug absorption are the drug's permeability across the GI membrane, its dissolution in the GI environment, and the administered dose. These key parameters are defined in the BCS by three dimensionless parameters [7]:

1) Absorption number (A_n)
2) Dissolution number (D_n)
3) Dose number (D_0)

These parameters incorporate both physicochemical and physiological factors and are essential to the oral drug absorption process.

The absorption number (A_n) represents the ratio between the effective permeability (P_{eff}) and the intestinal radius (R), multiplied by the residence time (t_{res}). It can also be understood as the product of the effective absorption rate constant (t_{abs}^{-1}) and the residence time.

$$A_n = \text{Absorption number} = \frac{P_{eff}}{R} t_{res} = t_{abs}^{-1} t_{res} \tag{8.2}$$

The dissolution number (D_n) represents the ratio between the t_{res} and the drug's dissolution time (t_{Diss}). This dissolution time depends on factors such as the drug's equilibrium solubility (C_s), its diffusivity (D), density (ρ), the initial particle radius (r_0), and the intestinal residence time (t_{res}).

$$D_n = \text{Dissolution number} = \frac{DC_s}{r_0} = \frac{4\pi r_0^2}{\left(\frac{4}{3}\right)\pi r_0^3 \rho} t_{res} = \frac{3t_{res}DC_s}{\rho r_0^2} = \frac{t_{res}}{t_{Diss}} \tag{8.3}$$

Finally, **the dose number (D_0)** is defined as the ratio of dose to dissolved drug. This dissolution time depends on factors such as the drug's equilibrium solubility (C_s), dose (M_0), and uptake volume of water (usually defined as 250 ml) (V_0).

$$D_0 = \text{Dose number} = \frac{M_0/V_0}{C_s} \tag{8.4}$$

A summary of these parameters is detailed in Table 8.1.

Table 8.1 Summary of dimensionless parameters and their meanings.

Dimensionless parameter	Meaning
A_n	The absorption rate is high when A_n exceeds 1. However, lower intestinal permeability leads to longer absorption time, which decreases the absorption number.
D_n	The dissolution rate is considered high when D_n exceeds 1. However, an increase in particle size or density prolongs the dissolution time and consequently reduces the ratio.
D_0	High solubility is indicated when D_0 is less than 1. Higher dose increases this ratio and may reduce drug absorption.

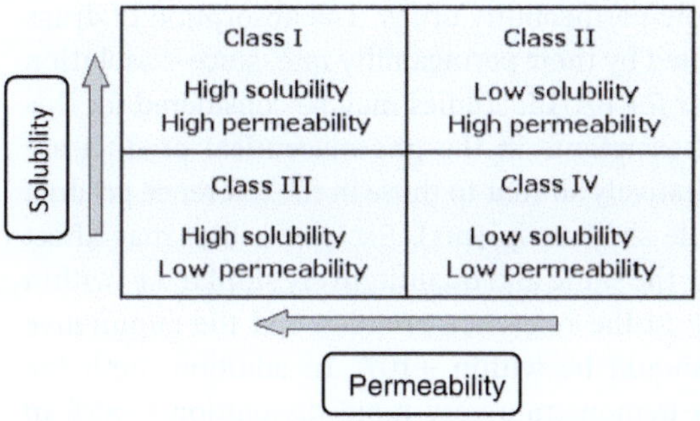

Figure 8.1 The BCS classifies drug substances based on their solubility (volume of aqueous media required to dissolve the highest dose strength across the physiological pH range) and permeability (the extent of absorption derived from human pharmacokinetic studies and also can be assessed by validated and standardized in vitro methods using Caco-2 cells) characteristics.

According to the BCS developed by Amidon et al. [6], drugs are classified into four categories based on permeability and solubility (Figure 8.1). Examples of drugs classified by the BCS are shown in Table 8.2.

- **BCS Class I:** High-solubility, high-permeability drugs. These drugs are expected to achieve complete absorption if at least 85% of the drug dissolves within 30 minutes across the physiological pH. Therefore, a waiver for BA/BE studies is granted for IR drug products in this group [13].
- **BCS Class II:** Low-solubility, high-permeability drugs. BCS Class II drugs are often limited by the dissolution/solubility rate. Therefore, these compounds are not eligible for biowaivers since they do not meet the requirements set by regulatory agencies. A biowaiver has been suggested for some Class IIa drugs in the BE criteria released by the European Medicines Agency (EMA) and World Health Organization (WHO) in addition to the recommendations from other research forums [13].

Table 8.2 Summary of Biopharmaceutical Classification System (BCS) with corresponding solubility, permeability, and example drugs.

Class	Solubility	Permeability	Meaning	Examples of drugs [9–12]
I	High	High	Maximum absorption	Acyclovir, atropine, diazepam
II	Low	High	Solubility limited	Atorvastatin, carbamazepine, digoxin
III	High	Low	Permeability limited	Amoxicillin, atenolol, bisphosphonates
IV	Low	Low	Poorly absorbed	Ciprofloxacin, furosemide, methotrexate

- **BCS Class III:** High-solubility, low-permeability drugs. The absorption of drugs in this class is expected to be limited by their permeability rate, since dissolution is likely to occur rapidly. A waiver for BA/BE studies may be considered for this class of drugs, provided that all excipients in the pharmaceutical product are qualitatively the same and quantitatively similar to those in the reference product (except for film coating or capsule shell excipients). Excipients that may affect absorption should be qualitatively the same and quantitatively similar, i.e. within ±10% of the amount of excipient in the reference product, and the cumulative difference for these excipients should be within ±10%. In addition, both the test product and reference must demonstrate very rapid dissolution (≥85% in ≤15 minutes) under standardized conditions (volume, temperature, pH range, etc.) [13–18].
- **BCS Class IV:** Low-solubility, low-permeability drugs. The absorption of drugs in this class is expected to be very low and generally considered unsuitable candidates for oral drug administration [19].

The BCS is an important tool to evaluate in vivo performance of drug substance based on in vitro permeability and solubility properties [20]. In this chapter, we discuss various aspects affecting intestinal absorption specific to BCS Class II drugs.

8.3 GI Physiological Variables Affecting BCS Class II Drug Dissolution

In most cases, drug absorption after oral administration depends on dissolution in GI tract milieu. For Class II compounds, dissolution or solubility is the rate-limiting factor in intestinal absorption [21–23]. The factors influencing dissolution rate $\left[\frac{dX_d}{dt}\right]$ are described by the equation mentioned below, derived from the Nernst–Brunner and Levich modifications of the Noyes–Whitney model [24–26]:

$$\frac{dX_d}{dt} = \frac{A \times D}{h}\left(C_s - \frac{X_d}{V}\right) \tag{8.5}$$

where

A = Surface area of the solid drug.

D = Diffusion coefficient of the drug.

h = Effective diffusion boundary layer thickness adjacent to the dissolving surface.

C_s = Saturation solubility of the drug.

V = Volume of the fluid available for dissolution.

X_d = Amount of drug already dissolved.

The dissolution rate is influenced by both the physicochemical properties of the drug and GI physiological factors, which will be reviewed in this section.

8.3.1 Bile Salts

Bile acids are synthesized and secreted by hepatocytes in the liver, stored in the gallbladder, and subsequently released into the small intestine. The major organic solutes of bile include bile acids, phospholipids (mainly lecithin), and cholesterol. Under fasting conditions, the intestinal concentrations of bile acids and phospholipids are relatively low (5 and 0.2 mM, respectively), compared to their elevated levels in fed conditions (15 and 4 mM, respectively) [27, 28].

Bile acids play a critical role in the solubility and dissolution of drugs through micelle formation and wetting effects [29, 30]. As such, they have a significant impact on the absorption processes of Class II compounds. When biliary secretions exceed their critical micelle concentration (CMC), they enhance the dissolution and facilitate the absorption of low water-soluble drugs by forming them into submicron mixed micelles, thereby promoting their transport to the enterocyte membranes [31–33]. Conversely, when bile concentrations are below the CMC [33, 34], the primary mechanism by which bile salts improve drug solubility is through wetting [35, 36]. For example, Bakatselou et al. demonstrated the ability of sodium taurocholate, a bile salt, in enhancing the solubilization and wetting of poorly soluble steroids (such as hydrocortisone, betamethasone, and dexamethasone), thereby improving their dissolution characteristics [29].

8.3.2 GI pH

As a drug product passes through the GI tract, it encounters a broad range of pH environments. Gastric pH is influenced by food intake, with reported values ranging from 1 to 8. In the fasting state, gastric pH is between 1.4 and 2.1 [37–39]. In contrast, the pH in the small intestine is consistently higher than in the stomach, less affected by food, and gradually increases from the proximal to the distal segments, covering a range of approximately 4.4–7.4 [21, 37, 40].

BCS Class II drugs include both non-ionizable substances (e.g. carbamazepine) and ionizable substances, which can be acids (e.g. ibuprofen) or bases (e.g. diazepam). For non-ionizable drugs, changes in pH along the GI tract usually don't affect the solubilization/dissolution process. However, for ionizable BCS Class II drugs, their absorption significantly depends on pH [30, 41].

Generally, the ionized form of a drug dissolves better in water because it can form more hydrogen bonds, but the unionized form passes through membranes more easily. Still, for Class II drugs, absorption is limited by dissolution process and not by membrane permeability. So, changes in pH may affect how much of the drug is ionized during dissolution and influence the absorption rate.

For BCS Class II weak base drugs (e.g. ketoconazole), ionization occurs in the stomach, resulting in rapid dissolution. However, as the drug passes from the stomach into the duodenum, the elevated pH causes a significant reduction in ionization, which may lead to drug precipitation [42, 43]. As a result, intestinal absorption becomes a multifactorial process, influenced by the degree of supersaturation, the solid form of the weak base, pH, fluid volume, viscosity, and bile salts [30, 44, 45].

In contrast, for BCS Class II weak acid drugs (e.g. ibuprofen, ketoprofen), the ionization process occurs mainly in the small intestine rather than the stomach. The intestinal pH is higher than the pK_a, so the solubility of the weak acid increases significantly. Therefore, the in vivo solubility and dissolution of these drugs would be high, causing them to behave more like Class I compounds (high solubility, high permeability) [34, 46, 47].

8.3.3 GI Transit

There are two important factors in GI transit: the gastric residence time, which is determined by gastric emptying, and the transit time through the small intestine. For BCS Class II drugs, dissolution is usually slower than gastric emptying, so it does not affect intestinal absorption. However, the transit time in the small intestine can influence absorption, as a longer time at the absorption site enhances drug absorption. On average, the small intestinal transit time is about 3–4 hours (Figure 8.2), is typically stable, and is not affected by food consumption or other physical characteristics [48, 49].

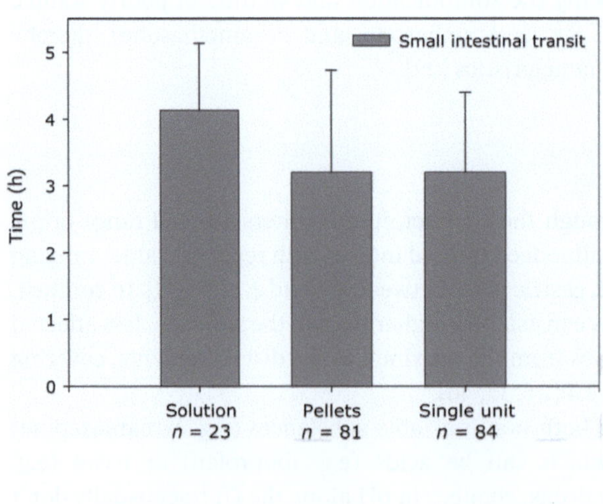

Figure 8.2 Human small intestinal transit time of different pharmaceutical dosage forms measured by g-scintigraphy. The intestinal transit time is typically stable and is not affected by food consumption or other physical characteristics. *Source:* Adapted from Davis et al. [37].

8.3.4 Drug Particle Size

Smaller particles provide a larger surface area exposed to the dissolution fluid, leading to a faster dissolution rate [50, 51]. This is especially important for Class II drugs, where the dissolution rate limits absorption. Therefore, increasing the surface area is expected to enhance the dissolution rate, making particle size an important factor in the formulation process.

Another factor that can influence dissolution is particle density and how well the particles are dispersed. The greater dispersion leads to improved dissolution [52, 53]. The effect of reducing particle size also appears to depend on food intake. Under fasting conditions, reducing particle size significantly improved both in vitro dissolution and in vivo oral absorption of DPC961, a BCS Class II reverse transcriptase inhibitor. However, under fed conditions, no effect was observed [54]. This difference likely results from variations in solubilization capacity between fasting and fed states.

Similarly, for spironolactone, smaller particle size improved dissolution rate in vivo but did not affect BA. The authors suggested that the lack of difference in BA between the two particle sizes may be due to insufficient washout of particles after perfusion, reabsorption of surface-active ingredients along the GI tract, relatively small differences in particle size, and large inter- and intra-individual variability in pharmacokinetic parameters [55].

It is important to note that what truly matters is the effective surface area in contact with the dissolution fluid (the wetted surface), rather than the actual particle size. This is especially relevant for highly hydrophobic drugs in dissolution media with poor wetting properties or in manufacturing processes that alter particle size during production. In such cases, reducing particle size may actually slow the dissolution rate [56, 57].

8.3.5 Volume Available for Dissolution

There are three main sources of fluids available in the GI tract for drug dissolution: the fluids taken with the drug, secretions into the GI lumen, and water flux across the intestinal wall. This factor is especially important for BCS Class II drugs, as a larger volume in GI tract may influence dissolution by simply presenting more liquid to the drug and creating a solvent drag effect. The total volume may vary by individual, and estimated volumes have been reported as follows:

- **Stomach:** Fasted state—500 ml (although the volume may be as low as 20–30 ml), fed state—800–900 ml.
- **Small intestine:** Fasted state—500 ml, fed state—900–1000 ml (although volumes in the upper small intestine in fed state can reach as much as 1.5 l).

In the fed state, estimated fluid volumes are higher than in the fasting state, both in the stomach and in the small intestine [21, 58–60].

8.4 In vitro Dissolution Tests for Class II Drugs

As mentioned earlier, the dissolution process in vivo is often the rate-limiting step in the oral absorption of Class II drugs [21–23]. Therefore, a well-designed dissolution test could predict in vitro–in vivo correlation (IVIVC).

8.4.1 Biorelevant Media

Choosing the appropriate medium is essential for achieving a good IVIVC. The dissolution process depends on many variables, such as surfactants, pH, buffer capacity, ionic strength, and the volume available for dissolution. Biorelevant dissolution media for BCS Class II drugs often fail to provide good IVIVC because they do not fully mimic these variables of the GI environment.

Examples of media modeling fasting conditions include simulated gastric fluid (SGF) and surfactants for the stomach and fasting-state simulated intestinal fluid (FaSSIF) for the proximal small intestine. For fed conditions, examples of media used include long-life milk (3.5% fat) for the stomach and fed-state simulated intestinal fluid (FeSSIF) for the proximal small intestine [21, 42, 59, 61]. The use of these media predicts good IVIVC for low-solubility drugs such as ketoconazole, danazol, and others [42, 59, 62–64].

In addition, the hydrodynamic conditions (mixing, peristaltic movements, fluid flow, etc.) are also important factors to consider. The use of the United States Pharmacopoeia (USP) paddle apparatus (50 rpm) and the basket apparatus (100 rpm) can mimic the hydrodynamic variations of the upper GI tract [65–67]. Nonetheless, variability in the results may still occur despite accurate simulation of the GI milieu.

8.4.2 Dynamic Lipolysis Model

Lipid-based drug delivery systems have the potential to enhance the dissolution, solubility, and oral BA of Class II compounds through various mechanisms [19, 68, 69]. One of the primary mechanisms involves stimulation of biliary and pancreatic secretions, which can enhance the dissolution and solubilization of coadministered lipophilic drugs [70].

Additional mechanisms to enhance the solubility of BCS II drugs using lipid-based formulations include [71–74]:

1) Administering the drug in a dissolved solution rather than in solid form.
2) Reducing metabolism and efflux transporter activity.
3) Physical modifications such as micronization and nanonization.
4) Changing the intra-enterocyte transport route by stimulation of the lymphatic transport pathway.

After oral administration, the lipid components of these formulations undergo enzymatic hydrolysis by pancreatic lipase and subsequently interact with endogenous bile salts and phospholipids. These interactions result in the formation of colloidal structures that promote the solubilization of low water-soluble drugs and prevent their precipitation within the aqueous GI milieu. In most cases, this is considered the principal mechanism by which lipid-based formulations improve the absorption of lipophilic drugs [28, 75]. Since lipid-based delivery systems depend on GI tract processing, release tests must include lipolytic enzymes that catalyze lipid digestion in vivo [76].

In recent years, a dynamic in vitro lipolysis model has been developed to mimic the in vivo lipid digestion process [77–79]. This model has been used to assess various lipid-based delivery systems, with studies demonstrating a correlation between in vitro dissolution and in vivo absorption for poorly water-soluble drugs, such as progesterone [80].

8.4.3 BCS-Based FDA/EMA Guidelines: Implications for Class II Drugs

The criteria for biowaiver are given by the United States Food and Drug Administration (FDA), WHO, and EMA.

A drug product is eligible for a BCS-based biowaiver provided that the drug substance(s) satisfy the criteria regarding solubility and permeability (Class I and Class III), the drug product is an IR oral dosage form with systemic action, and the drug product is the same dosage form and strength as the reference product [18, 81, 82].

BCS Class II drugs, characterized by low aqueous solubility and high permeability, have generally been excluded from biowaiver eligibility due to concerns that in vitro dissolution testing may not reliably predict in vivo absorption. However, emerging evidence suggests that certain weak acid Class II drugs may behave like Class I drugs under physiological conditions. As a result, proposals have been made to refine current guidelines and enable biowaiver extension to more Class II drugs.

8.5 Biowaiver Extension Potential

Over the past two decades, ongoing scientific research, case studies, and regulatory discussions have driven gradual updates to biowaiver eligibility guidelines. Notable revisions include, among others, narrowing the pH range required for solubility determination from 1.0–7.5 to 1.2–6.8, and lowering the permeability threshold from 90% to 85% of the oral fraction absorbed. Another consideration raised is the possible establishment of an intermediate solubility class for BCS Class II drugs with pH-dependent solubility characteristics [18, 83, 84].

BCS Class II drugs, characterized by low aqueous solubility and high permeability, are typically excluded from biowaiver eligibility due to concerns that in vitro dissolution data might not reliably predict in vivo absorption. However, as described in Section 8.3.2, the ionization of weakly acidic BCS Class II drugs ($pK_a \leq 5.0$) occurs primarily in the small intestine. These compounds usually exhibit poor solubility at acidic gastric pH, while their solubility can increase by 100-fold or more at the higher intestinal pH (~6.5) [85]. Consequently, the in vivo solubility and dissolution of these drugs from IR solid oral dosage forms become highly pH-dependent, allowing them to behave similarly to BCS Class I compounds (high solubility, high permeability) under certain physiological conditions.

Recent pharmacokinetic and biorelevant dissolution studies have supported this concept, demonstrating that several weakly acidic BCS Class II drugs, including non-steroidal anti-inflammatory drugs (NSAIDs), exhibit poor solubility in the stomach but show markedly improved solubility in the intestinal environment. These findings have generated increasing interest in the potential extension of biowaiver eligibility to the selected BCS Class II weak acids [86].

Another example of a BCS Class II weak acid is voriconazole, whose recent evaluations have highlighted limitations within the current International Council for Harmonization (ICH) M9 biowaiver framework. Although voriconazole demonstrates rapid dissolution at its highest marketed dose strength and achieves high BA, it remains ineligible for a biowaiver under ICH M9 due to the absence of dose-proportional pharmacokinetics across the therapeutic range. Interestingly, under the former FDA solubility criteria (based on the highest dose strength), voriconazole would have been classified as BCS Class I and thus eligible for a biowaiver.

These findings, combined with supportive pharmacokinetic and biorelevant dissolution data, suggest that the existing ICH M9 criteria may be overly conservative in certain cases. This has prompted proposals to revise current eligibility criteria and explore integration of refined classification systems, such as the refined Developability Classification System (rDCS), which better accounts for in vivo drug behavior. Such refinements could allow for a safe and scientifically justified extension of biowaiver approvals to additional BCS Class II weak acid drugs [87].

8.6 Advanced Simulation Models for FDA Biowaiver Submission

In recent years, regulatory agencies such as FDA, EMA, and international organizations have demonstrated the potential and the value of using simulation models in supporting regulatory decision-making, including granting biowaivers [88, 89]. Physiologically Based Pharmacokinetic (PBPK) models and Physiologically Based Biopharmaceutics Models (PBBM) are now widely used to support drug product development and to simulate drug disposition profiles in the body. These modeling and simulation approaches play a critical role in predicting human drug exposure and assessing BA and BE virtually, thereby reducing the effort, cost, and time associated with in vivo BE studies [90, 91].

Wu et al. discussed scientific considerations for extending biowaivers to BCS Class III through the use of modeling and simulation. In 2017, FDA issued guidance recommending the criteria for granting a BCS-based biowaiver. These include high solubility of the drug across the physiological pH range and ensuring that the formulation is qualitatively the same and quantitatively very similar to the reference product. These criteria were confirmed by the ICH M9 guideline (June 2018). The FDA has also used similar terminology referring to the BCS-based biowaiver option for BCS Class III drugs (highly soluble but poorly permeable). However, establishing IVIVC using conventional mass balance deconvolution approaches, such as Wanger–Nelson or Loo–Riegelman methods, which assume a single absorption compartment, is generally not feasible for very rapid dissolving dosage forms containing BCS Class III drugs. For compounds with low permeability, absorption is not restricted to a single region of the GI tract. The dissolved drug fraction may transit to different intestinal segments before entering systemic circulation. Therefore, when estimating the fraction absorbed, it is important to consider the heterogeneity of the small intestine, including enzyme expression, intestinal metabolism, and regional intestinal effective permeability. Unlike conventional mass balance deconvolution techniques, PBPK models are able to disentangle multiple processes that contribute to drug absorption, e.g. dissolution, GI transit, and permeation, and can establish IVIVC using variants of the compartmental absorption and transit model, supporting biowaiver for formulations containing BCS Class III drugs. However, knowledge gaps still need to be closed to enable the envisioned pathway to combine in vitro testing and physiologically based modeling as a promising biowaiver practice [92].

A notable example of using PBBM application in generic (test) product comes from Dr. Reddy's Laboratories (DRL), which has successfully applied PBBM to support regulatory biowaivers for BCS Class III drugs. The test product contained the same API in an IR formulation but as a different salt. A few excipients were not similar between reference and test products; however, none of the excipients were expected to impact permeability. Traditionally, the dissolution profiles of reference and test products are compared using similarity factor (f2) to assess their in vitro and in vivo similarity, both in generic industry and by regulatory agencies as part of regulatory submissions to support biowaivers. However, in certain markets, DRL's generic formulation exhibits faster dissolution than the local reference product. This case resulted in f2 values below 50 at pH 6.8 for both 50 and 100 mg strengths, which would normally prevent a biowaiver. Importantly, in the context of BCS Class III drugs, f2 can become over-discriminatory and clinically irrelevant, since absorption is governed by permeability rather than dissolution. To address this, DRL developed and validated a GastroPlus PBBM model and conducted virtual BE trials. These simulations demonstrated that, despite the observed dissolution mismatch, BE was achieved for both the 50 and 100 mg strengths. Furthermore, the modeling approach enabled the establishment of a dissolution safe space, which showed that drug release above 85% within 60 minutes had no clinical impact on BE. This strategy allowed DRL to obtain biowaivers even in cases where traditional f2 analysis failed, thereby highlighting the growing regulatory value of model-based evidence in generic drug development [91].

Another work presents an example of a PBPK model-based virtual BE (VBE) study that bridges differing formulations of Elagolix to support the waiver of BE studies. Elagolix (Orilissa®) is an orally active gonadotropin-releasing hormone (GnRH) receptor antagonist that inhibits endogenous GnRH signaling by binding competitively to GnRH receptors in the pituitary gland. In the Phase 3 clinical trials for endometriosis, treatment was administered as two separate tablets: an Elagolix tablet and an estradiol/norethindrone acetate (E2/NETA) 1/0.5 mg tablet. However, the commercial product was later designed as fixed-dose combination (FDC) capsules, consisting of a morning capsule (AM capsule, containing an Elagolix tablet and E2/NETA tablet) and an evening capsule (PM capsule, consisting of an Elagolix tablet). Dissolution testing revealed dissimilar in vitro profiles between the clinical trial tablets and the commercial capsules, which would traditionally mandate additional in vivo BE studies or BE waiver request. To address this, a previously verified PBPK model was updated with in vitro dissolution data to simulate virtual crossover BE trials. The simulations, further validated against relevant BE study data, confirmed that the AM and PM capsules met the regulatory BE criteria of 0.80–1.25 when compared with the reference tablets. This case represents a pioneering example where PBPK modeling, integrated with dissolution data, successfully supported a regulatory biowaiver, thereby avoiding large and unnecessary clinical BE studies [93].

These examples, among others, underscore a gradual paradigm shift, whereby advanced simulation models are increasingly recognized as an integral component of the regulatory evidence framework. There is a critical need for standardized protocols and robust regulatory guidelines to facilitate the integration of virtual BE methodologies into pharmaceutical practices [94].

8.7 Conclusions

This chapter reviewed the characteristics of BCS Class II drugs, including their physicochemical properties, physiological parameters, and dosage form variables. Understanding these and other factors, as well as their impact on the absorption process, is essential for the development of effective oral delivery systems for BCS Class II drug substances. Future research should focus on the development of advanced simulation models to reduce unnecessary human exposure in clinical trials, reduce development costs, and accelerate the availability of new drug products.

Abbreviations

API	active pharmaceutical ingredients
BA/BE	bioavailability/bioequivalence
BCS	Biopharmaceutics Classification System
CMC	critical micelle concentration
DRL	Dr. Reddy's Laboratories
E2/NETA	estradiol/norethindrone acetate
EMA	European Medicines Agency
FaSSIF	Fasting-State Simulated Intestinal Fluid
FDA	US Food and Drug Administration

FeSSIF	fed-state simulated intestinal fluid
GI	gastrointestinal
GnRH	gonadotropin-releasing hormone
ICH	International Council for Harmonization
IR	immediate release
IVIVC	in vitro–in vivo correlation
NSAIDs	non-steroidal anti-inflammatory drugs
PBBM	physiologically based biopharmaceutics models
PBPK	physiologically based pharmacokinetic
rDCS	refined developability classification system
SGF	simulated gastric fluid
VBE	virtual bioequivalence
WHO	World Health Organization

Symbols

A	surface area of the solid drug
A_n	absorption number
C_s	equilibrium solubility
C_W	concentration of the drug at the membrane
D	diffusivity
D_0	dose number
D_n	dissolution number
h	effective diffusion boundary layer thickness adjacent to the dissolving surface
J_W	mass transport across the GI wall
M_0	dose
P_{eff}	effective permeability
r_0	initial particle radius
R	intestinal radius
t_{abs}^{-1}	effective absorption rate constant
t_{Diss}	dissolution time
t_{res}	residence time
V_0	volume of water
X_d	amount of drug already dissolved

References

1 Sachan, N.K., Bhattacharya, A., Pushkar, S., and Mishra, A. (2009). Biopharmaceutical classification system: a strategic tool for oral drug delivery technology. *Asian J. Pharm.* 3 (2): 76–81.

2 Chavda, H.V., Patel, C.N., and Anand, I.S. (2010). Biopharmaceutics classification system. *Syst. Rev. Pharm.* 1 (1): 62–69.

3 Dahan, A. and Amidon, G. (2009). Gastrointestinal dissolution and absorption of class II drugs. In: *Drug Bioavailability: Estimation of Solubility, Permeability, Absorption and Bioavailability* (ed. H. van de Waterbeemd and B. Testa), 33–51. Weinheim: WILEY-VCH Verlag GmbH & Co. KGaA.

4 Lennernas, H. (1998). Human intestinal permeability. *J. Pharm. Sci.* 87 (4): 403–410.

5 Yu, L.X., Lipka, E., Crison, J.R., and Amidon, G.L. (1996). Transport approaches to the biopharmaceutical design of oral drug delivery systems: prediction of intestinal absorption. *Adv. Drug Deliv. Rev.* 19 (3): 359.

6 Amidon, G.L., Lennernas, H., Shah, V.P., and Crison, J.R. (1995). A theoretical basis for a biopharmaceutic drug classification: the correlation of in vitro drug product dissolution and in vivo bioavailability. *Pharm. Res.* 12 (3): 413.

7 Oh, D.M., Curl, R.L., and Amidon, G.L. (1993). Estimating the fraction dose absorbed from suspensions of poorly soluble compounds in humans: a mathematical model. *Pharm. Res.* 10 (2): 264.

8 Sinko, P.J., Leesman, G.D., and Amidon, G.L. (1991). Predicting fraction dose absorbed in humans using a macroscopic mass balance approach. *Pharm. Res.* 8 (8): 979.

9 Nagesh, X.B. and Murti, V. (2007). BCS class IV drugs: develop or discard? *AAPS Annual Meeting and Exposition.* San Diego Convention Centres, San Diego, CA, 11–15.

10 Bala, I., Bhardwaj, V., Hariharan, S. et al. (2005). Design of biodegradable nanoparticles: a novel approach to encapsulating poorly soluble phytochemical ellagic acid. *Nanotechnology* 16 (12): 2819.

11 Kommuru, T.R., Gurley, B., Khan, M.A., and Reddy, I.K. (2001). Self-emulsifying drug delivery systems (SEDDS) of coenzyme Q10: formulation development and bioavailability assessment. *Int. J. Pharm.* 212 (2): 233–246.

12 Sharma, P., Varma, M.V., Chawla, H.P., and Panchagnula, R. (2005). In situ and in vivo efficacy of peroral absorption enhancers in rats and correlation to in vitro mechanistic studies. *Farmaco (Societa Chimica Italiana : 1989)* 60 (11–12): 874–883.

13 Dhake, P.R., Kumbhar, S.T., and Gaikwad, V.L. (2024). Biowaiver based on biopharmaceutics classification system: considerations and requirements. *Pharm. Sci. Adv.* 2: 100020.

14 Becker, C., Dressman, J.B., Amidon, G.L. et al. (2008). Biowaiver monographs for immediate release solid oral dosage forms: pyrazinamide. *J. Pharm. Sci.* 97 (9): 3709–3720.

15 Blume, H.H. and Schug, B.S. (1999). The biopharmaceutics classification system (BCS): class III drugs–better candidates for BA/BE waiver? *Eur. J. Pharm. Sci.* 9 (2): 117.

16 Cheng, C.L., Yu, L.X., Lee, H.L. et al. (2004). Biowaiver extension potential to BCS Class III high solubility-low permeability drugs: bridging evidence for metformin immediate release tablet. *Eur. J. Pharm. Sci.* 22 (4): 297.

17 Jantratid, E., Prakongpan, S., Amidon, G.L., and Dressman, J. (2006). Feasibility of biowaiver extension to biopharmaceutics classification system class III drug products: cimetidine. *Clin. Pharmacokinet.* 45 (4): 385–399.
18 ICH Harmonised Guideline (2020). Biopharmaceutics classification system-based biowaivers M9. International Council for Harmonisation of Technical Requirements for Pharmaceuticals for Human Use.
19 Dahan, A. and Hoffman, A. (2006). Enhanced gastrointestinal. In: *Enhancement in Drug Delivery* (ed. E. Touitou and B.W. Barry), 111–127. CRC Press.
20 Yasir, M., Asif, M., Kumar, A., and Aggarval, A. (2010). Biopharmaceutical classification system: an account. *Int. J. PharmTech Res.* 2 (3): 1681–1690.
21 Dressman, J.B., Amidon, G.L., Reppas, C., and Shah, V.P. (1998). Dissolution testing as a prognostic tool for oral drug absorption: immediate release dosage forms. *Pharm. Res.* 15 (1): 11.
22 Fagerholm, U. (2007). Evaluation and suggested improvements of the biopharmaceutics classification system (BCS). *J. Pharm. Pharmacol.* 59: 751.
23 Wilding, I. (1999). Evolution of the biopharmaceutics classification system (BCS) to oral modified release (MR) formulations: what do we need to consider? *Eur. J. Pharm. Sci.* 8 (3): 157–159.
24 Levich, V. (1962). *Physicochemical Hydrodynamics*. Prentice-Hall.
25 Nernst, W. and Brunner, E. (1904). Theorie der Reaktionsgeschwindigkeit in heterogenen Systemen. *Z. Phys. Chem.* 47: 52–110.
26 Noyes, A. and Whitney, W. (1897). The rate of solution of solid substances in their own solutions. *J. Am. Chem. Soc.* 19: 930–934.
27 Persson, E., Gustafsson, A.-S., Carlsson, A. et al. (2005). The effects of food on the dissolution of poorly soluble drugs in human and in model small intestinal fluids. *Pharm. Res.* 22 (12): 2141.
28 Porter, C.J.H., Trevaskis, N.L., and Charman, W.N. (2007). Lipids and lipid-based formulations: optimizing the oral delivery of lipophilic drugs. *Nat. Rev. Drug Discov.* 6 (3): 231.
29 Bakatselou, V., Oppenheim, R.C., and Dressman, J.B. (1991). Solubilization and wetting effects of bile salts on the dissolution of steroids. *Pharm. Res.* 8 (12): 1461.
30 Sheng, J.J. (2007). Toward an in vitro bioequivalence test. PhD thesis. University of Michigan, Ann Arbor, MI.
31 Amidon, G.E., Higuchi, W., and Ho, N. (1982). Theoretical and experimental studies of transport of micelle-solubilized solutes. *J. Pharm. Sci.* 71 (1): 77–84.
32 Granero, G., Ramachandran, C., and Amidon, G.L. (2005). Dissolution and solubility behavior of fenofibrate in sodium lauryl sulfate solutions. *Drug Dev. Ind. Pharm.* 31 (9): 917–922.
33 Horter, D. and Dressman, J.B. (2001). Influence of physicochemical properties on dissolution of drugs in the gastrointestinal tract. *Adv. Drug Deliv. Rev.* 46 (1–3): 75.
34 Sheng, J.J., Kasim, N.A., Chandrasekharan, R., and Amidon, G.L. (2006). Solubilization and dissolution of insoluble weak acid, ketoprofen: effects of pH combined with surfactant. *Eur. J. Pharm. Sci.* 29 (3–4): 306.

35 Luner, P.E. (2000). Wetting properties of bile salt solutions and dissolution media. *J. Pharm. Sci.* 89 (3): 382–395.

36 Luner, P.E. and Kamp, D.V. (2001). Wetting behavior of bile salt–lipid dispersions and dissolution media patterned after intestinal fluids. *J. Pharm. Sci.* 90 (3): 348–359.

37 Davis, S.S., Hardy, J.G., and Fara, J.W. (1986). Transit of pharmaceutical dosage forms through the small intestine. *Gut* 27 (8): 886–892.

38 Lindahl, A., Ungell, A.L., Knutson, L., and Lennernäs, H. (1997). Characterization of fluids from the stomach and proximal jejunum in men and women. *Pharm. Res.* 14 (4): 497–502.

39 Oberle, R. and Amidon, G.L. (1987). The influence of variable gastric emptying and intestinal transit rates on the plasma level curve of cimetidine: an explanation for the double peak phenomenon. *J. Pharmacokinet. Biopharm.* 15 (5): 529–544.

40 Kalantzi, L., Goumas, K., Kalioras, V. et al. (2006). Characterization of the human upper gastrointestinal contents under conditions simulating bioavailability/bioequivalence studies. *Pharm. Res.* 23 (1): 165–176.

41 Balakrishnan, A., Rege, B.D., Amidon, G.L., and Polli, J.E. (2004). Surfactant-mediated dissolution: contributions of solubility enhancement and relatively low micelle diffusivity. *J. Pharm. Sci.* 93 (8): 2064–2075.

42 Galia, E., Nicolaides, E., Hörter, D. et al. (1998). Evaluation of various dissolution media for predicting in vivo performance of class I and II drugs. *Pharm. Res.* 15 (5): 698.

43 Kalantzi, L., Persson, E., Polentarutti, B. et al. (2006). Canine intestinal contents vs. simulated media for the assessment of solubility of two weak bases in the human small intestinal contents. *Pharm. Res.* 23 (6): 1373.

44 Russell, T.L., Berardi, R.R., Barnett, J.L. et al. (1994). pH-related changes in the absorption of dipyridamole in the elderly. *Pharm. Res.* 11 (1): 136.

45 Zhou, R., Moench, P., Heran, C. et al. (2005). pH-dependent dissolution in vitro and absorption in vivo of weakly basic drugs: development of a canine model. *Pharm. Res.* 22 (2): 188.

46 Granero, G., Ramachandran, C., and Amidon, G.L. (2006). Rapid in vivo dissolution of ketoprofen: implications on the biopharmaceutics classification system. *Pharmazie* 61 (8): 673–676.

47 Jinno, J., Oh, D.M., Crison, J.R., and Amidon, G.L. (2000). Dissolution of ionizable water-insoluble drugs: the combined effect of pH and surfactant. *J. Pharm. Sci.* 89 (2): 268–274.

48 Yu, L.X. and Amidon, G.L. (1999). A compartmental absorption and transit model for estimating oral drug absorption. *Int. J. Pharm.* 186 (2): 119.

49 Yu, L.X., Crison, J.R., and Amidon, G.L. (1996). Compartmental transit and dispersion model analysis of small intestinal transit flow in humans. *Int. J. Pharm.* 140 (1): 111.

50 Dressman, J.B., Vertzoni, M., Goumas, K., and Reppas, C. (2007). Estimating drug solubility in the gastrointestinal tract. *Adv. Drug Deliv. Rev.* 59 (7): 591.

51 Lobenberg, R., Kramer, J., Shah, V.P. et al. (2000). Dissolution testing as a prognostic tool for oral drug absorption: dissolution behavior of glibenclamide. *Pharm. Res.* 17 (4): 439.

52 Dressman, J. and Fleisher, D. (1986). Mixing-tank model for predicting dissolution rate control or oral absorption. *J. Pharm. Sci.* 75 (2): 109–116.

53 Sinko, C.M., Yee, A.F., and Amidon, G.L. (1990). The effect of physical aging on the dissolution rate of an anionic polyelectrolyte. *Pharm. Res.* 7 (6): 648.

54 Aungst, B.J., Nguyen, N.H., Taylor, N.J., and Bindra, D.S. (2002). Formulation and food effects on the oral absorption of a poorly water-soluble, highly permeable antiretroviral agent. *J. Pharm. Sci.* 91 (6): 1390–1395.

55 Bonlokke, L., Hovgaard, L., Kristensen, H.G. et al. (2001). Direct estimation of the in vivo dissolution of spironolactone, in two particle size ranges, using the single-pass perfusion technique (Loc-I-Gut®) in humans. *Eur. J. Pharm. Sci.* 12 (3): 239.

56 Kostewicz, E.S., Brauns, U., Becker, R., and Dressman, J.B. (2002). Forecasting the oral absorption behavior of poorly soluble weak bases using solubility and dissolution studies in biorelevant media. *Pharm. Res.* 19 (3): 345.

57 Takano, R., Sugano, K., Higashida, A. et al. (2006). Oral absorption of poorly water-soluble drugs: computer simulation of fraction absorbed in humans from a mini-scale dissolution test. *Pharm. Res.* 23 (6): 1144.

58 Custodio, J.M., Wu, C.Y., and Benet, L.Z. (2008). Predicting drug disposition, absorption/elimination/transporter interplay and the role of food on drug absorption. *Adv. Drug Deliv. Rev.* 60 (6): 717.

59 Dressman, J.B. and Reppas, C. (2000). in vitro–in vivo correlations for lipophilic, poorly water-soluble drugs. *Eur. J. Pharm. Sci.* 11 (Suppl. 2): S73.

60 Fordtran, J. and Locklear, T. (1966). Ionic constituents and osmolality of gastric and small intestinal fluids after eating. *Am J Dig Dis* 11 (7): 503–521.

61 Macheras, P., Koupparis, M., and Tsaprounis, C. (1986). Drug dissolution studies in milk using the automated flow injection serial dynamic dialysis technique. *Int. J. Pharm.* 33 (1–3): 125.

62 Galia, E., Horton, J., and Dressman, J.B. (1999). Albendazole generics–a comparative in vitro study. *Pharm. Res.* 16 (12): 1871.

63 Nicolaides, E., Symillides, M., Dressman, J.B., and Reppas, C. (2001). Biorelevant dissolution testing to predict the plasma profile of lipophilic drugs after oral administration. *Pharm. Res.* 18 (3): 380.

64 Vertzoni, M., Dressman, J., Butler, J. et al. (2005). Simulation of fasting gastric conditions and its importance for the in vivo dissolution of lipophilic compounds. *Eur. J. Pharm. Biopharm.* 60 (3): 413.

65 Scholz, A., Abrahamsson, B., Diebold, S.M. et al. (2002). Influence of hydrodynamics and particle size on the absorption of felodipine in labradors. *Pharm. Res.* 19 (1): 42.

66 Scholz, A., Kostewicz, E., Abrahamsson, B., and Dressman, J.B. (2003). Can the USP paddle method be used to represent in-vivo hydrodynamics? *J. Pharm. Pharmacol.* 55: 443.

67 Sheng, J.J., Sirois, P.J., Dressman, J.B., and Amidon, G.L. (2008). Particle diffusional layer thickness in a USP dissolution apparatus II: a combined function of particle size and paddle speed. *J. Pharm. Sci.* 97 (11): 4815–4829.

68 Charman, W.N. (2000). Lipids, lipophilic drugs, and oral drug delivery – some emerging concepts. *J. Pharm. Sci.* 89 (8): 967–978.

69 Gershanik, T. and Benita, S. (2000). Self-dispersing lipid formulations for improving oral absorption of lipophilic drugs. *Eur. J. Pharm. Biopharm.* 50 (1): 179–188.

70 Fleisher, D., Li, C., Zhou, Y. et al. (1999). Drug, meal, and formulation interactions influencing drug absorption after oral administration: clinical implications. *Clin. Pharmacokinet.* 36 (3): 233–254.

71 Dintaman, J.M. and Silverman, J.A. (1999). Inhibition of P-glycoprotein by D-α-tocopheryl polyethylene glycol 1000 succinate (TPGS). *Pharm. Res.* 16 (10): 1550.

72 Nerurkar, M.M., Burton, P.S., and Borchardt, R.T. (1996). The use of surfactants to enhance the permeability of peptides through Caco-2 cells by inhibition of an apically polarized efflux system. *Pharm. Res.* 13 (4): 528.

73 Dahan, A. and Hoffman, A. (2005). Evaluation of a chylomicron flow blocking approach to investigate the intestinal lymphatic transport of lipophilic drugs. *Eur. J. Pharm. Sci.* 24 (4): 381–388.

74 Gershkovich, P. and Hoffman, A. (2007). Effect of a high-fat meal on absorption and disposition of lipophilic compounds: the importance of degree of association with triglyceride-rich lipoproteins. *Eur. J. Pharm. Sci.* 32 (1): 24.

75 Tso, P. (1985). Gastrointestinal digestion and absorption of lipid. *Adv. Lipid Res.* 21: 143–186.

76 Dahan, A. and Hoffman, A. (2008). Rationalizing the selection of oral lipid-based drug delivery systems by an in vitro dynamic lipolysis model for improved oral bioavailability of poorly water-soluble drugs. *J. Control. Release* in press.

77 Porter, C.J.H. and Charman, W.N. (2001). in vitro assessment of oral lipid-based formulations. *Adv. Drug Deliv. Rev.* 50 (Suppl. 1): S127.

78 Sek, L., Porter, C.J.H., Kaukonen, A.M., and Charman, W.N. (2002). Evaluation of the in vitro digestion profiles of long and medium chain glycerides and the phase behaviour of their lipolytic products. *J. Pharm. Pharmacol.* 54: 29.

79 Zangenberg, N.H., Mullertz, A., Gjelstrup Kristensen, H., and Hovgaard, L. (2001). A dynamic in vitro lipolysis model. II. Evaluation of the model. *Eur. J. Pharm. Sci.* 14 (3): 237.

80 Dahan, A. and Hoffman, A. (2006). Use of a dynamic in vitro lipolysis model to rationalize oral formulation development for poor water-soluble drugs: correlation with in vivo data and the relationship to intra-enterocyte processes in rats. *Pharm. Res.* 23 (9): 2165–2174.

81 Guidance for Industry (2000). Waiver of in vivo bioavailability and bioequivalence studies for immediate-release solid oral dosage forms based on a biopharmaceutics classification system. Department of Health and Human Services, US Food and Drug Administration, Rockville, MD. https://www.gmp‐compliance.org/files/guidemgr/UCM070246.pdf.

82 WHO (2015). Multisource (generic) pharmaceutical products: guidelines on registration requirements to establish interchangeability. In: *WHO Expert Committee on Specifications for Pharmaceutical Preparations, Forty-Ninth Report*, vol. 992, 134–184. Geneva: World Health Organization.

83 Yu, L.X., Amidon, G.L., Polli, J.E. et al. (2002). Biopharmaceutics classification system: the scientific basis for biowaiver extensions. *Pharm. Res.* 19 (7): 921–925. doi: 10.1023/a:1016473601633.

84 Guidance for Industry (2000). *Waiver of in vivo Bioavailability and Bioequivalence Studies for Immediate-Release Solid Oral Dosage Forms Based on a Biopharmaceutics Classification System*. Food and Drug Administration (FDA).

85 Tsume, Y., Mudie, D.M., Langguth, P. et al. (2014). The biopharmaceutics classification system: subclasses for in vivo predictive dissolution (IPD) methodology and IVIVC. *Eur. J. Pharm. Sci.* 57: 152–163. doi: 10.1016/j.ejps.2014.01.009.

86 Li, M., Zhang, X., Wu, D. et al. (2021). Understanding in vivo dissolution of immediate release (IR) solid oral drug products containing weak acid BCS class 2 (BCS class 2a) drugs. *AAPS J.* 23 (6): 113. doi: 10.1208/s12248-021-00639-0.

87 Beran, K., Abrahamsson, B., Charoo, N. et al. (2025 Feb). Biowaiver monographs for immediate-release solid oral dosage forms: Voriconazole. *J. Pharm. Sci.* 114 (2): 660–680. doi: 10.1016/j.xphs.2024.10.055.

88 USFDA (2020). The use of physiologically based pharmacokinetic analyses—biopharmaceutics applications for oral drug product development, manufacturing changes, and controls. https://www.fda.gov/media/142500/download (accessed 15 June 2022).

89 EMA (2019). Guideline on the reporting of physiologically based pharmacokinetic (PBPK) modelling and simulation. https://www.ema.europa.eu/en/documents/scientific-guideline/guideline-reporting-physiologically-based-pharmacokinetic-pbpk-modelling-simulation_en.pdf (accessed 15 June 2022).

90 Pawestri, S.A. (2023). Application in silico modeling simulation in bioequivalence studies: a review. *J. Food Pharm. Sci.* 11 (1): 763–769. doi: 10.22146/jfps.5893.

91 Bhattiprolu, A.K., Kollipara, S., Ahmed, T. et al. (2022). Utility of physiologically based biopharmaceutics modeling (PBBM) in regulatory perspective: application to supersede f2, enabling biowaivers & creation of dissolution safe space. *J. Pharm. Sci.* 111 (12): 3397–3410. doi: 10.1016/j.xphs.2022.09.003.

92 Wu, F., Cristofoletti, R., Zhao, L., and Rostami-Hodjegan, A. (2021). Scientific considerations to move towards biowaiver for biopharmaceutical classification system class III drugs: how modeling and simulation can help. *Biopharm. Drug Dispos.* 42 (4): 118–127. doi: 10.1002/bdd.2274.

93 Mukherjee, D., Chen, M.J., Shao, X. et al. (2023). Virtual bioequivalence assessment of elagolix formulations using physiologically based pharmacokinetic modeling. *AAPS J.* 25 (3): 30. doi: 10.1208/s12248-023-00794-6.

94 Alotaiq, N. and Dermawan, D. (2024). Advancements in virtual bioequivalence: a systematic review of computational methods and regulatory perspectives in the pharmaceutical industry. *Pharmaceutics* 16 (11): 1414. doi: 10.3390/pharmaceutics16111414.

9

Cyclodextrin-Based Formulations to Improve Solubility of Poorly Soluble Drugs

Ganesh Shevalkar[1], Shradhha Tiwari[2], Deepak Bari[3], Chandrakantsing Pardeshi[3], and Sarika Kamble[4]

[1]*Department of Pharmaceutics, Sandip Institute of Pharmaceutical Sciences, Affiliated to Savitribai Phule Pune University (SPPU, Pune), Trambak Road, Mahiravani, Nashik, Maharashtra 422213, India*
[2]*Department of Pharmaceutics, School of Pharmacy, MIT Vishwaprayag University, Solapur - Pune Hwy, Kegaon, Maharashtra, Solapur 413255, India*
[3]*Department of Pharmaceutics, R. C. Patel Institute of Pharmaceutical Education and Research, Affiliated to Kavayitri Bahinabai Chaudhari North Maharashtra University (KBCNMU, Jalgaon), Karwand Naka, Shirpur, Maharashtra 425405, India*
[4]*Department of Pharmacology, SMBT College of Pharmacy, Affiliated to Savitribai Phule Pune University (SPPU, Pune), Nandi Hills, Dhamangaon, Nashik, Maharashtra 422403, India*

9.1 Introduction

9.1.1 Overview of Cyclodextrins (CDs)

Cyclodextrins (CDs) are a family of cyclic oligosaccharides produced from starch by enzymatic conversion. They possess a unique truncated cone-shaped structure with a hydrophobic central cavity and a hydrophilic outer surface, enabling them to form inclusion complexes with a wide range of guest molecules. This ability to encapsulate poorly soluble drugs enhances their solubility, stability, and bioavailability, making CDs valuable excipients in modern drug delivery systems. Depending on the number of glucose units, CDs are classified mainly as α-, β-, and γ-CDs, each with distinct cavity sizes and pharmaceutical applications.

9.1.2 Definition and Structural Properties

CDs are cyclic oligosaccharides of α-D-glucopyranose units linked by α-1,4-glycosidic bonds. These naturally occurring molecules are generated by the enzymatic degradation of starch using CD glycosyltransferase [1]. Their unique truncated cone shape arises from the chair conformation of glucose residues. The internal cavity of CDs is relatively hydrophobic, while the external surface is hydrophilic due to the hydroxyl groups projecting outward. This dual character enables CDs to form non-covalent inclusion complexes with a wide range of hydrophobic drug molecules, significantly altering the physicochemical properties of the included compounds [2].

9.1.3 Historical Perspective of CDs in Drug Formulations

CDs were formally discovered in 1891 as Antoine Villers studied the enzymatic degradation of potato starch in bacteria [3–5]. Villers isolated two compounds (most likely α-CD and β-CD) with properties similar to cellulose (i.e. resistance to acid hydrolysis and non-reducing properties). He consequently named them "cellulosines" [6]. Franz Schardinger, the so-called Founding Father of CD chemistry, renamed these two compounds dextrins, from which the modern name of "cyclodextrins" originates [7]. In 1939, Karl Freudenberg and his co-workers published a full description of the two separated compounds. In 1954, Friedrich Cramer focused on separating and purifying CDs naturally and studied CD–guest inclusion complexes in different states [1].

Even though CDs were discovered well over a century ago, the number of papers about CDs from 1955 to 1975 was minimal. This low number is partially due to the (erroneous) statement that CDs were toxic. Dexter French reported this in a critical review of CDs in 1957 [7]. He and B. H. Thomas fed rats small quantities of highly purified CD, and oddly enough, all subsequently died. These results, however, were unpublished, and the authors did not explain the cause of death. The methods French and Thomas used are questionable and have since been proven incorrect; other factors such as solubility, chemical modification, and route of administration (oral, intravenous, topical, etc.) must be taken into account when considering CDs' toxicity, which varies even among the native CDs [1].

9.1.4 Importance in Drug Delivery

CDs have become essential excipients in drug formulation, particularly for enhancing the solubility of poorly water-soluble compounds. Their ability to encapsulate hydrophobic drug molecules within their cavities leads to improved dissolution, chemical stability, and protection from enzymatic degradation [8]. Moreover, inclusion complexation often results in enhanced permeability across biological membranes and, consequently, improved oral bioavailability [9]. CDs can also help in masking unpleasant tastes or odors and enable the development of more patient-friendly dosage forms [10].

9.2 CD Structure and Types

9.2.1 Alpha, Beta, and Gamma CDs

CDs are biocompatible and biodegradable materials produced by the enzymatic degradation of starch [1]. They consist of 6 to 12 (α-1, 4)-linked α-D-glucopyranose units. Due to the chain conformation of the glucopyranose units, they have the shape of a truncated cone with a slightly hydrophobic central cavity provided by the skeletal carbon atoms of glucose and a hydrophilic exterior provided by the hydroxyl groups of glucose [2]. Their physicochemical properties are listed in Table 9.1. Most prominent are α-, β-, and γ-CD, as illustrated in Figure 9.1. In addition to these pharmaceutically approved CDs, larger cyclic oligosaccharides with 9–12 glucose

Table 9.1 Physical–chemical parameters of pharmaceutically approved CDs [2, 11].

	Number of glucopyranose units	Degree of substitution	Molecular weight (Da)	Solubility in water (g/l, 25 °C)	Internal diameter (Å)	Internal volume (nm^3)
α-CD	6	—	972	145	4.7–5.3	0.174
β-CD	7	—	1135	18.5	6.0–6.5	0.262
γ-CD	8	—	1297	232	7.5–8.3	0.427
HP-β-CD	7	0.65	1400	>600	6.0	0.262
RM-β-CD	7	1.8	1312	>500	5.8–6.5	0.262
SB-β-CD	7	0.9	2163	>500	6.0–6.5	NA
HP-γ-CD	8	0.6	1576	>500	8.0	0.427

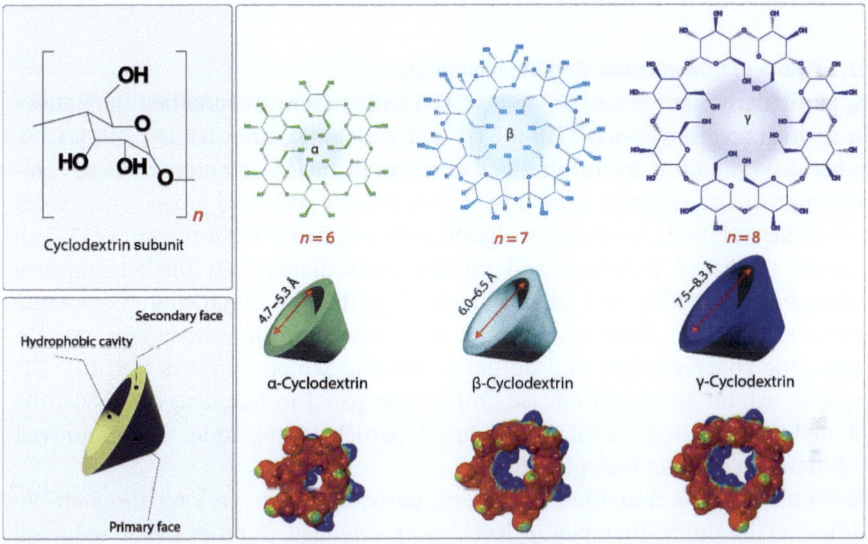

Figure 9.1 The structure of the anhydroglucose repeating unit of CDs (upper left), with a schematic representation of the 3D structure (lower left) and chemical and 2D structures, as well as computational surface analyses of α-, β-, and γ-CDs (right-hand side).

units, named δ-, ε-, κ-, and μ-CDs, respectively, are available. The terminology can be theoretically even continued till ω-CD with 29 repeats [8].

9.2.2 CDs for Enhancing Bioavailability of Pharmaceutical Compounds

The therapeutic efficacy of an active pharmaceutical ingredient (API) is not solely dependent on its inherent pharmacological activity but is significantly influenced by the formulation's excipients. These excipients can enhance drug solubility, stability, and release kinetics, thereby modifying the onset, intensity, and duration of therapeutic action. In this context, CDs—cyclic oligosaccharides capable of forming inclusion complexes—have emerged as pivotal excipients for improving drug bioavailability [12].

Bioavailability is defined as the proportion of an administered dose of an API that reaches the systemic circulation in an unaltered form. While intravenous administration provides 100% bioavailability, other routes often result in reduced availability due to various barriers, including poor aqueous solubility, limited membrane permeability, enzymatic degradation, and first-pass metabolism.

Several factors contribute to the reduced bioavailability of orally administered drugs, which include the following:

- **Physicochemical barriers**, such as crystalline nature, poor aqueous solubility, and slow dissolution rate.
- **Chemical instability**, including hydrolysis, oxidation, or photodegradation under physiological conditions.
- **Physiological factors**, like enzymatic degradation and limited permeability across biological membranes.

9.2.2.1 Role of CDs in Enhancing Bioavailability

CDs, particularly natural (α-, β-, and γ-CDs) and chemically modified derivatives such as hydroxypropyl-β-CD (HP-β-CD) and sulfobutyl ether-β-CD (SBE-β-CD), enhance drug solubility by forming non-covalent inclusion complexes. These complexes improve the apparent solubility of hydrophobic APIs and protect them from chemical degradation, resulting in significantly improved bioavailability [13–15]. Moreover, CD-based polymers and protein-encapsulated CDs further improve membrane permeability and chemical stability, thereby enhancing therapeutic effectiveness [16, 17]. Such systems can modulate the drug release rate, sustain plasma drug concentration, and optimize pharmacokinetic profiles [18, 19]. CD complexes exhibit enhanced release profiles compared to free drugs. This can be attributed to increased wettability, reduced particle aggregation, and improved dissolution dynamics in biological fluids.

Furthermore, CDs and their polymeric derivatives can prolong the half-life of drugs, maintaining therapeutic levels over extended durations and reducing dosing frequency [20, 21]. CD-based nanomaterials, including nanosponges, polymeric nanoparticles, graphene-based nanocarriers, and silica nanoparticles, offer advanced delivery options that combine solubility enhancement with controlled and targeted drug release [22].

These nanostructures not only improve pharmacokinetics but also enhance drug stability under physiological conditions. However, while CDs are known for their biodegradability and low toxicity [9], it is essential to assess the biocompatibility and safety of the nanoparticulate carriers used in such systems to ensure their suitability for clinical use.

9.2.3 Derivatives and Modified CDs

Modification of CDs is feasible through the three hydroxyl groups in the anhydroglucose repeats, as illustrated in Figures 9.2–9.4. The –OH in the C-6 position is the

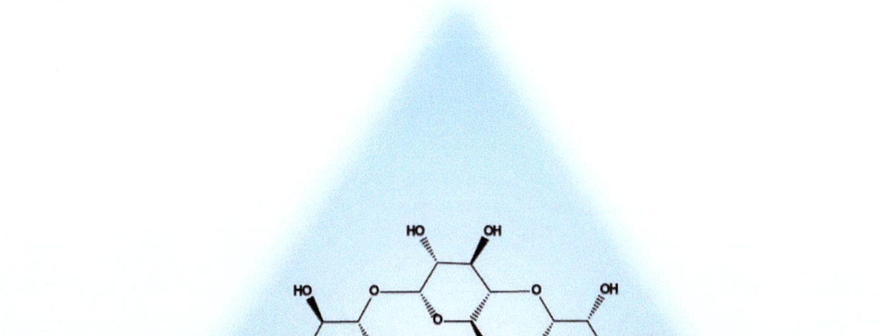

Figure 9.2 Synthetic pathways for various substituted CD derivatives, where R is the desired functionality, while R? is R or H.

most basic and most available for modifications [23]. The hydroxyl in the C-2 position is the most acidic, while the C-3 position is the most sterically hindered. CDs can be selectively modified due to these differences in the reactivity of the three hydroxyl groups [24]. Nonionic hydrophilic, slightly lipophilic, and ionic CDs are commercially available for various applications [25]. In most cases, CDs are modified to improve their solubility, but there are also examples of modifications aimed at other goals [26].

9.2.3.1 Substitution Reactions

CDs can be derivatized using various substitution reactions. One of the most widely applied methods is converting hydroxyls to other functionalities, resulting in deoxy-functionalized CDs. Another widespread approach is the reaction with hydroxyl groups, leading to the substitution of the oxygen of the hydroxyl moiety [27]. Monosubstitution and persubstitution of CDs at a specific position (i.e. C-2, C-3, or C-6) are important in order to prevent the formation of different isomers with

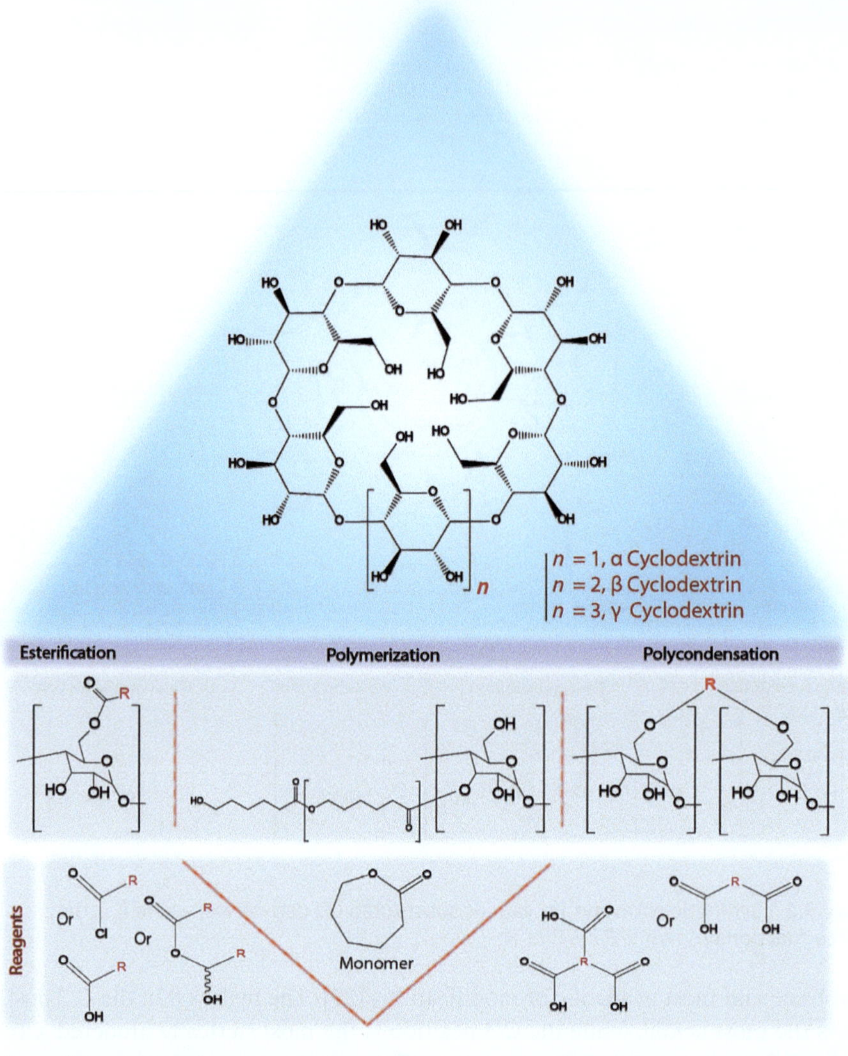

Figure 9.3 Schematic representation of the various esterification reactions of CDs, where R is the desired functionality or crosslinker.

various degrees of substitution [28]. Otherwise, these isomers have to be separated via chromatographic methods. The secondary hydroxyl groups at the C-2 and C-3 positions are difficult to modify individually [29]. Even though C-2 hydroxyl is the most acidic one, attempts to modify it in the presence of an equivalent amount of base resulted in unsatisfactory yield and uniformity of the product. The C-3 position is most inaccessible but can be selectively substituted via the inclusion complex method, where the reagent forms a host–guest complex with the CD, and the reactive site is oriented to the C-3 hydroxyl to react with it [8]. Finally, the C-6 position on the primary site is the most accessible and most nucleophilic hydroxyl group, freely available for several substitution reactions (Figure 9.2).

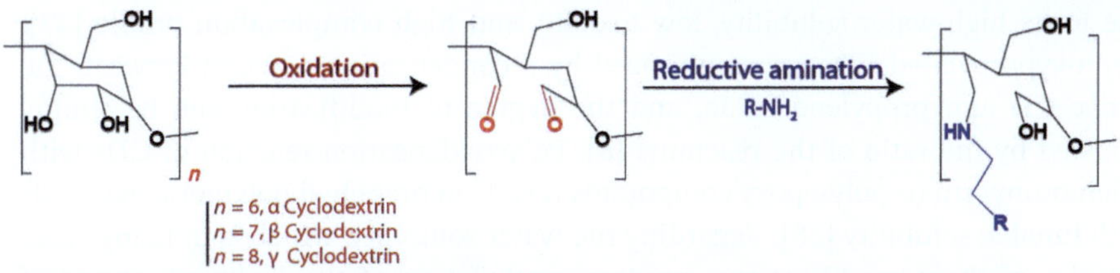

Figure 9.4 Schematic reaction pathway of oxidation of CDs followed by reductive amination, where R is the desired functionality.

9.2.3.2 Deoxy CDs

Mono-6-substituted CDs bear only one functional group per ring; consequently, these derivatives lack the symmetry of the parental CD. For the synthesis of such materials, 6-*O*-aryl/alkyl-sulfonated CDs are formed in the first step and modified further [27]. Predominantly, 6-deoxy-6-halogenated CDs are synthesized because of their high reactivity, but the 6-*O*-aryl/alkyl-sulfonated CDs can also be processed further to target products like monoazides, amines, or thiols [27, 30]. These monosubstituted CDs can be directly used [31] or modified further, for example, with click chemistry or oxidation [27]. Di-6-deoxy-substituted CDs are also synthesized with high selectivity through a protection–selective deprotection step, resulting in di-6-deoxy-6-thiolated α-CD forming intra- and intermolecular disulfides [8]. Per-6 modified CDs offer high symmetry and more functionalities than monosubstituted ones. Per-6-halogenated-per-6-deoxy CDs are synthesized in the first step and can be further modified with another substitution reaction, for example, with thiourea, leading to per-6-thiolated CDs [32]. Randomly substituted CDs bear several functional moieties in various positions (C-2, C-3, and/or C-6) [30]. The reaction of native CDs with thiourea and the subsequent hydrolysis of the thiouronium anion can be enhanced by using a microwave reactor to achieve a high degree of random thiolation [33]. The per-modification of CDs results in the derivatization of all hydroxyl groups. It is mostly used to increase solubility [30]. A recent study used phosphorus pentasulfide to gain per-deoxy-thiolated β-CD in a one-step reaction [34].

9.2.3.3 O-Substituted CDs

Among O-substituted CDs, methylated and hydroxypropylated ones are likely the most important derivatives due to their high-water solubility and complexing ability. Per-6-*O*-(*tert*-butyldimethylsilylated) CDs are protected intermediates for selective modification of the C-2 and C-3 positions, followed by a deprotection step [30]. This method is widely used for per-6 methylation of β-CD [35]. A simpler one-pot reaction was recently established to synthesize per-6-methyl β-CD using $CuSO_4$-complexed oligosaccharide [36]. Besides the selective modifications, random substitution of the hydroxyl groups at various positions is also feasible. Among others, randomly methylated β-CD is one of the most prominent methylated derivatives. The hydroxyl groups of CDs, especially on the primary face, can be utilized for condensation reactions. The most important CDs, modified by condensation reactions, are 2-hydroxypropylated CDs. Among them, 2-HP-β-CD is the most investigated variant due to its high-water solubility, low toxicity, and high complexation profile [37]. Hydroxypropylated CDs are synthesized by a condensation reaction between the native CD and propylene oxide, and the degree of modification can be simply adjusted by the ratio of the reactants [8]. Polycondensation reaction of CDs with epichlorohydrin or poly-epoxy compounds results in branched polymeric materials with tunable solubility [38]. Regarding the water solubility, which is in many cases the aim of these modifications, random methylation of the hydroxyl groups of β-CD increases its aqueous solubility up to 500 g/l, while hydroxypropylation of the hydroxyl groups at the C-6 position up to above 600 g/l. Introducing ionic groups into the CD structure can further increase aqueous solubility [11].

9.2.3.4 Esterification

A widely used modification method for CDs is their esterification with carboxylic acids, also a kind of O-substitution. This reaction is discussed separately due to the importance of esterification, mostly used to conjugate drugs or targeting moieties to the CDs, to form CD-based surfactants [39] or to enhance the solubility of some CD-based supramolecular structures [40]. The possible esterification reactions are summarized in Figure 9.3. Per-acetylated CDs with high solubility in supercritical carbon dioxide or organic solvents can form host–guest complexes in non-aqueous media [8]. Esterification of CDs can be achieved by enzymatic trans-esterification [41] or chemically, using carbodiimide chemistry or applying the acyl chloride form of the carboxylate [42]. Another option is to use CDs to initiate the ring-opening polymerization of lactide or ε-caprolactone monomers, resulting in a fully biodegradable macromolecular CD-ester [43]. Polycondensation is achievable by the esterification of CDs with multiple carboxylic acid or anhydride moiety-bearing molecules, such as citric acid, poly(acrylic acid), or pyromellitic dianhydride [44]. Depending on the conditions, soluble branched polymers or insoluble gels/nanosponges are formed; both structures are of particular interest for pharmaceutical and environmental applications [45].

9.2.3.5 Oxidation

The oxidation of CDs can lead to aldehyde functionalities by opening some anhydroglucose repeating units with $NaIO_4$. The formed aldehydes showed enhanced water solubility. Furthermore, broad-spectrum antimicrobial activity was found for these derivatives [46]. The aldehyde groups can be subjected to reductive amination, resulting in cationic substructures that provide high-water solubility [47], as depicted in Figure 9.4.

9.3 Approved CD-Based Pharmaceutical Formulations

Most of the pharmaceutical companies utilize β-CD with the prime aim of improving the bioavailability of the drugs with poor aqueous solubility. It has been revealed from the literature that CDs can alter the physicochemical and biological properties of drugs by forming inclusion complexes through host–guest interactions.

Table 9.2 Distribution of APIs in commercial drug products formulated with different types of CDs, including parent and chemically modified derivatives.

Types of CDs	α-CD	β-CD	γ-CD	n.s.
Parent	3	53	4	5
HP	—	36	4	2
SBE	—	14	—	—
RM	—	3	—	—
Sulfolipo	—	1	—	—
Total	3	111	8	7

n.s.: Not specified.
Source: Adapted from Puskás et al. [25]

9.3.1 CD-Based Products Containing Blockbuster Drugs

Japan-based Ono Pharmaceutical Co. (Osaka, Japan) was the first to commercialize a CD-based injectable formulation in 1976, which contained an α-CD/prostaglandin E1 complex. This was later followed by a tablet formulation comprising a β-CD/prostaglandin E2 complex [48]. In 1988, Chiesi Farmaceutici (Italy) introduced Brexin®, a formulation containing piroxicam complexed with β-CD, offering improved gastrointestinal tolerance and faster onset of action. In the United States, the first CD-based pharmaceutical product was launched in 1997 under the brand name Sporanox®, a solution of itraconazole complexed with HP-β-CD, developed by Janssen Pharmaceutical Co. (Belgium) [25]. Since then, the number of CD-based drug products on the market has continued to increase steadily.

Table 9.2 summarizes the number of APIs incorporated into commercially marketed drug products formulated with various CDs. The data indicate that the majority of these products utilize β-CD or its derivatives. Among the different β-CD derivatives employed, the most commonly used are HP-β-CD, hydroxypropyl-γ-CD, SBE-β-CD, randomly methylated β-CD (RAMEB), and sulfolipo-β-CD. These derivatives are selected based on their enhanced solubility, safety profiles, and suitability for specific routes of administration [25].

9.3.2 Successful CD-Based Formulations and Their Commercial Impact

CDs are widely employed in pharmaceutical product development, primarily to enhance the solubility, stability, and bioavailability of APIs. These improvements are achieved through the formation of inclusion complexes, where CDs encapsulate APIs within their hydrophobic cavities, thereby protecting them from degradation and increasing their aqueous solubility. Common dosage forms for CD-based formulations include tablets, parenteral solutions, and other solid and liquid forms. For instance, **Nitropen®**, a sublingual tablet used in the treatment of angina, incorporates β-CD to form an inclusion complex with nitroglycerin. This complex significantly reduces the volatility of nitroglycerin, resulting in improved stability. Additionally, the tablet was well-tolerated, with no reported sensation of a foreign body in the sublingual area, and maintained consistent nitroglycerin content during storage. **Olmetec®** orodispersible tablets, containing olmesartan medoxomil, utilize β-CD to effectively mask the unpleasant odor of the drug, enhancing patient compliance. Similarly, **Brivlera®**, a formulation of brivaracetam used for epilepsy treatment, employs β-CD to improve the drug's dissolution profile, thereby potentially enhancing its therapeutic efficacy [49].

9.3.2.1 Veklury®: A Success Story

Several studies have demonstrated that CD and CD derivative-based formulations can be effectively utilized to enhance the solubility and stability of antiviral agents, thereby improving their therapeutic performance in the treatment of various viral infections, including Coronavirus Disease 2019 (COVID-19) [50, 51]. In addition to drug delivery, CDs have also been explored for their utility as cryopreservatives and adjuvants in vaccine formulations.

Veklury®, a lyophilized powder for reconstitution and intravenous infusion, is the first commercially available antiviral formulation to utilize a CD derivative—SBE-β-CD—for complexing with remdesivir. SBE-β-CD, with an average degree of substitution of 6.5 sulfobutylether groups (SBE = 6.5), significantly improves the solubility and stability of remdesivir and is a critical excipient in the Veklury® formulation. Developed by the U.S.-based Gilead Sciences, Veklury® received the US Food and Drug Administration (FDA) approval in 2020 for the treatment of COVID-19 in hospitalized patients aged 12 years and older, weighing at least 40 kg. It remains the only antiviral treatment for COVID-19 currently recommended by the National Institutes of Health (NIH). Veklury® exerts its antiviral effect by directly inhibiting SARS-CoV-2 replication through the suppression of RNA-dependent RNA polymerase (RdRp), a key enzyme essential for viral RNA synthesis. The formulation has also demonstrated in vitro efficacy against multiple clinical isolates of SARS-CoV-2 variants [52].

9.4 Drug Delivery Applications of CD-Based Formulations

CD-based formulations have demonstrated broad applicability for the delivery of therapeutic APIs through various routes of administration, including oral, parenteral, transdermal, ocular, nasal, and rectal routes, among others. Furthermore, CD-based formulations can be developed as sustained or controlled-release drug delivery systems. Additionally, the surface of CDs can be functionalized with specific targeting ligands to develop targeted drug delivery systems designed for particular cells or tissues.

9.4.1 Oral Delivery

Poor gastric solubility, inadequate intestinal permeability, and pre-systemic degradation present significant challenges in the oral delivery of many drugs, often leading to poor therapeutic outcomes and compliance issues, particularly in chronic diseases. CD-based drug delivery systems, including nanosystems, offer a promising platform for enhancing the oral bioavailability of drugs by improving their ability to navigate the gastrointestinal tract. The stomach serves as a site for the absorption of drugs susceptible to degradation in high pH environments, while the ileum facilitates lymphatic drug transport, which is modulated by Peyer's patches in the gut-associated lymphoid tissue (GALT). Additionally, the duodenum, jejunum, and ileum are involved in receptor-mediated transport of vitamin B_{12}. Therefore, surface modification of nanosystems with vitamin B_{12} could enhance the intestinal uptake of drugs [53].

Itraconazole, a Biopharmaceutical Classification System (BCS) Class II drug with an aqueous solubility of approximately 1 ng/ml, was formulated as CD-based nanosponges, resulting in a 27-fold increase in its aqueous solubility. The in vitro dissolution rate of the CD-based formulation was faster compared to the commercially available itraconazole, highlighting the potential of CD-based nanosponges in improving the pharmacokinetic profile of drugs whose bioavailability is limited by poor solubility [54].

Torne et al. [55] evaluated the therapeutic efficacy of tamoxifen, an anticancer drug, using CD-based nanosponges. After oral administration via gastric intubation, tamoxifen-loaded CD-based nanosponges achieved significantly higher plasma concentrations ($p < 0.01$) compared to commercially available tamoxifen citrate, which was originally developed to enhance the drug's aqueous solubility and therapeutic efficacy. Cytotoxicity studies on breast cancer MCF-7 cells showed concentration and incubation time-dependent inhibition of cell proliferation, with the CD-based nanosponges formulation demonstrating superior inhibitory effects [55].

In another study, Hu et al. [56] developed polymer vesicles (polymersomes) based on β-CD-centered triarm star polymers (methoxy poly(ethylene glycol)-*b*-polylactide) (Figure 9.5a) to modify the oral absorption pathway of doxorubicin, a chemotherapeutic agent. Although doxorubicin is a BCS Class III drug with high solubility, its oral absorption is hindered by poor permeability through paracellular transport (Figure 9.5b). In this case, β-CD increased drug loading into the nanosystem, while the polymersomes altered the absorption pathway from paracellular to active transcellular transport. This modification resulted in a seven-fold increase in oral bioavailability and an eight-fold increase in anticancer efficacy of doxorubicin in a mouse sarcoma model. Additionally, the polymersome formulation helped mitigate the cardiac and gastrointestinal toxicities commonly associated with doxorubicin [56].

9.4.1.1 Parenteral Delivery

CDs are among the most widely used multifunctional pharmaceutical excipients, capable of forming hydrophilic host–guest inclusion complexes with a wide range of small molecules. This property enhances the apparent water-solubility, chemical stability, and bioavailability of various drug molecules, making CDs an ideal choice for the development of parenteral dosage forms with improved characteristics [57]. Figure 9.6 summarizes the potential drawbacks and benefits associated with CD-based parenteral formulations.

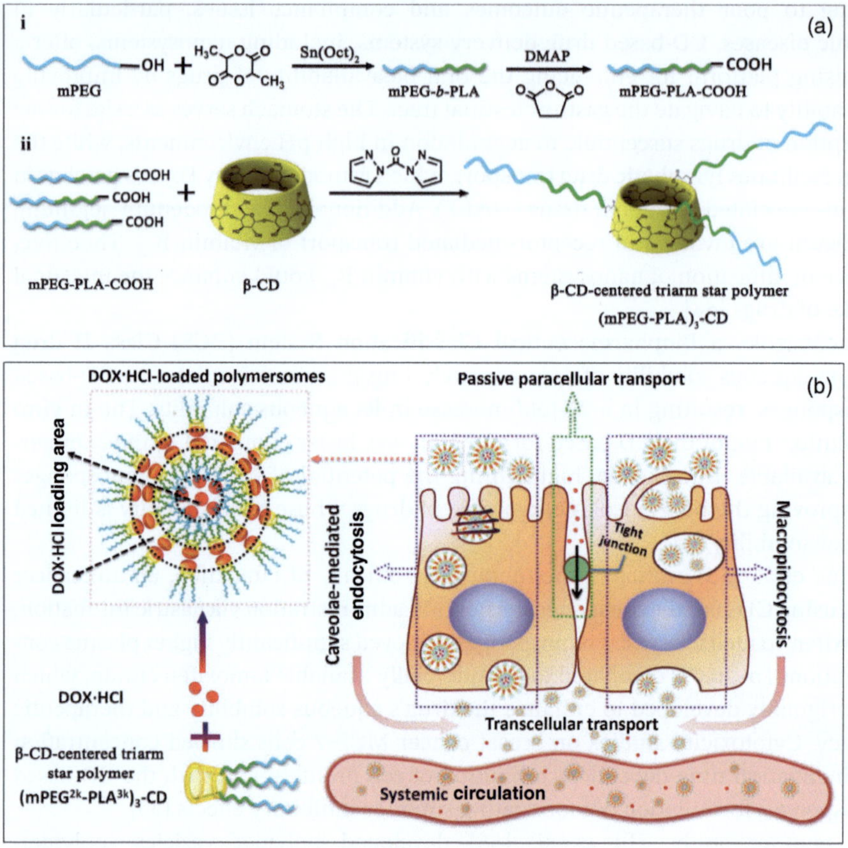

Figure 9.5 (a) Scheme of synthesis of β-CD-centered triarm star polymer, (b) probable mechanisms of oral absorption of doxorubicin following oral administration of CD-based polymersomes of doxorubicin. *Source:* Hu et al. [56] / with permission of American Chemical Society.

Trapani et al. [58] evaluated two novel CD-based injectable formulations of 6-[^{19}F]Fluoro-LDOPA in vitro and in vivo, utilizing SBE-β-CD and HP-β-CD as delivery vehicles for parenteral administration, naming the formulations FA and FB, respectively. The CDs were selected as delivery vehicles based on the hypothesis that adverse reactions at the injection site could result from low drug solubility and subsequent precipitation in the bloodstream. in vitro experiments assessed the effects of FA and FB formulations on mouse skeletal muscle fibers, and on human neuroblastoma SH-SY5Y and embryonal kidney tsA201 cell lines. The cell viability assay revealed that the FA formulation outperformed 6-fluoro-L-3,4-dihydroxyphenylalanine (F-DOPA) solutions. in vivo tolerability tests in adult male rats showed no significant effects on body weight or changes in organ weight. The authors concluded that [18F]F-L-DOPA, formulated as FA, represents a promising dosage form for positron emission tomography (PET)–computed tomography (CT) diagnosis of neuroendocrine tumors and Parkinson's disease [58].

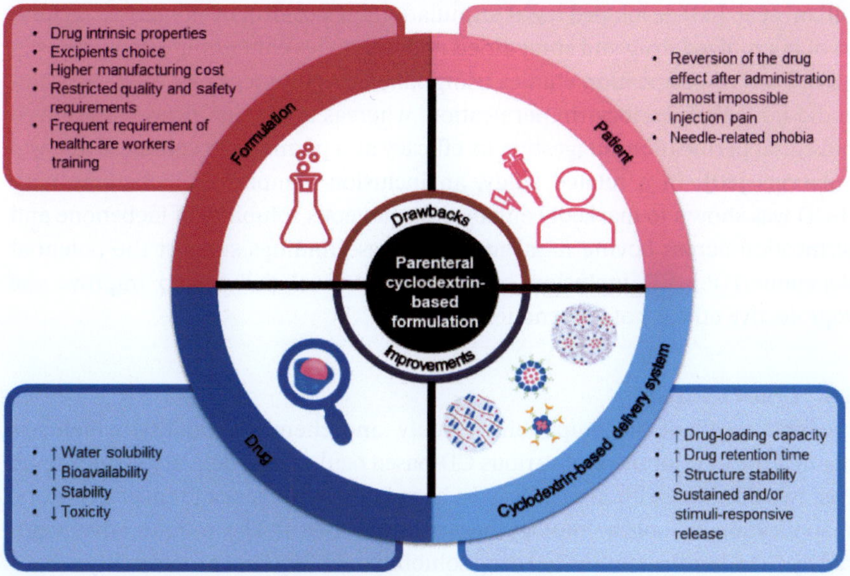

Figure 9.6 Parenteral CD-based formulation: drawbacks and improvements. *Source:* Hu et al. [57] / with permission of Elsevier.

9.4.1.2 Transdermal Delivery

CD-based drug delivery systems offer significant potential for topical therapeutic applications by encapsulating active APIs within their hydrophobic cavities and facilitating their delivery to the skin via various formulations. The beneficial outcomes primarily stem from CDs' ability to enhance drug solubility, stability, and site-specific targeting, while also potentially minimizing toxicity.

Hu et al. developed inclusion complexes of saikosaponin-d (SSD) with HP-β-CD using the solvent evaporation method. Three different molar ratios of SSD to HP-β-CD (1:1, 1:5, and 1:10) were prepared. These inclusion complexes enhanced the aqueous solubility of SSD by 350-fold, 1074-fold, and 906-fold, respectively. The cytotoxic potential of the complexes was evaluated using the 3-(4,5-dimethylthiazol-2-yl)-5-(3-carboxymethoxyphenyl)-2-(4-sulfophenyl)-2H-tetrazolium (MTS) assay in human cutaneous squamous cell carcinoma (HSC-1) cells. Among the formulations, SSD–HP-β-CD (1:5) demonstrated superior cytotoxic activity compared to the 1:1 and 1:10 complexes, which may be attributed to its optimal solubility and drug loading profile. The study concluded that SSD–HP-β-CD (1:5) is a promising candidate for future development as a topical formulation for the treatment of skin cancer via transdermal delivery [59].

9.4.1.3 Nasal Delivery

CDs can be employed in nasal drug delivery systems either as absorption enhancers or as functional excipients within nasal formulations. Their ability to enhance drug solubility and stability, along with their capacity to reduce the irritating effects of certain drugs on the nasal mucosa by encapsulating them within their hydrophobic cavities, makes them particularly advantageous for this route of administration.

Giuliani et al. [60] developed nasal formulations of ribavirin by preparing agglomerates of micronized drug and spray-dried microparticles containing either chitosan or α-CD. Ex vivo permeation studies using rabbit nasal mucosa revealed that chitosan did not enhance ribavirin permeation, whereas α-CD significantly improved its transnasal permeation, suggesting its efficacy as a permeation enhancer in nasal formulations [60]. In a related study, an inclusion complex of idebenone with HP-β-CD was shown to markedly enhance the aqueous solubility of idebenone and its permeation across bovine nasal mucosa. These findings support the potential of idebenone/HP-β-CD inclusion complexes for nasal delivery to improve the neuroprotective efficacy of idebenone [61].

9.4.1.4 Ocular Delivery

CDs exhibit excellent biocompatibility, safety, and chemical stability, which has accelerated the development of various CD-based ocular drug delivery systems over the past two decades. Although CDs have been known to science for over a century, their application in ophthalmology began in the 1990s. The unique advantages offered by CDs—such as enhanced drug solubility, stability, and permeability—have significantly contributed to the evolution of CD-based ocular delivery platforms [62].

Wang et al. [63] developed brinzolamide (BRZ)-HP-β-CD inclusion complexes and further encapsulated them into liposomes using the thin-film dispersion method (Figure 9.7a) to explore their potential for glaucoma treatment. The researchers evaluated the formulation by assessing in vitro drug release, ex vivo corneal permeability, and in vivo intraocular pressure (IOP) reduction efficacy. The results supported the potential of this nanoliposomal system for effective ocular delivery of BRZ. The BRZ/HP-β-CD nanoliposomes (BCL) demonstrated sustained drug release in vitro, as evaluated using the dialysis method (Figure 9.7b). When tested at a concentration of 1 mg/ml of BRZ, the BCL formulation exhibited a 9.36-fold increase in the apparent permeability coefficient compared to the commercial suspension formulation (BRZ-Sus, 10 mg/ml BRZ). Furthermore, BCL achieved sustained and enhanced IOP reduction efficacy in white New Zealand rabbits, significantly outperforming the commercial formulation (Figure 9.7c) [63].

9.4.1.5 Rectal Delivery

Rectal drug delivery offers a promising alternative to the oral route by reducing systemic side effects and bypassing hepatic first-pass metabolism [64]. Balakrishnan et al. formulated a clotrimazole/β-CD suppository, demonstrating that the resulting inclusion complex could serve as a potential strategy to enhance the bioavailability of hydrophobic drugs like clotrimazole [65].

More recently, Lei et al. developed a meloxicam (MLX) and HP-β-CD inclusion complex-based thermosensitive in situ gel (MLX/HP-β-CD-ISG) for rectal delivery. This formulation significantly improved the absorption and bioavailability of MLX in rats and extended rectal residence time without any incidence of rectal irritation. These findings suggest that MLX/HP-β-CD-ISG holds substantial promise for achieving enhanced therapeutic efficacy via rectal administration [66].

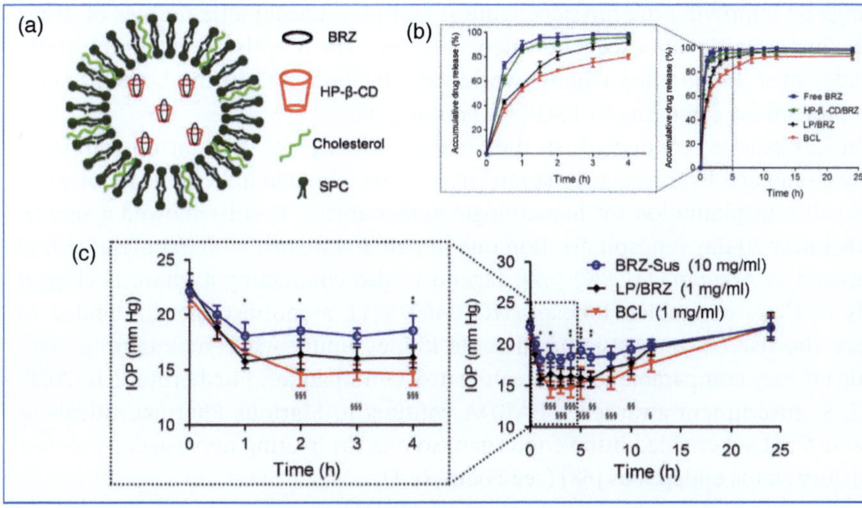

Figure 9.7 (a) Schematic microstructure of BRZ/HP-β-CD complexes loaded in liposomes. (b) in vitro release profiles of BRZ from free BRZ, HP-β-CD/BRZ, LP/BRZ, and BCL at 37 °C in simulated tear fluid. (c) in vivo IOP profiles of white New Zealand rabbits treated with BRZ-Sus, LP/BRZ, and BCL. *Source:* Wang et al. [63] / Frontiers Media SA / CC BY 4.0.

9.5 Current Landscape Post Approval of Captisol®

Captisol® (SBE-β-CD) was first approved by the FDA in Sweden in 2001, and currently, 15 Captisol-enabled drug products are commercially available in over 50 countries. It is used in various pharmaceutical formulations, including ready-to-use parenteral solutions, concentrates, and oral liquids, for treating conditions such as viral infections, cancer, fungal diseases, and psychiatric disorders. Developed by Ligand Pharmaceuticals, Captisol® is supported by a robust global patent portfolio. It has gained broad regulatory acceptance, with approvals from multiple FDA divisions and other international regulatory agencies. Numerous Captisol®-based drug candidates are currently in various stages of clinical development [67, 68].

Captisol® has a well-established safety profile and is an excipient in several FDA-approved drugs, including Nexterone®, Abilify®, Kyprolis®, Geodon®, VFend®, and Veklury®. With over two decades of clinical use, regulatory agencies worldwide have accumulated substantial experience with Captisol®-enabled products. Ligand Pharmaceuticals maintains two Drug Master Files (DMFs) with the FDA—Type IV (CMC) and Type V (regulatory safety)—along with DMFs filed in Canada, China, and Japan. Captisol® has been evaluated and approved for use in over 500 clinical trials globally.[1]

Although thousands of patents are filed annually on solubility and stability enhancement, only a small fraction reach clinical application or commercialization. Captisol® has successfully translated its potential into clinical and commercial

1 https://www.captisol.com/.

settings by improving the physicochemical and pharmacokinetic profiles of drugs while maintaining an excellent safety margin. The Captisol® patent portfolio includes over 400 granted patents worldwide, including 37 in the United States, with protections extending to 2041 (see Footnote 1).

Clinical studies conducted at the Medical College of Wisconsin compared Captisol-enabled Evomela® to Alkeran® in patients who had undergone autologous stem cell transplantation for hematologic malignancies. Results showed a significantly lower 30-day rehospitalization rate in patients treated with Evomela® (6.8%) compared to Alkeran® (17.9%) [69]. Ligand is also conducting a Phase II clinical study of Captisol-enabled Iohexol (NCT04627831), a contrast agent intended to reduce the risk of contrast-induced acute kidney injury while maintaining diagnostic efficacy comparable to GE Healthcare's Omnipaque®. Furthermore, in 2020, the U.S. government awarded a BARDA contract to Marinus Pharmaceuticals to develop Captisol-enabled intravenous ganaxolone for treating nerve agent-induced refractory status epilepticus [68] (see Footnote 1).

9.6 Regulatory and Clinical Landscape of Cyclodextrin-Based Biomaterials

CDs have received regulatory recognition for safety, with the FDA granting Generally Recognized As Safe (GRAS) status to α-, β-, and γ-CD under Notices No. 000155, 000074, and 000046, respectively. This regulatory endorsement supports their continued use in pharmaceutical, nutraceutical, diagnostic, food, and biomedical applications.

CD derivatives have demonstrated significant clinical utility in drug delivery and solubilization, with several formulations either approved or under investigation. For instance, Captisol® SBE-β-CD is included in multiple FDA-approved injectable formulations, most notably remdesivir, underscoring its effectiveness in antiviral therapy [68]. Similarly, CRLX101, a camptothecin–CD polymer conjugate, has been evaluated in clinical trials for solid tumor therapy, highlighting the role of CD-based systems in oncology [70]. Furthermore, HP-β-CD has received orphan drug designation from the FDA for its therapeutic potential in Niemann–Pick Type C disease, a rare lysosomal storage disorder [71]. These examples reflect the expanding application of CDs in controlled release, targeted therapy, and biomaterial design. Table 9.3 provides a summary of key clinical trials exploring CD-based biomaterials.

9.6.1 Regulatory Challenges and Considerations

Despite their promise, the regulatory landscape for CD-based biomaterials remains complex. Approval requires comprehensive toxicological, pharmacokinetic, and biocompatibility studies. Key concerns include the following:

- **Toxicity and dosage safety:** Preclinical assessments are necessary to establish safe dose ranges and mitigate risks associated with systemic accumulation or organ-specific toxicity.

Table 9.3 Selected clinical trials involving cyclodextrin-based biomaterials.

Clinical trial identifier	Study title	Phase	Status	Cyclodextrin type	Condition
NCT03879655	A Phase 2b/3 Open-label Trial of VTS-270 in Subjects with Neurologic Manifestations of Niemann–Pick Type C1 Disease Previously Treated Under Protocol VTS301	Phase 2b/3	Discontinued	2-HP-β-CD	Niemann–Pick Type C1 Disease
NCT05343156	DX211 Clinical Trial Protocol	Phase 1	Completed	CD-based nanoparticles	Not specified
NCT02055716	Sulforadex in Healthy Human Males MAD	Phase 1	Completed	α-CD	Prostate cancer
NCT03931057	The Use of ADV6209 for Premedication in Pediatric Anesthesia	Phase 4	Completed	γ-CD	Preanesthetic medication
NCT01612546	Pilot Trial of CRLX101 in Treatment of Patients With Advanced or Metastatic Stomach, Gastroesophageal, or Esophageal Cancer	Phase 2	Completed	CD-based polymer–camptothecin (CRLX101)	Adenocarcinoma of the esophagus
NCT00333502	Study of CRLX101 (NLG207) in the Treatment of Advanced Solid Tumors	Phase 1	Completed	Camptothecin conjugated to a linear CD-based polymer	Solid tumor
NCT01131299	Evaluation of Oral Alpha-Cyclodextrin for Decreasing Serum Cholesterol	Phase 2	Completed	α-CD	Cardiovascular disease

Source: Adapted from Bin Lee et al. [72].

- **Biodegradability and clearance:** Understanding the metabolic fate of CDs in vivo is crucial, particularly for modified CDs intended for long-term or high-dose use.
- **Immunogenicity:** Evaluation of potential immune responses is critical, especially for implantable or long-circulating biomaterials [73].

Another barrier is the classification ambiguity of CD-based materials. Depending on their design and therapeutic purpose, they may be regulated as drug delivery systems, medical devices, or combination products, each falling under distinct regulatory pathways. Establishing clear classification guidelines would facilitate clinical translation and streamline approval processes.

9.6.2 Manufacturing and Scale-Up Challenges

Achieving manufacturing consistency presents significant hurdles, particularly for chemically modified CDs, where batch-to-batch variability can arise from synthetic and purification inefficiencies. Variations in purity, substitution degree, structural integrity, and the presence of residual impurities or degradation products can impact both safety and efficacy [74].

To address this, there is a need for:

- Optimized purification protocols (e.g. chromatography, dialysis) to improve product uniformity.
- Scalable and cost-effective synthesis techniques, as current methods are often labor-intensive and reliant on expensive reagents.

Cost and scalability remain key limitations to broader industrial adoption. Overcoming these barriers through technological advancements and regulatory clarity will be essential to fully realize the potential of CD-based biomaterials in therapeutic and commercial settings.

9.7 Toxicological Considerations of CDs with Emphasis on Renal Safety

CDs have become integral to pharmaceutical and biomedical applications due to their ability to form inclusion complexes that enhance the solubility and stability of therapeutic agents. Although CDs are generally considered safe at therapeutic concentrations [75], their toxicological profile is influenced by several factors including dose, route of administration, degree of substitution, and nature of the guest molecule.

9.7.1 Route-Dependent and Dose-Related Toxicity

Orally administered native CDs, such as α-, β-, and γ-CDs, exhibit low toxicity due to their limited gastrointestinal absorption and enzymatic degradation by amylase into non-toxic glucose derivatives [76]. These characteristics contribute to their favorable safety profile in oral drug delivery and food applications. However, toxicity concerns increase significantly with parenteral administration, particularly in the case of β-CD. When administered via non-oral routes, CDs bypass enzymatic degradation, potentially leading to systemic accumulation. β-CD has been associated with renal toxicity, with studies showing nephrotoxic effects due to its accumulation in kidney tissues. Notably, Frank et al. observed nephrotoxicity in rats receiving subcutaneous injections of α- and β-CDs, attributing this to the formation of acicular microcrystals and giant vacuoles in renal tissue [72]. Such findings underscore the importance of considering both dose and physical form of CD in formulation design. To address the nephrotoxic limitations of β-CD, derivatives such as HP-β-CD have been developed. These modified CDs display enhanced aqueous solubility, lower renal accumulation, and improved safety profiles in parenteral applications [77].

9.7.2 Inclusion Complexes and Drug-Associated Toxicity

Beyond the inherent toxicity of CDs, their interaction with guest molecules in inclusion complexes can further influence safety. CDs can significantly enhance the bioavailability of poorly soluble drugs, which, if not properly controlled, may result in drug levels exceeding therapeutic thresholds, leading to organ-specific toxicity such as nephrotoxicity or hepatotoxicity [78, 79]. In particular, for drugs with narrow therapeutic indices, the altered distribution and release kinetics resulting from complexation with CDs can result in increased accumulation in specific organs, including the kidneys and liver. Environmental conditions such as pH and the presence of competing substances can also modulate drug release, potentially causing local irritation or systemic toxicity [80].

9.7.3 Hepatic and Metabolic Considerations

While CDs are generally well-tolerated, hepatotoxicity has been reported in long-term animal studies, where mild elevations in liver enzymes were observed after extended oral administration in rats and dogs [81]. These changes were reversible upon discontinuation and typically occurred at high doses, indicating a dose-dependent hepatic stress.

Moreover, certain CDs possess strong lipid-binding affinities, which can disrupt lipid metabolism [82]. Suzuki et al. reported that high doses of CDs interfered with lipid absorption in the gastrointestinal tract, occasionally causing indigestion and metabolic disturbances [83]. While this property may be beneficial in applications such as cholesterol reduction, it necessitates careful dose management to avoid adverse effects.

In summary, CDs exhibit a generally favorable safety profile when used within therapeutic limits, particularly for oral applications. Nevertheless, their toxicological behavior is strongly influenced by dose, route of administration, degree of chemical modification, and interactions with guest molecules. Among safety concerns, nephrotoxicity has been well-documented, most notably with β-CD in parenteral applications [72] due to its limited renal excretion capacity. Although derivatives such as HP-β-CD and SBE-β-CD (Captisol®) were developed to improve aqueous solubility and reduce toxicity, renal safety is not entirely risk-free. Both HP-β-CD and Captisol® can accumulate in renal tissue at high doses or in patients with impaired kidney function, leading to potential nephrotoxic effects [77]. Other potential adverse effects, including hepatotoxicity [81] and disruption of lipid metabolism [83], further emphasize the importance of controlled dosing and formulation design. Continued preclinical evaluation and rigorous safety assessments are essential to ensure the safe use of CDs in biomedical applications.

Therefore, careful dose selection, monitoring of renal function, and ongoing development of next-generation CD derivatives with improved elimination profiles remain crucial for minimizing toxicity risks in clinical use.

9.8 Patent Scenario Regarding the Use of CD

CDs have garnered significant attention in the pharmaceutical, food, cosmetic, and environmental sectors due to their unique ability to form inclusion complexes, enhance solubility, improve drug stability, and facilitate targeted delivery. This versatility has translated into a robust and evolving patent landscape, with numerous filings globally related to their synthesis, modification, and application [84].

9.8.1 Areas of Patent Focus

The patenting activity concerning CDs is concentrated in several key areas:

- **Inclusion complexes:** Many patents claim the use of CDs to improve the solubility, stability, and bioavailability of poorly water-soluble drugs through host–guest complexation. These include specific drug–CD complexes and processes for their preparation [85].
- **Modified CDs:** Chemically altered CDs such as HP-β-CD, SBE-β-CD, and methylated β-CDs have been patented for their improved safety and pharmacokinetic properties. These derivatives show better aqueous solubility and reduced toxicity, especially for parenteral use [77].
- **Drug delivery systems:** Patents have been granted for CD-based drug delivery vehicles including liposomes, micelles, nanoparticles, nanosponges, hydrogels, and implantable scaffolds, where CDs act as functional excipients or structural units to control drug release or improve targeting [86].
- **Polymer conjugates and advanced therapeutics:** CD–drug conjugates, such as CRLX101 (a camptothecin–CD polymer conjugate), represent innovative strategies in targeted cancer therapy. These are covered under patents protecting the polymeric structure, synthesis methods, and medical indications [87].
- **Non-pharmaceutical uses:** Beyond pharmaceuticals, patents cover CD applications in environmental remediation (e.g. pollutant capture), food preservation (flavor stabilization), and cosmetics (fragrance encapsulation and sebum control) [88].

9.8.2 Notable Patented Products and Technologies

Several CD-based technologies have achieved commercial success and regulatory approval:

- **Captisol® (SBE-β-CD):** Patented by CyDex Pharmaceuticals (a Ligand Pharmaceuticals subsidiary), Captisol has been incorporated into several FDA-approved injectables with strong global sales records. Notable examples include Voriconazole (Vfend®; Pfizer), a blockbuster antifungal formulation that demonstrated Captisol's ability to enhance solubility and stability, as well as Carfilzomib (Kyprolis®; Amgen) for multiple myeloma, and Ziprasidone mesylate (Geodon®; Pfizer) for schizophrenia. These successes highlight Captisol's regulatory acceptance and clinical utility across diverse therapeutic areas. While remdesivir (Veklury®; Gilead) showcased Captisol's adaptability during the COVID-19 pandemic, the broader market impact has been driven by antifungal, oncology, and antipsychotic drug products [68].
- **CRLX101:** Developed by Cerulean Pharma, CRLX101 is a nanoparticle–drug conjugate comprising a CD–polymer–camptothecin complex. It has undergone multiple clinical trials for various solid tumors and is protected by patents covering its formulation and method of treatment [89].
- **HP-β-CD for Niemann–Pick disease:** HP-β-CD has received orphan drug designation from the FDA and the European Medicines Agency (EMA) for the treatment of Niemann–Pick Type C1 disease. The product VTS-270, based on HP-β-CD, was developed by Vtesse Inc. and is protected by patents for its therapeutic application and administration protocols [90].

9.8.3 Patentability Considerations

Securing patent protection for CDs can be challenging due to their natural origin and extensive prior art. As a result, newer patents often focus on the following:

- Novel derivatives or modifications
- Combination formulations
- New routes of administration or therapeutic indications
- Synthesis and purification processes

The classification of CD-based technologies can vary (e.g. as drugs, excipients, medical devices, or biomaterials), influencing the applicable regulatory and intellectual property frameworks [91].

9.8.4 Manufacturing and Patent Challenges

Despite the promising patent landscape, challenges remain in ensuring the manufacturing consistency of modified CDs, especially on a commercial scale. Variability in synthesis, difficulty in purification, and high production costs can impact reproducibility and scalability—factors that are crucial both for regulatory approval and sustaining patent exclusivity. Furthermore, freedom-to-operate (FTO) analyses are critical before commercialization, given the dense network of overlapping patents, especially in the United States, Europe, Japan, and China [92].

9.8.5 Future Directions

Emerging areas of CD-related patent filings include the following aspects:

- Green synthesis methods and environmentally friendly production.
- CD-based vaccines and gene delivery systems.
- Personalized medicine platforms using CD carriers for patient-specific drug delivery.

To accelerate clinical translation, there is a pressing need for standardized regulatory guidelines that clearly define the classification, safety testing, and quality standards for CD-based biomaterials [92]. In addition, the development of structurally modified CDs remains an important avenue, as chemical modifications can not only enhance the aqueous solubility and complexation efficiency of poorly soluble drugs but also reduce inherent toxicity concerns associated with certain native CDs (e.g. β-CD). Tailored functionalization strategies may therefore open up safer and more effective CD-based delivery systems for next-generation therapeutics.

9.9 Conclusion

CDs have established themselves as versatile excipients in modern pharmaceutical formulations, owing to their ability to enhance solubility, stability, and bioavailability of poorly soluble drugs. Their successful integration into a wide range of marketed products across oral, parenteral, ocular, nasal, transdermal, and rectal routes underline their translational potential. Despite this progress, safety concerns such as nephrotoxicity, particularly with β-CD and even with derivatives like HP-β-CD and SBE-β-CD under certain conditions, highlight the need for cautious dose selection and patient monitoring. Manufacturing challenges, including batch-to-batch variability and the risk of impurities or degradation products, also demand stringent quality control. Regulatory acceptance, with GRAS status for parent CDs and established DMFs for derivatives like Captisol®, provides a strong foundation for further development. Looking ahead, structurally modified CDs with improved safety and solubilizing properties, along with greener synthesis methods, will accelerate clinical translation. Their potential applications extend beyond drug delivery to diagnostics, nutraceuticals, and vaccine technologies, ensuring that CDs will remain at the forefront of pharmaceutical innovation in the coming decades.

References

1 Poulson, B.G., Alsulami, Q.A., Sharfalddin, A. et al. (2021). Cyclodextrins: structural, chemical, and physical properties, and applications. *Polysaccharides* 3: 1–31. doi: 10.3390/polysaccharides3010001.

2 Arruda, T.R., Marques, C.S., and Soares, N.F.F. (2021). Native cyclodextrins and their derivatives as potential additives for food packaging: a review. *Polysaccharides* 2: 825–842. https://doi.org/10.3390/polysaccharides2040050.

3 Szente, L., Szemán, J., and Sohajda, T. (2016). Analytical characterization of cyclodextrins: history, official methods and recommended new techniques. *J. Pharm. Biomed. Anal.* 130: 347–365. doi: 10.1016/j.jpba.2016.05.009.

4 Crini, G. (2014). Review: a history of cyclodextrins. *Chem. Rev.* 114: 10940–10975. doi: 10.1021/cr500081p.

5 Crini, G. (2020). The contribution of Franz Schardinger to cyclodextrins: a tribute on the occasion of the centenary of his death. *J. Incl. Phenom. Macrocycl. Chem.* 97: 19–28. doi: 10.1007/s10847-020-00990-3.

6 Szejtli, J. (1998). Introduction and general overview of cyclodextrin chemistry. *Chem. Rev.* 98: 1743–1754. doi: 10.1021/cr970022c.

7 French, D. (1957). *The Schardinger Dextrins*, 189–260. Elsevier. doi: 10.1016/S0096-5332(08)60209-X.

8 Kali, G., Haddadzadegan, S., and Bernkop-Schnürch, A. (2024). Cyclodextrins and derivatives in drug delivery: new developments, relevant clinical trials, and advanced products. *Carbohydr. Polym.* 324: 121500. doi: 10.1016/j.carbpol.2023.121500.

9 Á. Sarabia-Vallejo, M del M. Caja, A.I. Olives, M.A. Martín, J.C. Menéndez, Cyclodextrin inclusion complexes for improved drug bioavailability and activity: synthetic and analytical aspects, *Pharmaceutics* 15 (2023) 2345. 10.3390/pharmaceutics15092345s.

10 Adamkiewicz, L. and Szeleszczuk, Ł. (2023). Review of applications of cyclodextrins as taste-masking excipients for pharmaceutical purposes. *Molecules* 28: 6964. doi: 10.3390/molecules28196964.

11 Saokham, P., Muankaew, C., Jansook, P., and Loftsson, T. (2018). Solubility of cyclodextrins and drug/cyclodextrin complexes. *Molecules* 23: 1161. doi: 10.3390/molecules23051161.

12 Adepu, S. and Ramakrishna, S. (2021). Controlled drug delivery systems: current status and future directions. *Molecules* 26: 5905. doi: 10.3390/molecules26195905.

13 Kim, D.-H., Lee, S.-E., Pyo, Y.-C. et al. (2020). Solubility enhancement and application of cyclodextrins in local drug delivery. *J. Pharm. Investig.* 50: 17–27. doi: 10.1007/s40005-019-00434-2.

14 Rasheed, A. (2008). Cyclodextrins as drug carrier molecule: a review. *Sci. Pharm.* 76: 567–598. doi: 10.3797/scipharm.0808-05.

15 Vyas, A., Saraf, S., and Saraf, S. (2008). Cyclodextrin based novel drug delivery systems. *J. Incl. Phenom. Macrocycl. Chem.* 62: 23–42. doi: 10.1007/s10847-008-9456-y.

16 Jansook, P., Ogawa, N., and Loftsson, T. (2018). Cyclodextrins: structure, physicochemical properties and pharmaceutical applications. *Int. J. Pharm.* 535: 272–284. doi: 10.1016/j.ijpharm.2017.11.018.

17 Topuz, F. and Uyar, T. (2022). Advances in the development of cyclodextrin-based nanogels/microgels for biomedical applications: drug delivery and beyond. *Carbohydr. Polym.* 297: 120033. doi: 10.1016/j.carbpol.2022.120033.

18 Liao, Y., Zhang, X., Li, C. et al. (2016). Inclusion complexes of HP-β-cyclodextrin with agomelatine: preparation, characterization, mechanism study and in vivo evaluation. *Carbohydr. Polym.* 147: 415–425. doi: 10.1016/j.carbpol.2016.04.022.

19 Morina, D., Sessevmez, M., Sinani, G. et al. (2020). Oral tablet formulations containing cyclodextrin complexes of poorly water soluble cefdinir to enhance its bioavailability. *J. Drug Deliv. Sci. Technol.* 57: 101742. doi: 10.1016/j.jddst.2020.101742.

20 Lahiani-Skiba, M., Hallouard, F., Bounoure, F. et al. (2018). Enhanced dissolution and oral bioavailability of cyclosporine a: microspheres based on αβ-cyclodextrins polymers. *Pharmaceutics* 10: 285. doi: 10.3390/pharmaceutics10040285.

21 Gabr, M.M., Mortada, S.M., and Sallam, M.A. (2018). Carboxylate cross-linked cyclodextrin: a nanoporous scaffold for enhancement of rosuvastatin oral bioavailability. *Eur. J. Pharm. Sci.* 111: 1–12. doi: 10.1016/j.ejps.2017.09.026.

22 Real, D.A., Bolaños, K., Priotti, J. et al. (2021). Cyclodextrin-modified nanomaterials for drug delivery: classification and advances in controlled release and bioavailability. *Pharmaceutics* 13: 2131. doi: 10.3390/pharmaceutics13122131.

23 Scriba, G.K.E., Konjaria, M., and Krait, S. (2023). Cyclodextrins. In: *Chiral Separation Stereochemical Elucidation: Fundamentals, Methods, and Applications* (ed. Q.B. Cass, M.E. Tiritan, J.M.B. Junior, and J.C. Barreiro), 273–323. Wiley. doi: 10.1002/9781119802280.ch8.

24 Köse, K., Tüysüz, M., Aksüt, D., and Uzun, L. (2022). Modification of cyclodextrin and use in environmental applications. *Environ. Sci. Pollut. Res.* 29: 182–209. doi: 10.1007/s11356-021-15005-y.

25 Puskás, I., Szente, L., Szőcs, L., and Fenyvesi, É. (2023). Recent list of cyclodextrin-containing drug products, period. *Polytech. Chem. Eng.* 67: 11–17. doi: 10.3311/PPch.21222.

26 Jain, M., Nowak, B.P., and Ravoo, B.J. (2022). Supramolecular hydrogels based on cyclodextrins: progress and perspectives. *ChemNanoMat* 8: doi: 10.1002/cnma.202200077.

27 Kasal, P. and Jindřich, J. (2021). Mono-6-substituted cyclodextrins—synthesis and applications. *Molecules* 26: 5065. doi: 10.3390/molecules26165065.

28 Zultanski, S.L., Kuhl, N., Zhong, W. et al. (2021). Mechanistic understanding of a robust and scalable synthesis of per(6-deoxy-6-halo)cyclodextrins, versatile intermediates for cyclodextrin modification. *Org. Process Res. Dev.* 25: 597–607. doi: 10.1021/acs.oprd.0c00249.

29 García, A., Priotti, J., Codina, A.V. et al. (2019). Synthesis and characterization of a new cyclodextrin derivative with improved properties to design oral dosage forms. *Drug Deliv. Transl. Res.* 9: 273–283. doi: 10.1007/s13346-018-0591-8.

30 Řezanka, M. (2019). Synthesis of substituted cyclodextrins. *Environ. Chem. Lett.* 17: 49–63. doi: 10.1007/s10311-018-0779-7.

31 Tang, W. and Ng, S.-C. (2008). Facile synthesis of mono-6-amino-6-deoxy-α-, β-, γ-cyclodextrin hydrochlorides for molecular recognition, chiral separation and drug delivery. *Nat. Protoc.* 3: 691–697. doi: 10.1038/nprot.2008.37.

32 Asim, M.H., Nazir, I., Jalil, A. et al. (2020). Per-6-thiolated cyclodextrins: a novel type of permeation enhancing excipients for BCS class IV drugs. *ACS Appl. Mater. Interfaces* 12: 7942–7950. doi: 10.1021/acsami.9b21335.

33 Grassiri, B., Knoll, P., Fabiano, A. et al. (2022). thiolated hydroxypropyl-β-cyclodextrin: a potential multifunctional excipient for ocular drug delivery. *Int. J. Mol. Sci.* 23: 2612. doi: 10.3390/ijms23052612.

34 Kali, G., Haddadzadegan, S., Laffleur, F., and Bernkop-Schnürch, A. (2023). Per-thiolated cyclodextrins: nanosized drug carriers providing a prolonged gastrointestinal residence time. *Carbohydr. Polym.* 300: 120275. doi: 10.1016/j.carbpol.2022.120275.

35 Bálint, M., Darcsi, A., Benkovics, G. et al. (2019). Synthesis of the chiral selector heptakis(6-O-methyl)-β-cyclodextrin by phase-transfer catalysis and hydrazine-mediated transfer-hydrogenation. *Electrophoresis* 40: 1941–1950. doi: 10.1002/elps.201900065.

36 Bucur, S., Niculaua, M., Ciobanu, C.I. et al. (2021). A simple synthesis route for selectively methylated β-cyclodextrin using a copper complex sandwich protecting strategy. *Molecules* 26: 5669. doi: 10.3390/molecules26185669.

37 D'Aria, F., Pagano, B., and Giancola, C. (2022). Thermodynamic properties of hydroxypropyl-β-cyclodextrin/guest interaction: a survey of recent studies. *J. Therm. Anal. Calorim.* 147: 4889–4897. doi: 10.1007/s10973-021-10958-1.

38 Gidwani, B. and Vyas, A. (2014). Synthesis, characterization and application of epichlorohydrin-β-cyclodextrin polymer. *Colloids Surf. B Biointerf.* 114: 130–137. doi: 10.1016/j.colsurfb.2013.09.035.

39 Schlüter, F., Bela, M.M., Glikman, D. et al. (2020). A cyclodextrin surfactant for stable emulsions with an accessible cavity for host–guest complexation. *Chem. Commun.* 56: 15434–15437. doi: 10.1039/D0CC06657E.

40 Kato, K., Hori, A., and Ito, K. (2018). An efficient synthesis of low-covered polyrotaxanes grafted with poly(ε-caprolactone) and the mechanical properties of its cross-linked elastomers. *Polymer (Guildf).* 147: 67–73. doi: 10.1016/j.polymer.2018.05.072.

41 Putaux, J.-L., Lancelon-Pin, C., Legrand, F.-X. et al. (2017). Self-assembly of amphiphilic biotransesterified β-cyclodextrins: supramolecular structure of nanoparticles and surface properties. *Langmuir* 33: 7917–7928. doi: 10.1021/acs.langmuir.7b01136.

42 Blaj, D.-A., Kowalczuk, M., and Peptu, C. (2023). Mass spectrometry of esterified cyclodextrins. *Molecules* 28: 2001. doi: 10.3390/molecules28052001.

43 Meimoun, J., Phuphuak, Y., Miyamachi, R. et al. (2022). Cyclodextrins initiated ring-opening polymerization of lactide using 4-dimethylaminopyridine (DMAP) as catalyst: study of DMAP/β-CD inclusion complex and access to new structures. *Molecules* 27: 1083. doi: 10.3390/molecules27031083.

44 Caldera, F., Tannous, M., Cavalli, R. et al. (2017). Evolution of cyclodextrin nanosponges. *Int. J. Pharm.* 531: 470–479. doi: 10.1016/j.ijpharm.2017.06.072.

45 Escobar, K., Garrido-Miranda, K.A., Pulido, R. et al. (2023). Coatings of cyclodextrin/citric-acid biopolymer as drug delivery systems: a review. *Pharmaceutics* 15: 296. doi: 10.3390/pharmaceutics15010296.

46 Ye, Y., Ren, H., Zhu, S. et al. (2017). Synthesis of oxidized β-cyclodextrin with high aqueous solubility and broad-spectrum antimicrobial activity. *Carbohydr. Polym.* 177: 97–104. doi: 10.1016/j.carbpol.2017.08.123.

47 Hussain Asim, M., Ijaz, M., Rösch, A.C., and Bernkop-Schnürch, A. (2020). Thiolated cyclodextrins: new perspectives for old excipients. *Coord. Chem. Rev.* 420: 213433. doi: 10.1016/j.ccr.2020.213433.

48 Inaba, K., Wakuda, T., and Uekama, K. (1984). Prostaglandins and their cyclodextrin complexes. *J. Incl. Phenom.* 2: 467–474. doi: 10.1007/BF00662213.

49 Nicolaescu, O.E., Belu, I., Mocanu, A.G. et al. (2025). Cyclodextrins: enhancing drug delivery, solubility and bioavailability for modern therapeutics. *Pharmaceutics* 17: 288. doi: 10.3390/pharmaceutics17030288.

50 Almeida, B., Domingues, C., Mascarenhas-Melo, F. et al. (2023). The role of cyclodextrins in COVID-19 therapy—a literature review. *Int. J. Mol. Sci.* 24: 2974. doi: 10.3390/ijms24032974.

51 Zokaei, E., Badoei-dalfrad, A., Ansari, M. et al. (2019). Therapeutic potential of DNAzyme loaded on chitosan/cyclodextrin nanoparticle to recovery of chemosensitivity in the MCF-7 cell line. *Appl. Biochem. Biotechnol.* 187: 708–723. doi: 10.1007/s12010-018-2836-x.

52 Anitha, A., Rajamohan, R., Murugan, M., and Seo, J.H. (2024). Inclusion complexation of remdesivir with cyclodextrins: a comprehensive review on combating coronavirus resistance—current state and future perspectives. *Molecules* 29: 4782. doi: 10.3390/molecules29194782.

53 Adeoye, O. and Cabral-Marques, H. (2017). Cyclodextrin nanosystems in oral drug delivery: a mini review. *Int. J. Pharm.* 531: 521–531. doi: 10.1016/j.ijpharm.2017.04.050.

54 Swaminathan, S., Vavia, P.R., Trotta, F., and Torne, S. (2007). Formulation of betacyclodextrin based nanosponges of itraconazole. *J. Incl. Phenom. Macrocycl. Chem.* 57: 89–94. doi: 10.1007/s10847-006-9216-9.

55 Torne, S., Darandale, S., Vavia, P. et al. (2013). Cyclodextrin-based nanosponges: effective nanocarrier for tamoxifen delivery. *Pharm. Dev. Technol.* 18: 619–625. doi: 10.3109/10837450.2011.649855.

56 Hu, M., Shen, Y., Zhang, L., and Qiu, L. (2016). Polymersomes via self-assembly of amphiphilic β-cyclodextrin-centered triarm star polymers for enhanced oral bioavailability of water-soluble chemotherapeutics. *Biomacromolecules* 17: 1026–1039. doi: 10.1021/acs.biomac.5b01676.

57 Ferreira, L., Campos, J., Veiga, F. et al. (2022). Cyclodextrin-based delivery systems in parenteral formulations: a critical update review. *Eur. J. Pharm. Biopharm.* 178: 35–52. doi: 10.1016/j.ejpb.2022.07.007.

58 Trapani, A., Tricarico, D., Mele, A. et al. (2017). A novel injectable formulation of 6-fluoro- l -DOPA imaging agent for diagnosis of neuroendocrine tumors and Parkinson's disease. *Int. J. Pharm.* 519: 304–313. doi: 10.1016/j.ijpharm.2017.01.038.

59 Hu, S.C.-S., Lai, Y.-C., Lin, C.-L. et al. (2019). Inclusion complex of saikosaponin-d with hydroxypropyl-β-cyclodextrin: improved physicochemical properties and anti-skin cancer activity. *Phytomedicine* 57: 174–182. doi: 10.1016/j.phymed.2018.11.012.

60 Giuliani, A., Balducci, A.G., Zironi, E. et al. (2018). in vivo nose-to-brain delivery of the hydrophilic antiviral ribavirin by microparticle agglomerates. *Drug Deliv.* 25: 376–387. doi: 10.1080/10717544.2018.1428242.

61 Venuti, V., Crupi, V., Fazio, B. et al. (2019). Physicochemical characterization and antioxidant activity evaluation of idebenone/hydroxypropyl-β-cyclodextrin inclusion complex †. *Biomolecules* 9: 531. doi: 10.3390/biom9100531.

62 Wang, Q., Zhang, A., Zhu, L. et al. (2023). Cyclodextrin-based ocular drug delivery systems: a comprehensive review. *Coord. Chem. Rev.* 476: 214919. doi: 10.1016/j.ccr.2022.214919.

63 Wang, F., Bao, X., Fang, A. et al. (2018). Nanoliposome-encapsulated brinzolamide-hydropropyl-β-cyclodextrin inclusion complex: a potential therapeutic ocular drug-delivery system. *Front. Pharmacol.* 9: doi: 10.3389/fphar.2018.00091.

64 Bialik, M., Kuras, M., Sobczak, M., and Oledzka, E. (2021). Achievements in thermosensitive gelling systems for rectal administration. *Int. J. Mol. Sci.* 22: 5500. doi: 10.3390/ijms22115500.

65 Balakrishnan, P., Song, C.K., Cho, H.-J. et al. (2012). Inclusion complex effect on the bioavailability of clotrimazole from poloxamer-based solid suppository. *Arch. Pharm. Res.* 35: 1169–1175. doi: 10.1007/s12272-012-0707-5.

66 Lei, X., Zhang, G., Yang, T. et al. (2023). Preparation and in vitro and in vivo evaluation of rectal in situ gel of meloxicam hydroxypropyl-β-cyclodextrin inclusion complex. *Molecules* 28: 4099. doi: 10.3390/molecules28104099.

67 Kfoury, M., Pipkin, J.D., Antle, V., and Fourmentin, S. (2017). Captisol®: an efficient carrier and solubilizing agent for essential oils and their components. *Flavour Fragr. J.* 32: 340–346. doi: 10.1002/ffj.3395.

68 Pardeshi, C.V., Kothawade, R.V., Markad, A.R. et al. (2023). Sulfobutylether-β-cyclodextrin: a functional biopolymer for drug delivery applications. *Carbohydr. Polym.* 301: 120347. doi: 10.1016/j.carbpol.2022.120347.

69 Monahan, K., Kleman, A., Thapa, B. et al. (2020). Propylene glycol-free melphalan versus PG-melphalan as conditioning for autologous hematopoietic cell transplantation for myeloma. *Biol. Blood Marrow Transplant.* 26: 2229–2236. doi: 10.1016/j.bbmt.2020.08.030.

70 Weiss, G.J., Chao, J., Neidhart, J.D. et al. (2013). First-in-human phase 1/2a trial of CRLX101, a cyclodextrin-containing polymer-camptothecin nanopharmaceutical in patients with advanced solid tumor malignancies. *Invest. New Drugs* 31: 986–1000. doi: 10.1007/s10637-012-9921-8.

71 Ottinger, E., Kao, M., Carrillo-Carrasco, N. et al. (2014). Collaborative development of 2-hydroxypropyl-β-cyclodextrin for the treatment of Niemann-Pick type C1 disease. *Curr. Top. Med. Chem.* 14: 330–339. doi: 10.2174/1568026613666131127160118.

72 Bin Lee, Y., Kyun, M.-L., Lee, Y.J. et al. (2025). Cyclodextrins as multifunctional tools for advanced biomaterials in tissue repair and regeneration. *Bioact. Mater.* 49: 627–651. doi: 10.1016/j.bioactmat.2025.03.018.

73 Thind, M. and Kowey, P. (2020). The role of the food and drug administration in drug development: on the subject of proarrhythmia risk. *J. Innov. Card. Rhythm Manag.* 11: 3958–3967. doi: 10.19102/icrm.2020.110103.

74 Taylor, C.J., Pomberger, A., Felton, K.C. et al. (2023). A brief introduction to chemical reaction optimization. *Chem. Rev.* 123: 3089–3126. doi: 10.1021/acs.chemrev.2c00798.

75 Esteso, M.A. and Romero, C.M. (2024). Cyclodextrins: properties and applications. *Int. J. Mol. Sci.* 25: 4547. doi: 10.3390/ijms25084547.

76 Loftsson, T. and Brewster, M.E. (2010). Pharmaceutical applications of cyclodextrins: basic science and product development. *J. Pharm. Pharmacol.* 62: 1607–1621. doi: 10.1111/j.2042-7158.2010.01030.x.

77 Gould, S. and Scott, R.C. (2005). 2-Hydroxypropyl-β-cyclodextrin (HP-β-CD): a toxicology review. *Food Chem. Toxicol.* 43: 1451–1459. doi: 10.1016/j.fct.2005.03.007.

78 Kraszni, M., Ágh, F., Horváth, D. et al. (2023). Effect of substitution degree and homogeneity on cyclodextrin-ligand complex stability: comparison of fenbufen and fenoprofen using CD and NMR spectroscopy. *Int. J. Mol. Sci.* 24: 7544. doi: 10.3390/ijms24087544.

79 Skuredina, A.A., Tychinina, A.S., Le-Deygen, I.M. et al. (2022). Cyclodextrins and their polymers affect the lipid membrane permeability and increase levofloxacin's antibacterial activity in vitro. *Polymers (Basel).* 14: 4476. doi: 10.3390/polym14214476.

80 Ghitman, J. and Voicu, S.I. (2023). Controlled drug delivery mediated by cyclodextrin-based supramolecular self-assembled carriers: from design to clinical performances. *Carbohydr. Polym. Technol. Appl.* 5: 100266. doi: 10.1016/j.carpta.2022.100266.

81 Bellringer, M.E., Smith, T.G., Read, R. et al. (1995). β-Cyclodextrin: 52-week toxicity studies in the rat and dog. *Food Chem. Toxicol.* 33: 367–376. doi: 10.1016/0278-6915(94)00149-I.

82 Kovacs, T., Nagy, P., Panyi, G. et al. (2022). Cyclodextrins: only pharmaceutical excipients or full-fledged drug candidates? *Pharmaceutics* 14: 2559. doi: 10.3390/pharmaceutics14122559.

83 Suzuki, M. and Sato, A. (1985). Nutritional significance of cyclodextrins: Indigestibility and hypolipemic effect of ALPHA-cyclodextrin. *J. Nutr. Sci. Vitaminol. (Tokyo).* 31: 209–223. doi: 10.3177/jnsv.31.209.

84 Musuc, A.M. (2024). Cyclodextrins: advances in chemistry, toxicology, and multifaceted applications. *Molecules* 29: 5319. doi: 10.3390/molecules29225319.

85 Boczar, D. and Michalska, K. (2022). Cyclodextrin inclusion complexes with antibiotics and antibacterial agents as drug-delivery systems—a pharmaceutical perspective. *Pharmaceutics* 14: 1389. doi: 10.3390/pharmaceutics14071389.

86 Matsuda, H. (1999). Cyclodextrins in transdermal and rectal delivery. *Adv. Drug Deliv. Rev.* 36: 81–99. doi: 10.1016/S0169-409X(98)00056-8.

87 Ekladious, I., Colson, Y.L., and Grinstaff, M.W. (2019). Polymer–drug conjugate therapeutics: advances, insights and prospects. *Nat. Rev. Drug Discov.* 18: 273–294. doi: 10.1038/s41573-018-0005-0.

88 Yousaf, R., Abdul Razzaq, F., Asghar, S. et al. (2023). Cyclodextrins: an overview of fundamentals, types, and applications. In: *Cyclodextrins – Core Concepts New Front (ed. Rashid Ali).* IntechOpen. doi: 10.5772/intechopen.108078.

89 Young, C., Schluep, T., Hwang, J., and Eliasof, S. (2011). CRLX101 (formerly IT-101) – a novel nanopharmaceutical of camptothecin in clinical development. *Curr. Bioact. Compd.* 7: 8–14. doi: 10.2174/157340711795163866.

90 Sharma, R., Hastings, C., Staretz-Chacham, O. et al. (2023). Long-term administration of intravenous Trappsol® Cyclo™ (HP-β-CD) results in clinical benefits and stabilization or slowing of disease progression in patients with Niemann-Pick disease type C1: results of an international 48-week phase I/II trial. *Mol. Genet. Metab. Reports* 36: 100988. doi: 10.1016/j.ymgmr.2023.100988.

91 Rincón-López, J., Almanza-Arjona, Y.C., Riascos, A.P., and Rojas-Aguirre, Y. (2021). Technological evolution of cyclodextrins in the pharmaceutical field. *J. Drug Deliv. Sci. Technol.* 61: 102156. doi: 10.1016/j.jddst.2020.102156.

92 Kaushik, R., Verma, R., Budhwar, V., and Kaushik, D. (2023). An overview of recent patents and future perspective based on cyclodextrin complexation. *Recent Adv. Drug Deliv. Formul.* 17: 31–46. doi: 10.2174/2667387817666230123114114.

10

Cosolvent-Based Formulations to Improve Solubility and Bioavailability

Nilam Bhusare[1,*], *Chetana Eknure*[2,*], *Yogeshwar Bachhav*[2], *and Maushmi S. Kumar*[1]

[1]*Somaiya Vidyavihar University, Somaiya Institute for Research and Consultancy, Center for Drug Design and Discovery, Vidyavihar (E), Mumbai, Maharashtra 400077, India*
[2]*Adex Pharma*^{TM}*, Tilak Nagar, Chembur, Mumbai, Maharashtra 400089, India*

10.1 Background of the Cosolvent-Based Formulations

Solubility is a critical physicochemical property. It governs the formulation, therapeutic efficacy, and analytical performance of drug molecules. According to the United States Pharmacopoeia (USP), solubility is defined as the volume of solvent required to dissolve one part of solute under specified conditions [1]. It refers to the maximum quantity of solute that can dissolve in a solvent to form a saturated solution under defined temperature and pressure conditions. Quantitatively, solubility can be expressed in concentration units such as molarity, molality, mole fraction, percentage, and milliequivalents. Qualitatively, it shows the spontaneous molecular interaction between the solute and the solvent, resulting in homogeneous dispersion [2]. The most persistent challenge in drug development is the poor aqueous solubility of the active pharmaceutical ingredients (API). This limitation significantly impacts the drug's efficacy and dose estimations and necessitates enabled formulation development [3]. The advent of high-throughput screening has led to the identification of multiple new chemical entities (NCEs) that are high-molecular-weight and lipophilic. These compounds are often difficult to transform into final dosage forms due to their low water solubility [1].

Solubility plays an important role in the bioavailability of orally administered drugs, specifically for those that are classified as Biopharmaceutical Classification System (BCS) Class II and IV. For a drug to get absorbed into systemic circulation, it should dissolve in the gastrointestinal (GI) fluids. Poor aqueous solubility leads to incomplete dissolution and becomes the rate-limiting step in the absorption of the drug. It directly impacts the fraction of the administered dose that reaches the systemic circulation, hence reducing therapeutic efficacy. Therefore, optimizing solubility is not merely a physicochemical prerequisite but is essential in

*Nilam Bhusare and Chetana Eknure contributed equally to this work.

pharmaceutical development to ensure consistent and effective drug delivery [4]. In addition, solubility enhancement is important to enhance analytical method sensitivity, formulation stability, and patient safety. From the various techniques available to date, cosolvency is one of the most effective and widely adopted strategies [5].

The solubility of a drug substance is affected by physicochemical properties and other external conditions, which makes it a critical determinant in formulation development. Solute-related factors such as molecular size, shape, surface area, ionization constant (p*K*a), and polymorphic form directly affect the dissolution rate and equilibrium solubility. These characteristics mainly influence the drug's interaction with solvents and its ability to transform into a bioavailable form [6]. Polarity, pH, and volume are solvent-related properties that are equally important and affect the ionization state and solvation capacity. Environmental conditions, such as temperature and pressure, affect solubility by altering molecular motion and intermolecular forces between the solute and solvent [7]. A detailed understanding of these variables is crucial for the rational design of pharmaceutical formulations, particularly while addressing the challenges due to poorly water-soluble compounds [5]. Solubilization of drugs involves a series of molecular interactions that enhance the solubility of poorly aqueous-soluble drugs. The interactions include dipole–dipole, ion–dipole attractions, and hydrogen bonding. It causes the incorporation of hydrophobic compounds into solvent systems [5]. It is known for its simplicity, scalability, and versatility across a wide range of techniques [8].

10.2 Cosolvency

Cosolvency refers to the phenomenon wherein a solute exhibits enhanced solubility in a mixture of solvents as compared to any single solvent. This principle is widely used in pharmaceutical formulation to address the challenges arising due to poorly water-soluble drugs. A cosolvent is a water-miscible and low-molecular-weight organic solvent such as alcohol, ketone, and ester. It is added to aqueous systems to modify the polarity and dielectric constant of the medium. This alteration into the system facilitates improved solute–solvent interactions, hence increasing the apparent solubility of hydrophobic drugs [9]. Frequently used cosolvents include ethanol, propylene glycol, glycerin, and polyethylene glycols (PEG 300 and PEG 400). These are generally recognized as safe and compatible with a wide range of dosage forms. Other supplementary solvents such as dimethyl sulfoxide (DMSO), dimethyl formamide, and dimethylacetamide are used for specialized applications. The selection and combination of cosolvents play a vital role in optimizing drug solubility, stability, and bioavailability [5, 10].

Cosolvency is mostly employed in the formulation of liquid dosage forms. It includes elixirs, syrups, injections, and aerosols, where the enhancement of solubility is critical for achieving therapeutic efficacy. This technique involves the critical alteration of the solvent environment via the addition of small quantities of a water-miscible organic solvent. With the changes in the polarity and the reduction of the dielectric constant of the medium, cosolvency improves the solute–solvent interactions. Hence, it enables the dissolution of hydrophobic or weakly ionizable drug molecules. The resulting increase in the solubility allows the development of stable and bioavailable liquid formulations. However, these do not necessitate chemical modification of the API [5].

Cosolvents disrupt the hydrogen bonding network of water. Hence, it reduces self-association and enhances its capacity to accommodate nonpolar solutes. This disruption lowers the cohesive energy of the aqueous phase. It creates a more permissive environment for solubilization. Cosolvents exhibit a dual molecular character. It possesses both hydrophilic hydrogen bonding groups and hydrophobic hydrocarbon regions [10]. This amphiphilic nature enables them to interact favorably with both polar and nonpolar domains in the system. This effectively modifies the solvent polarity. By destabilizing the structured lattice of water, cosolvents facilitate the incorporation of hydrophobic compounds into solution in a phenomenon often described as "squeezing out" nonpolar solutes from exclusion zones into the bulk medium. This physicochemical modification is central to the success of cosolvency in enhancing drug solubility, particularly for compounds with limited aqueous solubility [11].

In comparison to their intrinsic aqueous solubility, cosolvents enhance the solubility of poorly water-soluble compounds by several 1000-fold. Thus, making cosolvency a powerful tool in formulation development. This increase is achieved by modification of the solvent polarity and disrupting water's hydrogen bonding network. It facilitates the incorporation of hydrophobic molecules into solution. However, the use of cosolvents must be carefully optimized due to potential drawbacks. These include adverse effects such as precipitation upon dilution and instability caused by solvent evaporation [2]. These risks require precise control over solvent ratios and formulation conditions, mainly in parenteral and pediatric applications. Cosolvency is most suitable for minor solvent adjustments. It offers a cost-effective and scalable solution for early stage development. Nonetheless, its broader applicability may be limited by regulatory expectations and safety profiles, especially when high cosolvent concentrations are needed for adequate solubilization of the drugs [12].

10.2.1 Theory

The exact mechanisms behind cosolvency remain unclear, yet multiple theories have been put forth to account for its solubilizing effect. One of the theories suggests that cosolvents improve solubility and molecular dispersion by lowering the interfacial tension between hydrophobic solutes and aqueous solutions. Another theory suggests that cosolvents work by modulating the polarity gradient between the aqueous media and the drug, creating a more compatible solvent environment that facilitates solute incorporation. The observed increase in apparent solubility for nonpolar and weakly ionizable compounds is probably the result of these mechanisms working in harmony [10, 13]. It's interesting to note that a compound's solubility in a solvent blend frequently differs from predictions made based on how soluble it is in each solvent alone. The intricacy of cosolvent interactions is highlighted by this nonlinear behavior. Additionally, cosolvency affects buffer systems and can change both the pH and ionic strength of pharmaceutical formulations, which can impact drug stability and compatibility. It works well, especially with weak electrolytes and nonpolar molecules, whose intrinsic solubility in water is restricted due to unfavorable polarity and ionization profiles [14]. Cosolvents facilitate the dissolution of such compounds by reducing the polarity of the aqueous phase and create a more accommodating environment for the dissolution. The solubility of polar substances, like amino acids, which depend on hydrogen bonds and ionic interactions, may be negatively impacted by this same polarity shift [15].

Poor aqueous solubility is becoming increasingly prominent in formulation science as drug candidates continue to develop in terms of structural complexity and lipophilicity. Bioavailability, therapeutic efficacy, and patient compliance are frequently impeded by these physicochemical constraints [16]. Cosolvency offers a scientifically robust and practically viable solution, which enables formulators to overcome solubility barriers without altering the chemical structure of the API. Cosolvency facilitates the incorporation of the hydrophobic compounds into liquid systems by modifying the solvent environment. Hence, it supports the development of stable, effective, and patient-compliant drug products. Its use in solvent-based formulations not only enhances solubility of drugs but also aligns with the evolving priorities of pharmaceutical development [5].

10.2.2 Concept of Solubility

The solubility of the water depends on the properties of the crystal lattice (melting point and molar heat of fusion) and its molecular structure (activity coefficient). These two factors are crucial in determining how easily a drug dissolves in the solvent [17]. Yalkowsky and Valvani [18] introduced the general solubility equation (GSE) as a predictive model [1]. It is used for estimating the aqueous solubility of the organic nonelectrolytes. It incorporates key physicochemical parameters, melting point, activity coefficient, and crystal lattice energy, providing a thermodynamically grounded framework for solubility prediction. By linking solid-state properties with behavior, an equation provides insight into dissolution profiles of drug candidates, supporting rational formulation design [6, 17].

The aqueous solubility (X_a) can be generally expressed as:

$$\log Xa = \log X_i - \log \gamma \tag{10.1}$$

where

X_i = ideal solubility
γ = activity coefficient

Thus, aqueous solubility can be improved by

1) Increasing X_i, the ideal solubility of the compound, which reflects its thermodynamic propensity to dissolve.
2) Decreasing γ, the activity coefficient, which minimizes nonideal interactions and enhances solute–solvent compatibility.
3) Applying both strategies simultaneously, if feasible, to achieve a synergistic improvement in solubility.

The ideal solubility depends only on the crystal lattice and molecular structure of the solute, without any influence from the solubilizing medium. It can be expressed as:

$$-\log X_i = \frac{\Delta H_m}{2.303R}\left(\frac{T_{m'} - T}{TT_m}\right) \tag{10.2}$$

where

X_i = ideal solubility (mole fraction, X_2^{ideal})
T_m = melting point (K)
ΔH_m = enthalpy of melting (cal/mol)
R = gas constant (1.98 cal/deg/mol)
T = temperature of solution (K)

Since the ratio $\Delta H_m/T_m$ corresponds to the entropy of melting (ΔS_m), Walden's rule approximates it as 13.5 cal/deg/mol for rigid molecules. Therefore, the ideal solubility at room temperature can be estimated as:

$$\log X_i = -0.01(T_{m'} - 298) \tag{10.3}$$

or

$$\log X_i = -0.01(T_m - 25) \tag{10.4}$$

(when T_m is measured in °C)

This shows that ideal solubility depends exclusively on the melting point (which represents crystal lattice energy) and remains constant across different solvents. Notably, a 10-fold solubility enhancement can be achieved for every 100 °C reduction in melting point. The activity coefficient (γ) represents the decrease in solubility caused by intermolecular interactions between solute and solvent molecules. In other words, it reflects the strength of intermolecular forces that must be overcome to transfer a molecule from the solute to the solvent [17].

It is defined as:

$$\log \gamma = \frac{(\delta_1 - \delta_2)^2 \phi_1^2}{2.303RT} \tag{10.5}$$

Yalkowsky and Valvani [18] further demonstrated that the activity coefficient also depends on melting point and the octanol–water partition coefficient ($K_{o/w}$):

$$\log \gamma = \log K_{o/w} + 0.94 \tag{10.6}$$

Combining (10.1), (10.4), and (10.6), the aqueous solubility becomes:

$$\log Xa = -0.001(T_{m'} - 25) - K_{o/w} - 0.94 \tag{10.7}$$

If a solute exhibits a low melting point and a high octanol–water partition coefficient, its molecular structure, particularly its lipophilicity and limited hydrogen bonding capacity, is the primary barrier to aqueous solubility. In such cases, modifying the solvent medium, for example, through cosolvency or surfactant systems, may enhance solubilization without altering the compound itself [17]. Conversely, if the solute has a high melting point and a low partition coefficient, its crystal lattice energy is the dominant limiting factor. Here, structural modification of the solute, such as salt formation, prodrug design, or amorphization, may be necessary to disrupt lattice stability and improve dissolution [17].

The role of cosolvents can also be described mathematically in terms of activity coefficients:

$$\log \gamma = W_{11} - W_{22} - 2W_{12} \tag{10.8}$$

where

W_{11} = cohesive energy of solvent
W_{22} = cohesive energy of solute
W_{12} = adhesive energy between solute and solvent

When W_{11} is large, solubility in water is poor, leading to a higher activity coefficient and lower solubility. Upon addition of a cosolvent, W_{11} becomes closer to W_{22}, reducing the disparity and thereby lowering γ, which results in improved solubility [6, 17] (Figure 10.1).

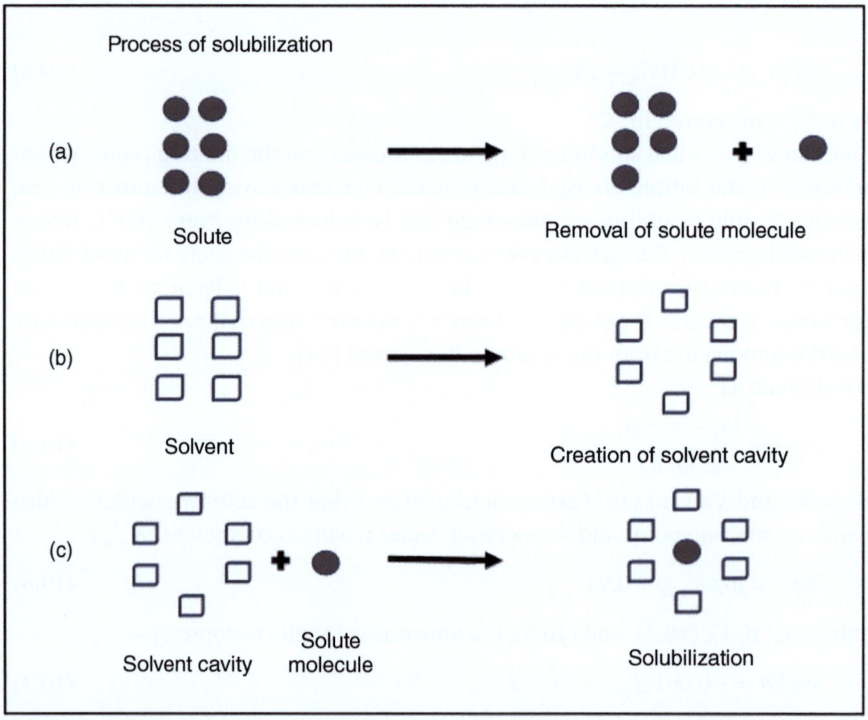

Figure 10.1 Process of solubilization. (a) Removal of solute from the crystal lattice (cohesive energy of the solute), (b) creation of solvent cavity (cohesive energy of the solvent), and (c) placement of solute molecule in the solvent cavity (adhesion energy). *Source:* Adapted from Jain et al. [17].

In general, the solubility of a drug in a cosolvent can be described as:

$$\log S_{mix} = \log S_W + \sigma f_c \tag{10.9}$$

where

S_{mix} = solubility in cosolvent mixture
S_w = solubility in water
f_c = fraction of cosolvent
σ = solubilizing power of cosolvent (a function of activity coefficient)

This equation shows that on a linear scale, solubility increases exponentially with the fraction of cosolvent, highlighting the efficiency of cosolvents in improving aqueous solubility of poorly soluble drugs.

The cosolvent approach is suitable for drug candidates that have specific physicochemical limitations [17]. These include the following:

(a) Non-ionizable compounds: Cosolvency is a more suitable alternative for the drugs that lack an ionizable functional group, which are inherently unsuitable for the pH-based solubilization strategy.
(b) Moderately lipophilic molecules: Moderately lipophilic molecules have log P values between 1 and 3. These molecules have low affinity for solubilization via surfactants or oil-based systems. Hence, it limits the effectiveness of micellar or emulsification techniques. Cosolvent system offers a practical route to enhance the aqueous solubility of such compounds without relying on ionization or lipid partitioning. Examples of cosolvent-based parenteral products illustrate the practical application of cosolvency in enhancement of drug solubility and stability. These include chlordiazepoxide injection (2.5% propylene glycol), diazepam injection (10% ethanol), and digoxin pediatric elixir (10% ethanol with 40% propylene glycol). A notable commercial example of calcium channel blocker is nimodipine IV injection (Nimotop, Bayer). These formulations underscore the versatility of cosolvency in parenteral and liquid dosage forms, particularly for drugs that have limited aqueous solubility [11].

Cosolvent plays an important role in enhancing the solubility of poorly water-soluble drugs, especially in oral solutions and soft gelatin capsules. Their inclusion not only improves solubilization but also facilitates emulsification in aqueous environments. For example, the oral solution and soft gelatin capsule formulation of amprenavir (Agenerase) utilizes a combination of PEG 400, propylene glycol, and d-alpha tocopheryl PEG 1000 succinate (vitamin E TPGS) to achieve effective solubilization [19, 20]. Similarly, the self-emulsifying soft gelatin capsule product Sandimmune incorporates corn oil, glycerol, sorbitol, and ethanol, which collectively promote emulsification through diffusion and solubilizing mechanisms [21]. As the drug transitions into the aqueous phase, it forms fine oil droplets, expanding the formulation and improving bioavailability. Recent advances have introduced novel cosolvent excipients such as diethylene glycol monoethyl ether

(DEGEE) (Transcutol P) as promising solubilizers and permeation enhancers across various administration routes [22, 23].

Saquinavir, another antiretroviral drug, is formulated in a self-emulsifying soft gelatin capsule, which contains gelatin, glycerol, propylene glycol, and polyethoxylated castor oil 35 [24]. In this formulation, cosolvents not only improve solubility but also facilitate the emulsification in aqueous environments, thus enhancing the bioavailability. Nevertheless, the use of propylene glycol is contraindicated in neonates because of its reported toxicity [25].

10.2.3 Advantages of Cosolvency

Cosolvency is a widely adopted technique, offering significant formulation benefits. It allows simple and rapid formulation by blending water-miscible solvents, which reduces the need for complex processing steps. This method is effective for nonpolar and non-ionizable drugs, as it works by altering the overall polarity of the solvent system. A major advantage is the high solubility enhancement of drugs, which in turn enables higher drug loading into the formulation. It contributes to enhancing the palatability by masking the unpleasant taste or odor of the drug. Cosolvency is versatile, applying to use in topical, oral, and parenteral preparations.

10.2.4 Disadvantages of Cosolvency

Cosolvency, despite the benefits, presents several notable risks. A primary concern is that many cosolvents can pose toxicity risks, which is particularly critical in pediatric and parenteral preparations. There is also a risk of API precipitation upon dilution, which directly compromises the safety and efficacy of the drug. There is a possibility of the incomplete solubilization of the API particles, reducing the bioavailability. Furthermore, solvent evaporation may lead to the instability of the formulation. It can also cause an alteration of pH and buffer strength in the system (in case of mixed formulations with buffers). The technique often introduces scale-up challenges when transitioning to large-scale industrial production.

Despite its limitations, cosolvency remains a powerful and widely used technique in pharmaceutical formulation. Its application in approximately 13% of the United States Food and Drug Administration (FDA)-approved parenteral products underscores its clinical relevance and regulatory acceptance [13].

10.3 Types of Cosolvents Allowed and Their IIG Limits

Cosolvents are the essential excipients used to enhance the solubility of APIs. Their inclusion is governed by safety thresholds outlined in the FDA's Inactive Ingredient Guide (IIG), which vary by route of administration and formulation type. Table 10.1 summarizes the maximum allowable potencies per unit dose for commonly used cosolvents, supporting safe and compliant formulation design [26].

Table 10.1 Cosolvents allowed and their IIG limits (cited 30 October 2025).

Cosolvent	Dosage form	Maximum potency per unit dose	Route of administration
Alcohol	Mouthwash	200 mg/ml	Buccal
Alcohol	Gel, paste	1.8% w/w	Dental
Alcohol	Injection	1491 mg	Intramuscular
Alcohol	Injection, solution	6405 mg	Intravenous
Alcohol	Solution/drops	10 mg/ml	Nasal
Alcohol	Solution	6624 mg	Oral
Glycerin	Solution	63.64% w/v	Auricular (otic)
Glycerin	Mouthwash	100 mg/ml	Buccal
Sorbitol	Paste	14.6% w/w	Dental
Glycerin	Injection	1.6% w/v	Intradermal
Propylene glycol	Injection	7459 mg	Intramuscular
PEG 400	Injection, emulsion	28,216 mg	Intravenous
Sorbitol	Injection	45% w/v	Intra-articular
Sorbitol	Injection	45% w/v	Intrasynovial
Glycerin	System	168.1 mg	Iontophoresis
PEG 3350	Solution	40,000 mg/100 ml	Nasal
PEG 400	Solution	46 mg	Ophthalmic
PEG 400	Solution	70,200 mg	Oral
Propylene glycol	Suspension	90,400 mg	Rectal
Propylene glycol	Solution	25% w/w	Respiratory (inhalation)
PEG 3350	Injection, suspension	2.88% w/v	Subcutaneous
Glycerin	System	18,144 mg	Topical
Dipropylene glycol	Film, extended release	296 mg	Transdermal
Sorbitol solution	Cream	1999 mg	Vaginal
PEG 1450	Suppository	9.75 mg	Urethral
PEG 3350	Injection, suspension	87 mg	Parenteral
	Other supplementary solvents		
Cosolvent	Dosage form	Maximum potency per unit dose	Route of administration
Benzyl alcohol	Injection	1620 mg	Intravenous
Benzyl alcohol	Injection	1000 mg	Intramuscular

(Continued)

Table 10.1 (Continued)

Cosolvent	Dosage form	Maximum potency per unit dose	Route of administration
Benzyl alcohol	Lotion	4000 mg	Topical
Benzyl alcohol	Gel	185 mg	Vaginal
Benzyl alcohol	Suspension	129 mg	Rectal
Benzyl alcohol	Spray	5 mg/1 ml	Nasal
Benzyl alcohol	Cream	160 mg	Topical
Benzyl alcohol	Solution	2% w/w	Topical
Benzyl alcohol	Injection, suspension	40 mg	Intra-articular
Benzyl alcohol	Injection, solution	0.95% w/v	Subcutaneous
DMSO	Solution	8008 mg	Topical
DMSO	Lotion	45.5% w/w	Topical
DMSO	Implant	104 mg	Subcutaneous
N, *N*-Dimethylacetamide	Injection	11,583 mg	Intravenous
N, *N*-Dimethylacetamide	Injection, solution	9891 mg	Intravenous
N, *N*-Dimethylacetamide	Spray	0.01 ml	Nasal

Source: Adapted from [26].

10.4 Possible Adverse Events Due to the Use of Cosolvents—Case Studies

Using cosolvents is still a primary method for getting drugs to dissolve, but this strategy comes with real risks. Even those solvents that are approved for pharmaceutical use have been linked to serious problems, from simple allergic reactions to life-threatening organ failure. The core issues leading to these toxicities are often an overload of the excipient, the creation of harmful metabolic byproducts, or simply being unsafe for children and other vulnerable patients. This history of adverse events makes one thing clear: drug formulation absolutely requires rigorous screening of all ingredients, careful safety evaluation based on the dose, and constant regulatory vigilance. The examples that follow show the clinical cost of choosing the wrong cosolvent and emphasize our urgent need for both safer alternatives and stricter oversight [27].

10.4.1 Diethylene Glycol Poisoning

One of the most notorious cases of excipient poisoning occurred in the United States in 1937, when diethylene glycol (DEG) was incorrectly used as a solvent in a liquid form of sulfanilamide. This catastrophic incident led to 107 deaths, including

34 children, with three dying directly from DEG toxicity. The tragedy proved to be a major turning point in drug regulation. It directly triggered the formation of the USFDA to strictly monitor drug safety. In response, the Federal Food, Drug, and Cosmetic Act (FFDC) of 1938 was enacted, shaping the regulatory framework that governs pharmaceutical safety even today [28].

DEG poisoning continued to be a deadly threat for decades. In 1969, South Africa experienced a major incident when five children died of kidney failure after taking an over-the-counter (OTC) sedative. Subsequent investigation proved DEG to be the toxic agent. The danger resurfaced tragically in 1990 at Jos University Teaching Hospital in Nigeria, where 41 children died from renal failure; the cause was confirmed to be DEG contamination in their paracetamol (acetaminophen) syrup. Similarly, devastating outbreaks later occurred in India, with DEG-tainted paracetamol syrup claiming 32 more lives. This chain of recurring global fatalities clearly emphasizes the lethal consequences of misusing DEG as a solvent in medicine [29].

10.4.2 Propylene Glycol Toxicity

Propylene glycol is a widely used cosolvent that has been clearly linked to toxicity in both adults and children. In adults, one notable case involved a patient treated with benzodiazepines using a propylene glycol formulation. The patient developed a severe illness, an iatrogenic (medically caused) condition featuring lactic acidosis, renal dysfunction, and ultimate multi-organ failure. Toxicity in children is an even greater concern. Propylene glycol exposure has been tied to serious problems, including central nervous system (CNS) depression, hypoactivity, seizures, metabolic acidosis, hyperosmolarity, coma, and lactic acidosis [30].

In a striking incident, a 15-month-old infant who received large doses of a vitamin D formulation containing propylene glycol quickly developed alarming symptoms: rapid breathing (tachypnea), fast heart rate (tachycardia), sweating (diaphoresis), irritability, and convulsions. These symptoms are consistent with propylene glycol toxicity, particularly because infants and young children have immature hepatic and renal pathways that slow down the metabolism and elimination of the excipient. The resulting accumulation of propylene glycol can quickly lead to CNS depression, hyperosmolality, and severe lactic acidosis, as their metabolic pathways become saturated [25].

10.5 Challenges in Selecting a Formulation in the Preclinical Studies Using Cosolvents

Selecting a formulation, especially for oral drugs' preclinical studies, presents major difficulties. Both preclinical and clinical data consistently show that using cosolvents carries a significant risk of drug precipitation. PEG 400 is a frequently used cosolvent because it effectively dissolves lipophilic drugs. However, its use demands caution—PEG 400 strongly attracts water and can draw it from surrounding tissues. When the cosolvent is diluted in the body, this shift in the solvent's composition can cause the drug to precipitate, severely compromising its solubility and bioavailability. This phenomenon leads to variable absorption profiles, which makes it challenging to accurately interpret pharmacokinetic data and reliably assess the drug's dose–response relationship [31].

This issue of water-induced dilution and API precipitation isn't limited to PEG 400; it can also occur with other common cosolvents like ethanol, propylene glycol, Transcutol P, and glycerol. The use of polymeric precipitation inhibitors (PPIs) can help to address this issue of API precipitation from the cosolvent-containing formulations to some extent [32].

Reports have consistently shown that PEG 400 can inflict irreversible toxic effects in animal models, making it a notably undesirable choice during the initial phases of drug development. A separate constraint arises with cosolvent concentrations—once levels surpass 60%, the resulting significant viscosity spike severely compromises absorption kinetics and leads to highly variable drug release [33]. Crucially, to mitigate the risk of local irritation and systemic toxicity, a safeguard often guided by existing preclinical safety data and the FDA IIG list, formulators generally restrict cosolvent levels in preclinical vehicles below 25%. Addressing these challenges is the inherent stability issue, as these cosolvents mix with and become diluted in aqueous environments, the drug's solubility tends to decrease exponentially [34].

Cosolvents like PEG 400 have demonstrated a dose-related tendency to cause laxative effects and GI upset, such as loose stools, particularly in animal models like dogs and monkeys, even when concentrations fall within typical formulation guidelines. These adverse reactions can severely complicate oral drug delivery and potentially skew the results of subsequent pharmacodynamic studies. Similarly, ethanol-based vehicles pose a different risk, having been observed to trigger drug precipitation within the GI tract. This precipitation impairs the drug's dissolution profile and ultimately reduces its bioavailability [35]. This unwelcome precipitation beyond the GI tract has been implicated in more serious systemic issues. In animal studies, peritoneal cavity adhesion, specifically involving the liver and spleen, has been linked to peritonitis. Precipitation also negatively affects the compound's stability and can prolong systemic circulation times, a phenomenon documented with formulations utilizing DMSO and PEG 400. Furthermore, this chemical instability has been associated with severe clinical findings, including cardiovascular abnormalities, thrombophlebitis, and a general diminishment of the formulation's integrity. The complexity of these side effects strongly emphasizes the need for meticulous excipient selection and vehicle design during the initial stages of drug formulation [36].

Cosolvent systems frequently fail to maintain drug solubilization upon dilution in the body, leading to supersaturation and eventual drug precipitation. For instance, PEG 400-based formulations were only effective up to a dose of 50–100 mg/g. At doses higher than this, the expected proportional increase in drug solubility and absorption didn't happen. This dose-dependent limitation significantly reduces the utility of these systems, especially in early clinical trials where researchers often need to test higher drug doses [37].

The poor dilution capacity of cosolvents restricts their use because precipitation often occurs when the formulations enter biological fluids. In intravenous infusions, this precipitation is a critical hazard, potentially causing therapeutic failure, tissue injury, or even organ damage. Additionally, toxicological concerns persist, such as hepatotoxicity associated with excipients like PEG. These challenges underscore the strong need for alternative methods to boost solubility. One prominent alternative is the use of a Self-Emulsifying Drug Delivery System (SEDDS). Although SEDDS still uses cosolvents, they have gained popularity because they offer more reliable solubilization and significantly reduce the risk of precipitation during administration *in vivo* [36].

10.6 Design of Formulations Using Cosolvents for First-in-Man Studies or Late Clinical Phases

While cosolvent systems effectively boost solubility at low concentrations, they frequently fail to maintain drug solubilization at higher doses. Studies show that once the dose surpasses the formulation's capacity, significant precipitation occurs, especially in the body's aqueous environments. For example, formulations using 100% PEG 400 might work for doses up to 50–100 mg (depending on the specific drug's solubility), but going beyond this range often makes the system unable to keep the drug dissolved through the absorption process. This failure severely disrupts the drug's bioavailability, as the precipitated drug forms can't be absorbed, regardless of how easily it can cross membranes (passive permeability). When this happens, the drug's limited solubility overrides its pharmacokinetic benefits, resulting in suboptimal therapeutic outcomes. Furthermore, undissolved drug can lead to GI side effects, an inefficient waste of the active ingredient, and reduced overall biological efficacy. These findings emphasize the critical need for dose-formulation matching and highlight the importance of pursuing alternatives like SEDDS or amorphous solid dispersions (ASDs), which can accommodate higher doses without compromising solubility or safety [38].

A critical hurdle for drug delivery is the high risk of precipitation upon dilution, particularly with intravenous (IV) and solubilized formulations. When a formulation enters the bloodstream or mixes with biological fluids, the concentration of both the drug and the cosolvent drops proportionally. If the drug concentration in this new medium exceeds its inherent solubility, the solution becomes supersaturated, causing the drug to precipitate out. This is a dangerous phenomenon that can lead to therapeutic failure, especially during intravenous infusions, and may cause tissue damage or even organ toxicity. Adding to this challenge, the use of cosolvents like PEG 400 is also restricted by their known toxicity profiles and the strict limits on the maximum safe administered quantity [39].

10.7 Regulatory Aspects

The use of cosolvents like PEG 400, propylene glycol, and ethanol in drug formulations, which helps dissolve poorly water-soluble drugs for oral, IV, and topical routes, demands strict regulatory oversight. Their inclusion must be scientifically justified, toxicologically supported, and fully compliant with regulatory guidelines.

The safety of these excipients can't be assumed; it must be demonstrated through structured nonclinical studies and historical data proving safe use [40].

The emphasis on excipient safety is rooted in history. The infamous 1937 sulfanilamide tragedy, where an untested solvent caused over 100 deaths (many of them children), directly led to the passing of the FFDC of 1938. This foundational law established the mandatory requirement for pre-market safety testing of all drug components, including excipients. Today, this core principle remains central to regulatory review. Authorities like the USFDA and EMA require that cosolvents be chemically identified and characterized for purity and quality, backed by comprehensive toxicological profiling and supporting literature evidence [41, 42].

When developing injectable formulations, the chosen cosolvents must meet stringent criteria. They need to show compatibility with the final processes and components, specifically terminal sterilization and the container/closure system, as well as confirmed hemocompatibility. This safety profile must be backed by data addressing tissue toxicity and protein binding. The FDA 2005 guidance, Nonclinical Studies for the Safety Evaluation of Pharmaceutical Excipients, establishes a clear framework for assessing excipient safety. For any cosolvent that hasn't been previously qualified for the specific route of administration or intended exposure level, comprehensive acute and repeat-dose toxicity studies are mandated in both rodent and non-rodent species. Specifically, a 1-month repeat-dose study is recommended for preclinical submissions and must incorporate key endpoints: clinical pathology, histopathology, and toxicokinetic analysis. Further required evaluations may encompass genetic toxicology (such as the Ames test and chromosomal aberration assays), ADME profiling to confirm systemic exposure, reproductive toxicity for compounds intended for systemic or long-term use, and full safety pharmacology screens covering the cardiovascular, CNS, and respiratory systems [41].

Depending on the product type and route of administration, regulatory expectations vary. New Drug Applications/Biologics License Applications (NDAs/BLAs) must include full safety data or reference a drug master file. Abbreviated New Drug Applications (ANDAs) for topical, oral, parenteral, ophthalmic, or otic products must match the reference listed drug (RLD) in excipient type and quantity. OTC products must use excipients recognized as safe and maintain internal documentation. Changes to excipient composition post-approval may require a prior approval supplement (PAS) for major modifications or a CBE-30/Annual Report for minor adjustments [41].

Ultimately, cosolvent excipients offer powerful tools for solubility enhancement, but their regulatory acceptance hinges on rigorous safety evaluation, dose justification, and formulation stability.

10.8 Selected Examples of Cosolvent-Based Formulations Using 505(b)(2) Pathway and Intellectual Property (IP) Protection

The USFDA published draft guidelines on applications under the FFDC section 505(b)(2) in 1999. This was the first official launch of the 505(b)(2) approach, which allows organizations to submit NDAs that include data from a reference medicine that has already been approved. These NDAs typically focus on reformulations that improve properties such as administration, bioavailability, or solubility. Drugs of BCS Class II or IV that are poorly soluble in water are dissolved and delivered with the aid of these advances. When diluted in biological fluids, common cosolvents such as ethanol, propylene glycol, PEG, or mixes that improve drug dispersion, modify solvent polarity, and inhibit precipitation are used [43].

10.8.1 Cosolvent-Comprised Formulations for Hydrophobic Drug Solubilization

A well-known example is the formulation of an oral solution that contains a cannabinoid, ideally dronabinol, and uses a cosolvent system of ethanol (15–65% w/w), propylene glycol (0–50% w/w), and PEG 400 (0–50% w/w) with water (0–40% w/w) to solubilize the hydrophobic drug at concentrations of 0.05–20 mg/ml authorized as Syndros® (NDA 205525) in 2016. By aligning with the drug's partition coefficient, the formulation improves solubility, leading to a 20% increase in bioavailability, a quicker onset (T_{max} 0.51–2.13 hours), and a decrease in inter-subject variability. The drug's poor water solubility (5–20% oral bioavailability due to inadequate absorption and first-pass metabolism) was addressed by cosolvents in this approval, which referred to the original Marinol® capsules (NDA 018651). In comparison to capsules, the formulation enhances uniformity, onset, and absorption variability [44].

Furthermore, the patent US6136799A, "Cosolvent Formulations," presents a novel method for producing stable, self-preserved aqueous pharmaceutical formulations intended for parenteral delivery. Ethanol (15–50% v/v) and propylene glycol (15–35% v/v) in water (balanced to 100%) are used by therapeutic agents with poor water solubility, such as vitamin D compounds like paricalcitol and calcitriol (2–10 μg/ml), to overcome poor aqueous solubility (<1 μg/ml) without the need for antioxidants, buffers, or chelating agents [45]. This patent is continued in US6361758B1, which outlines a comparable cosolvent method for vitamin D analogs. When combined, the cosolvents increase solubility by up to 119 μg/ml and enable antimicrobial preservation that satisfies USP criteria. Stability is maintained at 30 °C for 24 months with 96–100% efficacy. Facilitating irritation-free intravenous distribution without the need for extra preservatives improves bioavailability [46]. To improve process efficiency rather than direct solubility, the patented lyophilized formulation of dantrolene sodium (0.1–50 mg/ml, ~100 mg/vial) uses a *tert*-butanol-based cosolvent solution (2–10%) in water with mannitol (~3 g/vial) and NaOH for pH 8.8–10.9. The cosolvent improves bioavailability through quick injection and cuts the lyophilization time by 33% (to about 50 hours) through porous ice structures and reconstitution time of <10 seconds, which is crucial for emergency IV use in malignant hyperthermia [47]. Likewise, a solid dispersion for hydrophobic drugs like raloxifene, modafinil, and glimepiride, using PEG (e.g. PEG 4500) as the primary carrier with surfactants (e.g. Polysorbate 80) and cosolvents like propylene glycol, was formulated to form melts or granules, enhancing dissolution in gastric/intestinal fluids. PEG dominates for wettability and dispersion, improving bioavailability by increasing aqueous solubility and absorption. These systems exemplify cosolvent-mediated entropy-driven solubilization, often combined with antioxidants (e.g. butylated hydroxyanisole at 0.01% w/w) to prevent oxidative degradation [48].

Docetaxel is a widely used anticancer drug, and the original formulations of Taxotere contain ethanol and Tween 80 to solubilize the drug (0.54 g polysorbate 80 and 0.395 g dehydrated alcohol) [49]. However, hypersensitivity reactions using this product were reported due to the surfactants in the formulation. Sandoz, Inc., Hospira Inc., and Apotex Inc. each have docetaxel-containing new drug products using 505(b)(2) pathway. The new formulations have PEG 300 as an additional cosolvent, and the Tween 80 content is significantly less than that of Taxotere. These formulations were claimed to be safer and stable than Taxotere [50–52]. Enabled drug delivery technology utilizing the cosolvent–surfactant approach has indeed proved vital in providing an effective treatment option for cancer patients.

10.8.2 pH-Modified Cosolvent Formulations for Ionizable Drugs

Several patents integrate pH adjusters with cosolvents to modulate ionization states, enhancing solubility for weakly acidic/basic actives. The innovative solvent system targets salts of weakly acidic, basic, or amphoteric pharmaceutical agents, such as naproxen sodium (10–50% w/w), by incorporating PEG (10–80% w/w) and water (1–20% w/w), along with a de-ionizing agent (0.2–1.0 molar eq., e.g. hydrochloric, lactic, or acetic acid) to achieve partial neutralization. This approach mitigates issues like precipitation from excessive ionization and unwanted ester formation with PEG. By controlling pH (2.5–7.5) and enhancing compatibility with capsule shells, this improves solubility for increased drug loading in soft gelatin capsules and improves bioavailability, producing stable, high-loading formulations perfect for oral delivery [53]. Additionally, by combining solvents comprising transcutol or polyglycerol oleate (10–80% w/w) with limited ionization via ions (0.05–1.5 mol per ionizable group) from hydroxide or hydrogen sources, along with cosolvents like PEG and Tween esters, a solution for acidic, basic, and amphoteric agents (10–80% w/w) in soft gelatin capsules was formulated. With either water (1–10% w/w) or propylene glycol (4–12% w/w), this improves oral bioavailability by improving dissolution and increasing solubility for larger concentrations in smaller capsules [54].

Boehringer Ingelheim-affiliated entities' innovations pertaining to angiotensin-II antagonists for the management of hypertension vary mainly in their scope and level of invention. New substituted benzimidazole formulations that contain cosolvent systems or pH modifiers to improve the solubility of telmisartan, a medication that is hard to dissolve in water, are disclosed by US5591762A. The method improves solubility and absorption in oral forms like Micardis® by adjusting pH with alkalis such as sodium hydroxide and meglumine [55]. The polymorphic forms of telmisartan are also covered by US6358986B1, which uses the same pH-modifying agents in cosolvent-like systems to increase water solubility from less than 0.01 mg/ml to more, improving oral bioavailability through steady absorption. The formulation is perfect for antihypertensive treatment since it provides stability and better solubility [56]. However, EP1545467A1 emphasizes solid dosage forms of telmisartan, like tablets or capsules, that contain basic agents (like meglumine), nonionic surfactants (like poloxamer 188), and diluents (like mannitol) in a cosolvent system to boost solubility and enable quick, pH-independent dissolution for improved bioavailability [57].

Similarly, a combination therapy of lacidipine and telmisartan, employing cosolvents or pH adjusters to improve solubility and bioavailability of these lipophilic drugs, reduces first-pass metabolism and variability in absorption. The system enhances systemic exposure for better antihypertensive effects [57]. Additionally, a pH modifier, meglumine, is used in cosolvent systems (e.g. USRE37035E1, US5312924A, US6143769A) to enhance the solubility and rapid-onset bioavailability of repaglinide for antidiabetic treatment. Use of meglumine facilitates rapid insulinotropic effects with minimized glycemic variability in type 2 diabetes regimens [58–60]. Similarly, US6677358B1 describes a regimen for non-insulin-dependent diabetes mellitus using repaglinide in cosolvent-enhanced formulations with pH modifiers like meglumine, improving solubility and bioavailability for better glycemic control [61].

10.8.3 Surfactant–Cosolvent Hybrids for Use in Oral and Transdermal Delivery

Hybrid systems incorporate surfactants for micellization alongside cosolvents. EP0179583A1 introduces a surfactant–cosolvent system for poorly soluble drugs, such as ivermectin, which combines surfactants (e.g. sodium lauryl sulfate, Polysorbate 80) with cosolvents (e.g. polyols, polyoxyethylene ethers) to form in situ micelles upon aqueous exposure, thereby enhancing dissolution and solubility for improved absorption. The formulation is prepared as a solid via solution evaporation or melt cooling, emphasizing uniform drug distribution for oral or other routes. Encouraging micellar encapsulation and lowering surface tension, it increases bioavailability [62]. US20010053383A1 also focuses on transdermal systems that have a polymer matrix of polysiloxane (10–90% w/w) and polyacrylate (10–90% w/w). Cosolvents such as propylene glycol or oleic acid are used to modify solubility parameters, thereby improving drug delivery (such as nitroglycerin and estradiol) and flow across skin. Through increased saturation concentrations and regulated release, it increases bioavailability [63].

10.8.4 Solid Dispersion for Solubility Enhancement

Through amorphous stabilization, the dispersion approaches simulate the effects of cosolvent. In order to achieve improved bioavailability and two to five times dissolution rates through supersaturation and precipitation inhibition, solid dispersions of cucurbitacin B with carriers such as hydroxypropyl methylcellulose and PEG derivatives, rotigotine with polyvinylpyrrolidone (weight ratio 9:3.5 to 9:6), cilnidipine with hydroxypropyl cellulose and hydroxypropyl methylcellulose phthalate, and heterocyclic compounds in polymers (e.g. Poloxamer 188 or PVP K30) are described in CN108042496A, US20180147154A1, CN108210472A, and US10004719B1 [64–67].

As demonstrated by patents offering improved solubility and bioavailability, the 505(b)(2) approach has enabled creative cosolvent-based reformulations for medications with low solubility.

10.9 Clinical Pipeline

A crucial tactic for improving the solubility and bioavailability of weakly water-soluble medications (BCS Class II/IV) is the use of cosolvent-based formulations, which are frequently incorporated into SEDDS. There are several approved products as of 2025, and many are in the testing phase. To create stable emulsions during dilution in the GI tract, these formulations usually combine cosolvents (such as ethanol, propylene glycol, and DMSO) with oils, surfactants, and emulsifiers. This promotes lymphatic uptake, inhibits P-glycoprotein (P-gp) efflux, and prevents API precipitation, leading to higher oral bioavailability and lower toxicity when compared to IV alternatives [68].

As of October 2025, the most well-known candidate is DHP107 (marketed as Liporaxel in Korea), an oral paclitaxel drug formulation invented by Daehwa Pharmaceutical. This lipid-based SEDDS combines oils (like monoolein) and surfactants with cosolvents like ethanol and propylene glycol to form nanoemulsions upon GI dilution, improving paclitaxel absorption via lymphatic pathways, avoiding first-pass metabolism, and achieving three to five times higher bioavailability than IV paclitaxel (Taxol) [69, 70]. In 2025, it finished Phase III (OPTIMAL-3, NCT03326102), showing non-inferior efficacy to IV paclitaxel with less neuropathy and hypersensitivity, and allowed for easy oral administration without the need for premedication. It is indicated for metastatic breast and gastric malignancies [71, 72].

To further improve intestinal absorption and bioavailability, Kaken Pharmaceutical's Oraxol (licensed from Hanmi) combines oral paclitaxel with encequidar (a P-gp inhibitor). As of 2025, active Phase I/II trials for angiosarcoma and metastatic breast cancer (e.g. NCT03165955 PK study, NCT06835400 for dose optimization) show anticancer activity, safety, and equivalent exposure to IV paclitaxel, allowing for variable dosing [73, 74]. Similarly, a tablet substitute for intravenous docetaxel is provided by ModraDoc006/r, an oral docetaxel tablet containing ritonavir (CYP3A4/P-gp inhibitor) from Modra Pharmaceuticals that improves solubility and bioavailability through metabolic inhibition. With superior tolerability, lower toxicity, and easy weekly dosage, Phase II trials (finished in 2024, NCT03136640) targeting metastatic castration-resistant prostate cancer demonstrate equivalent efficacy to IV, opening the door for Phase III in 2025 [75]. The macrocyclic peptide enlicitide chloride, manufactured by Merck & Co., uses lipid-based formulations with the surfactant/cosolvent Labrasol ALF to increase the solubility, permeability, and bioavailability of BCS Class III/IV drugs. Human clinical trials emphasize the potential for oral delivery of complex compounds by showing a two- to threefold increase in plasma concentration through improved absorption and lymphatic uptake with fewer dietary effects [76].

10.10 Future Prospects

Future developments to enhance the solubility of poorly water-soluble medications are anticipated to be revolutionary due to the need for sustainability, technological integration, and a better understanding of mechanisms. To improve therapeutic indices for BCS Class IV drugs, future research could concentrate on combining cosolvents with lipid nanoparticles or polymeric micelles to generate hybrid systems that target tissues, such as tumors or the brain, and provide controlled release profiles. This could potentially reduce off-target effects [77]. To lessen the toxicity risks associated with conventional options like propylene glycol or ethanol, investigate creating green, bio-based cosolvents, such as those derived from renewable sources like glycerol or plant-derived polyols. This will support regulatory demands for smaller ecological footprints in pharmaceutical manufacturing as well as environmental sustainability goals [78]. With preclinical models evaluating response kinetics and efficacy, cosolvent mixes that react to physiological inputs (such as pH, temperature, or enzymes) to initiate medication release can be engineered to provide smart delivery systems for chronic illnesses like diabetes or arthritis [7]. At the same time, recent ML applications in supercritical CO_2 systems for BCS Class II/IV drugs demonstrate how important AI and ML models are for forecasting solubility enhancements, predicting cosolvent–drug interactions, and optimizing mixture compositions. These models may speed up formulation design and minimize empirical trial-and-error in preclinical stages [79]. To enable customized formulations for complex biologics and natural products, molecular dynamics simulations and sophisticated spectroscopic techniques will further clarify cosolvency interactions at an atomic level, for example, cosolvent impact on cyclodextrin–drug complexation or micelle aggregation in SEDDS. Whereas hybrid systems that combine hydrotropes or ionic liquids with cosolvents could improve bioavailability and prevent precipitation upon dilution for oral delivery of peptides such as semaglutide [5, 9].

New developments in cosolvent platforms integrated with nanomedicine, like organic solvent nanofiltration for formulation purification or the use of cosolvents in bioinks for 3D-printed customized dosage forms, offer enhanced scalability, patient-centered delivery for cancer and other medications, and the possibility of decreased toxicity [80]. As the market is expected to grow to USD 5.89 billion by 2032, regulatory evolution will prioritize excipient safety analysis by digital twins and AI-driven toxicokinetics, encouraging developments in crystallization–cosolvent hybrids for therapeutics derived from natural products, and inevitably tackling missing gaps in disease management and personalized medicine [81].

10.11 Conclusion

Poor water solubility hinders bioavailability, formulation stability, and therapeutic efficacy, making it a crucial obstacle in pharmaceutical development, especially for BCS Class II and IV medications. Cosolvency is a flexible and scalable approach that uses organic solvents that are water-soluble, such as ethanol, propylene glycol, and PEG, to change the polarity of the solvent, break the hydrogen bond in water, and increase the solubility of the drug exponentially—often by thousands of times—without changing the active ingredient. Mechanistic linkages between crystal lattice energy, lipophilicity, and solute–solvent interactions are highlighted by theoretical frameworks like the GSE and activity coefficient models, which direct the creation of logical formulations. The effectiveness of cosolvency is demonstrated by practical uses in liquid dosage forms, parenteral injections, and self-emulsifying systems. Nimotop and Agenerase are examples of commercial pharmaceuticals that enhance dissolution kinetics and patient compliance. Patents that address the precipitation issues and allow for high-drug-loading capsules include US9693978B2—pH-modified PEG systems for naproxen—and US6136799A—synergistic ethanol-propylene glycol blends for vitamin D analogs. Despite the benefits of simplicity and cost-effectiveness, careful optimization is required due to restrictions such as toxicity (e.g. propylene glycol-induced acidosis), dilution-induced API precipitation, and regulatory obstacles. Future barriers should be addressed by combining green cosolvents, AI-driven forecasts, and hybrid nanocarrier systems, which will promote sustainable, customized medications that improve global health outcomes and conform to changing regulatory requirements.

Acknowledgment

The authors are grateful to Somaiya Vidyavihar University and Adex Pharma™ for the opportunity and the support provided while authoring the book chapter.

References

1 Savjani, K.T., Gajjar, A.K., and Savjani, J.K. (2012). Drug solubility: importance and enhancement techniques. *ISRN Pharm.* 2012: 195727.

2 Bhalani, D.V., Nutan, B., Kumar, A., and Singh Chandel, A.K. (2022). Bioavailability enhancement techniques for poorly aqueous soluble drugs and therapeutics. *Biomedicines* 10 (9): 2055.

3 L Hart, M. et al. (2013). Brief overview of various approaches to enhance drug solubility. *J. Dev. Drugs* 02 (03).

4 Hashmi, A.R., Sekar, M., Zahra, F. et al. (2025). Advanced drug delivery strategies to overcome solubility and permeability challenges: driving biopharmaceutical advancements toward commercial success. *ACS Omega* 10 (36): 40769–40792.

5 El Hamd, M.A., Obaydo, R.H., Nashed, D. et al. (2025). Hydrotropy and co-solvency: sustainable strategies for enhancing solubility of poorly soluble pharmaceutical active ingredients. *Talanta Open* 11: 100391.

6 Sinko, P.J. and Singh, Y. (eds.) (2024). *Martin's physical pharmacy and pharmaceutical sciences: physical chemical and biopharmaceutical principles in the pharmaceutical sciences*, 8the. Wolters Kluwer Health.

7 Xie, B., Liu, Y., Li, X. et al. (2024). Solubilization techniques used for poorly water-soluble drugs. *Acta Pharm Sin B* 14 (11): 4683–4716.

8 Zhu, G., Yu, H., Wang, Q. et al. (2024). Molecular mechanism of the cosolvency effect in organic solutes with similar structures. *Ind Eng Chem Res* 63 (28): 12733–12744.

9 He, M., Zheng, W., Wang, N. et al. (2022). Molecular dynamics simulation of drug solubilization behavior in surfactant and cosolvent injections. *Pharmaceutics* 14 (11): 2366.

10 Lachman, L., Lieberman, H.A., and Kanig, J.L. (1987). *The Theory and Practice of Industrial Pharmacy*, 3e. Varghese Publication House.

11 Almoazen, H. (2013). Felton L. Remington: Essentials of Pharmaceutics. Pharmaceutical Press; 2013, 772pp, $69.00 (softcover), ISBN 9780857111050. *Am. J. Pharm. Educ.* 77: 233.

12 Nyamba, I., Sombié, C.B., Yabré, M. et al. (2024). Pharmaceutical approaches for enhancing solubility and oral bioavailability of poorly soluble drugs. *Eur. J. Pharm. Biopharm.* 204.

13 Millard, J.W., Alvarez-Núñez, F.A., and Yalkowsky, S.H. (2002). Solubilization by cosolvents: establishing useful constants for the log–linear model. *Int. J. Pharm.* 245 (1–2): 153–166.

14 Samuelsen, L., Holm, R., Lathuile, A., and Schönbeck, C. (2019). Buffer solutions in drug formulation and processing: how pKa values depend on temperature, pressure and ionic strength. *Int. J. Pharm.* 560: 357–364.

15 Jouyban, A. (2008). Review of the cosolvency models for predicting solubility of drugs in water-cosolvent mixtures. *J. Pharm. Pharmaceut. Sci.* 11 (1): 32–58.

16 Fine-Shamir, N. and Dahan, A. (2024). Solubility-enabling formulations for oral delivery of lipophilic drugs: considering the solubility-permeability interplay for accelerated formulation development. *Exp. Opin. Drug Deliv.* 21 (1): 13–29.

17 Jain, S., Patel, N., and Lin, S. (2015). Solubility and dissolution enhancement strategies: current understanding and recent trends. *Drug Dev. Ind. Pharm.* 41 (6): 875–887.

18 Yalkowsky, S.H. and Valvani, S.C. (1980). Solubility and partitioning I: solubility of nonelectrolytes in water. *J. Pharm. Sci.* 69 (8): 912–922.

19 U.S. Food and Drug Administration. (2005) AGENERASE (amprenavir) Oral Solution Prescribing Information. NDA 21-039/SLR017. GlaxoSmithKline, NC.

20 U.S. Food and Drug Administration. (2005) AGENERASE (amprenavir) Capsules [Prescribing Information]; NDA 21-007/SLR017; GlaxoSmithKline, NC.

21 U.S. Food and Drug Administration. Sandimmune® (cyclosporine) prescribing information. East Hanover, NJ: Novartis Pharmaceuticals Corporation.

22 Musakhanian, J., Osborne, D.W., and Rodier, J.D. (2024). Skin penetration and permeation properties of Transcutol® in complex formulations. *AAPS PharmSciTech* 25 (7): 1–23.

23 Ha, E.S., Lee, S.K., Choi, D.H. et al. (2020). Application of diethylene glycol monoethyl ether in solubilization of poorly water-soluble drugs. *J. Pharm. Invest.* 50 (3): 231–250.

24 US20060217320A1 - Soft gel formulations for saquinavir - Google Patents.

25 Wilson, K.C., Reardon, C., Theodore, A.C., and Farber, H.W. (2005). Propylene glycol toxicity: a severe iatrogenic illness in ICU patients receiving IV benzodiazepines: a case series and prospective, observational pilot study. *Chest* 128 (3): 1674–1681.

26 U.S. Food and Drug Administration/FDA/Center for Drug Evaluation and Research. (2025). Inactive Ingredient Search for Approved Drug Products.

27 Saito, J., Nadatani, N., Setoguchi, M. et al. (2021). Potentially harmful excipients in neonatal medications: a multicenter nationwide observational study in Japan. *J. Pharm. Health Care Sci.* 7 (1): 23.

28 Ballentine, C. (1981). Sulfanilamide disaster- taste of raspberries, taste of death: the 1937 elixir sulfanilamide incident. *FDA Consumer Magazine.*

29 Schep, L.J., Slaughter, R.J., Temple, W.A., and Beasley, D.M.G. (2009). Diethylene glycol poisoning. *Clin. Toxicol.* 47 (6): 525–535.

30 Lim, T.Y., Poole, R.L., and Pageler, N.M. (2020). Propylene glycol toxicity in children. *J. Pediatr. Pharmacol. Ther.* 19 (4): 277–282.

31 Shah, S.M., Jain, A.S., Kaushik, R. et al. (2014). Preclinical formulations: insight, strategies, and practical considerations. *AAPS PharmSciTech* 15 (5): 1307–1323.

32 Parmar, T., Kadu, P., and Kale, P. (2022). Critical strategies for drug precipitation inhibition: a review with the focus on poorly soluble drugs. *Curr. Drug Delivery* 20 (5): 497–507.

33 Rytting, E., Lentz, K.A., Chen, X.Q. et al. (2004). A quantitative structure-property relationship for predicting drug solubility in PEG 400/water cosolvent systems. *Pharm. Res.* 21 (2): 237–244.

34 Pellegrini, G., Starkey Lewis, P.J., Palmer, L. et al. (2013). Intraperitoneal administration of high doses of polyethylene glycol (PEG) causes hepatic subcapsular necrosis and low-grade peritonitis with a rise in hepatic biomarkers. *Toxicology* 314 (2–3): 262–266.

35 Thomazini, V.C., da Cunha, G.M., Guimarães, N.M. et al. (2023). Impact of concerning excipients on animal safety: insights for veterinary pharmacotherapy and regulatory considerations. *DARU J. Pharm. Sci.* 32 (1): 289.

36 Neervannan, S. (2006). Preclinical formulations for discovery and toxicology: physicochemical challenges. *Expert Opin. Drug Metab. Toxicol.* 2 (5): 715–731.

37 Tønsberg, H., Holm, R., Bisgaard, J. et al. (2010). Effects of polysorbate 80 on the in-vitro precipitation and oral bioavailability of halofantrine from polyethylene glycol 400 formulations in rats. *J. Pharm. Pharmacol.* 62 (1): 63–70.

38 Fine-Shamir, N., Beig, A., Miller, J.M., and Dahan, A. (2020). The solubility, permeability and the dose as key factors in formulation development for oral lipophilic drugs: maximizing the bioavailability of carbamazepine with a cosolvent-based formulation. *Int. J. Pharm.* 582: 119307.

39 Narazaki, R., Sanghvi, R., and Yalkowsky, S.H. (2007). Estimation of drug precipitation upon dilution of pH-cosolvent solubilized formulations. *Chem. Pharm. Bull (Tokyo)* 55 (8): 1203–1206.

40 Gullapalli, R.P. and Mazzitelli, C.L. (2015). Polyethylene glycols in oral and parenteral formulations - a critical review. *Int. J. Pharm.* 496 (2): 219–239.

41 U.S. Department of Health and Human Services, Food and Drug Administration, Center for Drug Evaluation and Research (CDER), and Center for Biologics Evaluation and Research (CBER) (2005). Guidance for industry nonclinical studies for the safety evaluation of pharmaceutical excipients.

42 European Medicines Agency. (2007). *Guideline on excipients in the dossier for application for marketing authorisation of a medicinal product* (Revision 2, EMEA/CHMP/QWP/396951/2006). https://www.ema.europa.eu/en/excipients-dossier-application-marketing-authorisation-medicinal-product-scientific-guideline.

43 Chen, J., Zhao, Z., Wang, X., and Huang, J. (2023). Utilizing 505(b)(2) regulatory pathway for new drug applications: an overview on the advanced formulation approach and challenges. In: *Drug Repurposing - Advances, Scopes and Opportunities in Drug Discovery*. IntechOpen.

44 US8222292B2 (2011) Liquid cannabinoid formulations. Google Patents.

45 US6136799A (2000) Cosolvent formulations. Google Patents.

46 US6361758B1 (2002) Cosolvent formulations. Google Patents.

47 EP2219605B1 (2015) Co-solvent compositions and methods for improved delivery of dantrolene therapeutic agents. Google Patents.

48 Ray, A.K., Nandi, I., Palaniswamy, S. et al. (2005). *Pharmaceutical composition for solubility enhancement of hydrophobic drugs*. World Intellectual Property Organization (WIPO) Patent No. WO2005004917A2.

49 RxList. (2023). Taxotere (docetaxel for injection): Side effects, uses, dosage, interactions, warnings. https://www.rxlist.com/taxotere-drug.htm.

50 FDA and Cder (2012). Docetaxel injection 40 mg/mL (20 mg/0.5 mL and 80 mg/2 mL).

51 U.S. Food and Drug Administration. (2011). *NDA Approval Letter: Docetaxel Injection (NDA 022234)*. https://www.accessdata.fda.gov/drugsatfda_docs/nda/2011/022234Orig1s000Approv.pdf (accessed 13 April 2026).

52 Center for Drug Evaluation and Research (2011). Clinical pharmacology and biopharmaceutics review(S). Application Number: 201525Orig1s000

53 US9693978B2 (2017) Solvent system for enhancing the solubility of pharmaceutical agents. Google Patents.

54 US5376688A (1994) Enhanced solubility pharmaceutical solutions. Google Patents.

55 US5591762A (1997) Benzimidazoles useful as angiotensin-11 antagonists. Google Patents.

56 US6358986B1 (2002) Polymorphs of telmisartan. Google Patents.

57 EP1545467A1 (2005) Solid pharmaceutical formulations comprising telmisartan. Google Patents.

58 US6143769A (2000) Phenylacetic acid benzylamides - Google Patents.

59 US5312924A (1994) Phenylacetic acid benzylamides. Google Patents.

60 USRE37035E1 (2001) Phenylacetic acid benzylamides. Google Patents.

61 US6677358B1 (2004) NIDDM regimen. Google Patents.

62 EP0179583A1 (1986) A system for enhancing the water dissolution rate and solubility of poorly soluble drugs. Google Patents.

63 US20010053383A1 (2001) Solubility parameter based drug delivery system and method for altering drug saturation concentration. Google Patents.

64 US10004719B1 (2018) Solid dispersion formulation. Google Patents.

65 CN108210472A (2018) A kind of Cilnidipine solid dispersions tablet and preparation method thereof. Google Patents.

66 US20180147154A1 (2018) Polyvinylpyrrolidone for the stabilization of a solid dispersion of the non-crystalline form of rotigotine. Google Patents.

67 CN108042496A (2018) A kind of Cucurbitacin B solid dispersions and preparation method thereof. Google Patents.

68 Bagade, P.V., Kulkarni, N.S., Chaudhari, P.D. et al. (2025). Exploring modern techniques for solubility enhancement in drug formulations. *Int. J. Curr. Pharm. Res.* 17: 31–41.

69 Moon, B.S., Park, H.S., Sunwoo, J. et al. (2021). Tissue pharmacokinetics of DHP107, a novel lipid-based oral formulation of paclitaxel, in mice and patients by positron emission tomography. *Clin. Transl. Sci.* 14 (5): 1747–1755.

70 Study Details. NCT04675528. (2024) Food Effect on Pharmacokinetics and Safety of DHP107 (Liporaxel®) FEEL Study. ClinicalTrials.gov.

71 Kim, S.-B., Zhang, Q., Sun, T., Seo, J.H., Lee, K.S., Kim, T.-Y., Tong, Z., Park, K.H., Moon, Y.W., Wang, S., Li, W., Yang, Y., Wang, J., Wang, X., Choi, J., Lee, J.E., Yoon, K.E., Chung, S., Xu, B., and Sohn, J. (2020) [OPTIMAL 3] A phase III trial to evaluate the efficacy and safety of DHP107 (Liporaxel, oral paclitaxel) compared to Taxol (IV paclitaxel) as first line therapy in patients with recurrent or metastatic HER2 negative breast cancer (BC) (NCT03315364). *Journal of Clinical Oncology*, 38 (15_suppl), TPS1106–TPS1106.

72 NCT03326102 (2020) A Multi-center, Open-label, Phase 2 Clinical Trial to Evaluate the Efficacy, Safety and Pharmacokinetics of DHP107 (Liporaxel®, Oral Paclitaxel) Compared to IV Paclitaxel in Patients With Recurrent or Metastatic Breast Cancer(OPERA). *My Cancer Genome.*

73 Study Details | NCT06835400 | Oral Paclitaxel + Encequidar vs IV Paclitaxel in Treatment of HER2 Negative Metastatic Breast Cancer | ClinicalTrials.gov.

74 Yu, Y., Wang, Y., Mao, L. et al. (2025). Phase I clinical trial to assess safety and efficacy of Oraxol, a novel oral paclitaxel chemotherapy agent, in patients with previously treated metastatic breast cancer. *MedComm. (Beijing)* 6 (3): e70097.

75 Vaishampayan, U.N., Keessen, M., Dreicer, R. et al. (2024). A global phase II randomized trial comparing oral taxane ModraDoc006/r to intravenous docetaxel in metastatic castration resistant prostate cancer. *Eur. J. Cancer* 202: 114007.

76 Merck (2025). Merck's Investigational Oral PCSK9 Inhibitor Enlicitide Decanoate Met all Primary and Key Secondary Endpoints in Adults with Hypercholesterolemia in Pivotal CORALreef Lipids Study. Merck.com.

77 Araman, C., Beck-Broichsitter, M., Erwin, N. (2022). Seven trends shaping the future of pharmaceutical formulation development. *Pharmtech.*

78 Future Market Insights (2025). Pharmaceutical solvents market insights - size, share & industry growth 2025 to 2035. *Future Market Insights.*

79 Alamoudi, J.A. (2024). Recent advancements toward the incremsent of drug solubility using environmentally-friendly supercritical CO_2: a machine learning perspective. *Front Med. (Lausanne)* 11.

80 Xiao, H., Feng, Y., Goundry, W.R.F., and Karlsson, S. (2024). Organic solvent nanofiltration in pharmaceutical applications. *Org. Process Res. Dev.* 28 (4): 891–923.

81 Coherent Market Insights (2025). Pharmaceutical solvents market size and analysis, 2025-2032. *Coherent Market Insights.*

11

First-in-Animal Studies and Formulation Considerations

Dnyaneshwar N. Kapote[1] and Vandana Gawande-Kapote[2]

[1]*Aenova Holding GmbH, Global Pharmaceutical Development, Science and technology department Berger Strasse 8-10, Starnberg 82319, Germany*
[2]*Saanvi Pharma Consulting UG (haftungsbeschränkt), Siemensstrasse 40, Nuremberg 90459, Germany*

11.1 Introduction

Over the years, with the advent of high-throughput screening and combinatorial chemistry, pharmaceutical industry has witnessed that the number of new chemical entities (NCEs) having poor aqueous solubility has been exponentially increased. It has been estimated that more than 40% of the marketed active pharmaceutical ingredients (APIs) and even 70% to 90% of the NCEs are poorly water soluble compounds [1–3]. The early formulation development plays an important role in drug discovery and development. These formulations are produced in the discovery phase and the preclinical phase, and are dosed to the animals in different forms such as solution or suspension. An example of a simple suspension formulation of NCE dosed to mice via oral gavage for efficacy study is reported by Ananthula et al [4]. These early formulations are termed as preclinical formulations. Preclinical formulations are developed with major objectives as follows:

(a) Optimum *in vivo* exposure at the target site which is important to evaluate biological or pharmacological response.
(b) Accurate dosing is imperative to derive dose–response relationship and it is essential that the formulation is physically and chemically stable throughout preparation, dosing as well as short-term storage.
(c) Minimizing the excipient toxicity: The given vehicle or formulation must have maximum *in vivo* tolerability [5].

The goal of preclinical studies is to evaluate the safety of the new investigational compound to progress into the first-in-human trials. The results of preclinical studies are validated by conducting clinical studies (phase 1 onward). Most often, the preclinical studies are extrapolated to predict clinical outcomes. The preclinical studies involve the administration of the starting dose and checking dose frequency to obtain the desired pharmacological response. It also aims to have information on safe dose escalation regimen that permits differentiation of response as a function of drug exposure, and an understanding of potential toxicity which may supersede potential pharmacological benefit [6]. The purpose of animal models (preclinical model) is to establish and provide evidence for the non-clinical "proof-of-concept" with respect to the safety, efficacy, and rationale for the development of specific new drug molecules [7]. The data obtained from the use of animal models supports the decision whether to develop or not the investigational drug candidate (e.g. termination of further development). Rodents, canines, and nonhuman primates are common preclinical models due to the established understanding of these animals and their underlying physiology [8–10].

Preclinical models are conducted to gain insight into therapeutic agents and disease physiology. It's also essential to conduct discovery and development of new drugs with greater precision and following principles of Good Laboratory Practice (GLP). It is also recommended that the accuracy of animal models is improved using three components: improving the predictive power of existing models, fine-tuning the means in which animal models are used in decision-making, and investment in basic animal model research with the focus on development of clinically relevant, highly predictive models [11]. It is necessary to record the similarities and differences between nonhuman and human physiological systems to deduce information useful in design of a preclinical program. The results obtained from the designated preclinical studies will be the basis for every decision to make predictions on human exposure, safety, and efficacy.

The preclinical formulation scientist plays a vital role in the drug development process. The scientist needs to have thorough understanding of various formulation aspects, toxicology, pharmacokinetics (PKs), analytical techniques, and physicochemical characterization. The NCE is still in its early development phase with limited knowledge of its physicochemical properties. This information gap can be filled by applying advanced thermal and mechanical techniques like differential scanning calorimetry (DSC), thermogravimetric analysis (TGA), and dynamic mechanical analysis, as it provides valuable information on the thermal behavior, stability, mechanical, and assess material properties of the NCEs. A detailed physicochemical characterization of the NCE can provide good understanding of physical and chemical stability of the NCE and drug-like aspects. Formulation scientists can interact with medicinal chemists in lead optimization. There is also possibility to interact with analytical chemists to recognize if there is any stability issue which can easily derail or save an NCE at a later stage. The formulation scientist should also be aware of the dose to be administered and expectations of the preclinical pharmacokinetist regarding the exposure. Focus should also be on the specific needs of drug administration and safety/tolerability of composition used in formulation. The preclinical formulator plays a crucial role in driving the NCE from wet chemistry lab to preclinical testing phase [12].

The development of preclinical formulations pose significant challenges for the formulator such as (i) increase in number of low solubility NCEs suffering from poor permeability and/or solubility in gastric/intestinal fluids; (ii) appropriate selection of excipients with low toxicity potential and less negative impact on NCE performance; (iii) tailoring the formulation based on animal model under consideration, i.e. small volume for rodent, large volumes for canines and non-primates; (iv) shorter timeframe for development; and (v) limited options to solubilization that can be used to develop enabled formulations.

This chapter summarizes literature-based data and the author's experience regarding preclinical formulations of small-molecule NCEs. It also focuses on various formulation aspects to develop preclinical formulation and consideration for first-in-animal studies.

In the following section, we summarize (i) steps in preclinical studies, (ii) focus on pre-formulation aspects to develop prototypes, and (iii) formulation considerations for preclinical studies with specific focus on conventional approach and enabling approach.

In the final section, we present the rationale/strategy to bridge the gap between first-in-animal studies and first-in-human studies. The study of drug PKs and the bioanalysis for preclinical formulation are out of scope of this section.

11.2 Steps in Preclinical Studies

To anticipate a better outcome of the proposed preclinical studies, it's important to consider both the scope of animal studies and the early stage formulations used (including its tolerability). Following are the three typical preclinical studies in the course of drug discovery.

11.2.1 Pharmacological/Pharmacodynamic Study

The pharmacological/pharmacodynamic study is carried out to understand the effect of a drug on the body, i.e. to determine the pharmacological or physiological impact of the drug on the body, the mechanism of action, and the correlation between the drug concentration and its effect. Various studies performed in this category are

- disease model development
- activity or efficacy studies with respect to specificity, selectivity, and tolerability to target organ and/or tissue
- safety monitoring for the follow-up studies
- dose-related activity in validated disease model
- activity/efficacy studies in comparison to comparator drug molecules

Typically, the formulations used in this study are solutions and suspensions. It's important to note that the vehicle used in this study must be well-tolerated by the animal under investigation [5].

11.2.2 Pharmacokinetic Study

The PK study is used to evaluate effect of body on the drug, i.e. the fate of drug after administration, which is also known as absorption, distribution, metabolism, and excretion (ADME) in humans and animals. It is conducted in rodents and non-rodent species using different routes of administration. For the early PK studies, solution form is mainly used to measure the exposure and identify the mechanism of absorption. Specifically, it can also help to estimate the drug absorption, which could be dissolution, solubility, or permeability limited. The suspension is generally used for both early phase (PK) and late-phase (PK) studies. It supports the rationale and basis for late-stage oral solid dosage form development based on preclinical studies data.

11.2.3 Toxicokinetic Study

The term toxicokinetics refers to evaluating the preclinical toxicity of a drug compound. It is used to study the adverse effects of the drug compound in animal models. The results of these studies will help to select tolerated clinical dose range. The selection of a robust vehicle that can provide desired exposure to support the toxicological program is a challenging part.

The doses used in toxicological studies are higher compared to the PK/efficacy studies; hence, the concentration of the excipients used are also high, which leads to the incidence of the adverse events [13]. The toxicokinetic data assists scientists in establishing the safe starting dose and maximum dose for phase I clinical study.

11.3 Pre-Formulation Considerations for First-in-Animal Studies (Preclinical Formulations)

Pre-formulation is the first key step for the success of early drug development programs. One needs to harness various information related to the study purpose, dose, administration route, formulation requirement, animal species, etc. As per the United States Food and Drug Administration (USFDA) the NCE is defined as a "drug that contains no active moiety that has been approved by FDA in any other application submitted under section 505(b) of the Act. 9" [14].

Pre-formulated NCEs need to be administered in suitable dosage form. To study its *in vivo* effects, it is essential to determine the physicochemical properties of NCE. The various physicochemical properties such as solubility, pK_a, log P, melting point, polymorphic form, crystalline/amorphous nature, and particle size distribution (PSD) are to be measured accurately. This information needs to be carefully synced to come up with a precise pre-formulation/formulation development program for preclinical studies. These details are also important for medicinal chemists who can incorporate this data and attempt any optimization efforts which can contribute to getting a molecule which is favorable for formulation development (as a backup candidate). During the early phase development, cost of synthesis of a drug substance is high, and amount of the drug substance available is limited, which is one of the limiting factors to come up with enabled formulations [15]. Hence, there should be attempts to develop material sparring techniques which utilize lesser amounts of drug substance. Various tools are needed to be used for investigation linked to the pre-formulation/formulation program for first-in-animal studies. The solubility of ionizable compounds, pK_a, and log P can be determined using Sirius T3 from PION Inc., which consumes a few milligrams of active substance [16]. It is important to investigate each lab-scale batch and also the batch-to-batch variability of drug substance. The analytical investigation such as DSC, X-ray powder diffraction (XRPD), and PSD needs 10–50 mg of drug substance. It provides valuable information and ensures that API batches used for PK and toxicokinetic studies display comparable morphology and particle size. Thus, it can support holistic data comparison to understand the batch-to-batch consistency. Figure 11.1 represents the schematic flowchart/decision tree for early formulations used in animal studies. It depicts that for early formulation in animal studies, it is important to assess the physicochemical properties of the NCE under consideration, such as solubility and permeability, which support in deciding the Biopharmaceutical Classification System (BCS) for the respective NCEs. In case of BCS Class I, the NCEs/APIs have high solubility and high permeability and can be solubilized by pH adjustment or with use of surfactant/polymer, so the conventional approach can be followed, and APIs can be formulated into aqueous solution and oral suspension

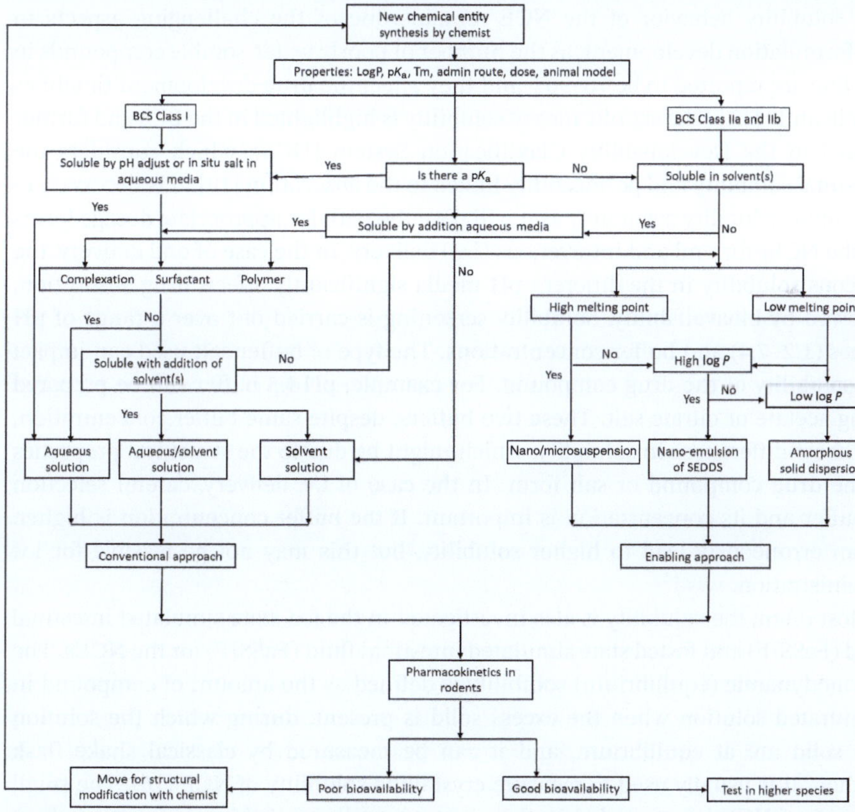

Figure 11.1 Schematic decision tree for early formulations used in animal studies. *Source:* Adapted from [17].

dosage form. The APIs are tested in animals (e.g. rodents) for PKs. If they show poor bioavailability, it implies the necessity to modify the chemical structure of the NCE under investigation, but in case of good bioavailability, then further studies in higher animal species can be performed. In case of BCS Class IIa compounds, the drug's dissolution rate is limited, it can be improved using particle size reduction technique such as jet mill or hammer mill. But for the BCS Class IIb, the drug's solubility rate is limited which needs enabling approach like amorphous solid dispersion (ASD) prepared by hot melt extrusion (HME), which is applicable for high-melting-point (thermostable) drugs, spray drying (SD) for thermolabile drugs, nanosuspension for high-melting-point and high drug load drugs, or lipid-based system for low-melting-point drugs and more lipophilic drugs. This approach is also applicable for BCS Class IV drugs which suffer from poor solubility and poor permeability.

The different aspects involved in preclinical formulations that need to be considered are as follows.

11.3.1 Solubility Studies

The solubility behavior of the NCE remains one of the challenging aspects in pre-formulation development, as the number of poorly water-soluble compounds in pipeline are reported to be 70–90% and they affect the drug development timelines significantly [18]. The significance of solubility is highlighted in the BCS and further refined in the Developability Classification System (DCS), which considers the intestinal solubility and permeability linked to the absorption [19]. It is necessary to determine solubility accurately as it will pave the path for appropriate dosage forms for the NCEs for oral and intravenous (i.v.) delivery. In the case of oral delivery, the aqueous solubility in the different pH media significantly affects drug absorption, followed by bioavailability. Solubility screening is carried out over a range of pH values (1.2–7.4) and buffer concentrations. The type of buffer salt used can impact the solubility of the drug compound. For example, pH 4.5 buffer can be prepared using acetate or citrate salt. These two buffers, despite same buffer concentration, can show differences in solubility, which might be due to the structural properties of the drug compound or salt form. In the case of i.v. delivery, careful selection of buffer and its concentration is important. If the buffer concentration is higher, it can erroneously lead to higher solubility, but this may not be feasible for i.v. administration.

Most often, the solubility is also investigated in the fed state simulated intestinal fluid (FeSSIF) and fasted state simulated intestinal fluid (FaSSIF) for the NCEs. The thermodynamic (equilibrium) solubility is defined as the amount of compound in a saturated solution when the excess solid is present, during which the solution and solid are at equilibrium, and it can be measured by classical shake flask method. It is usually used to measure crystalline solubility of NCE [16]. The small amount of NCE is suspended in the aqueous media, and this suspension is kept in temperature-controlled bath at 37 °C for 24 hours. Once the equilibrium is reached, the NCE is separated from the undissolved media by using centrifugation

Table 11.1 Tools for estimation of equilibrium solubility and kinetic solubility.

Company	Technique/method	Remarks
PION	Potentiometric titration/CheqSol	Less amount of NCE required but applicable mostly to ionizable compounds
Roche	Solubility and residual solid screening (SORESOS)	Small-scale screening of drug solubility (Wyttenbach, Grassmann)
Roche	Lyophilization solubility assay (LYSA)	The solvent stock solutions are evaporated, resulting in solid samples used for solubility assay
Waters	UPLC–MS/MS	Solubility screening of various compounds can be determined in different buffer media with minimum amounts of active ingredient

or membrane filtration. The sample is subsequently analyzed by ultraviolet (UV) spectroscopy or high-performance liquid chromatography (HPLC) having the diode array detector (DAD) or UV detector. The equilibrium solubility is much broader value applicable and feasible for all pH values. It is often termed as true solubility of a compound and is one of the crucial parts of information for formulation development.

In addition to equilibrium solubility, the kinetic solubility measurement is quite common. It is defined as the solubility at the time when first appearance of the induced precipitation in a solution is observed. Kinetic solubility values are typically higher than the corresponding thermodynamic (equilibrium) solubility values [20]. Table 11.1 highlights the application of various solubility determination tools useful to screen equilibrium solubility and kinetic solubility of NCEs.

For the ionizable drug molecule, the state of ionization varies with pH as response to protonation and de-protonation equilibria. Thus, the difference in the concentration of the neutral species will determine the shape of the pH solubility curve [21]. The Figure 11.2, represents the dissociation of telmisartan (zwitterionic compound)

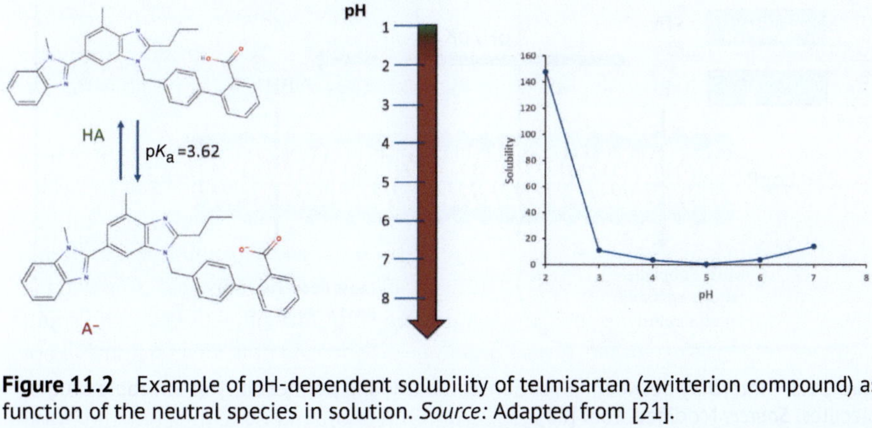

Figure 11.2 Example of pH-dependent solubility of telmisartan (zwitterion compound) as function of the neutral species in solution. *Source:* Adapted from [21].

in water which has acidic and basic ionizable groups, the direction of the shift is likely related to nature of ionizable group involved in the establishment of a certain equilibrium. It shows the pH-dependent solubility.

11.3.2 Lipophilicity and *in vitro* Permeability

Lipophilicity is one of the important considerations in the progress of right NCE candidate for preclinical animal testing. Lipophilicity characterizes the affinity of a molecule or moiety to the lipophilic environment [22]. The biological cell membranes are composed of lipidic bilayers, which consist of phospholipids and cholesterol, where lower lipophilic molecular structure is unlikely to pass through this membrane by passive diffusion. Due to advent of high-throughput screening, a large number of NCEs identified at early stages tend to show high lipophilicity, which may result in poor aqueous solubility [15]. Figure 11.3 depicts the complex interplay of various properties for ionizable drugs. The poorly soluble drug molecules tend to show a high lipophilicity to enable a pharmacological response within a cell. While, in case of water-soluble drug, there is rapid dissolution, but it might be least lipophilic and low permeable. In case of ionizable drugs, the state of ionization defines the pH-dependent solubility and lipophilicity; in this case, the unionized form is absorbed rapidly compared to ionized form. It is one of the important factors for the gastrointestinal (GI) tract uptake. Thus, the pH gradient across the GI tract drives the complex interplay between dissolution, precipitation, and permeation, which is caused by the constant change in ionization, solubility, and lipophilicity of the drug molecule [23].

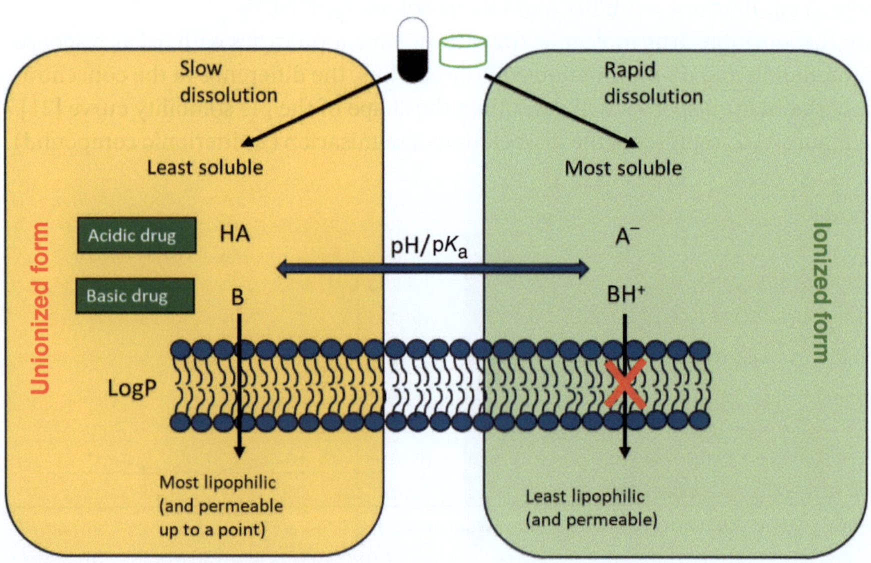

Figure 11.3 Interplay between ionization, solubility, and permeability of ionizable drug molecules. *Source:* Modified from [23].

The lipophilicity of a compound is defined by its partition coefficient (P). This concept was first introduced in 1872 by Berthelot and Jungfleisch [24]. It is the ratio of dissolved drug in an aqueous medium to partition into the organic lipophilic phase, when sink condition is applied at equilibrium. It is independent of pH and is concerned with partitioning of the neutral species. The partition coefficient defines the unionized drug concentration in two immiscible phases at equilibrium, as defined in Eq. (11.1).

$$P = \frac{\text{[unionized species] organic}}{\text{[unionized species] aqueous}} \tag{11.1}$$

The extent of partition depends on the lipophilicity of the compound. The partition coefficient between octanol and water (K_{ow}) is used as a reference to compare lipophilicity of compounds. For practical reasons, the partition coefficient is expressed in logarithmic scale ($\log P$).

The lipophilicity of the NCEs impacts their permeability through biological membrane [25]. To elicit its desired pharmacological and therapeutic effects, the potential drug molecule must cross different cellular barriers by passive and/or active transport-mediated uptake. Hence, membrane permeability is one of the key determinants in PK behavior (absorption, distribution, metabolism and excretion [ADME]) of the compound. The permeation of drugs across intestinal epithelial membrane can occur via passive transcellular pathway, paracellular pathway, or active transport which depends on physicochemical property and structural characteristics of the NCEs [26].

The permeability can be tested *in vitro* by using artificial membranes like phospholipid vesicle-based permeation assay (PVPA), which is helpful to predict the passive absorption of compounds [27]. While the parallel artificial membrane permeability assay termed as (PAMPA) is applicable for compounds that are absorbed by passive diffusion and not by active transport [28]. The most widely used cell culture monolayers from human colon adenocarcinoma are known as Caco-2 [29]. The Caco-2 measures the transport rate of a compound across Caco-2 cell monolayer membrane. The *in vivo* absorption of the compound across the gut wall can be well predicted with this procedure. It is a widely used tool to study permeation improvement of compounds belonging to BCS Class III and IV, possessing poor lipophilicity and poor permeability. It supports the evaluation of compounds that undergo passive diffusion and active uptake [12]. Table 11.2 gives examples of blockbuster drugs with Caco-2 cell permeability and *in vivo* bioavailability.

11.3.3 Ionization Constant (pK_a)

Most of the compounds have ionizable groups within their chemical structure. The strength of a compound to donate or accept protons (H+) in aqueous medium defines its ionization state as well as acidic or basic character. The dissociation behavior of such compound results in smaller ions, particles, etc., and is represented as dissociation constant, also known as ionization constant (pK_a) [6]. The pK_a describes the pH value at which 50% molecular species are ionized, while the

Table 11.2 Examples of blockbuster drugs with their Caco-2 cell permeability and *in vivo* bioavailability [30–33].

Blockbuster molecule	Brand name	Caco-2 cell permeability (P_{app})	*in vivo* bioavailability
Suzetrigine	Journavx ™	12.25×10^{-6} cm/s	69.5%
Lazertinib	Lazcluze®	$\sim 17 \pm 7 \times 10^{-6}$ cm/s	33.7–48.3%
Nirmatrelvir/ritonavir	Paxlovid®	1.76×10^{-6} cm/s (low) (Nirmatrelvir)	34–50% (Nirmatrelvir)
Enzatulamide	Xtandi®	$\geq 31 \times 10^{-6}$ cm/s (high)	>27.1%

other 50% is unionized. It is well-established that the ionization constant affects physicochemical properties of the compound, i.e. aqueous solubility, which further dictates the drug formulation approaches. More importantly, absorption, distribution, metabolism, excretion and toxicity (ADMET) are profoundly affected by the charged state of compounds under varying pH conditions [34]. The detailed studies of pH partition hypothesis from Brodie, Hogben, and co-workers investigated the effect of ionization state on rates of absorption of drugs from the small intestine. In this work, they have demonstrated that the acids with pK_a values below 3 and bases with pK_a above 8 were poorly absorbed [35, 36]. Thus, the pK_a value is an important parameter for absorption and permeability of the compound.

11.3.4 Physical State and Polymorphism

NCEs can exist in different physical states such as crystalline or amorphous forms. Crystalline compound might also display different polymorphic forms. Crystalline forms are the most stable and less soluble, while amorphous forms may show higher solubility but less stability.

The characterization of polymorphism is crucial in navigating the strategy for formulation development. It is defined as the phenomenon in which a compound has a different crystal structure but similar chemical composition; thus, due to their different network structures/molecular conformations, they possess different physicochemical properties [37]. Compound having various polymorphs impacts the properties like solubility, dissolution, and stability of the formulation. It is observed that the compound exists in different polymorphic form; its one polymorph is active while another polymorph is metastable which tends to convert to a more stable polymorph. Polymorphs can be characterized using DSC studies. DSC is usually coupled with XRPD to understand the difference in structural morphology of the compound. Precisely, XRPD provides information about physical state of compound, if it is in crystalline or amorphous form. DSC supports in determining the glass transition, melting point, and decomposition temperature of NCE, which may help to decide appropriate approach during pre-formulation/formulation development.

11.4 Formulation Considerations for First-in-Animal Formulations

The development of a new drug compound from discovery to regulatory approval usually takes approximately 10–12 years. It is estimated that the average costs to research and develop new compound are from $314 million to $2.8 billion [38]. Out of 5000–10,000 molecules that enter research and development pipeline, approximately 250 molecules enter preclinical program, only five molecules can manage to reach to clinical stage trials, and only one receives the marketing approval. Thus, there is significant amount of resources and investment needed to develop a new drug molecule. The reduction of costs and time to market is key priority for the drug discovery companies [39].

For this reason, it is important to focus on successful preclinical formulation development. A well-defined formulation development strategy should be established based on the preclinical data and knowledge of physicochemical properties available. The extensive description of the formulation development considerations that can be adopted for first-in-animal studies is mentioned in the following sections.

11.4.1 Biopharmaceutical Classification System (BCS) and Its Application

The introduction of the BCS in the mid-1990s by the USFDA guidelines represents a major step forward in the regulation of oral drug products. This guidance classifies drug substances into four categories according to their permeability and solubility properties, as shown in Table 11.3.

The drug is considered to be highly soluble when the entire dose of molecule is soluble in 250 ml or less of GI fluid/aqueous media [14]. It is reported that more than 50–70% NCEs belong to Class II, which are poorly soluble but have good permeability in the GI tract [41]. The BCS provides a framework to consider key parameters (dose, solubility, permeability, and dissolution rate) that might influence *in vivo* performance [19].

Butler and Dressman developed a modified BCS, also known as DCS, that systematically categorized Class II compounds into subcategories that depend on limiting factor of oral absorption [19, 42].

Class IIa: In this class, molecules show dissolution rate-limited absorption, in which the dissolution rate is too slow, and the absorption window is shorter as the time needed for its dissolution. One of the strategies used to overcome this challenge involves using particle size reduction as standard approach.

Class IIb: In this class, molecules show solubility-limited absorption. The GI fluid volume is insufficient to dissolve the entire administered drug dose, and the compounds will tend to be incompletely absorbed unless the drug is formulated in an already solubilized form and will therefore present a major challenge to the formulator [19]. Hence, for this class of compounds, an enabling formulation approach is needed to improve solubility and achieve higher oral absorption.

Table 11.3 Biopharmaceutical Classification System.

Class	Solubility	Permeability	Absorption pattern	Examples	Challenges in drug delivery
I	High	High	Well absorbed	Diltiazem, Propranolol, Metoprolol	No major challenges for immediate release forms, but controlled release formulation might be needed to limit drug release or dissolution since absorption of released drug is rapid
II	Low	High	Variable	Telmisartan, Itraconazole, Dipyridamole	Formulations are designed to overcome solubility or dissolution problems by using different strategies
III	High	Low	Variable	Insulin, Metformin, Cimetidine	Approaches are employed to enhance permeability
IV	Low	Low	Poorly absorbed	Taxol, Chlorthiazide	A combination of strategies used for Class II and III drugs is employed to improve dissolution and permeability

Source: Adapted from [40].

Hence, the rational selection of excipients and delivery route, deep understanding of physicochemical properties of molecule, and precise formulation strategy will support in improving solubility-driven bioavailability of the NCEs that belong to Class IIb and IV categories.

11.4.2 Choice of Excipients and Its Applications in First-in-Animal Studies

Excipients are an integral part of formulation development and are used to deliver the drug molecules, which are evaluated for safety. The excipients are used for one or more functions like viscosity reduction, pH buffering, solubilization, improving drug adsorption, tonicity modification, or long-term stabilization. A heterogeneous group of materials is available as excipients, and they differ in origin, chemistry, physicochemical properties, and functionality. The pharmaceutical regulations recommend that all excipients used in formulation, along with their decomposition products, are evaluated and are safe, i.e. devoid of toxicologic risk after consumption. It is crucial that the principle of non-toxicity to the species should be followed, and hence the excipient possesses no toxicity risk and is tolerable to the animal species. It is necessary to check the excipient status if it's already mentioned in inactive ingredient database, is used in marketed products with relevant route of administration, and is listed as generally recognized as safe (GRAS). The concentration of excipients used in formulation should be adequate to meet the overall objective. In the case of animal studies, the formulation containing fewer or no excipients would be good due to less complexity. However, it is practically impossible to formulate complex APIs without using excipients. The reasons for usage of excipients are different and are selected based on the intended purpose in the formulation. In view of the high emergence of poorly soluble NCEs, it's necessary to use functional excipients which enhance the compound's solubility. Sometimes a NCE is unstable and prone to oxidation; in such cases, it is necessary to add excipients having antioxidant property which act as oxygen scavengers [42]. If the NCE is poorly permeable, the formulation will need a permeability enhancer to improve permeability of the compound. Excipients may also be used to maintain the API in stable form during preparation, storage, and administration. A formulator must have understanding and rationale for selection of appropriate excipients in formulation development. The knowledge of the physicochemical properties of the excipients and their correlation with that of the drug molecules needs to be assessed. In the preclinical formulations, various types of excipients need to be used such as polymers, oils, surfactants, emulsifiers, antioxidants, cosolvents, solubilizers, inorganic salts, and pH modifiers. The most common dosage form for delivery of the NCEs in preclinical studies is the solution. It is necessary that the drug molecule is soluble in the aqueous solvent, which can be readily administered to the animals. The advantage of such formulation is that its simplicity to formulate, uniformity of content, and ease of administration. But in case of poorly soluble NCEs, which are predominant in the present pipeline, the alternate strategies are emulsion, suspension, cosolvent, and the reduction of the particle size. In case of solubility-limited bioavailability, it necessitates using solubilization techniques such as solid dispersions, lipid-based systems (self-emusifying drug delivery systems [SEDDS], self-microemulsifying drug delivery systems [SMEDDs], micellar solutions, liposomes), and nanosuspensions, to improve bioavailability of these poorly soluble compounds. The following sections provide a deep dive into the various approaches that can support the formulation scientists to develop rational formulation for animal studies.

11.4.3 Conventional Approach

The oral route of administration is a convenient, economical, and safe way for the dosing of investigational drugs to the animals. If the desired route of administration during the clinical studies is oral, then the preclinical evaluation in oral form will be preferred. It is necessary to know the bioavailability of the compound when administered orally. Usually, the oral PK data is used to compare against the single dose i.v. administration to determine absolute bioavailability from the active compound in an oral formulation.

11.4.3.1 Oral Gavage

It is used to deliver the fixed volume of the dose directly into the stomach, which is usually done in rodents and non-rodents. The formulation in the form of solution and suspensions can be orally administered in which the dose is 5 ml/kg for rats and 10 ml/kg for dogs [43]. An et al. investigated the oral bioavailability and PK properties of G7883 and G6893 for the three groups of mice that were dosed through oral gavage. The formulation for G6893 comprised of 0.6% methyl cellulose, 0.2% Tween 80 in water with pH adjusted to 2.5 at 30 mg/kg, and a dose volume of 5 ml/kg. The polymer and surfactant mixture is applied to improve suspendability of the compound [44].

11.4.3.2 Oral Solution

It is one of the preferred dosage forms for oral administration in animals. The drug molecule can be dissolved into aqueous media such as water or buffer forming a solution without undissolved particles (monophasic aqueous systems). But if the NCE is not sufficiently soluble in aqueous media, then it should be dissolved using the surfactants and cosolvents. Many surfactants are reported in literature to improve solubility and bioavailability of poorly water-soluble drugs through different mechanisms. For example, the nonionic surfactants like medium-chain mono–diglyceride of C8 and C10 chain, lecithin (diacyl glycerol phosphatidylcholine), and lauroyl macrogol-32 glycerides (Nakmode et al.), and cosolvents like propylene glycols; glycerol; diethylene glycol monoethyl ether; *N*-methyl pyrrolidone; 2-pyrrolidone; alcohol; different grades of polyethylene glycols like PEG 100, PEG 200, PEG 300, and PEG 400 [45, 46]. Often, the solubilizers are added to dissolve the poorly water-soluble NCEs. The various types of solubilizers that can be used to improve solubility are propylene glycol monocaprylate, propylene glycol monolaurate, linoleoyl polyoxyl-6 glycerides, 2-pyrrolidone, *N*-methyl pyrrolidone, soya oil, corn oil, and cyclodextrins [47]. The formulations prepared using such excipients exhibit bitter taste. To improve the palatability of such formulation, various flavors can be used, and to mask the bitter taste, taste masking agents can be added. Mahajan et al. reported higher oral bioavailability of tenofovir formulation when dosed as oral solution of SEDDS compared to marketed formulation as tablet. The formulation consists of 8 mg of tenofovir, 1.08 g of eucalyptus oil, 1.8 g of glycerol and polyoxyl 35 castor oil (Kolliphor® EL), and 1.2 g of medium-chain triglycerides (MCTs) (Kollisolv® MCT 70) [48].

11.4.3.3 Oral Suspension

Oral suspension may be used in preclinical studies if the investigational drug is not sufficiently soluble in solvent of choice and higher dose needs to be administered. Such type of formulation is used for dose determination and toxicity studies. In this case, the concentration of the compound that needs to be administered is higher than equilibrium solubility, and hence, the suspension formulation is a better option. Oral suspension is prone to flocculation and may exhibit the settling of drug particles upon storage. Hence, in such a scenario addition of good suspending agents and deflocculating agents is opted to ensure that the particles can be redispersed uniformly throughout the volume and remain suspended for sufficient time after shaking. Various surfactants and viscosity modifiers are used in such formulations. The commonly used surfactants are polysorbates, poloxamers, cremophors, and the viscosity modifiers like hydroxyl propyl methyl cellulose, sodium carboxyl methyl cellulose, and methyl cellulose are applied as they are safe, inert, and widely used in oral suspension formulation for animal studies [49]. Larson et al. prepared the oral suspension of danazol, cinnarizine, and halofantrine by mixing these compounds with 0.5% methyl cellulose solution. The formulation with 28 mg/kg of danazol, cinnarizine, and halofantrine was administered to male Sprague–Dawley rats as an aqueous suspension [50].

11.4.3.4 Oral Mini-Tablet/Mini-Capsule

Oral mini-tablet/capsule formulation is a convenient form to administer animals compared to oral gavage. The NCEs in its pure solid form can be filled into the capsule using capsule size 9. It has the advantage as pure compound will accurately predict the bioavailability without the interference coming from the addition of excipients into the formulation. Physical form of the investigational drug is vital in deciding the formulation type. If the drug molecule to be investigated is in semi-solid/gel form, it needs to be dissolved and adsorbed onto an inert carrier which is suitable for tablet or capsule filling. One such adsorbent is mesoporous silica. Krome et al. formulated liquisolid of corallopyronin (Coro A), in which they used propylene carbonate and the mesoporous silica Syloid® XDP 3050 [51]. For low-dose NCEs, i.e. dose <10 mg of total unit, one strategy is to adsorb the compound on inert carrier and further process into granules or powder, which can be compressed into minitablets or filled into capsule size 9. But if the dose is higher for semi-solid/gel form of NCEs, then it will be convenient to formulate it into oral solution or suspension dosage form. Maghrabia et al. encapsulated the poorly oral bioavailable Ceftriaxone (CTX) into nanoparticles using two natural biopolymers (chitosan and shellac), by a simple technique (ionic crosslinking). The optimized nanoparticle formulation was freeze-dried with nanoparticle equivalent to 50 mg of CTX filled into hard gelatin capsule size 4 and administered to male New Zealand rabbits. The absolute bioavailability of the hard gelatin capsules loaded with optimized formulation obtained was 92% compared to tablets, which was 80% [52]. Frank et al. investigated the crystalline formulation containing 20% BIXX drug load for three formulations prepared using fluid bed granulation and high shear granulation. While the one amorphous formulation consists of 30% BIXX prepared by HME, having copovidone. All the four formulations were filled into capsule size 3 that was administered to beagle dog. It was observed that the PK- parameters of amorphous formulation showed slightly better *in vivo* performance compared to other three formulations. But, there was no major difference between four formulations.

11.4.4 Specialized/Enabling Formulations Approach

It is often important to categorize the investigational drugs into the brick dust and grease ball category. The brick dust compounds are characterized by tight crystal lattice, high melting point $T_m > 200\,°C$, practically insoluble in water but possess low solubility in lipids, and are rather hydrophobic than lipophilic. While, in case of grease ball compounds, they are poorly water soluble but may show good lipid solubility (due to high log P) and low melting point. Due to these typical physicochemical properties, these molecules are unable to form strong bonds with water molecules and possess extremely low solubility [54, 55]. In general, grease balls are preferred over brick dust compounds due to higher chances of achieving higher bioavailability. To improve solubility-driven bioavailability of such compounds, it is necessary to use enabling approaches such as ASDs, lipid-based systems, nanosuspensions, and cyclodextrins.

11.4.4.1 Amorphous Solid Dispersions (ASDs)

The solid dispersions are the systems in which one or more actives are dispersed or incorporated into the inert carrier/matrix at the solid state, prepared by solvent evaporation (SD) or melting (HME) methods [56]. In case of solid dispersion, the crystalline drug is converted to amorphous form in which the API particles are molecularly dispersed in the carrier, and dissolved into the polymer-rich diffusion layer together with the carrier. Due to the higher surface area of the API particles and possibility of improved wetting and decreased agglomeration, this may lead to considerable improvements in dissolution, compared to conventional dosage forms. The rationale design via decision tree for ASD preparation using HME and SD is presented in Figure 11.4. It explains the initial assessment for selection of synthetic polymers or mixtures with API for ASDs based on literature and experience.

In the next step, the identification and characterization of polymer/mixture is carried out to understand the melting point and glass transition temperature of the mixture. Once these mixtures are selected, it is necessary to assess their solubility in organic solvents, which are common for API and polymer. If the APIs are soluble in desired ratio, solvent casting can be initiated which consumes low amounts of API. Depending on the solubility, if API is soluble, the further fine-tuning of the drug and loading of API/polymer can be performed after conducting analytical characterization of film casts and achieving satisfactory results. Solvent evaporation technique (SD) can be one lead for ASDs preparation. In case if its insoluble in organic solvent, and also if the polymer/API are insoluble in film casting, its unsuitable for solvent evaporation technique, and melt fusion technique can be applied. The API/polymer suitability can be assessed via vacuum compression molding (VCM) using Melt Prep®. If the mixture is miscible, it can be evaluated for HME method. Huang et al.

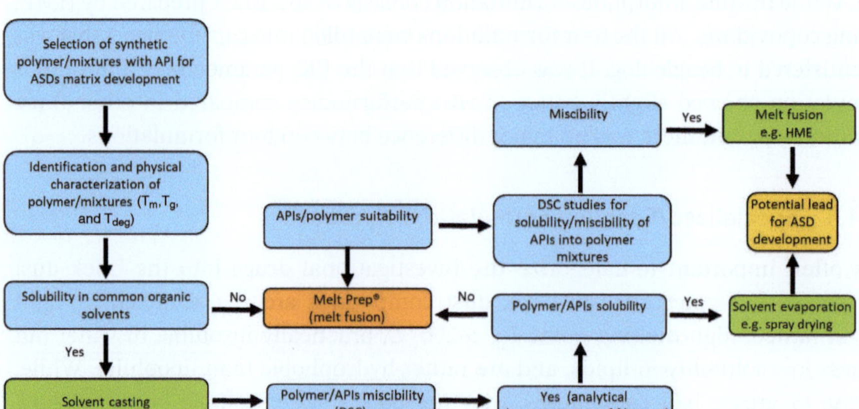

Figure 11.4 Decision tree for rationale of ASDs preparation via HME and spray drying.

used Affinisol® HPMC HME, a novel grade of HPMC class with low glass transition temperature and melt viscosity compared to other grades of Hypromellose. The formulations containing polymer and carbamazepine (CBZ) were extruded using 16 mm twin screw extruder, and the effect of temperature, screw speed, and feed rate was investigated. The authors concluded that Affinisol® HPMC HME is an interesting polymer for the HME process to increase wettability and dissolution property of poorly soluble drugs [57]. In the ASDs, there is a hydrogen bonding interaction of drug molecules with the carrier. The IR frequency shifts are generally consistent with the amorphous form except for the 374 cm^{-1} (lattice vibration) and 1160 cm^{-1} (C–C ring stretch/C–N–C asymmetric stretch) bands for pure CBZ. In these cases, shifts are substantially greater than the amorphous form, indicating the polymer had substantial interactions with CBZ. Also, the Affinisol® HPMC HME has a higher T_g (115 °C), which is advantageous for improving the physical stability of CBZ during storage, and it helps to prevent recrystallization of the drug molecule, which leads to drug release in a supersaturated state and results in faster dissolution rates. Zecevic et al. designed solid dispersion formulations containing different ratios of APIs such as dipyridamole, griseofulvin with hydroxypropyl methylcellulose acetate succinate (HPMC-AS) and hydroxy propyl cellulose Super Special Low viscosity (HPC-SSL) prepared by HME for site-specific dissolution improvement or maintenance of a potent supersaturation over the entire GI pH range for a reliable drug therapy [58]. The solid dispersions can be tailored into the solid oral dosage form, resulting in increased patient compliance compared to liquid dosage produced by solubilization methods.

Krome et al. have reported that Coro A, a potent novel antibiotic, which is poorly water-soluble and unstable at room temperature was formulated into ASD using SD method. They used povidone and copovidone polymers as carriers and ethanol as a solvent for SD. The SD formulation was administered to BALB/c mice for PKs. The Coro A showed high C_{max} of 64.3 μg/ml at second time point at 10 minutes and AUC values of 67.8 μg h/ml, showing overall good oral bioavailability (F_{abs}) of 59% as determined. It was concluded that the ASDs' approach was suitable for stability and solubility enhancement of oral formulation which was used in preclinical studies [59]. One of the important limitations of ASDs is the physical stability of drug molecules in the solid dispersion formulation. The drug molecules in the ASD tend to recrystallize during storage, leading to decreased dissolution rate, affecting bioavailability. Suryanarayanan et al. investigated the correlation between molecular mobility and physical stability and characterized molecular mobility in ASD of itraconazole (ITZ) with each polyvinylpyrrolidone (PVP) and HPMC-AS using SD. The investigators observed that ASDs of ITZ with PVP and HPMC-AS, only HPMC-AS acted as anti-plasticizer of global mobility. The results further suggested that global molecular mobility found good correlation to crystallization onset and growth rate, indicating the role of cooperative motions in physical instability with storage temperature. HPMC-AS was more effective than PVP in inhibiting ITZ crystallization [60].

The inhibition of recrystallization of the amorphous drug molecule in ASD can be achieved in two ways: (i) by decreasing the molecular mobility of the system and (ii) by favoring the solubility of the API into the polymer matrix. The favorable options among the two depends mainly on the physicochemical properties of the drug molecule under investigation [61–67]. The drug polymer miscibility needs to be evaluated to obtain stable ASD formulations, which can be done using DSC. The molecular state of drug and solid dispersion is characterized by the following techniques: DSC [68], modulated differential scanning calorimetry (mDSC) [69], X-ray diffraction (XRD) [70], and confocal Raman spectroscopy [57]. While the other instrumental technique like Fourier transform infrared spectroscopy [71], solid-state nuclear magnetic resonance [72], dynamic vapor sorption [73], and TGA are used to assess the chemical stability and molecular interaction of drug in polymer matrix. Other techniques such as polarized light microscopy [74], scanning electron microscopy (SEM) [75], transmission electron microscopy (TEM) [76], atomic force microscopy (AFM) [77], and confocal laser microscopy are useful to characterize the crystalline states of the drug, surface morphology, molecular miscibility, and the phase separation of solid dispersions. Transmission electron microscopy has been recently used to study even small degrees of drug crystallites in solid dispersion, in which interactions between the electron beam and the sample produce both a real-space image and an electron diffraction pattern of the sample. Moseson et al. reported that 16 drug products were manufactured using HME and 27 using SD as ASD methods have been approved by USFDA in 12-year period from 2012 to 2023. Thus, ASD formulations demonstrate their clinical success to deliver poorly soluble drugs [78]. It will be important to know how the ASDs improve the bioavailability of poorly soluble drugs. The recent bibliographic data contributes to better understanding of drug release kinetics of ASDs in dissolution and drug absorption. The mechanisms underpinning the drug release from the ASDs are often concerned but poorly understood. There are mainly two mechanisms for the drug release from solid dispersions: carrier- or drug-controlled.

The carrier-controlled release is governed by the property of the carrier. The carrier may be medium-soluble or medium-insoluble in the different pH aqueous media. Sun and Lee et al. prepared amorphous dispersions of 20% indomethacin using medium-soluble carriers such as HPMC-AS. They observed that the drug release and supersaturation follow a dissolution-controlled mechanism. These might be attributed to the rapidly dissolved or dispersed carrier in the dissolution medium, which leads to rapid liberation of the amorphous API and generates a highly supersaturated drug solution. They found that this early surge of drug supersaturation is followed by a rapid decline in the concentration of the drug, which may be due to precipitation triggered by rapid buildup of supersaturation. To improve the dissolution profile, it is essential to know the release mechanism of solid dispersions [79]. Thus, the focus should not only be on the polymorphic states of the API, but also on the essential carrier properties such as solubility, viscosity, ability to maintain supersaturation, and crystallization inhibition; the ratio of drug–carrier is the key factor affecting the dissolution profile and should be regarded [80].

If a drug-dependent release predominates, the rate-determining step is the dissolution of the poorly water-soluble API. This drug-controlled dissolution is known for crystalline glass suspension. Consequently, the dissolution is not associated with the polymer but is primarily determined by the API's solubility properties. In case of glassy solid solutions, a carrier-controlled dissolution is observed. The API particles are molecularly dispersed in the carrier and dissolved into the polymer-rich diffusion layer together with the carrier. The hydrodynamics of dissolution process may also play a role in determining the mechanism, in that more rapid stirring speed may favor drug-controlled dissolution by enhancing the rate of polymer dissolution into the bulk in relation to drug dissolution into the diffusion layer [81]. The increase in drug absorption is due to increased concentrations of molecularly dissolved API and facilitated diffusion through unstirred layer by drug-rich particles. The effective drug release from the drug-rich particles is clear advantage of ASDs compared to other enabling approaches [82]. The concept of drug uptake from ASDs is represented in Figure 11.5.

In ASDs the API is embedded into the polymer matrix using SD or HME technique. This embedded API undergoes the first step of dissolution, in which the drug-rich particles containing polymers, micelles, and crystals are observed. Later, this complex mixture of API occurs in the solution after dissolution, and some colloidal API forms also emerged in the solution. This dissolved form of drug is uptaken into the intestinal membrane [82]. Thus, the potential change in API solid state (crystalline to amorphous) results in increased wettability of the APIs by using water-soluble polymers, better separation of API particles, micellar solubilization of API by using high concentrations of surfactants/polymers, and hydrogen bonding interaction between polymers and APIs at the molecular level could be observed in ASDs. Thus, these together would lead to an increase in rate and extent of dissolution of poorly soluble APIs with improved bioavailability [83–85].

The marketed ASD-approved product of Jaypirca® (Pirtobrutinib) indicated as antineoplastics was prepared by using HPMC-AS polymer and the SD method.

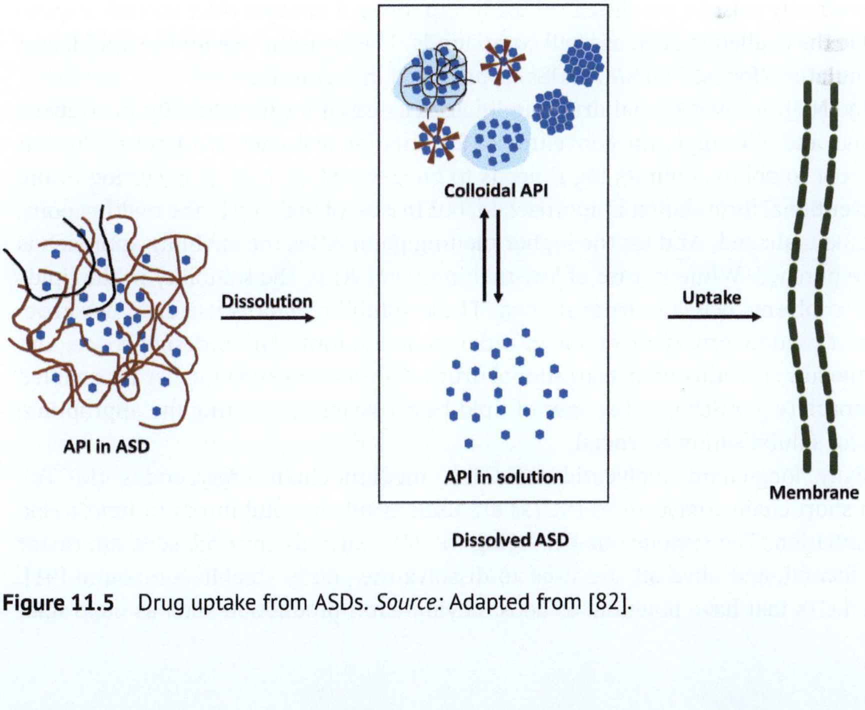

Figure 11.5 Drug uptake from ASDs. *Source:* Adapted from [82].

Repeat-dose toxicology studies up to 3 months were conducted in rats and dogs, using an oral route of administration. Overall, toxicities were consistent with BTK inhibition and included the following: B-cell suppression, increase in activated T cells, pancreatic findings (e.g. inflammation, hemorrhage, and fibrosis in rats). It was reported that the pancreatic findings could be species- and strain-specific, as observed with administration of other BTK inhibitors [86, 87]. Another market-approved product Alvaiz® (Eltrombopag), indicated for thrombocytopenia, consists of copovidone polymer prepared by HME technique. Renal toxicity was observed in studies up to 14 days of duration in mice and rats at exposures that were generally associated with morbidity and mortality [88]. One of the approved ASD formulations is Qulipta®/Aquipta® (Atogepant) which is indicated for migraine treatment. This ASD formulation contains copovidone as the main polymer and is prepared using HME technique. The repeat-dose toxicity program was conducted in three species (mice, rats, and monkeys). The following were reported mainly in rodents (mice, rats) studies: post-dose salivation, decreased body weight gain, reversible epithelial vacuolation in the small intestine, and slight elevation of liver ALT transaminase. Along with these no neoplastic changes were observed in the intestines in the carcinogenicity studies in rats or mice. Additionally, these findings were not observed in rhesus monkeys [89].

11.4.4.2 Lipid-Based Systems

Lipid-based formulation approach is used for the compounds representing the category of grease balls (molecules with high log *P* and low melting point). This type of formulation helps to dissolve poorly soluble compounds using oils, surfactants, or cosolvent mixtures. The compound is formulated into a molecular solution, which circumvents the slower dissolution of poorly water-soluble compounds as they already exist in pre-dissolved form. Lipid-based systems offer several ways to tackle the challenge of grease ball compounds. The decision tree for the lipid-based formulation (focused on SMEDDSs) is proposed in Figure 11.6.

The NCE/investigational drug candidate is assessed for its solubility in aqueous media, and if it's high, the conventional formulation approach is followed. But if it has low solubility, then its log *P* needs to be assessed. In case of lower log *P*, the conventional formulation is approached, but in case of high log *P*, the melting point is to be evaluated. And for the higher melting point APIs, the enabling approach is to be pursued. While in case of low-melting-point APIs, the solubility in the lipids and cosolvents needs to be evaluated. These solubility data is useful in construction of pseudo-ternary diagrams in order to define formulation design space, and for the low solubility of investigational drugs, they are screened further for enabled approach (e.g. ASDs). In the case of lipid-based systems, selecting the appropriate oils for solubilization is crucial.

Many long-chain triglycerides (LCTs), medium-chain triglycerides (MCTs), and short-chain triglycerides (SCTs) are used as oils or solubilizers in lipid-based formulation. The various oils belonging to LCTs, such as corn oil, soya oil, castor oil, fish oil, and olive oil, are used to dissolve the poorly soluble compound [91]. The LCTs that have potential to boost chylomicron production such as oleic acid,

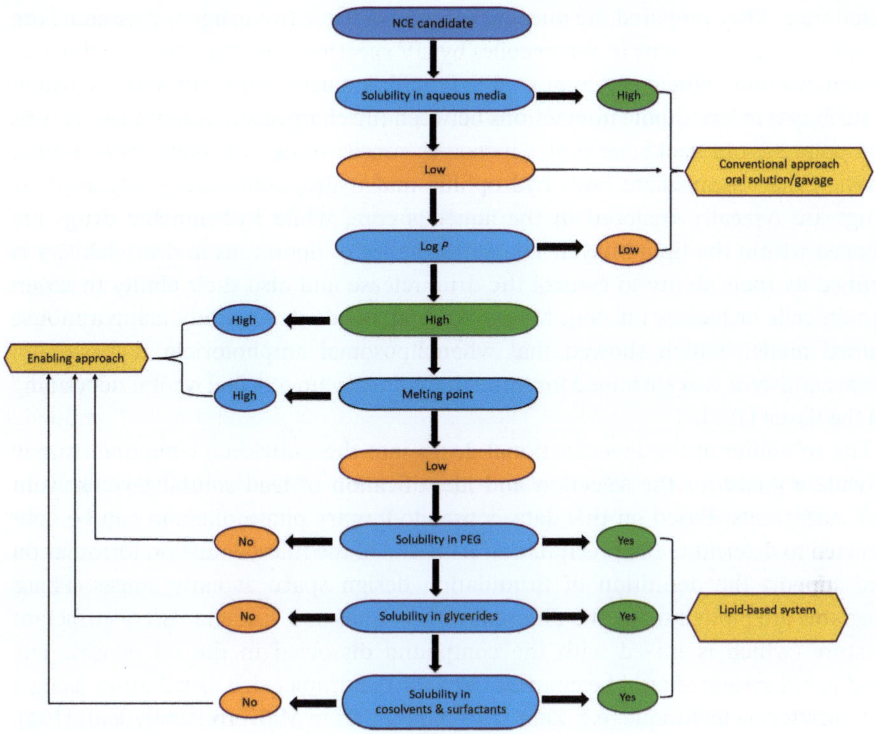

Figure 11.6 Decision tree for rationale design of lipid-based system. *Source:* Adapted from [90].

linoleic acid, and their derivatives are applied as they improve lymphatic uptake, avoid first-pass metabolism, and improve overall bioavailability of compounds [92]. MCTs and SCTs are preferred by formulation scientists due to their higher solubilization potential compared to LCTs. The MCTs and SCTs such as lauroyl polyoxyl-6-glycerides (Labrafil® M2125), mixture of medium-chain mono- and diglycerides (Capmul®MCM), and MCT (Labrafac™ lipophile WL1349) excipients are selected based on their physicochemical properties and impact on drug solubilization, formulation stability, and bioavailability improvement [93]. The investigational compounds can be dissolved in these oils to obtain the simple solution. The formulations are easy to prepare and are comprised of GRAS excipients, but they show limitations due to slow digestion and slow release of the compound. The compounds not soluble enough in oils can be solubilized by using the combination of oils and surfactants.

The surfactants with hydrophilic lipophilic balance (HLB < 12) are preferred in this type of lipid-based system. The oil-based formulation may self-emulsify into oil-in-water emulsions when they encounter GI fluids, and is known as SEDDS. The SMEDDS consists of oils, water-soluble surfactants with HLB > 12, and cosolvent or solubilizer [91, 94]. Micellar solubilization enhances the solubility of the drug by utilizing surfactants and other components to improve dissolution. Vinarov et al. reported the micellar solubilization of poorly soluble drugs such as danazol and fenofibrate. They prepared the micellar solution of these two drugs and assessed the drug solubilization locus in the micelles by UV spectroscopy. The observed danazol solubilized more efficiently compared to fenofibrate using ionic surfactants, which is attributed to ion–dipole interactions between the charged surfactant head groups and polar steroid backbone [95]. Liposomes consist of one or more phospholipid bilayers that encapsulate both hydrophilic and hydrophobic drugs. Hydrophilic drugs are typically enclosed in the aqueous core, while hydrophobic drugs are trapped within the lipid bilayer. The performance of liposomes in drug delivery is defined by their ability to control the drug release and also their ability to target certain cells or tissues [96–99]. Moore AJ et al. conducted a study using a mouse animal model, which showed that when liposomal amphotericin B was given intravenously, it was sustained for more than one day up to a few weeks, depending on the tissue [100].

The solubility of the investigational drugs into the individual components may provide a guide for the selection and identification of lead solubilizer/cosolvent and surfactants. Based on this data, a pseudo-ternary phase diagram can be constructed to determine the precipitation potential of the microemulsion formulation and support the definition of formulation design space at early stages. Phase diagrams are plotted using the different ratios of oils and surfactants/co-surfactant mixture (which is mixed with the compound dissolved in the oil phase). The lipid-based system should be characterized for the droplet size distribution using a light-scattering technique (e.g. Zeta-sizer nano ZS from Malvern Panalytical) [101]. If the emulsion has a smaller droplet size, it has larger surface area, which can result in higher solubility and faster absorption of the compound. But care needs to be taken so that the concentration of surfactants/co-surfactants should not exceed the permissible dose of excipients in a given animal species. Also, physical stability of the formulation can be evaluated using zeta-sizer, which determines the zeta potential. If the zeta potential is negative, it indicates the formulation is physically stable during storage [102]. The simple lipid formulation will be preferred, which uses a minimum number of excipients to achieve the desired drug loading, which will generate an emulsion on dispersion with GI fluids without a high amount of precipitation risk [90]. It will be desirable to have a formulation that is simple, scalable, stable and improves solubility. Lee et al. developed a novel apixaban-loaded solid SEDDS formulation and administered it to rats. The liquid SEDDS consists of Peceol/labrasol/Cremophor EL in volume ratio of 15/35/50, and it was suspended in distilled water and SD using 0.75 g of Aerosil 200 as mesoporous carrier using fluid bed granulator to prepare apixaban-loaded S-SEDDS. The formulations were administered to rats and showed 3.2-fold improved bioavailability [103]. Kok et al. developed the self-nanoemulsifying drug delivery system (SNEDDS) for cannabidiol (CBD), which is poorly water-soluble and highly lipophilic. They prepared two oil-based formulations of CBD: one with MCT oil formulation (MCT-SNEDDS) composed of formulation 1 that consists of 45% w/w of Brij O10, 18% w/w of captex 355, and 27% w/w of propylene glycol that produced largest droplet size; formulation 2 consists of 36% w/w of Brij O10, 18% w/w of captex 355, and 36% w/w of propylene glycol produced smallest droplet size, and other containing sesame oil in formulation (SE-SNEDDS) that comprised 82% w/w of sesame oil, 8% w/w ethanol, 0.05% w/w sucralose, and 0.02% w/w of ethyl-3-methyl-3-phenylglycidate, which is similar to market-approved Epidiolex® oral solution. It is indicated for the treatment of seizures associated with Lennox–Gastaut and Dravet syndromes, as well as tuberous sclerosis complex. The CBD-SNEDDS were prepared by mixing the various excipients until clear solution was obtained, and SNEDDS with CBD was prepared by dissolving CBD crystals in the drug-free SNEDDS to a concentration of 10% w/w. The mixture is stirred at 500 rpm and 37 °C to dissolve CBD. Later, it was cooled to room temperature prior to use. The two oil-based formulations of MCT-SNEDDS were administered to healthy Sprague–Dawley rats via oral gavage (20 mg/kg CBD). The median T_{max} values of CBD-SNEDDS are within 1 hour, while the T_{max} for MCT-CBD is substantially greater (6 hours). While in comparison to administration of the MCT-CBD formulation, the $AUC_{0\text{–}10h}$ values showed that administration of the SNEDDS formulations resulted in significantly greater exposure to CBD, suggesting that the F1 and F2 formulations improved the bioavailability of CBD by 2.2- and 2.8-fold relative to MCT-CBD; however, no improvement was found compared to Epidiolex®. Hence, it is important to have a formulation that improves bioavailability and is stable during storage [104]. Xtandi® soft capsule (Enzalutamide), which is indicated for prostate cancer, was prepared using capryl caproyl macrogol-8 glycerides, butyl hydroxyanisole, and butylhydroxytoluene. The administration of enzalutamide to pregnant mice showed increased incidence of embryo–fetal deaths with external and skeletal changes. In studies involving mice (4 weeks), rats (4 and 26 weeks), and dogs (4, 13, and 39 weeks), changes in the reproductive organs associated with enzalutamide showed a decrease in organ weight with atrophy of the prostate and epididymis. In a 6-month study in transgenic rasH2mice, enzalutamide did not show carcinogenic potential (absence of neoplastic findings) at doses up to 20 mg/kg per day [105].

11.4.4.3 Nanosuspensions

Nanotechnology-based formulations are one of the most lucrative and fastest emerging technologies for the delivery of therapeutics. It is one of the important modes of drug delivery when it comes to preclinical studies. Nanosuspensions are preferred formulations for delivering investigational drugs, especially in safety studies, where the amount of drug delivered is relatively high. The nanosuspensions can be produced using a top-down approach, where the particle size reduction is done from the micron scale to the sub-micron scale using milling technology. The investigational drugs can be nanosized to sub-micron size up to 100–200 nm in a reproducible manner using wet milling techniques such as media milling (nanomilling), microfluidization, and high-pressure homogenization [106–108]. Nanosuspensions can be prepared by adding surfactants and polymer into the aqueous media containing drug molecules by adding zirconium balls. For example, Aprepitant is a basic compound with pK_a of 9.7 and melting point of 254 °C, suggesting high melting point. Its free base aqueous solubility (3–7 μg/ml) is very low in pH range of 2–10, with log P of 4.8 at pH 7.0, indicating high lipophilicity. The particle size reduction techniques could increase the surface area with better solubilization and faster absorption. Hence, nanosuspension using top-down approach was used to achieve enhanced dissolution rate and exposure. The *in vivo* dog studies of nanosuspensions of aprepitant indicated threefold increase in exposure with decrease in particle size from 5 μm to 150 nm [109].

Another approach used for preparing nanosuspension is bottom-up approach in which the drug molecules are dissolved in the organic solvent and then precipitated using anti-solvent in presence of stabilizer. Skypharma's solubilization technology Dissocubes™ uses high-pressure piston gap homogenizer to reduce the drug particle size in the presence of surface modifiers [110]. Nanoform developed a proprietary technology which uses controlled expansion of supercritical solutions (CESS) that produces pure homogeneous drug particles from solutions in an excipient-free process. The particle size reduction of 10 μm to 50 nm can be achieved using this technology [111].

Nanosuspension tends to undergo agglomeration and hence tends to show stability issues. Agglomeration of the particles can be prevented by using anionic and nonionic surfactants like poloxamers, phospholipids, and polysorbates 80 [112]. PSD of nanosuspension is a critical parameter, as it will affect the stability of nanosuspension formulation during storage. As the larger particles tend to grow, this will lead to large variation in particle size. The typical phenomenon observed in case of nanosuspensions is where small particles tend to dissolve and diffuse through the gradient concentration to end up accumulating on larger particle sizes, resulting in particle growth overtime and formation of microparticles phenomenon known as Ostwald ripening [113, 114]. Thus, solubilization of API in nanosuspension will lead to Ostwald ripening, which should be avoided. It can occur due to large variation in particle size of suspension, which can be decreased by reducing the drug PSD [115]. Thus, a rational selection of viscosity modifier or stabilizer surfactants such as carboxy methyl cellulose sodium will support the successful manufacturing of a stable nanosuspension [116]. Ryanodex® (Dantrolene sodium) is a lyophilized powder which is reconstituted to yield nanosuspension using sterile water for injection. Sprague–Dawley female rats fed dantrolene sodium for 18 months at dosage levels of 15, 30, and 60 mg/kg/day showed an increased incidence of benign and malignant mammary tumors compared with concurrent controls. A 24-month study in HaM/ICR mice revealed no evidence of carcinogenic activity. Dantrolene sodium, when administered to male and female rats at dose levels up to 45 mg/kg/day (approx. 1.4 times the maximum recommended daily human dose on a mg/m^2 basis), showed no adverse effects on fertility or general reproductive performance [117].

Rapamune® (Sirolimus) is an oral solution containing propylene glycol, soya oil, polysorbate 80, phosal 50 PG (phosphatidylcholine, propylene glycol), mono- and diglycerides, ethanol, soya fatty acids, and ascorbyl palmitate. The carcinogenicity studies performed in mouse and rats showed increased incidences of lymphomas (male and female mice), hepatocellular adenoma and carcinoma (male mice), and granulocytic leukemia (female mice). While the reproduction toxicity studies showed decreased fertility in male rats, partly reversible reductions in sperm counts were reported in a 13-week rat study. Testicular weight reduction and/or histological lesions were observed in rats and in a monkey study [118].

11.5 Bridging the Gap of First-in-Animal Studies with First-in-Human Studies

The first-in-animal studies form an integral part of drug discovery process and preclinical testing as they provide information about the safety, efficacy, target of interest, and importantly, preclinical proof-of-concept studies. The following section highlights the general gaps between first-in-animal studies and the first-in-human studies with strategies bridging those gaps.

Preclinical formulations are tested using various animal models, and the associated preclinical data are often not always predictable in human studies, leading to higher failure rates. The possible reasons for the failures are the lack of translation between preclinical data and clinical data, ineffective molecular targets, undesirable PK/PD profiles, differences in intra- and interspecies, and poor drug–dose relationships [119]. Hence, there is need to establish translational research and models between preclinical and first-in-human studies.

The use of Han (Wistar) or Sprague–Dawley rats and predominantly beagle dogs is widely reported in literature for the evaluation of investigational drugs [120]. It is important to consider that variation in the animal species will impact the outcome of studies, which might be due to different GI physiology between the animal species, i.e. rats and dogs. The human GI physiology, like gastric and intestinal pH, gastric emptying, intestinal transit times, and motility, as well as intestinal permeability, possesses insufficient predictability compared to animal species [121]. This gap can be filled by performing the systematic analysis of databases covering the use of various formulations used in different animal species. This analysis should also investigate the effect of these formulations, relationship with the physicochemical properties of the drug, and implementation of screening tools or prediction models in humans. Various biopharmaceutical tools are developed, which cover the physiological relevance and are designed to cover the complexity of human GI tract [122]. These tools support to describe drug performance in the human GI tract. One such tool was developed, known as TNO GI model (TIM), which represents human GI tract in realistic manner. Minekus et al. developed this model to investigate digestive process *in vitro*. It consists of four separate chambers, which represent stomach, duodenum, jejunum, and ileum. Recently, a fully automated bench-top model (tiny-TIM) that enables better simulation of the anatomy, peristalsis, and gastric emptying of human stomach when compared to previous models is also developed [123]. Lopez et al. investigated the application of tiny-TIM as preclinical model to mechanistically study the effect of meal intake and gastric pH modification on performance of oral drug products in the human GI tract for commercially available ITZ formulations (Sporanox® capsule). The tiny-TIM model gave insights into interplay between physiological conditions, the drug, and the formulation generated. The comparison with clinical data revealed a good correlation between *in vitro* bioaccessibility and *in vivo* bioavailability [124].

It was highlighted that the canine GI tract, the luminal characteristics for large dogs (body weight: 30 kg), showed various similarities with humans. Mainly, the canine stomach is anatomically similar to humans, e.g. volume of 0.5–1 l (living beagle) [125]. It is reported that dogs possess a well-developed small intestine that is consistent with a diet which is low in fiber but high in fat and protein, having simple colon. The major difference when compared to humans is shorter small intestine, especially for small dogs, the much shorter large intestine, and about one-fourth of human colon [126, 127]. Important distinctions from human GI tract relate to the higher bile concentrations and higher solubility/dissolution of BCS Class II drugs, higher absorption of BCS Class III drugs, and differences in colonic characteristics. A detailed understanding of enzyme (both luminal and mucosal) and transporter activities in dogs would boost the use of this model in drug absorption and API/formulation development studies [121].

The *in vitro* potencies in animals and humans are integrated with other preclinical data, e.g. free fractions in plasma and *in vivo* preclinical data, to understand the target for efficacy in humans. Human PK is forecasted to using physiologically based PK modeling, while the safety margins are determined by comparing exposures at the no observable effect level (NOAEL) to the predicted human efficacious exposure [128]. There is no single animal species which alone is a good surrogate for human subjects, and it needs to be evaluated based on situation and case-to-case basis [129].

The current knowledge regarding the predictability of *in vivo* effects of excipients which are used in formulation of oral drug products and the interaction of these excipients with NCEs and/or investigational drugs is still under development. There is a need to progress in this area and understand the pharmacological action of excipients to avoid undesirable side effects. Preclinical studies should be designed in a systematic way to understand the safety of the excipients in order to rule out any adverse effects for both animals and humans.

11.6 Conclusions

Early formulations play an important role in translating the investigational drugs from preclinical to clinical stages, which could successfully enter the market. It adds value, as it enables the selection of rational formulation approach, excipients based on their physicochemical properties, and desired target. A pre-formulation scientist needs to consider the challenges linked to PKs, toxicology, and stability aspects of investigational drugs. The focus should also be to use minimum amounts of excipient to avoid undesirable side effects due to the excipient. Conventional approach is one of the preferred ways for early formulation due to ease of preparation, administration, less complexity, less expensive/economic, and less time-consuming. However, it's driven by the properties of investigational drugs, especially solubility and permeability, and/or grease ball or brick dust category. Enabled formulation approach is used for poorly water-soluble drugs in which specialized technologies are needed to improve solubility and overall absorption of the NCE.

While developing first-in-animal formulation, it is essential to consider biopharmaceutical tools or prediction models. The decision trees for general formulation design and for enabled formulation are proposed, which serve as the guide for early formulation selection and development. It also helps to allow resource planning and budget provisions for clinical candidate selection. The decision tree makes use of the existing knowledge in the scientific community. Nowadays, to assist pre-formulation scientists about selection of appropriate formulation strategy, more digital tools are emerging such as Zoomlab®. It was introduced by BASF SE, Germany, which uses advanced algorithms to predict the suitability of formulation. It supports formulators to identify suitable excipients, their concentrations/amounts and their type (functional/non-functional excipients) in order to tailormade them into a suitable dosage form. One of the main applications of this tool for poorly water-soluble APIs is the calculation of Hansen solubility parameters (HSPs) for any API or NCEs, which can be compared to HSPs of excipients, excipient selection, and drug loading estimation for solid dispersions, predict theoretical solubility and miscibility of NCEs in polymers/excipients and lipids, and lastly assessment of formulation route (i.e. HME and/or SD and/or SMEDDS) [130]. Thus, these early predictions of suitable formulation strategy will save time and costs for early development. The potential of such digital tools for preclinical formulation needs to be explored in more holistic way. It is assumed that with more data generated from the studies in future, it will be possible to take assistance of machine learning and artificial intelligence to screen appropriate excipients and NCEs, encouraging application of digital pharmaceutics.

References

1 Benet, L.Z. (2013). The role of BCS (biopharmaceutics classification system) and BDDCS (biopharmaceutics drug disposition classification system) in drug development. *J. Pharm. Sci.* 102 (1): 34–42.

2 Zecevic, D.E. and Wagner, K.G. (2013). Rational development of solid dispersions via hot-melt extrusion using screening, material characterization, and numeric simulation tools. *J. Pharm. Sci.* 102 (7): 2297–2310.

3 Pham-The, H., Garrigues, T., Bermejo, M. et al. (2013). Provisional classification and in silico study of biopharmaceutical system based on caco-2 cell permeability and dose number. *Mol. Pharm.* 10 (6): 2445–2461.

4 Ananthula, H.K., Parker, S., Touchette, E. et al. (2018). Preclinical pharmacokinetic evaluation to facilitate repurposing of tyrosine kinase inhibitors nilotinib and imatinib as antiviral agents. *BMC Pharmacol. Toxicol.* 19 (1): 80.

5 Li, P. and Zhao, L. (2007). Developing early formulations: practice and perspective. *Int. J. Pharm.* 341 (1–2): 1–19.

6 Shah, A.K. and Agnihotri, S.A. (2011). Recent advances and novel strategies in pre-clinical formulation development: an overview. *J. Control. Release* 156 (3): 281–296.

7 McGonigle, P. and Ruggeri, B. (2014). Animal models of human disease: challenges in enabling translation. *Biochem. Pharmacol.* 87 (1): 162–171.

8 Vos, J.G. (1980). Immunotoxicity assessment screening and function studies. *Arch. Toxicol.* 4: 95–108.

9 Penta, J.S., Rosencweig, M., and Guarino, A.M. (1979). Mouse and large animal toxicology studies of 12 antitumor agents: relevance of starting doses for phase I clinical trials. *Cancer Chemother. Pharmacol* 3: 97–101.

10 Parkinson, C. and Grasso, P. (1993). The use of the dog in toxicity testing of pharmaceutical compounds. *Hum. Exp. Toxicol* 12: 99–109.

11 Singh, V.K. and Seed, T.M. (2021). How necessary are animal models for modern drug discovery? *Expert Opin. Drug Discov.* 16 (12): 1391–1397.

12 Shah, S.M., Jain, A.S., Kaushik, R. et al. (2014). Preclinical formulations: insight, Strategies, and Practical Considerations. *AAPS PharmSciTech* 15 (5): 1307–1323.

13 Higgins, J., Cartwright, M.E., and Templeton, A.C. (2012). Progressing preclinical drug candidates: strategies on preclinical safety studies and the quest for adequate exposure. *Drug Discov. Today* 17 (15–16): 828–836.

14 Loftsson, T. and Brewster, M.E. (2010). Pharmaceutical applications of cyclodextrins: basic science and product development. *J. Pharm. Pharmacol.* 62 (11): 1607–1621.

15 Leuner, C. and Dressman, J. (2000). Improving drug solubility for oral delivery using solid dispersions. *Eur. J. Pharm. Biopharm.* 50 (1): 47–60.

16 Box, K.J., Völgyi, G., Baka, E. et al. (2006). Equilibrium versus kinetic measurements of aqueous solubility, and the ability of compounds to supersaturate in solution—a validation study. *J. Pharm. Sci.* 95 (6): 1298–1307.

17 Huang, J. (2019). Formulation forum – formulation research strategy for discovery-stage new drug candidates. *Drug Dev. Deliv.* 19 (2).

18 Hall, D. (2023) Trends and challenges in the evolving small molecule development pipeline.

19 Butler, J.M. and Dressman, J.B. (2010). The developability classification system: application of biopharmaceutics concepts to formulation development. *J. Pharm. Sci.* 99 (12): 4940–4954.

20 Sou, T. and Bergström, C.A.S. (2018). Automated assays for thermodynamic (equilibrium) solubility determination. *Drug Discov. Today: Technol.* 27: 11–19.

21 Kádár, S., Csicsák, D., Tőzsér, P. et al. (2022). Understanding the pH dependence of supersaturation state—a case study of Telmisartan. *Pharmaceutics* 14 (8): 1635.

22 van de Waterbeemd, H., Carter, R.E., Grassy, G. et al. (1997). Glossary of terms used in computational drug design (IUPAC Recommendations 1997). *Pure Appl. Chem.* 69 (5): 1137–1152.

23 López Mármol, Álvaro: Supersaturation and solubilization to enhance the oral bioavailability of poorly soluble drug molecules: A mechanistic understanding of their potential and limitations. - Bonn, 2021. - Dissertation, Rheinische Friedrich-Wilhelms-Universität Bonn. Online-Ausgabe in bonndoc: https://nbn-resolving.org/urn:nbn:de:hbz:5-63300.

24 Berthelot, M.P.E. and Jungfleisch, E.C. (1872). Sur les lois qui président au partage d'un corps entre deux dissolvants (expériences). *Ann. Chim. Phys.* 4: 396–407.

25 Camenisch, G., Alsenz, J., van de Waterbeemd, H., and Folkers, G. (1998). Estimation of permeability by passive diffusion through Caco-2 cell monolayers using the drugs' lipophilicity and molecular weight. *Eur. J. Pharm. Sci.* 6: 313–319.

26 Liu, X., Testa, B., and Fahr, A. (2011). Lipophilicity and its relationship with passive drug permeation. *Pharm. Res.* 28 (5): 962–977.

27 Flaten, G.E., Awoyemi, O., Luthman, K. et al. (2009). The phospholipid vesicle-based drug permeability assay: 5. Development toward an automated procedure for high-throughput permeability screening. *JALA: J. Assoc. Lab. Autom.* 14 (1): 12–21.

28 Mensch, J., Melis, A., Mackie, C. et al. Evaluation of various PAMPA models to identify the most discriminating method for the prediction of BBB permeability. *Eur. J. Pharm. Biopharm.* 74 (3): 495–502.

29 Yee, S. (1997). in vitro permeability across Caco-2 cells (Colonic) can predict in vivo (small intestinal) absorption in man–fact or myth. *Pharm Res* 14 (6): 763–766.

30 USFDA (2025). Integrated review of Journavx. https://www.accessdata.fda.gov/drugsatfda_docs/nda/2025/219209Orig1s000IntegratedR.pdf (accessed 1 July 2025).

31 USFDA (2024). Assessment report of Lazertinib. https://www.ema.europa.eu/en/documents/assessment-report/lazcluze-epar-public-assessment-report_en.pdf (accessed 1 July 2025).

32 Lorenz, D.A., Konagurthu, S., Wald, R.J. and Jason, A. (2019). Formulations of enzalutamide. United States Patent Office, filed 4 January 2019.

33 Eng, H., Dantonio, A.L., Kadar, E.P. et al. (2022). Disposition of Nirmatrelvir, an orally bioavailable inhibitor of SARS-CoV-2 3C-like protease, across animals and humans. *Drug Metab. Dispos.* 50 (5): 576–590.

34 Manallack, D.T., Prankerd, R.J., Yuriev, E. et al. (2013). The significance of acid/base properties in drug discovery. *Chem. Soc. Rev.* 42 (2): 485–496.

35 Shore, P.A., Brodie, B.B., and Hogben, C.A.M. (1957). The gastric secretion of drugs: a pH Partition hypothesis. *J. Pharmacol. Exp. Therap.* 119: 361–369.

36 Hogben, C.A.M., Tocco, D.J., Brodie, B.B., and Schanker, L.S. (1959). On the mechanism of intestinal absorption of drugs. *J. Pharmacol. Exp. Therap.* 125: 275–282.

37 Bhalani, D.V., Nutan, B., Kumar, A., and Singh Chandel, A.K. (2022). Bioavailability enhancement techniques for poorly aqueous soluble drugs and therapeutics. *Biomedicines* 10 (9): 2055.

38 Wouters, O.J., McKee, M., and Luyten, J. (2020). Estimated research and development investment needed to bring a new medicine to market, 2009–2018. *JAMA* 323 (9): 844.

39 Thomas Hughes, M.B. and Dahlem, A. (2007). Drug discovery and development, understanding the R&D process. Pharmaceutical Research and Manufacturers of America, Pharmaceutical Industry Profile. (Washington, DC: PhRM. February 2007), 1–10.

40 LoÈbenberg, R. and Amidon, G.L. (2000). Modern bioavailability, bioequivalence and biopharmaceutics classification system. New scientific approaches to international regulatory standards. *Eur. J. Pharm. Biopharm.* 50: 3–12.

41 Miller, J.M., Beig, A., Carr, R.A. et al. (2012b). The solubility–permeability interplay when using cosolvents for solubilization: revising the way we use solubility-enabling formulations. *Mol. Pharm.* 9: 581–590.

42 Gabrič, A., Hodnik, Ž., and Pajk, S. (2022). Oxidation of drugs during drug product development: problems and solutions. *Pharmaceutics* 14 (2): 325.

43 Zane, P., Gieschen, H., Kersten, E. et al. (2019). in vivo models and decision trees for formulation development in early drug development: a review of current practices and recommendations for biopharmaceutical development. *Eur. J. Pharm. Biopharm.* 142: 222–231.

44 An, L., De Bruyn, T., Pang, J. et al. (2024). Early stage preclinical formulation strategies to alter the pharmacokinetic profile of two small molecule therapeutics. *Pharmaceuticals* 17 (2): 179.

45 Nakmode, D., Bhavana, V., Thakor, P. et al. (2022). Fundamental aspects of lipid-based excipients in lipid-based product development. *Pharmaceutics* 14 (4): 831.

46 Kim, Y.-H., Kim, S.-B., Choi, S.-H. et al. (2023). Development and evaluation of self-microemulsifying drug delivery system for improving oral absorption of poorly water-soluble Olaparib. *Pharmaceutics* 15 (6): 1669.

47 Deshmukh, K., Shrivastava, A., and Bambarkar, S. (2021). Lipid excipients: carving out a niche in pharmaceuticaL drug delivery. *Pharma Times* 53 (12): 15–18.

48 Mahajan, N., Mujtaba, M.A., Fule, R. et al. (2024). Self-emulsifying drug delivery system for enhanced oral delivery of tenofovir: formulation, physicochemical characterization, and bioavailability assessment. *ACS Omega* 9: acsomega.3c08565.

49 Gopinathan, S., O'Neill, E., Rodriguez, L.A. et al. (2013). in vivo toxicology of excipients commonly employed in drug discovery in rats. *J. Pharmacol. Toxicol. Methods* 68 (2): 284–295.

50 Larsen, A.T., Holm, R., and Müllertz, A. (2017). Solution or suspension – does it matter for lipid based systems? in vivo studies of chase dosing lipid vehicles with aqueous suspensions of a poorly soluble drug. *Eur. J. Pharm. Biopharm.* 117: 308–314.

51 Krome, A.K., Becker, T., Kehraus, S. et al. (2022). Corallopyronin A: antimicrobial discovery to preclinical development. *Nat. Prod. Rep.* 39 (9): 1705–1720.

52 Maghrabia, A., Boughdady, M., and Meshali, M. (2021). Design and optimization of new enteric nanoparticles of ceftriaxone for oral delivery: in vitro and in vivo assessments. *Int. J. Nanomed.* 16: 5937–5953.

53 Frank, K.J., Locher, K., Zecevic, D.E. et al. (2014). in vivo predictive mini-scale dissolution for weak bases: advantages of pH-shift in combination with an absorptive compartment. *Eur. J. Pharm. Sci.* 61: 32–39.

54 Fagerberg, J.H., Karlsson, E., Ulander, J. et al. (2015). Computational prediction of drug solubility in fasted simulated and aspirated human intestinal fluid. *Pharm Res* 32 (2): 578–589.

55 Bergström, C.A.S. and Larsson, P. (2018). Computational prediction of drug solubility in water-based systems: qualitative and quantitative approaches used in the current drug discovery and development setting. *Int. J. Pharm.* 540 (1–2): 185–193.

56 Chiou, W.L. and Riegelman, S. (1971). Pharmaceutical applications of solid dispersion systems. *J. Pharm. Sci.* 60 (9): 1281–1302.

57 Huang, S., O'Donnell, K.P., Keen, J.M. et al. (2016). A new extrudable form of hypromellose: AFFINISOL™ HPMC HME. *AAPS PharmSciTech* 17 (1): 106–119.

58 Zecevic, D.E., Meier, R., Daniels, R., and Wagner, K.-G. (2014). Site specific solubility improvement using solid dispersions of HPMC-AS/HPC SSL – mixtures. *Eur. J. Pharm. Biopharm.* 87 (2): 264–270.

59 Krome, A.K., Becker, T., Kehraus, S. et al. (2020). Solubility and stability enhanced oral formulations for the anti-infective corallopyronin A. *Pharmaceutics* 12 (11): 1105.

60 Bhardwaj, S.P., Arora, K.K., Kwong, E. et al. (2014). Mechanism of amorphous itraconazole stabilization in polymer solid dispersions: role of molecular mobility. *Mol. Pharm.* 11 (11): 4228–4237.

61 Marsac, P.J., Konno, H., and Taylor, L.S. (2006). A comparison of the physical stability of amorphous felodipine and nifedipine systems. *Pharm. Res.* 23 (10): 2306–2316.

62 Yoshioka, M., Hancock, B.C., and Zografi, G. (1995). Inhibition of indomethacin crystallization in poly (vinylpyrrolidone) coprecipitates. *J. Pharm. Sci.* 84 (8): 983–986.

63 Tian, Y., Jones, D.S., and Andrews, G.P. (2015). An investigation into the role of polymeric carriers on crystal growth within amorphous solid dispersion systems. *Mol. Pharm.* 12 (4): 1180–1192.

64 Aso, Y., Yoshioka, S., and Kojima, S. (2001). Explanation of the crystallization rate of amorphous nifedipine and phenobarbital from their molecular mobility as measured by 13C nuclear magnetic resonance relaxation time and the relaxation time obtained from the heating rate dependence of the glass transition temperature. *J. Pharm. Sci.* 90 (6): 798–806.

65 Van den Mooter, G., Wuyts, M., Blaton, N. et al. (2001). Physical stabilisation of amorphous ketoconazole in solid dispersions with polyvinylpyrrolidone K25. *Eur. J. Pharm. Sci.* 12 (3): 261–269.

66 Zhou, D., Zhang, G.G.Z., Law, D. et al. (2002). Physical stability of amorphous pharmaceuticals: importance of configurational thermodynamic quantities and molecular mobility. *J. Pharm. Sci.* 91 (8): 1863–1872.

67 Kothari, K., Ragoonanan, V., and Suryanarayanan, R. (2015). The role of polymer concentration on the molecular mobility and physical stability of nifedipine solid dispersions. *Mol. Pharm.* 12 (5): 1477–1484.

68 Nart, V., França, M.T., Anzilaggo, D. et al. (2015). Ball-milled solid dispersions of BCS Class IV drugs: impact on the dissolution rate and intestinal permeability of acyclovir. *Mater. Sci. Eng.: C* 53: 229–238.

69 Zhao, M., Barker, S.A., Belton, P.S. et al. (2012). Development of fully amorphous dispersions of a low Tg drug via co-spray drying with hydrophilic polymers. *Eur. J. Pharm. Biopharm.* 82 (3): 572–579.

70 Gangurde, A.B., Kundaikar, H.S., Javeer, S.D. et al. (2015). Enhanced solubility and dissolution of curcumin by a hydrophilic polymer solid dispersion and its insilico molecular modeling studies. *J. Drug Deliv. Sci. Technol.* 29: 226–237.

71 Al-Obaidi, H., Ke, P., Brocchini, S., and Buckton, G. (2011). Characterization and stability of ternary solid dispersions with PVP and PHPMA. *Int. J. Pharm.* 419 (1–2): 20–27.

72 Yang, F., Su, Y., Zhu, L. et al. (2016). Rheological and solid-state NMR assessments of copovidone/clotrimazole model solid dispersions. *Int. J. Pharm.* 500 (1–2): 20–31.

73 Lamm, M.S., Simpson, A., McNevin, M. et al. (2012). Probing the effect of drug loading and humidity on the mechanical properties of solid dispersions with nanoindentation: antiplasticization of a polymer by a drug molecule. *Mol. Pharm.* 9 (11): 3396–3402.

74 Sarode, A., Wang, P., Cote, C., and Worthen, D.R. (2013). Low-viscosity hydroxypropylcellulose (HPC) grades SL and SSL: versatile pharmaceutical polymers for dissolution enhancement, controlled release, and pharmaceutical processing. *AAPS PharmSciTech* 14 (1): 151–159.

75 Newa, M., Bhandari, K., Li, D. et al. (2007). Preparation, characterization and in vivo evaluation of ibuprofen binary solid dispersions with poloxamer 188. *Int. J. Pharm.* 343 (1–2): 228–237.

76 Ricarte, R.G., Lodge, T.P., and Hillmyer, M.A. (2015). Detection of pharmaceutical drug crystallites in solid dispersions by transmission electron microscopy. *Mol. Pharm.* 12 (3): 983–990.

77 Fule, R. and Amin, P. (2014). Development and evaluation of lafutidine solid dispersion via hot melt extrusion: investigating drug-polymer miscibility with advanced characterisation. *Asian J. Pharm. Sci.* 9 (2): 92–106.

78 Moseson, DT., Tran, T.B., Karunakaran, B., Ambardekar, R., and Hiew, T.N. (2024). Trends in amorphous solid dispersion drug products approved by the U.S. Food and Drug Administration between 2012 and 2023. *Int. J. Pharm.: X*, 7 (100259), 1–17.

79 Sun, D.D. and Lee, P.I. (2015). Probing the mechanisms of drug release from amorphous solid dispersions in medium-soluble and medium-insoluble carriers. *J. Control. Release* 211: 85–93.

80 Vo, C.L.-N., Park, C., and Lee, B.-J. (2013). Current trends and future perspectives of solid dispersions containing poorly water-soluble drugs. *Eur. J. Pharm. Biopharm.* 85 (3): 799–813.

81 Craig, D.Q.M. (2002). The mechanisms of drug release from solid dispersions in water-soluble polymers. *Int. J. Pharm.* 231 (2): 131–144.

82 Schittny, A., Huwyler, J., and Puchkov, M. (2020). Mechanisms of increased bioavailability through amorphous solid dispersions: a review. *Drug Deliv.* 27 (1): 110–127.

83 Maniruzzaman, M., Snowden, M.J., Bradely, M.S., and Douroumis, D. (2015). Studies of intermolecular interactions in solid dispersions using advanced surface chemical analysis. *RSC Adv.* 5 (91): 74212–74219.

84 Kapote, D.N. and Wagner, K.G. (2021). Influence of shellac on the improvement of solubility and supersaturation of loratadine amorphous solid dispersion using a new grade of HPMC. *J. Drug Deliv. Sci. Technol.* 61: 102116.

85 Srinarong, P., Hämäläinen, S., Visser, M.R. et al. (2011). Surface-active derivative of inulin (Inutec® SP1) is a superior carrier for solid dispersions with a high drug load. *J. Pharm. Sci.* 100 (6): 2333–2342.

86 Erickson, R.I., Schutt, L.K., Tarrant, J.M. et al. (2017). Bruton's tyrosine kinase small molecule inhibitors induce a distinct pancreatic toxicity in rats. *J. Pharmacol. Exp. Ther.* 360 (1): 226–238.

87 Bhaskaran, M., Cornwell, P.D., Sorden, S.D. et al. (2018). Pancreatic effects of a Bruton's tyrosine kinase small-molecule inhibitor in rats are strain-dependent. *Toxicol. Pathol.* 46 (4): 460–472.

88 USFDA (2023). ALVAIZ^{TM} (eltrombopag tablets), for oral use. https://www.accessdata.fda.gov/drugsatfda_docs/label/2023/216774s000lbl.pdf (accessed 2 July 2025).

89 EMEA (2023) Committee for Medicinal Products for Human Use (CHMP) (2023) Assessment report for Aquipta. https://www.ema.europa.eu/en/medicines/human/EPAR/aquipta (accessed 1 July 2025).

90 Kuentz, M., Holm, R., Kronseder, C. et al. (2021). Rational selection of bio-enabling oral drug formulations – a PEARRL commentary. *J. Pharm. Sci.* 110 (5): 1921–1930.

91 Holm, R., Kuentz, M., Ilie-Spiridon, A.-R., and Griffin, B.T. (2023). Lipid based formulations as supersaturating oral delivery systems: From current to future industrial applications. *Eur. J. Pharm. Sci.* 189: 106556.

92 O'Driscoll, C.M. and Griffin, B.T. (2008). Biopharmaceutical challenges associated with drugs with low aqueous solubility—the potential impact of lipid-based formulations. *Adv. Drug Deliv. Rev.* 60 (6): 617–624.

93 Bolko, K., Zvonar, A., and Gašperlin, M. (2014). Mixed lipid phase SMEDDS as an innovative approach to enhance resveratrol solubility. *Drug Dev. Ind. Pharm.* 40 (1): 102–109.

94 Pouton, C.W. (2006). Formulation of poorly water-soluble drugs for oral administration: physicochemical and physiological issues and the lipid formulation classification system. *Eur. J. Pharm. Sci.* 29 (3–4): 278–286.

95 Vinarov, Z., Katev, V., Radeva, D. et al. (2018). Micellar solubilization of poorly water-soluble drugs: effect of surfactant and solubilizate molecular structure. *Drug Dev. Ind. Pharm.* 44 (4): 677–686.

96 Guimarães, D., Cavaco-Paulo, A., and Nogueira, E. (2021). Design of liposomes as drug delivery system for therapeutic applications. *Int. J. Pharm.* 601: 120571.

97 Nsairat, H., Khater, D., Sayed, U. et al. (2022). Liposomes: structure, composition, types, and clinical applications. *Heliyon* 8 (5): e09394.

98 Liu, P., Chen, G., and Zhang, J. (2022). A review of liposomes as a drug delivery system: current status of approved products, regulatory environments, and future perspectives. *Molecules* 27 (4): 1372.

99 Agrawal, S.S., Baliga, V., and Londhe, V.Y. (2024). Liposomal formulations: a recent update. *Pharmaceutics* 17 (1): 36.

100 Adler-Moore, J., Lewis, R.E., Brüggemann, R.J.M. et al. (2019). Preclinical safety, tolerability, pharmacokinetics, pharmacodynamics, and antifungal activity of liposomal amphotericin B. *Clin. Infect. Dis.* 68 (Supplement_4): S244–S259.

101 Qian, K., Lin, Y., Zhao, B., and Liu, X. (2025). Droplet size reduction of self-emulsifying drug delivery system (SEDDS) using the hybrid of medium and long-chain triglycerides. *Pharmaceutics* 17 (7): 822.

102 Thakkar, H., Nangesh, J., Parmar, M., and Patel, D. (2011). Formulation and characterization of lipid-based drug delivery system of raloxifene-microemulsion and self-microemulsifying drug delivery system. *J. Pharm. Bioall. Sci.* 3 (3): 442.

103 Lee, H.I., Woo, M.R., Din, F.U. et al. (2025). Development of a novel apixaban-loaded solid self-emulsifying drug delivery system for oral administration: physicochemical characterization and pharmacokinetics in rats. *J. Pharm. Invest.* 55: 575–588.

104 Kok, L.Y., Bannigan, P., Sanaee, F. et al. (2022). Development and pharmacokinetic evaluation of a self-nanoemulsifying drug delivery system for the oral delivery of cannabidiol. *Eur. J. Pharm. Sci.* 168: 106058.

105 EMEA (2013). Product summary of Xtandi® capsules. https://www.ema.europa.eu/en/documents/product-information/xtandi-epar-product-information_en.pdf (accessed 1 July 2025).

106 Pınar, S.G., Oktay, A.N., Karaküçük, A.E., and Çelebi, N. (2023). Formulation strategies of nanosuspensions for various administration routes. *Pharmaceutics* 15 (5): 1520.

107 Karakucuk, A., Celebi, N., and Teksin, Z.S. (2016). Preparation of ritonavir nanosuspensions by microfluidization using polymeric stabilizers: I. A design of experiment approach. *Eur. J. Pharm. Sci.* 95: 111–121.

108 Oktay, A.N., Ilbasmis-Tamer, S., and Celebi, N. (2019). The effect of critical process parameters of the high-pressure homogenization technique on the critical quality attributes of flurbiprofen nanosuspensions. *Pharm. Dev. Technol.* 24 (10): 1278–1286.

109 Olver, I., Shelukar, S., and Thompson, K.C. (2007). Nanomedicines in the treatment of emesis during chemotherapy: focus on aprepitant. *Int. J. Nanomed.* 2 (1): 13–18.

110 Keck, C.M. and Müller, R.H. (2006). Drug nanocrystals of poorly soluble drugs produced by high pressure homogenisation. *Eur. J. Pharm. Biopharm.* 62 (1): 3–16.

111 Haeggström, E., Yliruusi, J., Falck, K., Räikkönen, H., Pessi, J., Lassila, I., and Meriläinen, A. (2017). A method and a device for producing nanoparticles. US

Patent office US 2017/0231914 A1, filed 5 October 2015 and issued 17 August 2017.

112 Wong, J., Brugger, A., Khare, A. et al. (2008). Suspensions for intravenous (IV) injection: a review of development, preclinical and clinical aspects. *Adv. Drug Deliv. Rev.* 60 (8): 939–954.

113 Lindfors, L., Skantze, P., Skantze, U. et al. (2006). Amorphous drug nanosuspensions. 1. Inhibition of Ostwald ripening. *Langmuir* 22 (3): 906–910.

114 Behrens, M.A., Franzén, A., Carlert, S. et al. (2025). On the Ostwald ripening of crystalline and amorphous nanoparticles. *Soft Matter* 21 (12): 2349–2354.

115 Swarbirck, J., Rubino, J., and Rubino, J. (2006). *Remington: The Science and Practice of Pharmacy*. Baltimore: Lippincott Williams & Wilkins.

116 Van Eerdenbrugh, B., Van den Mooter, G., and Augustijns, P. (2008). Top-down production of drug nanocrystals: nanosuspension stabilization, miniaturization and transformation into solid products. *Int. J. Pharm.* 64 (1): 64–75.

117 Breder C.D. USFDA (2014). NDA of Ryanodex suspension for injection/Dantrolene sodium. https://www.accessdata.fda.gov/drugsatfda_docs/nda/2014/205579Orig1s000CrossR.pdf (accessed 1 July 2025).

118 EMEA (2018). Product summary of Rapmune® oral solution. https://www.ema.europa.eu/en/documents/product-information/rapamune-epar-product-information_en.pdf (accessed 1 July 2025).

119 Mak, I.W., Evaniew, N., and Ghert, M. (2014). Lost in translation: animal models and clinical trials in cancer treatment. *Am. J. Transl. Res.* 6: 114–118.

120 Butler, L., Guzzie-Peck, P., Hartke, J. et al. (2017). Current nonclinical testing paradigms in support of safe clinical trials: an IQ Consortium DruSafe perspective. *Regul. Toxicol. Pharmacol.* 87 (S3): S1–S15.

121 Sjögren, E., Abrahamsson, B., Augustijns, P. et al. (2014). in vivo methods for drug absorption - comparative physiologies, model selection, correlations with in vitro methods (IVIVC), and applications for formulation/API/excipient characterization including food effects. *Eur. J. Pharm. Sci.* 57: 99–151.

122 Butler, J., Hens, B., Vertzoni, M. et al. (2019). in vitro models for the prediction of in vivo performance of oral dosage forms: Recent progress from partnership through the IMI OrBiTo collaboration. *Eur. J. Pharm. Biopharm.* 136: 70–83.

123 Minekus, M. (2015). The TNO gastro-intestinal model (TIM). In: *The Impact of Food Bioactives on Health: in vitro and Ex Vivo Models* (ed. K. Verhoeckx et al.). New York, NY: Springer International Publishing.

124 López Mármol, Á., Fischer, P.L., Wahl, A. et al. (2022). Application of tiny-TIM as a mechanistic tool to investigate the in vitro performance of different itraconazole formulations under physiologically relevant conditions. *Eur. J. Pharm. Sci.* 173: 106165.

125 Martinez, M., Amidon, G., Clarke, L. et al. (2002). Applying the biopharmaceutics classification system to veterinary pharmaceutical products. *Adv. Drug Deliv. Rev.* 54 (6): 825–850.

126 de Zwart, L.L. (1999). Anatomical and physiological differences between various species used in studies on the pharmacokinetics and toxicology of xenobiotics. A review of literature. *Natl. Inst. Public Health Environ.* https://

rivm.openrepository.com/entities/publication/c269962d-b115-4ccb-ae2a-0ff9aadd500b/full.

127 Kararli, T.T. (1995). Comparison of the gastrointestinal anatomy, physiology, and biochemistry of humans and commonly used laboratory animals. *Biopharm. Drug Disp.* 16 (5): 351–380.

128 Namdari, R., Jones, K., Chuang, S.S. et al. (2021). Species selection for nonclinical safety assessment of drug candidates: examples of current industry practice. *Regulat. Toxicol. Pharmacol.* 126: 105029.

129 Li, A.P. (1999). Overview: hepatocytes and cryopreservation – a personal historical perspective. *Chem.-Biol. Interact.* 121: 1–5.

130 Decision support for solubilization using Zoomlab™. https://pharma.basf.com/about-basf-pharma/digitalization/zoomlab (accessed 4 May 2025).

12

Improving Aqueous Solubility of Drug Candidates Through Structural Modifications

Kamal Kumar

Smartbax GmbH, Ingolstädter Landstraße 1, 85764 Oberschleißheim, Germany

12.1 Introduction

Over the past 30 years, drug discovery practices have evolved significantly, driven by a deeper and more sophisticated understanding of the characteristics of successful drugs [1]. These advancements aim to improve compound drugability and overall success rates, though drug candidate failure remains the norm in development programs. In 1990, these failures were evenly attributed to toxicity, lack of efficacy, and poor pharmacokinetics (PK). Enhanced preclinical profiling has substantially reduced PK-related failures by improving human PK predictions. Despite this, drug discovery output over the past two to three decades has remained low, with toxicity and lack of efficacy now the primary reasons for candidate failure. A growing focus on physical properties, developability, and side effects has led to the adoption of diverse in vitro and in vivo assays to eliminate high-risk compounds early and reduce the likelihood of unacceptable clinical outcomes.

The significance of physical properties has thereby re-emerged as a central focus in modern medicinal chemistry [1]. Physical properties like size, weight, charge, and lipophilicity are key metrics for defining "drug-like space" and predicting molecular behavior [2]. These descriptors have culminated in tools such as the quantitative estimate of drug-likeness (QED), which help quantify a molecule's drug-like potential [3, 4].

A better aqueous solubility of a drug candidate ensures systemic exposure through physiological fluids (e.g. bloodstream) and effectively reaches its target. Additionally, its physical characteristics must be complementary to the target for successful interaction. The aqueous environments of the digestive tract, bloodstream, and intracellular spaces contrast with the more lipophilic environments at drug action sites, where recognition depends on a precise balance of polar and hydrophobic interactions [5]. Historically, the use of "small hydrophilic

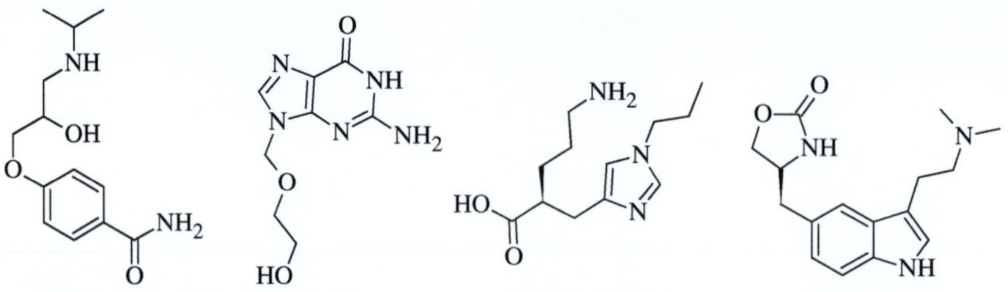

Atenolol	Acyclovir	UK-396,082	Zolmitriptan
clog $D_{7.4}$ = −1.6	clog $D_{7.4}$ = −1.57	clog $D_{7.4}$ = −2.7	clog $D_{7.4}$ = −0.4
MW = 266	MW = 225	MW = 239	MW = 287

Figure 12.1 Hydrophilic oral drugs absorbed by paracellular route.

molecules" has facilitated the discovery of numerous effective oral agents by enabling paracellular intestinal absorption. This process occurs through the tight aqueous spaces between the cells lining the small intestine [6]. To achieve this mode of absorption, the molecules are typically small (molecular weight <250 or calculated molar refraction <10) and are often basic or zwitterionic in nature, with a distribution coefficient below one (negative log D values), making them more water-soluble. Some well-known examples of such molecules are depicted in Figure 12.1, for instance, the β1-antagonist atenolol [7], antiviral acyclovir [8, 9], UK-396,082, a thrombin-activatable fibrinolysis inhibitor [10], and zolmitriptan, an anti-migraine serotonin $1\beta/1\delta$ receptor agonist (Figure 12.1) [6].

Poor aqueous solubility is a major cause of the failure of many drug candidates in clinical trials, even when they demonstrate strong pharmacological activity in animal studies [11]. To a greater extent, the low aqueous solubility stems from high- to medium-throughput synthesis that tends to produce more hydrophobic molecules, and the high-throughput screening (HTS) keep affording numerous hit compounds which exhibit low bioavailability, primarily due to their poor solubility [12, 13]. The pursuit of higher potency and selectivity has shifted drug discovery toward larger, more hydrophobic molecules, often at the expense of aqueous solubility [14, 15]. Nonetheless, adequate dissolution remains critical, particularly for oral drugs. Human physiology has evolved to favor the clearance of lipophilic compounds via solubilizing mechanisms such as metabolism and conjugation [16, 17]. While some marketed drugs lie at the extremes of physicochemical properties, most fall within a defined range that balances hydrophilicity with pharmacological efficacy [18].

Unfavorable physical properties, particularly low aqueous solubility, increasingly hinder drug development by affecting pharmacokinetics and pharmacology [1, 14, 16, 19–24]. While the determinants of solubility are complex and sometimes debated, both empirical and rational design strategies offer practical solutions. This chapter compiles recent medicinal chemistry approaches and structural modifications that effectively address solubility challenges.

12.2 Enhancing Drug Solubility and Bioavailability: A Key to Effective Therapy

Solubility generally refers to dissolving a solute in a solvent, essentially to produce a homogeneous system [25–27]. Solubility is typically measured in aqueous media to determine thermodynamic solubility, where the solute concentration in the equilibrium state is assessed. Kinetic solubility, on the other hand, evaluates the tendency of a pre-dissolved compound to precipitate. It is important to distinguish between dissolution and solubility. Dissolution refers to the process by which a solute transitions from any phase—gas, liquid, or solid—into a solvent to form a solution. Solubility, however, describes the maximum concentration of solute that can dissolve in a solvent at a specific temperature [28].

For orally administered drugs in solid dosage forms, disintegration into smaller particles facilitates molecular dissolution in the gastrointestinal tract (GIT) and permeates the intestinal barrier [29]. Since physiological fluids are water-based, aqueous solubility is a critical factor for achieving therapeutic drug concentrations in systemic circulation. Poor solubility hinders dissolution in GIT fluids, reducing drug permeability and bioavailability. Low bioavailability necessitates higher drug doses, which increases formulation costs and may also lead to toxicity [30, 31].

The solubility of a drug is directly linked to its dissolution rate, both of which are critical factors in determining oral bioavailability [32]. As such, solubility, dissolution rate, and bioavailability together shape a drug's overall in vivo performance [33, 34]. Bioavailability refers to the proportion of the drug that reaches systemic circulation and is influenced by both the drug's solubility and the route of administration. The oral route remains the most convenient and preferred method, making oral bioavailability a top priority in pharmaceutical development. However, compounds belonging to BCS Class II (low solubility, high permeability) and Class IV (low solubility, low permeability) often pose significant challenges and typically require structural modifications to enhance solubility and improve therapeutic outcomes [35, 36].

Given that approximately 60% of the human body is water—distributed across blood, cells, and tissues—understanding how water interacts with drug molecules is vital during drug design. Water not only acts as a medium for drug transport but also interacts with proteins, lipids, and carbohydrates, which are common drug targets. Solubility, therefore, is central to a drug's pharmacokinetic and pharmacodynamic behavior, influencing processes such as absorption, metabolism, receptor binding [37], and the effectiveness of HTS in early drug discovery [38, 39].

Aqueous solubility refers to the extent to which a substance dissolves in water, reflecting the interaction between water molecules and solute molecules. There are different types of aqueous solubility measurements. Intrinsic solubility refers to the solubility of the un-ionized (neutral) form of a compound in a specific solvent—typically water—at a given temperature and pressure. It represents the baseline solubility of the molecule without the influence of pH or any ionization processes. In other words, intrinsic solubility is: *The maximum concentration of the uncharged form of a compound that can dissolve in water under equilibrium conditions*. This concept is especially important for weak acids and bases, whose total solubility in solution often depends on the pH of the environment. The total solubility includes both the ionized and un-ionized forms, but only the intrinsic solubility reflects the true solubility of the neutral species, which is often the form that permeates biological membranes.

In pharmaceutical literature, solubility is commonly described using two method-dependent terms: kinetic solubility and thermodynamic solubility [40]. *Kinetic solubility* refers to the concentration of a solute in a solvent at the point when a precipitate first begins to form in the solution. In contrast, *thermodynamic solubility* is defined as the concentration of a compound in solution when the system is at equilibrium with the undissolved solute present. While thermodynamic solubility addresses the question, "How much does the substance dissolve?," kinetic solubility answers, "How much does the molecule precipitate?"[40]. Thermodynamic solubility is particularly crucial for poorly soluble drugs, as it directly affects dissolution rates and subsequent drug absorption. Thermodynamic solubility is regarded as the "true" solubility and is influenced by several factors, including pH, temperature, ionic strength, salt/buffer effects, and phase separation. The key difference between the two methods lies in the experimental approach: thermodynamic solubility is determined by adding the compound in its solid form to the aqueous medium, whereas kinetic solubility is measured using a pre-dissolved compound [41].

Kinetic and thermodynamic solubility serve distinct roles throughout drug discovery. Kinetic solubility is particularly useful in early stages, such as HTS and lead identification, due to its rapid, convenient assessment from dimethylsulfoxide (DMSO) stock solutions. While it may overestimate true solubility by not reflecting equilibrium conditions, it remains valuable for efficiently prioritizing hits for further development [42, 43].

For the later drug development, thermodynamic solubility is more relevant. In later stages, like lead optimization, preclinical development, and formulation, a deeper understanding of a compound's true solubility is needed. Thermodynamic solubility reflects the equilibrium solubility, which is critical for bioavailability, formulation development, and regulatory submissions. It is essential for developing robust dosage forms and ensuring consistent performance across batches. A comparison of kinetic and thermodynamic solubility is presented in Table 12.1.

Recent literature highlights the development of various advanced techniques for solubility determination, particularly those aligned with HTS methodologies [44–46]. Based on high-throughput assay platforms, several methods have been employed to predict equilibrium solubility, including solid-state characterization techniques such as polarized light microscopy, Raman spectroscopy, powder X-ray diffraction, ultraperformance liquid chromatography (UPLC) [45], and polychromatic turbidimetry [47]. In addition, quantitative structure–activity/property relationship (QSAR/QSPR) models have emerged as cost-effective computational tools for predicting solubility [48, 49].

Table 12.1 Comparison of thermodynamic and kinetic solubility of drugs.

Aspect	Thermodynamic solubility	Kinetic solubility
Definition	The maximum amount of a drug that can dissolve in a solvent under equilibrium conditions (fully saturated solution at a given temperature and pH).	The solubility of a drug before reaching equilibrium is usually determined as the concentration at which precipitation begins in a supersaturated solution.
Measurement method	• Drug incubated in aqueous medium (water/buffer) at fixed pH and temp. • Stirred and equilibrated for 24–48 hours. • Solution filtered; dissolved drug measured by HPLC or UV spectrophotometry.	• Drug dissolved in organic solvent (e.g. DMSO) and rapidly diluted into aqueous medium to form a supersaturated solution. • Precipitation monitored visually or by nephelometry/light scattering. • Highest concentration before precipitation is recorded as kinetic solubility.
Conditions	• Conducted at equilibrium conditions. • Typically requires longer incubation times (24–48 hours). • pH and temperature are carefully controlled.	• Measured under non-equilibrium conditions. • Conducted over shorter timescales (minutes to hours). • Involves supersaturation, which may not be stable over time.
Importance in drug discovery	• Crucial for later stages of drug development, including formulation design and predicting drug absorption. • Helps in assessing long-term solubility, stability in biological fluids and formulations. • Used in bioavailability studies to ensure drugs can dissolve effectively in the gastrointestinal (GI) tract.	• Widely used in early stage drug screening to quickly rank compounds based on their solubility. • Helps in identifying poorly soluble compounds early, allowing medicinal chemists to modify structures accordingly. • Important for assessing the risk of precipitation in the GI tract.
Advantages	• Provides an accurate and reproducible measurement of solubility. • Directly correlates with long-term formulation stability. • Helps in designing appropriate dosage forms and bioavailability predictions.	• Faster and more practical for high-throughput screening. • Requires smaller amounts of the drug (ideal for early stage screening). • Can identify supersaturable drugs that may be stabilized using formulation strategies.

(Continued)

Table 12.1 (Continued)

Aspect	Thermodynamic solubility	Kinetic solubility
Disadvantages	• Time-consuming and requires large amounts of the compound. • Not ideal for early stage screening due to lengthy equilibration times.	• May overestimate solubility, leading to misleading results. • Less reliable for formulation development as supersaturated states may not be stable over time.
Use in drug optimization	• Used for selecting drug candidates with sufficient solubility for formulation and bioavailability. • Helps in designing prodrugs, salts, or excipients to enhance solubility. • Essential for determining solubility-limited absorption in pharmacokinetics.	• Used early in drug discovery to prioritize soluble compounds. • Guides decisions on solubility enhancement strategies (e.g. solid dispersions, cyclodextrins). • Helps predict supersaturation potential for formulations.

Source: Adapted from Könczöl and Dargó [43].

Globally, considerable efforts are directed toward developing innovative strategies to enhance the solubility and dissolution rates of lead compounds while maintaining their biological activity. These strategies span a variety of approaches, including:

- **Physical methods:** Particle size reduction, crystal habit modification, drug dispersion in carriers, and complexation.
- **Chemical methods:** Salt formation, co-crystallization, derivatization, buffer use, micronization, nanonization, and co-solvency.

This chapter focuses specifically on medicinal chemistry strategies employing chemical modifications to address solubility and related performance challenges in early stage or advanced drug candidates.

12.3 Chemical Modification as a Strategy to Enhance Solubility

Improving the solubility of poorly water-soluble lead compounds continues to pose a major obstacle in drug development. For medicinal chemists and formulation scientists, chemical modification offers a promising and adaptable approach to address this challenge without compromising—and often enhancing—bioactivity [50]. A variety of structural modification techniques are used to boost solubility. These include altering functional groups to adjust the compound's acid–base profile, modifying molecular structures to strike an optimal hydrophilic–lipophilic balance (as reflected by $\log P$), and introducing hydrogen bond donors or acceptors to promote interactions with aqueous environments. Additional approaches involve incorporating heterocyclic ring systems or replacing hydrophobic groups with hydrophilic bioisosteres to improve aqueous solubility.

Interestingly, solubility limitations are not always attributed to hydrophobicity alone. In many cases, high melting points significantly hinder solubility. These elevated melting points are often associated with tightly packed crystal lattices, high molecular planarity, and symmetrical structures that confer greater stability in the solid phase [51, 52]. Therefore, disrupting molecular symmetry and reducing planarity has become a valuable tactic in modern lead optimization efforts to improve solubility [53].

Other alternate strategies to enhance drug solubility (other than chemical modifications) include various formulation and molecular strategies like salt formation, cocrystals, nanotechnology-based systems (e.g. nanoparticles, liposomes), solid dispersions, cyclodextrin inclusion complexes, pH adjustment, lipid-based formulations, and amorphous drug design. Additionally, polymeric micelles and other excipient-based methods are widely used to improve solubility. These strategies, however, are not covered in this chapter. This chapter focuses specifically on structure-based modifications (Figure 12.2), while the formulation-based strategies and other techniques are addressed in detail in other chapters of this book.

12.3.1 Structure (Chemical) Modification to Enhance Solubility

Structural modification is a key strategy employed to address solubility limitations in drug candidates, aiming to enhance aqueous solubility without compromising

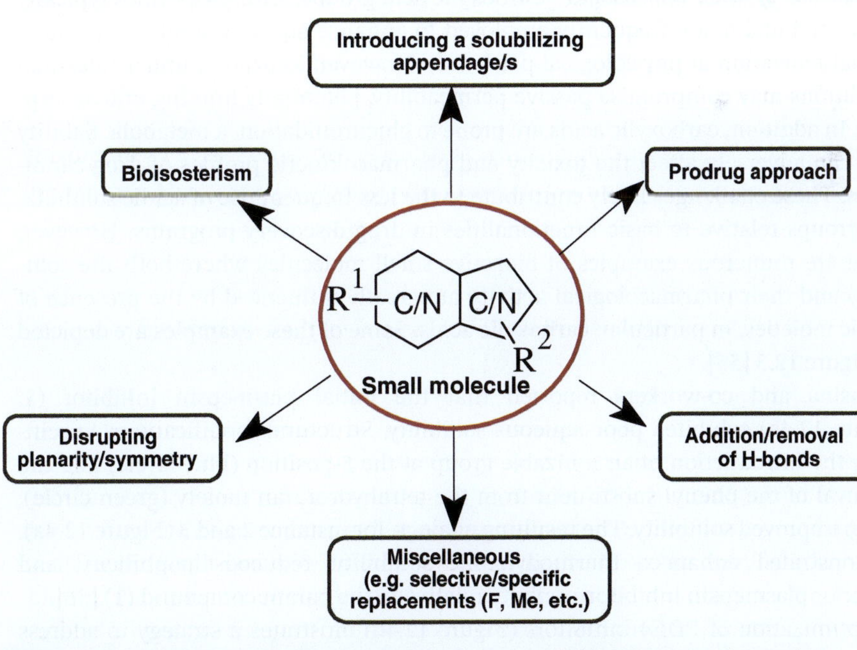

Figure 12.2 Medchem approaches through structural modifications to enhance the solubility of sparingly soluble drug candidates or early discovery hits.

pharmacological efficacy. Rational design approaches focus on increasing molecular hydrophilicity through the introduction of polar groups, reduction of hydrophobic regions, or implementation of prodrug strategies. These modifications must be carefully optimized to maintain target affinity, selectivity, and favorable pharmacokinetic properties. Given the significant impact of solubility on drug absorption and bioavailability, structural optimization remains a critical consideration throughout the drug discovery and development process. The following Sections (12.3.1.1–12.3.1.6) will detail established methodologies and representative examples illustrating successful applications of this approach.

12.3.1.1 Introduction of Polar or Ionizable Appendages

The strategic attachment of solubilizing appendages is an effective method for enhancing the aqueous solubility of drug candidates without perturbing target engagement. The functional groups such as typically acidic (e.g. carboxylic acids), neutral (e.g. polar hydrogen bond donors/acceptors), or basic (e.g. protonatable amines) are linked to the drug core via a flexible tether. Optimal attachment sites are selected to orient the appendage away from the target-binding interface, thereby preserving biological activity. Structural data, particularly high-resolution drug–target complexes, are critical for guiding both the positioning of the appendage and the design of the linker. This approach has enabled the solubilization of even highly hydrophobic molecules, such as C_{60} fullerenes, highlighting its broad applicability in drug design [54].

Introduction of Acidic Appendages Carboxylic acid groups, with p*K*a values typically between 4 and 6, are frequently employed to enhance aqueous solubility through partial ionization at physiological pH (~7.0). However, ionization under intestinal conditions may compromise passive permeability, potentially limiting oral absorption. In addition, carboxylic acids are prone to glucuronidation, a metabolic liability that can adversely affect the toxicity and pharmacokinetic profiles of drug candidates. These challenges likely contribute to the less frequent use of acidic solubilizing groups relative to basic functionalities in drug discovery programs. However, there are numerous examples of bioactive small molecules where both the solubility and their pharmacological activity are greatly influenced by the presence of acidic moieties, in particular, carboxylic acids. Some of these examples are depicted in Figure 12.3 [55].

Rasina and co-workers reported that the initial plasmepsin inhibitor (**1**, Figure 12.4a) exhibited poor aqueous solubility. Structural modifications, specifically the introduction of an ionizable group at the 5-position (blue circle) and the removal of the phenyl substituent from the tetrahydrofuran moiety (green circle), led to improved solubility. The resulting analogs, for instance **2** and **3** (Figure 12.4a), demonstrated enhanced thermodynamic solubility, reduced lipophilicity, and superior plasmepsin inhibitory activity relative to the parent compound (1) [56].

Optimization of PDE4 inhibitors (Figure 12.4b) illustrates a strategy to address poor solubility (e.g. lead compound **4**, solubility 0.3 μg/ml, $\text{clog}\,P = 5.20$) [57]. X-ray crystallography guided the introduction of linear carboxylic acid side chains

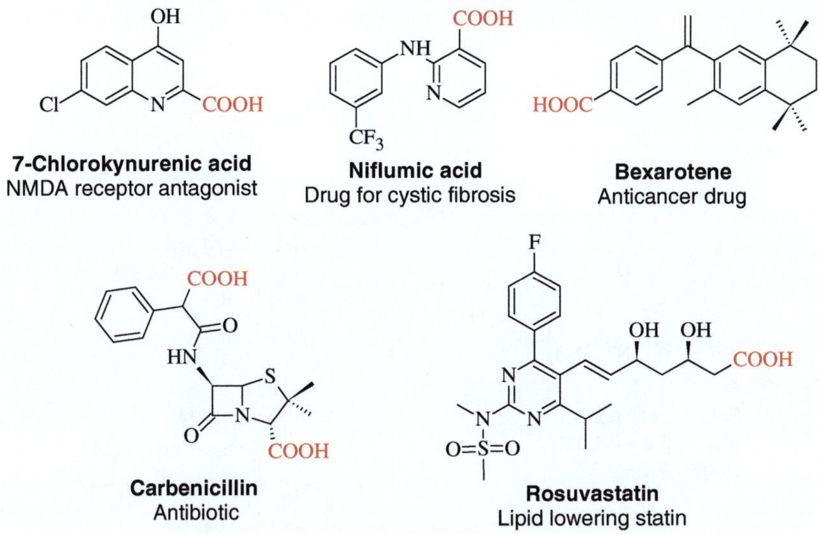

Figure 12.3 Selected examples of bioactive carboxylic acid small molecules.

($-(CH_2)_nCO_2H$), resulting in marked solubility improvements. Increasing linker length from $n = 4$ to 6 raised solubility from 500 to 1400 μg/ml, with compound 5 ($n = 6$) also showing enhanced potency. Longer chains reduced solubility, likely due to increased hydrophobicity. Potency required at least six methylene units, suggesting optimal spacing of the ionizable group is critical for maintaining target engagement.

Jorgensen and colleagues encountered a low-solubility issue during the development of quinolinyltriazole inhibitors targeting the tautomerase activity of human macrophage migration inhibitory factor (MIF) [58]. The low solubility in this class resulted from its near-planar structure and an intermolecular hydrogen bond, as revealed in a small-molecule X-ray structure. Molecular modeling suggested that substituents at the 6- and 7-positions of the quinoline ring should be solvent-exposed. Introducing polar substituents at these positions yielded promising results. While the methoxyethoxy group (**6**, Figure 12.4c) improved solubility to some extent, the addition of a carboxymethyl group (**7**, Figure 12.4c) significantly enhanced solubility, while maintaining the binding affinities for both compounds [58].

Together, these observations highlight both the potential and the challenges of employing acidic groups to improve solubility, emphasizing the need to carefully balance physicochemical optimization with target engagement considerations in lead design.

Introduction of Neutral Appendages Neutral polar groups are frequently employed to enhance aqueous solubility through hydrogen bonding interactions with water. This is typically achieved by incorporating functionalities with high hydrogen bond basicity, such as heterocycles, amides, or ureas. However, in some cases, strong

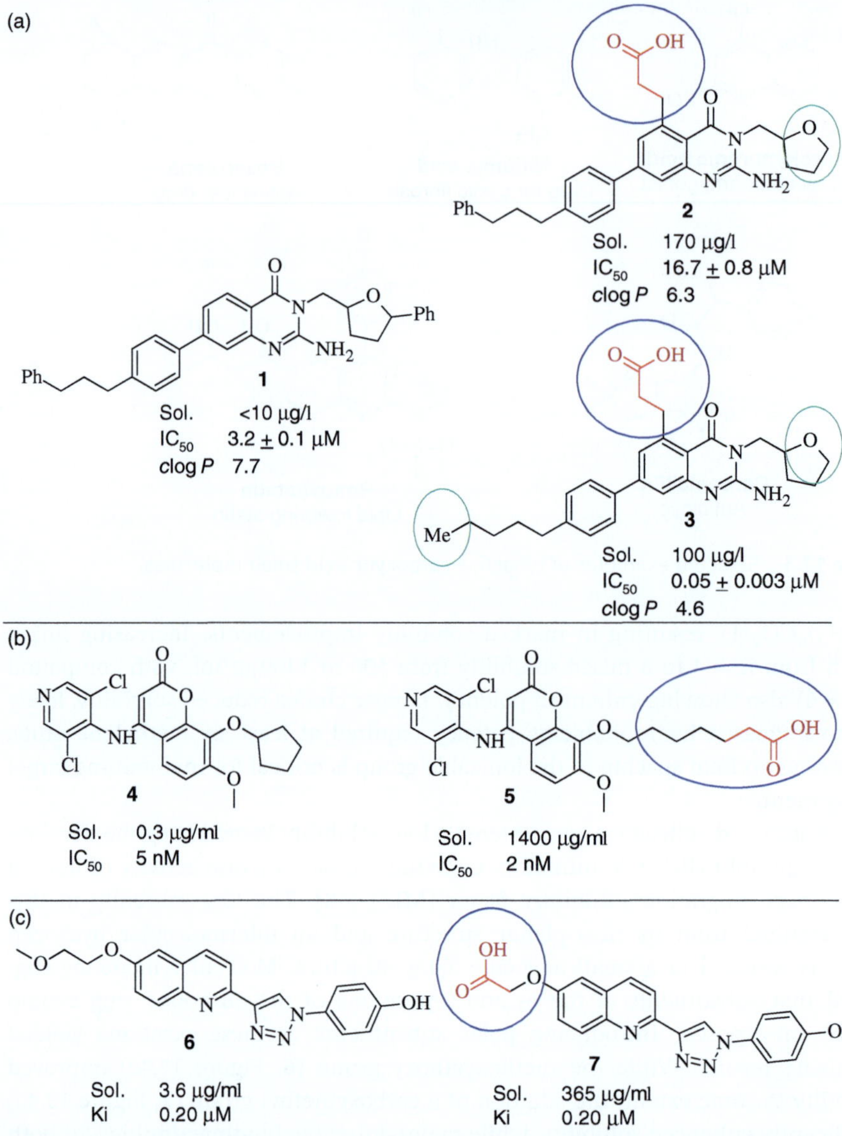

Figure 12.4 (a)–(c) Solubility improvement of poorly soluble lead compounds by introducing carboxylic acid moiety.

hydrogen-bonding capacity can compromise passive membrane permeability. As an alternative, weaker hydrogen bond acceptors, including ethers, have been investigated. Nonetheless, their application remains limited, as achieving sufficient solubility often necessitates the inclusion of multiple weak acceptors, potentially leading to increased molecular size and suboptimal physicochemical properties.

Lesch et al. reported their studies targeting identification of early clinical candidates against hepatitis B virus (HBV) from their earlier set of capsid assembly

Figure 12.5 Lead CAM inhibitor against HBV.

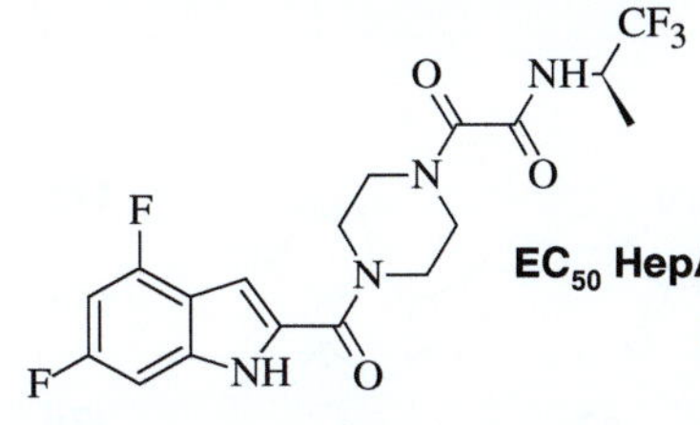

8

modulators (CAM). The initial lead compound **8** exhibited limited potency, good stability in mouse liver microsomes and permeability in Caco-2 cell lines, but their low solubility apparently did not favor dose–escalation studies, despite favorable rat PK profiles (Figure 12.5). Therefore, solubility and potency were the main optimization parameters [59].

Although small modifications on the indole ring brought necessary stability, the potency and solubility problems were addressed by introduction of neutral appendages, in particular, small heterocyclic ones (Figure 12.6). Modulating the stereocenter (**9**) or introduction of cyclobutane appendage (**10**) improved solubility. Adding an oxetane appendage (**11**) or a larger pyran (**12**) led to more soluble but relatively less potent molecules. The potency could be reinstated by introducing oxadiazole **11**, but this resulted in a loss in microsomal stability. The latter issue, as well as the hERG (human Ether-á-go-go-Related Gene) challenge, was addressed using 4-ethyl-6-fluoro indole motif and delivering potential clinical candidates (**14–16**).

An article describing the discovery of novel Akt kinase inhibitors provides an illustrative example of improved solubility achieved by means of a strong H-bond

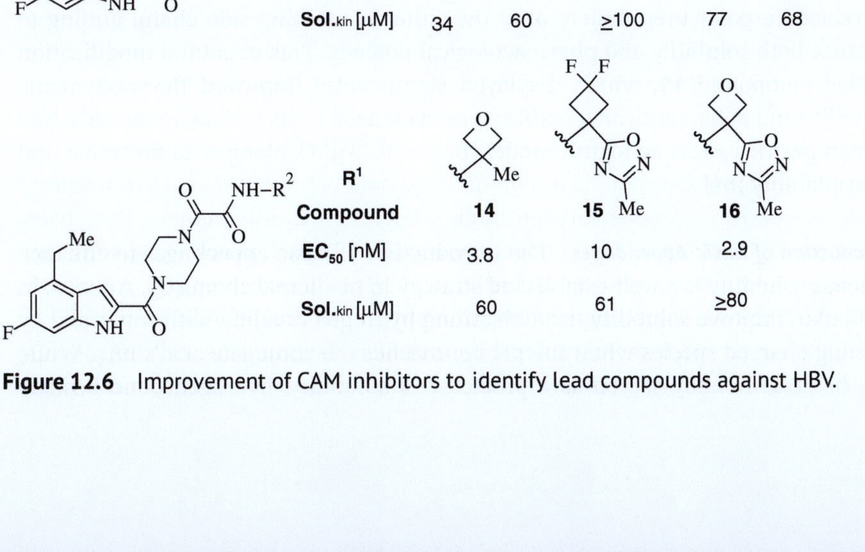

Compound	9	10	11	12	13
EC_{50} [nM]	127	9.8	52	18	13
Sol.kin [μM]	34	60	≥100	77	68

Compound	14	15	16
EC_{50} [nM]	3.8	10	2.9
Sol.kin [μM]	60	61	≥80

Figure 12.6 Improvement of CAM inhibitors to identify lead compounds against HBV.

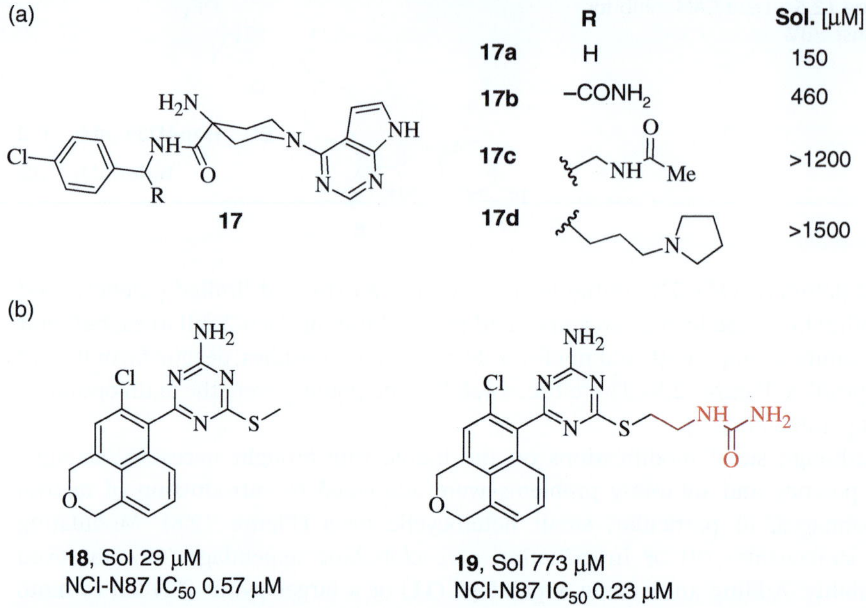

Figure 12.7 (a,b) Improvement in solubility of Akt inhibitors by structural modifications.

acceptor (Figure 12.7a) [60]. Amide groups (**17b** provided a more than twofold improvement in solubility over the unsubstituted parent compound **17a**, while the reverse amide **17c** resulted in a ≥10-fold increase. Interestingly, **17c** with a neutral amide appendage exhibited solubility comparable to analogues containing basic amine-solubilizing appendages, such as **17d** (Figure 12.7a).

The attachment of large polyethylene glycol (PEG) groups (MW ≥ 1000) is often employed in the development of protein therapeutics but is rarely pursued as a means for improving the solubility of small molecules [61–63].

In a related investigation, a series of benzoisochromene triazines emerged as promising Hsp90 inhibitors. The lead compound (**18**, Figure 12.7b) demonstrated notable target engagement but exhibited suboptimal oral bioavailability, attributed to its limited aqueous solubility (29 μM). To address these limitations, Suda et al. introduced a polar urea moiety onto the sulfur-containing side chain, aiming to enhance both solubility and pharmacological potency. This structural modification yielded compound **19**, which displayed significantly improved thermodynamic solubility and potent anticancer efficacy, as evidenced by its activity in the NCI-N87 human gastric cancer xenograft model (IC_{50} = 0.23 μM), along with favorable oral bioavailability [64].

Introduction of Basic Appendages The introduction of basic appendages to enhance aqueous solubility is a well-established strategy in medicinal chemistry. Amines, in particular, improve solubility through strong hydrogen bonding with water and by forming charged species when the pH approaches the conjugate acid's pK_a. While pK_a has traditionally served as a predictor of solubility, hydrogen bond basicity

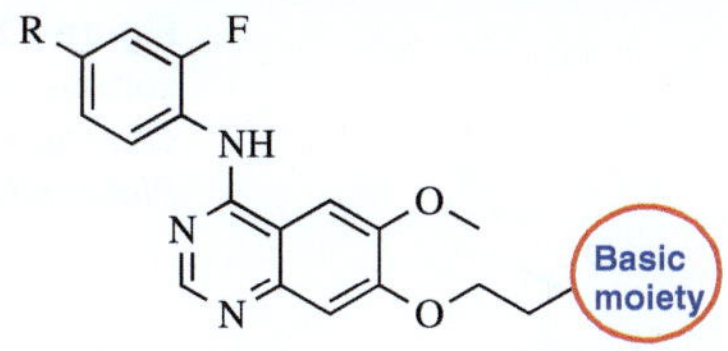

Compound	Basic moiety	R	pK_{BHX}	Sol. (μM)	Dose (mg/kg)	Plasma levels (μM) at 6 h	at 24 h
20	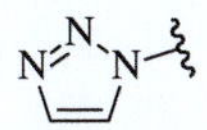	Br	—	0.7	50	14	0.2
21	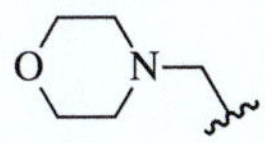	Cl	1.78	20	100	4.0	0.8
22	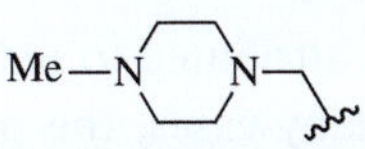	Cl	1.88	60	50	5.0	3.0
23	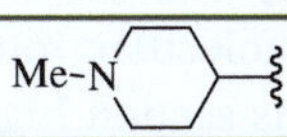	Br	**2.11**	**330**	50	**5.5**	**3.0**

Figure 12.8 Significance of basicity of appendages in modulating solubility and dose of a drug.

(pK_{BHX}) may, in some cases, provide a more accurate correlation. Both pK_a and hydrogen bond basicity are tuneable via electronic and steric effects, offering access to a wide variety of amine structures with diverse physicochemical profiles. This versatility enables precise optimization of solubility while managing associated risks, such as phospholipidosis and QT prolongation [21, 65].

As illustrated in Figure 12.8, the solubility of a series of 4-anilinoquinazolines correlates with their basicity, as indicated by pK_{BHX} [66]. Notably, under non-protonating conditions (pH 12.7), the piperidinyl compound **23** remained 17-fold more soluble than triazole **20**, highlighting the importance of hydrogen bonding in the solubility of neutral species. Moreover, solubility also appears to influence drug plasma concentrations at longer durations.

The poorly soluble lead compound (**24**, Figure 12.9) was found to be active against a chloroquine-sensitive *Plasmodium falciparum* strain (3D7). To enhance water solubility, Baragana and co-workers developed compound **25** with better thermodynamic solubility by replacing the pyridine ring with pyrrolidine and adding a morpholine nucleus [67]. These findings highlight the multifaceted influence of basicity and hydrogen bonding on solubility optimization and emphasize the importance of tailored physicochemical tuning during drug design.

12.3.1.2 Role of Small Key Modifications Including H-Bonding Modifiers to Improve Aqueous Solubility

The incorporation of hydrogen bond donors and acceptors—such as hydroxyl or amino functionalities—has long been recognized as a strategy to enhance aqueous solubility, primarily by promoting favorable interactions with surrounding water molecules. Paradoxically, the deliberate removal of such hydrogen-bonding groups

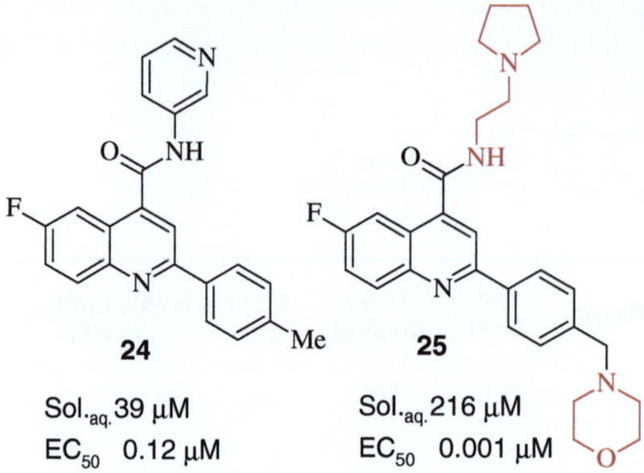

Figure 12.9 Development of more soluble compounds against chloroquine-sensitive *Plasmodium falciparum* strain (3D7).

can also lead to improved solubility, a phenomenon often attributed to reductions in melting point that facilitate dissolution [68, 69]. In many cases, the presence of hydrogen-bonding moieties contributes to strong intermolecular interactions in the crystalline state, which can hinder solubility [70]. This section highlights a few illustrative examples that demonstrate how strategic modulation of hydrogen bonding, through either introduction or elimination of these groups, has been effectively employed to improve solubility. It is also worth noting that hydrogen bonding may contribute to some of the solubility-enhancing effects observed with the introduction of polar or ionizable substituents, as discussed in the preceding Section 12.3.1.1.

A noteworthy strategy to improve solubility by disrupting intramolecular interactions was demonstrated by Li et al. in 2007 [71]. The authors reported the design and synthesis of a novel class of imidazolin-2-ones as potent, orally bioavailable, and muscle-selective selective androgen receptor (AR) modulators (SARMs). Many of the compounds exhibited sub-nanomolar activity in both AR binding and functional assays. In their lead optimization efforts, compound **26** emerged as a highly active candidate but displayed poor aqueous solubility. Detailed analysis revealed that its limited solubility was primarily due to extensive intermolecular hydrogen bonding in the crystal lattice, which facilitated close packing and π–π stacking of co-planar aryl rings—features that contributed to its high melting point (Figure 12.10a). To address this, the authors modified the scaffold by replacing one of the carbonyl groups with a methyl substituent, effectively disrupting the hydrogen-bonding network and increasing the spatial separation of the aryl rings. This modification, exemplified in compound **27**, resulted in a significant reduction in melting point and a marked improvement in solubility.

The p110α isoform of phosphoinositide 3-kinase (PI3K) is frequently mutated in various cancer types and has emerged as a promising therapeutic target in oncology. Among the chemical scaffolds investigated, pyrazolopyridine derivatives—such as compound **28** (Figure 12.10b)—have demonstrated potent inhibitory activity against the p110α isoform. However, their clinical potential has been limited by poor aqueous solubility. To address this issue, Kendall and colleagues undertook a lead optimization campaign focused on improving the physicochemical properties

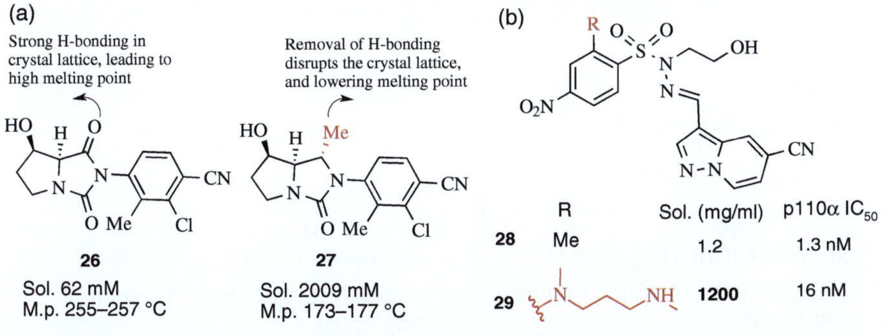

Figure 12.10 (a,b) Improvement of solubility by modulation of H-bonding.

of compound **28**. Introduction of hydrogen bond donor–acceptor diamino group on the nitrophenyl ring resulted in compound **29**, exhibiting a significant enhancement in thermodynamic aqueous solubility [72].

12.3.1.3 Improvement of Solubility by Disruption of Molecular Planarity and Symmetry

Aqueous solubility of solid compounds is governed by both molecular interactions with water and crystalline lattice stability [73]. Disruption of crystal packing, therefore, offers a viable strategy to enhance solubility. Molecular planarity and symmetry are key factors influencing packing efficiency and melting point. The correlation between reduced planarity and lower melting point is well-established, as exemplified by the effect of hydrogenation on unsaturated oils [74]. More recently, Lovering demonstrated that an increased fraction of sp^3-hybridized carbons in drug-like molecules correlates with reduced melting points [23]. These observations support the concept that decreasing molecular planarity may serve as a rational approach to improve aqueous solubility [74]. Wang et al. (Amgen Inc.) reported the development of TRPV1 antagonists [75], including the clinical candidate AMG 517 (**30**, Figure 12.11a), which exhibited poor thermodynamic aqueous solubility (<1 μg/ml in PBS or 0.01 M HCl). To address this, a second-generation analogue (**31**) was designed by partially saturating the 4-(trifluoromethyl)phenyl ring to reduce planarity and disrupt crystal packing [76, 77]. This modification led to a >13-fold increase in solubility (13 μg/ml in 0.01 M HCl), albeit with a 30-fold loss in potency in a capsaicin-mediated assay and a fourfold reduction in an acid-mediated assay. The higher melting point of **30** further supports the role of planarity in limiting solubility.

The aryl hydrocarbon receptor (AhR) plays a pivotal role in various physiological and toxicological processes, including digitoxin toxicity, hepatotoxicity, and immunotoxicity. A β-naphthoflavone derivative (**32**, Figure 12.11b) was identified as a potent AhR agonist; however, its poor aqueous solubility limited further development. To address this, Fujita and colleagues introduced fluorine atoms on the phenyl ring (compound **33**, Figure 12.11b), aiming to disrupt molecular planarity by increasing the dihedral angle of the biaryl system. This modification

(a)

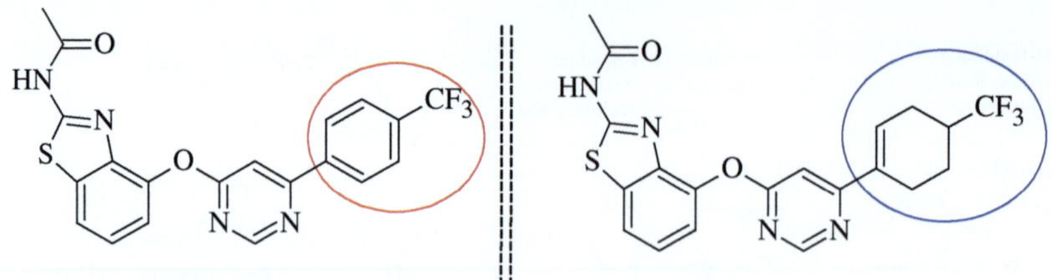

30
rat TRPV1 (CAP) IC_{50} : 0.9 nM
Solubility (0.01 M HCl): <1 μg/ml
clog *P*: 4.6
m.p.: 219–221 °C

31
rat TRPV1 (CAP) IC_{50}: 27 nM
Solubility (0.01 M HCl): 13 μg/ml
clog *P*: 3.7
m.p.: 130–131 °C

(b)

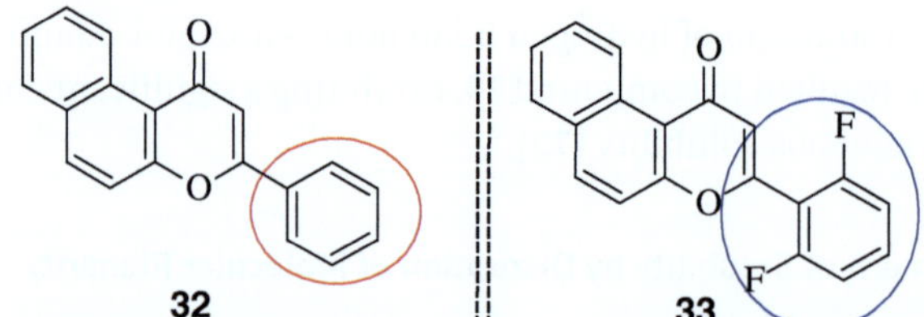

32
EC_{50}: 1.4 μM
Solubility: 84.6 μg/ml
Dihedral angle: 17.8°
m.p.: 166 °C

33
EC_{50}: 0.2 μM
Solubility: 153 μg/ml
Dihedral angle: 40.5°
m.p.: 150 °C

Figure 12.11 Improvement of aqueous solubility by: (a) removing aromaticity and (b) disruption of crystal packing.

led to reduced crystal packing efficiency, as evidenced by a lower melting point (150 °C) and improved thermodynamic solubility [53].

12.3.1.4 Bioisosterism to Address Aqueous Solubility

According to Burger, bioisosteric atoms or groups are those that possess near-equal molecular shapes and volumes, approximately the same distribution of electrons, and exhibit similar physical properties. Bioisosteres can be classified into two types, i.e. classical bioisosteres (monovalent, divalent, trivalent atom or groups, tetra substituted, and ring equivalents) and nonclassical bioisosteres (cyclic and noncyclic functional groups). Classical bioisosteres represent the results of an early appreciation of the concept and encompass structurally simple atoms or groups, for instance, the monovalent to tetravalent bioisosteres depicted in Figure 12.12. Nonclassical bioisosteres are structurally distinct, usually comprise different number of atoms and exhibit different steric and electronic properties, for instance, cyclic and noncyclic isosteres as well as the exchangeable group isosterism, in which the properties of discrete functional elements are emulated (Figure 12.12).

The development and application of bioisosteres have been adopted as a fundamental tactical approach useful to address a number of aspects associated with the design and development of drug candidates. Their applications are explored

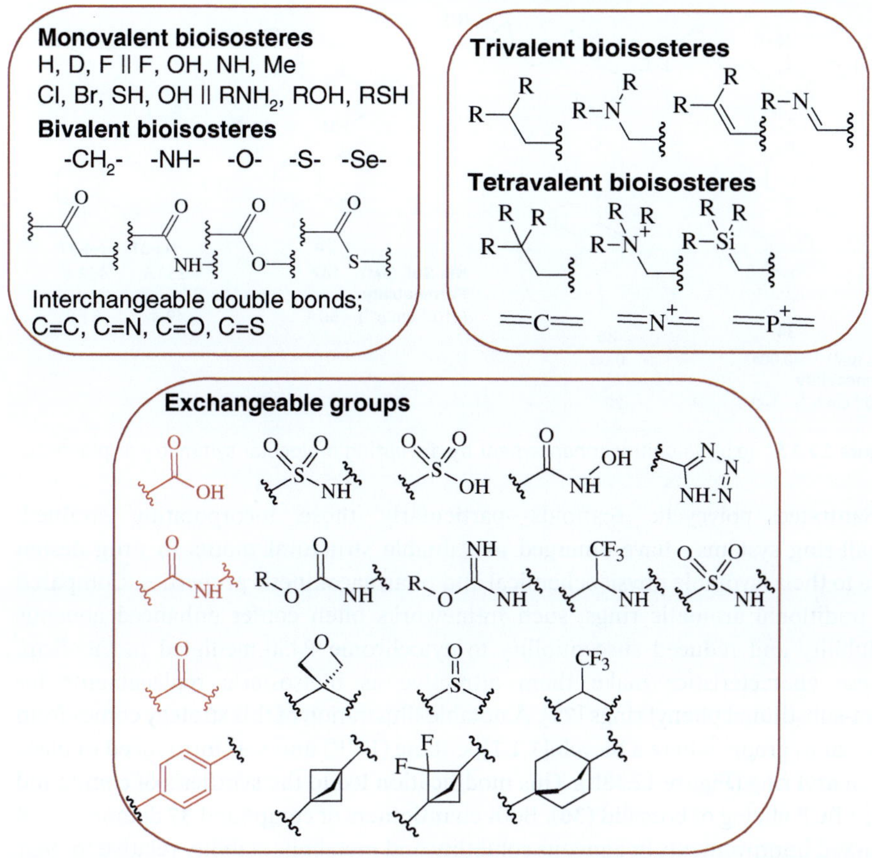

Figure 12.12 Depiction of selected set of bioisosteres of elements and functional groups.

to improve potency, enhance selectivity, alter physical properties like solubility, reduce or redirect metabolism, eliminate or modify toxicophores, and acquire novel intellectual property. The following are selected examples of how simple to complex bioisosteres modulate the physicochemical properties to advance a lead compound toward clinic.

Positive allosteric modulators (PAMs) of metabotropic glutamate receptor subtype 2 (mGlu2) have emerged as promising therapeutic agents for the treatment of various central nervous system (CNS) disorders. A representative advanced lead compound featuring a triazolopyridine core tethered to a piperidine side chain (compound **34**, Figure 12.13a) demonstrated potent activity, but was limited by poor aqueous solubility and suboptimal drug-like properties. In a series of lead optimization efforts, Cid and colleagues introduced a methylene spacer between the triazolopyridine scaffold and the cyclic amine, which significantly improved kinetic solubility without compromising functional activity. Replacing the piperidine moiety with a piperazine ring, and introduction of *o*,*p*-difluoro group on the aryl ring yielding compound **35**, endowed with a more favorable balance between solubility and receptor activity [78].

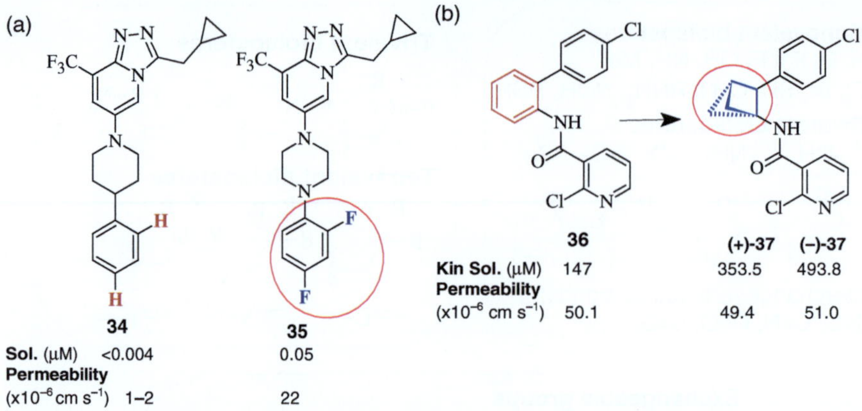

Figure 12.13 (a,b) Solubility enhancement by disrupting molecular symmetry or planarity.

Saturated polycyclic scaffolds—particularly those incorporating strained, small-ring systems—have emerged as valuable structural motifs in drug design due to their favorable physicochemical and pharmacokinetic properties. Compared to traditional aromatic rings, such frameworks often confer enhanced aqueous solubility and reduced susceptibility to cytochrome P450-mediated metabolism. These characteristics make them attractive as bioisosteric replacements for para-substituted phenyl rings [79]. A notable illustration of this strategy comes from the Baran group, where a bicyclo[1.1.1]pentane (BCP) unit was introduced in place of an aryl ring (Figure 12.13b). This modification led to the synthesis of compound **37**, a BCP analog of boscalid (**36**). Both enantiomers of compound **37** demonstrated marked improvements in aqueous solubility and metabolic stability relative to their aromatic predecessor, compound **36** (Figure 12.13b), highlighting the potential of BCP substitution to enhance drug-like properties [80].

The bioisosteric moiety could also modulate other undesired effects, such as off-targets or other pharmacokinetic parameters. One such bioisostere of a sulfonamide is the sulfoximine group that has impressed the medicinal chemistry community immensely in the last decade. Sulfoximines are isoelectronic with sulfones, but the introduction of the nitrogen creates asymmetry (Figure 12.14a).

Sulfoximines are stable compounds, and the nitrogen offers an additional point for substitution [81]. The nitrogen is also basic enough to allow metal ion coordination or salt formation. The heteroatoms bound to the sulfur are hydrogen bond acceptors and, in the case of NH sulfoximines (R^3 = H), the group has dual hydrogen bond donor/acceptor functionality. Moreover, structurally simple sulfoximines are readily soluble in protic solvents (Figure 12.14a) [82–84]. A combination of bioisosteric groups replacement, including introduction of a sulfoximine, is beautifully depicted in the following example.

The dose–escalation study in patients for one of the Bayer's molecules, ZK 304709 (**38**), was terminated due to dose-limited absorption at high doses, which was mainly attributed to its limited aqueous solubility of only 8 mg/l. Moreover, ZK 304709 was found to accumulate in the erythrocytes of patients due to an off-target

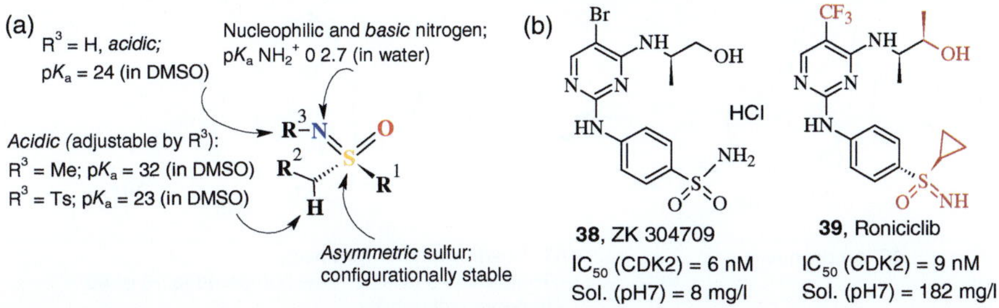

Figure 12.14 (a) General properties of sulfoximine group and (b) improvement of solubility and other parameters by introduction of sulfoximine moiety.

activity against carbonic anhydrases that was primarily due to the sulfonamide moiety. Small successive replacements of hydrophobic groups (Br to CF_3) and other H-bonding moieties (sulfonamide to sulfoximine and primary alcohol to secondary alcohol) led to the clinical compound roniciclib (**39**, Figure 12.14b). The aqueous solubility of roniciclib of 182 mg/l is more than 20-fold higher than that of the former clinical candidate. Furthermore, roniciclib does not inhibit carbonic anhydrases and does not accumulate in erythrocytes [85].

12.3.1.5 Prodrug Approach to Address Aqueous Solubility

A prodrug is a poorly active or inactive compound containing the parental drug that undergoes some in vivo biotransformation through chemical or enzymatic cleavage, enabling the delivery of the active molecule at efficacious levels [86, 87]. Prodrugs can be broadly classified into two major types: carrier-linked prodrugs and bioprecursors. Carrier-linked prodrugs are particularly useful for addressing poor solubility by covalently attaching a nontoxic pro-moiety to the parent drug, thereby modifying unfavorable physicochemical properties. Upon enzymatic or chemical cleavage, the active drug is released. These prodrugs may be bipartite (direct linkage to the drug) or tripartite (involving a spacer) [88]. Common carrier-linked examples include amide, phosphate, imine, carbonate, and ether prodrugs [86]. The design and advantages of such systems have been extensively reviewed [89, 90].

Bioprecursors, in contrast, involve structural modification of the active compound without a distinct carrier linkage [91]. Bioprecursor prodrugs can sometimes improve solubility, but they are generally designed for metabolic activation rather than solubility enhancement. Several FDA-approved prodrugs exemplify the success of this approach [86, 92].

Given the availability of comprehensive reviews on prodrug strategies, we highlight only select recent examples demonstrating improved solubility via prodrug design. Amprenavir (**40**, Figure 12.15a), an HIV-1 protease inhibitor approved in 1999, suffers from poor aqueous solubility, necessitating a softgel formulation. To address this, Dan Todd (GlaxoSmithKline plc, USA) presented the development of fosamprenavir (**41**), a phosphate prodrug with significantly improved solubility. Following the evaluation of 60 prodrug candidates, fosamprenavir was selected based on its high aqueous solubility, physicochemical stability, and efficient

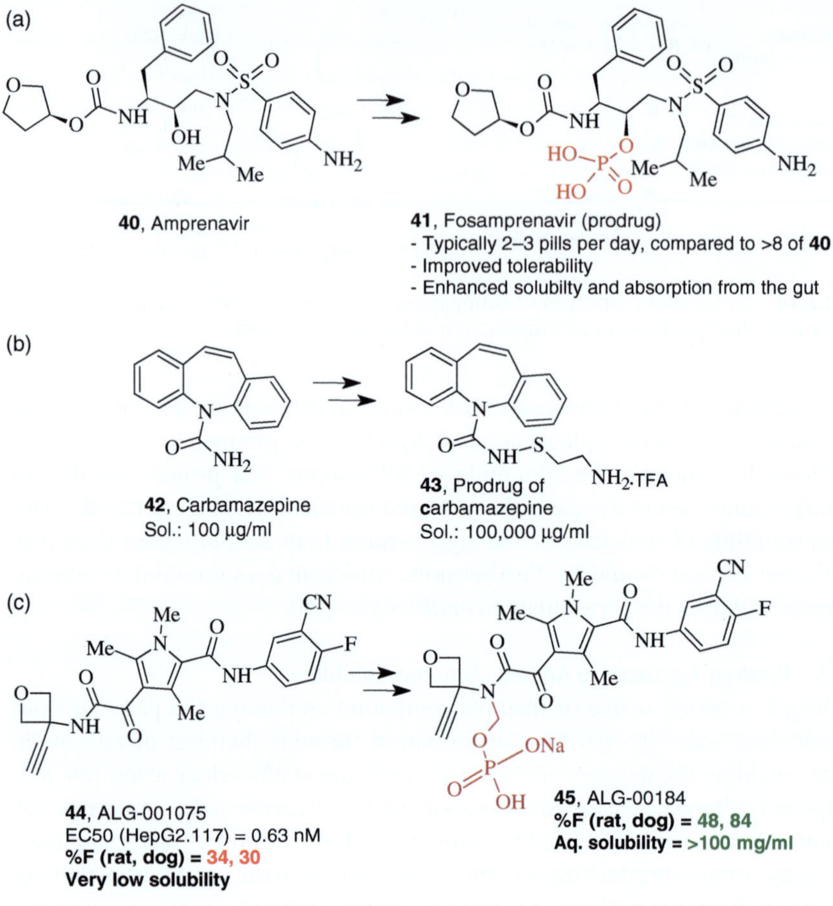

Figure 12.15 (a)–(c) Prodrug approach to enhance aqueous solubility of drug candidates.

enzymatic conversion to amprenavir at the epithelial surface. This enabled a solid oral formulation with a reduced pill burden (two tablets versus eight softgels). The reduced pill burden and smaller tablet size make treatment easier for patients to stick with, which is critical for HIV therapy, where strict adherence prevents resistance. Fosamprenavir received FDA approval in 2003 for use in combination antiretroviral therapy in adults with HIV infection.

Carbamazepine (CBZ, **42**, Figure 12.15b) is a clinically effective anticonvulsant with minimal sedative and cognitive side effects relative to other agents. However, its low aqueous solubility (~120 μg/ml) limits its suitability for intravenous (IV) administration [93]. To address this, a water-soluble prodrug (**43**) was synthesized (Figure 12.15b). The aminoethylthioamine prodrug **43** exhibited significantly enhanced aqueous solubility, exceeding 100 mg/ml at pH 2.6, and demonstrated superior chemical stability in solution compared to CBZ. Pharmacokinetic studies following IV administration confirmed rapid and efficient bioconversion of **43** to the active parent drug **42** in vivo [94].

A functional cure for chronic hepatitis B (CHB) remains elusive and a major therapeutic challenge. The current nucleoside analogue treatments rarely achieve this cure. CAMs offer a novel mechanism to suppress viral replication and reduce HBV DNA, RNA, and antigen levels. GLP-26, a potent CAM, was optimized for antiviral potency, safety, and drug-like properties, leading to the development of ALG-001075. To address solubility limitations, ALG-001075 (**44**, Figure 12.15c), an advanced and potent CAM, was identified as potential clinical candidate for further development to address the low solubility, along with some other minor issues. To this end, a hemiaminal phosphate salt, as the highly soluble prodrug ALG-000184 (**45**), was developed. **45** has demonstrated an excellent safety profile and superior efficacy based on reductions in viral markers (HBV DNA, HBV RNA) vs other CAM compounds at doses as low as 10 mg QD and is currently undergoing clinical evaluation in CHB patients [95].

12.3.1.6 Other Miscellaneous Approaches to Address Weak Aqueous Solubility

While the introduction of polar or ionizable groups remains a conventional strategy to enhance aqueous solubility, as discussed in Section 12.3.1.1, several non-traditional approaches have also proven effective. One such alternative involves disrupting molecular planarity or symmetry to improve solubility, as outlined in Section 3.1.2. An illustrative example of this approach was reported by Schröder and colleagues at Bristol-Myers Squibb, who developed selective Met kinase inhibitors [96]. In this study, an ethoxy substituent was introduced at the 4-position of the pyridone ring in the highly planar compound **46** (Figure 12.16a), yielding analog **47**. While maintaining comparable in vitro activity to **46**, compound **47** exhibited a >40-fold increase in aqueous solubility at pH 1.0 (400 μg/ml). This substantial enhancement in solubility was attributed to a disruption of

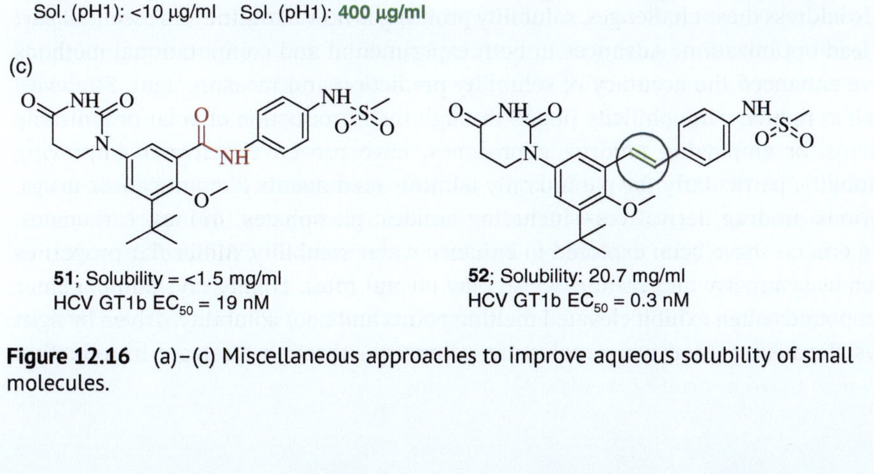

Figure 12.16 (a)–(c) Miscellaneous approaches to improve aqueous solubility of small molecules.

molecular planarity induced by the ethoxy substituent, which in turn reduced crystal packing efficiency [74].

Although fluorine is traditionally viewed as a hydrophobic substituent that may reduce solubility, emerging evidence suggests that, in specific structural contexts, fluorination can enhance aqueous solubility. While the incorporation of fluorine is often driven by its beneficial effects on potency and metabolic stability, certain examples demonstrate a rational use of fluorination to modulate solubility. For instance, in a series of diacylglycerol *O*-acyltransferase 1 (DGAT1) inhibitors, substitution at the terminal phenyl ring revealed a striking improvement in solubility for the bis-fluorinated analogue **50** relative to the parent compound **48** (Figure 12.16b). As this structural change did not induce any change in $\log D$ (2.5), this suggests that the enhanced solubility arises from disruption of crystal packing rather than changes in lipophilicity, as corroborated by the negligible solubility impact of Me and Cl substitutions (cf. **48–50**).

Substitution of rings or linkers can significantly influence solubility, alongside broader pharmacological and physicochemical properties. While amide bonds are commonly modified using bioisosteres to address physicochemical and drug metabolism and pharmacokinetics (DMPK) challenges, their replacement with olefins is relatively rare. In a notable example, a non-nucleoside HCV NS5B polymerase inhibitor (compound **51**, Figure 12.16c) exhibited poor aqueous solubility. Replacing the amide linkage with a trans-olefin yielded compound **52**, which demonstrated markedly improved solubility and permeability [97].

12.4 Conclusion

Aqueous solubility is a critical determinant of drug absorption, influencing both membrane permeability and systemic bioavailability. Inadequate solubility can compromise the entire drug development continuum—from initial hit identification through to preclinical evaluation—by reducing assay sensitivity, increasing data variability, and obscuring structure–activity relationships (SAR). Consequently, poorly soluble compounds often fail to progress into clinical candidates.

To address these challenges, solubility profiling is now a routine and essential part of lead optimization. Advances in both experimental and computational methods have enhanced the accuracy of solubility prediction and measurement. Strategies such as reducing lipophilicity ($\log P$) through the introduction of polar or ionizable groups, or employing prodrug approaches, have proven effective in improving solubility, particularly for parenterally administered agents like anticancer drugs. Various prodrug derivatives—including amides, phosphates, imines, carbonates, and ethers—have been explored to enhance water solubility. Molecular properties such as symmetry and planarity also play pivotal roles. Highly crystalline, planar compounds often exhibit elevated melting points and poor solubility, driven by tight crystal packing. Modifying molecular geometry—through strategic introduction or removal of functional groups—can disrupt crystallinity and improve solubility, offering a valuable design principle during optimization.

Heterocyclic modifications represent another widely adopted strategy. Heterocycles influence solubility through their p*K*a, polarity, and hydrogen-bonding capacity, and are central scaffolds in drug-like molecules. Substitution of aromatic systems with appropriate heterocycles or flexible linkers not only improves solubility but can also expand the physicochemical diversity of lead libraries. Nonetheless, these chemical interventions may impact potency, selectivity, or the integrity of the pharmacophore. Hence, careful balance between solubility and biological activity remains a key design consideration. Computational tools can guide such modifications, providing predictive insights into solubility-enhancing changes while preserving pharmacological properties.

In conclusion, structural modification remains one of the most effective approaches to enhance aqueous solubility in drug discovery. By integrating physicochemical understanding with computational design and synthetic innovation, medicinal chemists can better navigate the solubility challenges that frequently hinder lead progression.

References

1 Meanwell, N.A. (2011). Improving drug candidates by design: a focus on physicochemical properties as a means of improving compound disposition and safety. *Chem. Res. Toxicol.* 24: 1420–1456. doi: 10.1021/tx200211v.

2 van de Waterbeemd, H. and Gifford, E. (2003). ADMET *in silico* modelling: towards prediction paradise. *Nat. Rev. Drug Discov.* 2: 192–204. doi: 10.1038/nrd1032.

3 Bickerton, G.R., Paolini, G.V., Besnard, J. et al. (2012). Quantifying the chemical beauty of drugs. *Nat. Chem.* 4: 90–98. doi: 10.1038/nchem.1243.

4 Amidon, G.L., Lennernäs, H., Shah, V.P., and Crison, J.R. (1995). A theoretical basis for a biopharmaceutic drug classification: the correlation of in vitro drug product dissolution and in vivo bioavailability. *Pharm. Res.: Off. J. Am. Assoc. Pharm. Sci.* 12: 413–420. doi: 10.1023/A:1016212804288.

5 Hansch, C., Björkroth, J.P., and Leo, A. (1987). Hydrophobicity and central nervous system agents: on the principle of minimal hydrophobicity in drug design. *J. Pharm. Sci.* 76: 663–687. doi: 10.1002/jps.2600760902.

6 He, Y.L., Murby, S., Warhurst, G. et al. (1998). Species differences in size discrimination in the paracellular pathway reflected by oral bioavailability of poly(ethylene glycol) and D-peptides. *J. Pharm. Sci.* 87, 626–633. doi: 10.1021/js970120d.

7 Kirch, W. and Görg, K.G. (1982). Clinical pharmacokinetics of atenolol — A review. *Eur. J. Drug Metab. Pharmacokinet.* 7: 81–91. doi: 10.1007/BF03188723.

8 Wood, A.J.J., Whitley, R.J., and Gnann, J.W. (1992). Acyclovir: a decade later. *N. Eng. J. Med.* 327: 782–789.

9 Wei, Y.-P., Yao, L.-Y., Wu, Y.-Y. et al. (2021). Critical review of synthesis, toxicology and detection of acyclovir. *Molecules* 26: 6566.

10 Bunnage, M.E., Blagg, J., Steele, J. et al. (2007). Discovery of potent & selective inhibitors of activated thrombin-activatable fibrinolysis inhibitor for the treatment of thrombosis. *J. Med. Chem.* 50: 6095–6103. doi: 10.1021/jm0702433.

11 Dai, X.L., Yao, J., Wu, C. et al. (2020). Solubility and permeability improvement of allopurinol by cocrystallization. *Cryst. Growth Des.* 20: 5160–5168. doi: 10.1021/acs.cgd.0c00326.

12 Sou, T. and Bergström, C.A.S. (2018). Automated assays for thermodynamic (equilibrium) solubility determination. *Drug Discov. Today Technol.* 27: 11–19. doi: 10.1016/j.ddtec.2018.04.004.

13 Ting, J.M., Tale, S., Purchel, A.A. et al. (2016). High-throughput excipient discovery enables oral delivery of poorly soluble pharmaceuticals. *ACS Cent. Sci* 2 (10): 748–755. doi: 10.1021/acscentsci.6b00268.

14 Hill, A.P. and Young, R.J. (2010). Getting physical in drug discovery: a contemporary perspective on solubility and hydrophobicity. *Drug Discov. Today* 15: 648–655. doi: 10.1016/j.drudis.2010.05.016.

15 Leeson, P.D. and Springthorpe, B. (2007). The influence of drug-like concepts on decision-making in medicinal chemistry. *Nat. Rev. Drug Discov.* 6: 881–890. doi: 10.1038/nrd2445.

16 Faller, B., Ottaviani, G., Ertl, P. et al. (2011). Evolution of the physicochemical properties of marketed drugs: can history foretell the future? *Drug Discov. Today* 16: 976–984. doi: 10.1016/j.drudis.2011.07.003.

17 Roden, D.M. and George, A.L. (2002). The genetic basis of variability in drug responses. *Nat. Rev. Drug Discov.* 1: 37–44. doi: 10.1038/nrd705.

18 Zhao, H. (2011). Lead optimization in the nondrug-like space. *Drug Discov. Today* 16: 149–163. doi: 10.1016/j.drudis.2010.12.002.

19 Di, L., Fish, P.V., and Mano, T. (2012). Bridging solubility between drug discovery and development. *Drug Discov. Today* 17: 486–495. doi: 10.1016/j.drudis.2011.11.007.

20 Hann, M.M. (2011). Molecular obesity, potency and other addictions in drug discovery. *MedChemComm* 2: 349–355. doi: 10.1039/C1MD00017A.

21 Yang, Y., Engkvist, O., Llinàs, A., and Chen, H. (2012). Beyond size, ionization state, and lipophilicity: influence of molecular topology on absorption, distribution, metabolism, excretion, and toxicity for druglike compounds. *J. Med. Chem.* 55: 3667–3677. doi: 10.1021/jm201548z.

22 Ritchie, T.J. and Macdonald, S.J.F. (2009). The impact of aromatic ring count on compound developability – are too many aromatic rings a liability in drug design? *Drug Discov. Today* 14: 1011–1020. doi: 10.1016/j.drudis.2009.07.014.

23 Lovering, F., Bikker, J., and Humblet, C. (2009). Escape from flatland: increasing saturation as an approach to improving clinical success. *J. Med. Chem.* 52: 6752–6756. doi: 10.1021/jm901241e.

24 Lipinski, C.A. (2000). Drug-like properties and the causes of poor solubility and poor permeability. *J. Pharmacol. Toxicol. Methods* 44: 235–249. doi: 10.1016/S1056-8719(00)00107-6.

25 M. S. Patil, S. Z. Godse, R. B. Saudagar, ""Solubility enhancement by various techniques: an overview,"" *World J. Pharm. Pharm. Sci.* 2013, 2, 4558-4572.

26 Bogner, R.H., Murdande, S.B., Pikal, M.J., and Shanker, R.M. (2010). Solubility advantage of amorphous pharmaceuticals: II. Application of quantitative thermodynamic relationships for prediction of solubility enhancement in structurally diverse insoluble pharmaceuticals. *Pharm. Res.* 27: 2704–2714. doi: 10.1007/s11095-010-0269-5.

27 Rathi, P.B., Kale, M., Soleymani, J., and Jouyban, A. (2018). Solubility of etoricoxib in aqueous solutions of glycerin, methanol, polyethylene glycols 200, 400, 600, and propylene glycol at 298.2 K. *J. Chem. Eng. Data* 63: 321–330. doi: 10.1021/acs.jced.7b00709.

28 Lu, J.X., Tupper, C., and Murray, J. (2022). Biochemistry, dissolution and solubility. *StatPearls*.

29 Gabor, F., Fillafer, C., Neutsch, L. et al. (2010). Improving oral delivery. *Handb. Exp. Pharmacol.* 197: 345–398. doi: 10.1007/978-3-642-00477-3_12.

30 Joshi, J., Nainwal, N., and Saharan, V.A. (2019). A review on hydrotrophy: a potential approach for the solubility enhancement of poorly soluble drugs. *Asian J. Pharm. Clin. Res.* 12: 19–26. doi: 10.22159/ajpcr.2019.v12i10.34811.

31 Bhalani, D.V., Nutan, B., Kumar, A., and Chandel, A.K.S. (2022). Bioavailability enhancement techniques for poorly aqueous soluble drugs and therapeutics. *Biomedicines* 10: 2055. doi: 10.3390/biomedicines10092055.

32 Salehi, N., Al-Gousous, J., Mudie, D.M. et al. (2020). Hierarchical mass transfer analysis of drug particle dissolution, highlighting the hydrodynamics, pH, particle size, and buffer effects for the dissolution of ionizable and nonionizable drugs in a compendial dissolution vessel. *Mol. Pharm.* 17: 3870–3884.

33 Bremmell, K.E. and Prestidge, C.A. (2019). Enhancing oral bioavailability of poorly soluble drugs with mesoporous silica based systems: opportunities and challenges. *Drug Dev. Ind. Pharm.* 45: 349–358.

34 Kumar, R. (2020). Solubility and bioavailability of fenofibrate nanoformulations. *ChemistrySelect* 5: 1478–1490.

35 Tehler, U., Fagerberg, J.H., Svensson, R. et al. (2013). Optimizing solubility and permeability of a biopharmaceutics classification system (BCS) class 4 antibiotic drug using lipophilic fragments disturbing the crystal lattice. *J. Med. Chem.* 56: 2690–2694.

36 Charalabidis, A., Sfouni, M., Bergström, C., and Macheras, P. (2019). The biopharmaceutics classification system (BCS) and the biopharmaceutics drug disposition classification system (BDDCS): beyond guidelines. *Int. J. Pharm.* 566: 264–281.

37 van den Anker, J., Reed, M.D., Allegaert, K., and Kearns, G.L. (2018). Developmental changes in pharmacokinetics and pharmacodynamics. *J. Clin. Pharmacol.* 58: doi: 10.1002/jcph.1284.

38 Yan, A. (2018). Prediction of ADME properties. In: *Applied Chemoinformatics*, 333–357. Wiley.

39 Falcón-Cano, G., Molina, C., and Cabrera-Pérez, M.Á. (2020). ADME prediction with KNIME: development and validation of a publicly available workflow for the prediction of human oral bioavailability. *J. Chem. Inf. Model* 60: 2660–2667.

40 Saal, C. and Petereit, A.C. (2012). Optimizing solubility: kinetic versus thermodynamic solubility temptations and risks. *Eur. J. Pharm. Sci.* 47: 589–595.

41 Black, S., Dang, L., Liu, C., and Wei, H. (2013). On the measurement of solubility. *Org. Process. Res. Dev.* 17: 486–492.

42 Kerns, E., Di, L., and Carter, G. (2008). in vitro solubility assays in drug discovery. *Curr. Drug Metab.* 9: 879–885.

43 Könczöl, Á. and Dargó, G. (2018). Brief overview of solubility methods: recent trends in equilibrium solubility measurement and predictive models. *Drug Discov. Today Technol* 27: 3–10.

44 Alsenz, J. and Kansy, M. (2007). High throughput solubility measurement in drug discovery and development. *Adv. Drug Deliv. Rev.* 59: 546–567.

45 Lin, B. and Pease, J.H. (2016). A high throughput solubility assay for drug discovery using microscale shake-flask and rapid UHPLC–UV–CLND quantification. *J. Pharm. Biomed. Anal.* 122: 126–140.

46 Chai, Q., Shih, J., Weldon, C. et al. (2019). Development of a high-throughput solubility screening assay for use in antibody discovery. *MAbs* 11: 747–756.

47 Sugano, K., Kato, T., Suzuki, K. et al. (2006). High throughput solubility measurement with automated polarized light microscopy analysis. *J. Pharm. Sci.* 95: 2115–2122.

48 Wang, J. and Hou, T. (2011). Recent advances on aqueous solubility prediction. *Comb. Chem. High Throughput Screen* 14: 328–338.

49 Delaney, J.S. (2005). Predicting aqueous solubility from structure. *Drug. Discov. Today* 10: 289–295.

50 Di, L., Kerns, E., and Carter, G. (2009). Drug-like property concepts in pharmaceutical design. *Curr. Pharm. Des.* 15: 2184–2194.

51 Wenglowsky, S., Moreno, D., Rudolph, J. et al. (2012). Pyrazolopyridine inhibitors of B-RafV600E. Part 3: an increase in aqueous solubility via the disruption of crystal packing. *Bioorg. Med. Chem. Lett.* 22: 912–915.

52 Pinal, R. (2004). Effect of molecular symmetry on melting temperature and solubility. *Org. Biomol. Chem.* 2: 2692.

53 Fujita, Y., Yonehara, M., Tetsuhashi, M. et al. (2010). β-Naphthoflavone analogs as potent and soluble aryl hydrocarbon receptor agonists: improvement of solubility by disruption of molecular planarity. *Bioorg. Med. Chem.* 18: 1194–1203.

54 Troshina, O.A., Troshin, P.A., Peregudov, A.S. et al. (2006). Photoaddition of N-substituted piperazines to C_{60}: an efficient approach to the synthesis of water-soluble fullerene derivatives. *Chem. - Eur. J.* 12: 5569–5577.

55 Fan, Z., Zhang, Z., and Xi, C. (2020). Light-mediated carboxylation using carbon dioxide. *ChemSusChem* 13: 6201–6218.

56 Rasina, D., Stakanovs, G., Borysov, O.V. et al. (2018). 2-Aminoquinazolin-4(3H)-one based plasmepsin inhibitors with improved hydrophilicity and selectivity. *Bioorg. Med. Chem.* 26: 2488–2500.

57 Govek, S.P., Oshiro, G., Anzola, J.V. et al. (2010). Water-soluble PDE4 inhibitors for the treatment of dry eye. *Bioorg. Med. Chem. Lett.* 20: 2928–2932.

58 Cisneros, J.A., Robertson, M.J., Mercado, B.Q., and Jorgensen, W.L. (2017). Systematic study of effects of structural modifications on the aqueous solubility of drug-like molecules. *ACS Med. Chem. Lett.* 8: 124–127.

59 Lesch, B., Birkmann, A., Bonsmann, S. et al. (2025). Discovery of AIC263282, a hepatitis B virus capsid assembly modulator ready for candidate profiling. *J. Med. Chem.* 68: 6361–6371.

60 Addie, M., Ballard, P., Buttar, D. et al. (2013). Discovery of 4-amino- *N* -[(1 *S*)-1-(4-chlorophenyl)-3-hydroxypropyl]-1-(7 *H* -pyrrolo[2,3- *d*]pyrimidin-4-yl)piperidine-4-carboxamide (AZD5363), an orally bioavailable, potent inhibitor of Akt kinases. *J. Med. Chem.* 56: 2059–2073.

61 Greenwald, R.B., Choe, Y.H., McGuire, J., and Conover, C.D. (2003). Effective drug delivery by PEGylated drug conjugates. *Adv. Drug Deliv. Rev.* 55: 217–250.

62 Zalipsky, S., Gilon, C., and Zilkha, A. (1983). Attachment of drugs to polyethylene glycols. *Eur. Polym. J.* 19: 1177–1183.

63 Smith, J.L., Rossiter, K.I., Semko, C.M. et al. (2013). PEG conjugates of potent α4 integrin inhibitors, maintaining sustained levels and bioactivity in vivo, following subcutaneous administration. *Bioorg. Med. Chem. Let.t* 23: 4117–4119.

64 Suda, A., Kawasaki, K., Komiyama, S. et al. (2014). Design and synthesis of 2-amino-6-(1H,3H-benzo[de]isochromen-6-yl)-1,3,5-triazines as novel Hsp90 inhibitors. *Bioorg. Med. Chem.* 22: 892–905.

65 Fischer, H., Atzpodien, E.-A., Csato, M. et al. (2012). In silico assay for assessing phospholipidosis potential of small druglike molecules: training, validation, and refinement using several data sets. *J. Med. Chem.* 55: 126–139.

66 Hennequin, L.F., Stokes, E.S.E., Thomas, A.P. et al. (2002). Novel 4-anilinoquinazolines with C-7 basic side chains: design and structure activity relationship of a series of potent, orally active, VEGF receptor tyrosine kinase inhibitors. *J. Med. Chem.* 45: 1300–1312.

67 Baragaña, B., Norcross, N.R., Wilson, C. et al. (2016). Discovery of a quinoline-4-carboxamide derivative with a novel mechanism of action, multistage antimalarial activity, and potent in vivo efficacy. *J. Med. Chem.* 59: 9672–9685.

68 Di, L. and Kerns, E.H. (2015). Introduction and overview. In: *Blood-Brain Barrier in Drug Discovery*, 1–4. Wiley.

69 Kuhn, B., Mohr, P., and Stahl, M. (2010). Intramolecular hydrogen bonding in medicinal chemistry. *J. Med. Chem.* 53: 2601–2611.

70 Etter, M.C. (1990). Encoding and decoding hydrogen-bond patterns of organic compounds. *Acc. Chem. Res.* 23: 120–126.

71 J. J. Li, J. C. Sutton, A. Nirschl, Y. Zou, H. Wang, C. Sun, Z. Pi, R. Johnson, Stanley R. Krystek, R. Seethala, R. Golla, P. G. Sleph, B. C. Beehler, G. J. Grover, A. Fura, V. P. Vyas, C. Y. Li, J. Z. Gougoutas, M. A. Galella, R. Zahler, J. Ostrowski, L. G. Hamann, "Discovery of potent and muscle selective androgen receptor modulators through scaffold modifications," *J. Med. Chem.* 2007, 50, 3015–3025.

72 Kendall, J.D., Giddens, A.C., Tsang, K.Y. et al. (2017). Novel pyrazolo[1,5-a] pyridines with improved aqueous solubility as p110α-selective PI3 kinase inhibitors. *Bioorg. Med. Chem. Lett.* 27: 187–190.

73 Jain, N. and Yalkowsky, S.H. (2001). Estimation of the aqueous solubility I: application to organic nonelectrolytes. *J. Pharm. Sci.* 90: 234–252.

74 Ishikawa, M. and Hashimoto, Y. (2011). Improvement in aqueous solubility in small molecule drug discovery programs by disruption of molecular planarity and symmetry. *J. Med. Chem.* 54: 1539–1554.

75 Wang, H.-L., Katon, J., Balan, C. et al. (2007). Novel vanilloid receptor-1 antagonists: 3. The identification of a second-generation clinical candidate with improved physicochemical and pharmacokinetic properties. *J. Med. Chem.* 50: 3528–3539.

76 Brown, A., Brown, L., Brown, T.B. et al. (2008). Triazole oxytocin antagonists: identification of aryl ether replacements for a biaryl substituent. *Bioorg. Med. Chem. Lett.* 18: 5242–5244.

77 Kim, I.-H., Heirtzler, F.R., Morisseau, C. et al. (2005). Optimization of amide-based inhibitors of soluble epoxide hydrolase with improved water solubility. *J. Med. Chem.* 48: 3621–3629.

78 Cid, J.M., Tresadern, G., Vega, J.A. et al. (2016). Discovery of 8-trifluoromethyl-3-cyclopropylmethyl-7-[(4-(2,4-difluorophenyl)-1-piperazinyl)methyl]-1,2,4-triazolo[4,3- *a*]pyridine (JNJ-46356479), a selective and orally bioavailable mglu2 receptor positive allosteric modulator (PAM). *J. Med. Chem.* 59: 8495–8507.

79 Tsien, J., Hu, C., Merchant, R.R., and Qin, T. (2024). Three-dimensional saturated C(sp3)-rich bioisosteres for benzene. *Nat. Rev. Chem.* 8: 605–627.

80 Zhao, J.-X., Chang, Y.-X., He, C. et al. (2021). 1,2-Difunctionalized bicyclo[1.1.1]pentanes: long–sought-after mimetics for *ortho*/*meta*-substituted arenes. *Proc. Natl. Acad. Sci.* 118: doi: 10.1073/pnas.2108881118.

81 Arvidsson, P.I. (2025). Sulfilimines: an underexplored bioisostere for drug design? *J. Med. Chem.* 68: 4056–4058.

82 Lücking, U. (2019). Neglected sulfur(<scp>vi</scp>) pharmacophores in drug discovery: exploration of novel chemical space by the interplay of drug design and method development. *Org. Chem. Front.* 6: 1319–1324.

83 Frings, M., Bolm, C., Blum, A., and Gnamm, C. (2017). Sulfoximines from a medicinal chemist's perspective: physicochemical and in vitro parameters relevant for drug discovery. *Eur. J. Med. Chem.* 126: 225–245.

84 Lücking, U. (2013). Sulfoximines: a neglected opportunity in medicinal chemistry. *Angew. Chem. Int. Ed.* 52: 9399–9408.

85 Cho, B.C., Dy, G.K., Govindan, R. et al. (2018). Phase Ib/II study of the pan-cyclin-dependent kinase inhibitor roniciclib in combination with chemotherapy in patients with extensive-disease small-cell lung cancer. *Lung Cancer* 123: 14–21.

86 Jornada, D., Dos Santos Fernandes, G., Chiba, D. et al. (2015). The prodrug approach: a successful tool for improving drug solubility. *Molecules* 21: 42.

87 Sanches, B.M.A. and Ferreira, E.I. (2019). Is prodrug design an approach to increase water solubility? *Int. J. Pharm.* 568: 118498.

88 Niethammer, A., Gaedicke, G., Lode, H.N., and Wrasidlo, W. (2001). Synthesis and preclinical characterization of a paclitaxel prodrug with improved antitumor activity and water solubility. *Bioconjug. Chem.* 12: 414–420.

89 Gangrade, D. and Karande, R. (2017). Prodrugs to codrugs. *Curr. Drug Ther.* 12: 29–36.

90 Protti, Í.F., Rodrigues, D.R., Fonseca, S.K. et al. (2021). Do drug-likeness rules apply to oral prodrugs? *ChemMedChem* 16: 1446–1456.

91 Bremner, J. (2021). Design principles and development of prodrugs for multiply active antibacterials. In: *Multiple Action-Based Design Approaches to Antibacterials*, 121–158. Singapore: Springer.

92 Rautio, J., Kärkkäinen, J., and Sloan, K.B. (2017). Prodrugs – recent approvals and a glimpse of the pipeline. *Eur. J. Pharm. Sci.* 109: 146–161.

93 Chen, X.-Q. and Venkatesh, S. (2004). Miniature device for aqueous and non-aqueous solubility measurements during drug discovery. *Pharm. Res.* 21: 1758–1761.

94 Hemenway, J.N., Nti-Addae, K., Guarino, V.R., and Stella, V.J. (2007). Preparation, characterization and in vivo conversion of new water-soluble sulfenamide prodrugs of carbamazepine. *Bioorg. Med. Chem. Lett.* 17: 6629–6632.

95 Vendeville, S., Amblard, F., Bassit, L. et al. (2024). The discovery and preclinical profile of ALG-000184, a prodrug of the potent hepatitis B virus capsid assembly modulator ALG-001075. *J. Med. Chem.* 67: 21126–21142.

96 Schroeder, G.M., An, Y., Cai, Z.-W. et al. (2009). Discovery of *N* -(4-(2-Amino-3-chloropyridin-4-yloxy)-3-fluorophenyl)-4-ethoxy-1-(4-fluorophenyl)-2-oxo-1,2-dihydropyridine-3-carboxamide (BMS-777607), a selective and orally efficacious inhibitor of the met kinase superfamily. *J. Med. Chem.* 52: 1251–1254.

97 Randolph, J.T., Krueger, A.C., Donner, P.L. et al. (2018). Synthesis and biological characterization of aryl uracil inhibitors of hepatitis C virus NS5B polymerase: discovery of ABT-072, a trans-stilbene analog with good oral bioavailability. *J. Med. Chem.* 61: 1153–1163.

Index

a

d

e